Surveys on Recent Developments in Algebraic Geometry

Proceedings of Symposia in
PURE MATHEMATICS

Volume 95

Surveys on Recent Developments in Algebraic Geometry

Bootcamp for the
2015 Summer Research Institute
on Algebraic Geometry
July 6–10, 2015
University of Utah, Salt Lake City, Utah

Izzet Coskun
Tommaso de Fernex
Angela Gibney
Editors

American Mathematical Society
Providence, Rhode Island

2010 *Mathematics Subject Classification.* Primary 14H10, 14E30, 14E08, 14D07, 14N35, 14J60, 14G17, 13D02, 19D06, 37D40.

Library of Congress Cataloging-in-Publication Data

Names: Summer Research Institute on Algebraic Geometry (2015 : Salt Lake City, Utah) | Coskun, Izzet, 1978- editor. | De Fernex, Tommaso, 1970- editor. | Gibney, Angela, 1971- editor.

Title: Surveys on recent developments in algebraic geometry : bootcamp for the Summer Research Institute on Algebraic Geometry, July 6-10, 2015, University of Utah, Salt Lake City, Utah / Izzet Coskun, Tommaso de Fernex, Angela Gibney, editors.

Other titles: Bootcamp for the Summer Research Institute on Algebraic Geometry, July 6-10, 2015, University of Utah, Salt Lake City, Utah

Description: Providence, Rhode Island : American Mathematical Society, [2017] | Series: Proceedings of symposia in pure mathematics ; volume 95 | Includes bibliographical references.

Identifiers: LCCN 2017001962 | ISBN 9781470435578 (alk. paper)

Subjects: LCSH: Geometry, Algebraic–Congresses. | AMS: Algebraic geometry – Curves – Families, moduli (algebraic). msc | Algebraic geometry – Birational geometry – Minimal model program (Mori theory, extremal rays). msc | Algebraic geometry – Birational geometry – Rationality questions. msc | Algebraic geometry – Families, fibrations – Variation of Hodge structures. msc | Algebraic geometry – Projective and enumerative geometry – Gromov-Witten invariants, quantum cohomology, Gopakumar-Vafa invariants, Donaldson-Thomas invariants. msc | Algebraic geometry – Surfaces and higher-dimensional varieties – Vector bundles on surfaces and higher-dimensional varieties, and their moduli. msc | Algebraic geometry – Arithmetic problems. Diophantine geometry – Positive characteristic ground fields. msc | Commutative algebra – Homological methods – Syzygies, resolutions, complexes. msc | K-theory – Higher algebraic K-theory – Q- and plus-constructions. msc | Dynamical systems and ergodic theory – Dynamical systems with hyperbolic behavior – Dynamical systems of geometric origin and hyperbolicity (geodesic and horocycle flows, etc.). msc

Classification: LCC QA564 .S86 2015 | DDC 516.3/5–dc23

LC record available at https://lccn.loc.gov/2017001962

DOI: `http://dx.doi.org/10.1090/pspum/095`

♾ The paper used in this book is acid-free and falls within the guidelines
established to ensure permanence and durability.
Visit the AMS home page at `http://www.ams.org/`

10 9 8 7 6 5 4 3 2 1   22 21 20 19 18 17

Contents

Preface

The algebraic geometry community has a tradition of running a research institute every ten years. This important conference convenes a majority of the practitioners in the world to overview the developments of the past decade and to outline the most fundamental and far-reaching problems for the next. Previous institutes have included Woods Hole (1964), Arcata (1974), Bowdoin (1985), Santa Cruz (1995) and Seattle (2005). This past Algebraic Geometry Summer Institute took place at the University of Utah, in Salt Lake City, on July 13–31, 2015.

Increasingly, algebraic geometry has become very diverse and technically demanding. It is daunting for graduate students and postdocs to master the techniques and wide-range of activities in the field. In response, the community has been running a "Bootcamp" in the week preceding the institute with the aim of familiarizing the participants with a broad-range of developments in algebraic geometry in an informal setting. In 2015, from July 6 to 10, the Bootcamp took place at the University of Utah, Salt Lake City. Following tradition, the 150 graduate student participants were mentored by 15 postdocs and young assistant professors, helping to create a vibrant and informal atmosphere and allowing young researchers to form support networks. The mentors introduced the graduate students and postdocs to exciting new developments in the Minimal Model Program, Hodge theory, perfectoid spaces, positive characteristic techniques, Boij-Söderberg theory, p-adic Hodge theory, Bridgeland stability, tropical geometry and many more topics at the cutting edge of the field.

Activities at the Bootcamp included morning lectures given by the mentors, followed by afternoon working sessions. In this volume, in an attempt to make these excellent expositions more widely available, we have collected 10 survey papers that grew out of the lectures. Each paper discusses a different subfield of algebraic geometry that has seen significant progress in the last decade. The papers preserve the informal tone of the lectures and strive to be accessible to graduate students who have basic familiarity with algebraic geometry. They also contain many illuminating examples and open problems. We expect these surveys will become invaluable resources, not only for graduate students and postdocs, but also senior researchers starting along new directions.

We now summarize the contents of this volume in greater detail.

Higher dimensional birational geometry. The Minimal Model Program was initiated in the 1980s by Mori as a way of extending classification theorems for surfaces to higher dimensional varieties. In the last decade, several of the central conjectures in the field have been resolved. The fundamental and influential work of Birkar, Cascini, Hacon and McKernan showed the existence of minimal models for varieties of general type, and proved the finite generation of the canonical ring.

The consequences have ranged from the resolution of classical conjectures such as the Sarkisov program to the construction of new moduli spaces. The successes have also led to attempts at extending the theory to other contexts such as Kähler manifolds or characteristic p birational geometry. The paper *A snapshot of the Minimal Model Program* by Brian Lehmann describes the most important developments of the decade and gives a list of open problems and conjectures. The paper *Positive characteristic algebraic geometry* by Patakfalvi, Schwede and Tucker gives a detailed introduction to the use of the Frobenius morphism for defining birational and singularity invariants in characteristic p. This approach has led to tremendous advances including an extension of the Minimal Model Program to threefolds in characteristic $p > 5$. The paper contains many illustrative and instructional examples and exercises to familiarize the reader with the characteristic p techniques.

Progress on moduli spaces. Moduli spaces have been at the center of algebraic geometry and its applications to other areas of mathematics. Moduli spaces of curves, moduli spaces of abelian varieties and moduli spaces of sheaves appear in many guises in mathematics and have applications to number theory, combinatorics, mathematical physics and topology. In the last decade, there has been significant development in constructing new compactifications and new birational models of important moduli spaces. The developments were motivated in part by evolution in the Minimal Model Program and new techniques coming from derived categories and Bridgeland stability. As a result, our understanding of classical moduli spaces such as the moduli space of principally polarized abelian varieties and moduli spaces of Gieseker semistable sheaves on surfaces has vastly improved. The paper *The geometry of the moduli space of curves and abelian varieties* by Tommasi gives an overview of this progress. Tommasi's paper is notable for an accessible account of toroidal compactifications of the moduli spaces of abelian varieties. The paper *Birational geometry of moduli spaces of sheaves and Bridgeland stability* by Jack Huizenga gives a masterful introduction to the geometry of moduli spaces of sheaves on surfaces and Bridgeland stability conditions using the projective plane as a motivating example. In recent years, Bridgeland stability has revolutionized our understanding of the birational geometry of moduli spaces of sheaves on surfaces. Arcara, Bayer, Bertram, Coskun, Huizenga, Macrì and others have computed the ample and effective cones of moduli spaces of sheaves on certain surfaces and have given many geometric applications. Huizenga's paper also clearly illustrates the interactions between developments in the Minimal Model Program and moduli spaces.

New applications of moduli spaces and connections to dynamics. There has been significant expansion in the applications of moduli theory and in the interactions of moduli spaces with dynamics. Inspired by string theory, Gromov-Witten theory revolutionized classical enumerative geometry in the 1990s and early 2000s. In the last decade, the theory has matured and found new applications to the tautological ring of the moduli space of curves. Maulik, Nekrasov, Okounkov, Pandharipande conjectured the equivalence of various curve counting theories such as the Gromov-Witten and Donaldson-Thomas correspondence. Pandharipande, Pixton and others have resolved some of these conjectures. The paper *Gromov-Witten theory: From curve counts to string theory* by Clader, gives a succinct account of the vast advances that have taken place in Gromov-Witten theory in the last decade.

Clader masterfully summarizes the correspondence between Gromov-Witten and Donaldson-Thomas theory and applications of Gromov-Witten theory to the tautological ring of the moduli space of curves. She concludes with a set of central open problems in the field. The paper *Teichmüller dynamics in the eyes of an algebraic geometer* by Chen, introduces algebraic geometers to recent developments in Teichmüller dynamics following the fundamental work of Eskin, Kontsevich, Mirzakhani and Zorich. In his beautifully illustrated survey, Chen makes many analytic and dynamical concepts accessible to algebraic geometers. Chen emphasizes applications of the theory to the geometry of the Deligne-Mumford moduli spaces of curves and degenerations of abelian differentials.

Rationality of varieties. The last few years have seen an explosion in the study of rationality of varieties. Voisin introduced a new deformation technique based on the Chow theoretic decomposition of the diagonal to show that very general quartic double solids are not stably rational. Her approach was extended by Colliot-Thélène and Pirutka, and applied by them and many others such as Totaro, Hassett and Tschinkel to resolve long standing questions of rationality and stable rationality. In addition, Kuznetsov and others have suggested using the derived category and orthogonal decompositions to obtain new invariants for rationality. The survey *Unramified cohomology, derived categories and rationality* by Auel and Bernardara, gives a comprehensive introduction to these novel ideas and explores the interconnections between the two developments.

Hodge theory and degenerations. Hodge theory carries subtle transcendental information about the geometry of complex varieties. In the last decade, new insights have allowed a better understanding of degenerations of Hodge structures and the corresponding degenerations of varieties. The paper *Degenerations of Hodge structure* by Robles is an introduction to recent developments in Hodge theory, most notably to the classification of certain degenerations of Hodge structure pioneered by Green, Griffiths, Kerr, Robles and others.

Syzygies and cohomology tables. Betti tables of resolutions and cohomology tables of vector bundles are fundamental objects of study in commutative algebra and algebraic geometry. Eisenbud and Schryer's solution of the Boij-Söderberg conjecture and the resulting description of the cones of these tables have reshaped the theory in the last decade. The paper *Questions about Boij-Söderberg theory* by Erman and Sam surveys the developments and poses many fascinating and accessible further open problems.

Homotopy methods in algebraic geometry. The motivic, or $\mathbb{A}^1$ homotopy theory, introduced by Morel and Voevodsky, lies at the heart of recent progress, such as the classification of vector bundles on smooth complex affine varieties by Asok and Fasel. The paper *A primer for unstable motivic homotopy theory* by Antieau and Elmanto, gives an accessible introduction to this technical theory. Many key examples, useful exercises, enticing open problems and extensive references make this paper an indispensable reference for beginning practitioners of the subject.

Acknowledgements. We thank the mentors Ben Antieau, Bhargav Bhatt, Ana Caraiani, Dawei Chen, Angelica Cueto, Daniel Erman, Jack Huizenga, Brian Lehmann, Colleen Robles, Aaron Pixton, Giulia Saccà, Orsola Tommasi, Kevin Tucker, Yu-Jong Tzeng, Bianca Viray for giving excellent surveys of their areas of expertise and for their dedication to the younger generation of mathematicians. Without their hard work the Bootcamp would not have been possible. We are grateful to our other contributors Asher Auel, Marcello Bernardara, Emily Clader, Elden Elmanto, Zsolt Patakfalvi, Karl Schwede, Steven Sam for allowing us to include their excellent surveys. We thank Max Lieblich for all his help organizing the Bootcamp. We thank the University of Utah and the University of Utah Mathematics Department for hosting the Bootcamp and their support. We thank the National Science Foundation (DMS 1500652) and the National Security Agency (141006) for their financial support. Finally, we are greatful to the American Mathematical Society and especially Christine Thivierge and Sergei Gelfand for all their help in producing this volume.

Izzet Coskun
Tommaso de Fernex
Angela Gibney

A snapshot of the Minimal Model Program

Brian Lehmann

Abstract. We briefly review the main goals of the minimal model program. We then discuss which results are known unconditionally, which are known conditionally, and which are still open.

1. Introduction

This survey paper reviews the main goals and results of the Minimal Model Program (henceforth MMP). The paper has three parts. In Section 2, we give a very gentle introduction to the main ideas and conjectures of the MMP. We emphasize why the results are useful for many different areas of algebraic geometry.

In Sections 3-6, we take a "snapshot" of the MMP: we describe which results are currently known unconditionally, which are known conditionally, and which are wide open. We aim to state these results precisely, but in a manner which is as useful as possible to as wide a range of mathematicians as possible. Accordingly, this paper becomes more and more technical as we go. The hope is this paper will be helpful for mathematicians looking to apply the results of the MMP in their research.

In Section 7, we briefly overview current directions of research which use the MMP. Since the foundations of the MMP are discussed previously, this section focuses on applications. The choice of topics is not comprehensive and is idiosyncratically based on my own knowledge.

These notes are *not* intended to be a technical introduction to the MMP. There are many good introductions to the techniques of the MMP already: [**KM98**], [**Mat02**], [**HK10**], [**Kol13b**], and many others. These notes are also *not* intended to be a historical introduction. We will focus solely on the most recent results which are related to Principle 2.2: the existence of minimal models, termination of flips, and the abundance conjecture. Thus we will not cover in any length the many spectacular technical developments required as background. In particular, we will unfortunately omit most of the foundational results from the 1980's due to Kawamata, Kollár, Mori, Reid, Shokurov, and many others. We will also not cover the analytic side of the picture in any depth. To partially amend for this decision, we give a fairly complete list of references at the end.

Throughout we will work over $\mathbb{C}$ unless otherwise specified. Varieties are irreducible and reduced.

2010 *Mathematics Subject Classification.* Primary 14C20.

1.1. Acknowledgements. My sincere thanks to James McKernan, Calum Spicer, and Roberto Svaldi for many helpful suggestions and corrections.

2. Main idea of the MMP

WARNING 2.1. This entire section is filled with inaccuracies, imprecisions, over-simplifications, and outright falsehoods. The few terms which may be new to a general audience will be defined rigorously in the next section.

I will focus on the guiding principle:

PRINCIPLE 2.2. Let X be a smooth projective variety over a field. The geometry and arithmetic of X are governed by the "positivity" of the canonical bundle $\omega_X := \bigwedge^{\dim X} \Omega^1_X$.

We will participate in the traditional abuse of notation by letting K_X denote any Cartier divisor satisfying $\mathcal{O}_X(K_X) \simeq \omega_X$. Such a divisor is only unique up to linear equivalence, but since our statements are all linear-equivalence invariant this abuse is harmless in practice.

We first work out the case of dimension 1. Let C be a smooth integral projective curve with canonical bundle $\omega_C = \Omega_C$. The central feature of the curve is its genus:

$$g_C = \dim H^0(C, \Omega_C).$$

(Hodge theory shows that this coheres with the classical definition via Betti numbers.) By Riemann-Roch and Serre duality this is equivalent to saying that:

$$\deg(\Omega_C) = 2g_C - 2.$$

As is well-known, curves split into three categories based upon their genus or curvature. In line with Principle 2.2, it is most natural to describe this trichotomy in terms of the degree of the canonical bundle – only then do we see why the trichotomy is the right one.

$\deg(K_C)$	< 0	$= 0$	> 0
g_C	0	1	≥ 2
examples	$\mathbb{P}^1$	elliptic curves	smooth plane curves of degree > 3
universal cover/$\mathbb{C}$	$\mathbb{P}^1$	$\mathbb{C}$	$\mathbb{H}$
automorphisms	PGL_2	$\approx$ itself	finite
rational points over # field	dense after deg 2 ext	dense and thin after ext	finite

It would be very nice to have a similar trichotomy in higher dimensions. Of course this is too optimistic – a complex manifold of higher dimensions can have different curvatures in different directions – but we will soon see that there is some hope.

For arbitrary varieties X, we first need to decide what properties of K_X best correspond to the conditions $\deg(K_C) < 0$, $= 0$, > 0 for curves. Ample divisors are the most natural generalization of "positive degree" divisors for curves (and dually for antiampleness and "negative degree"). In the setting of the MMP, the best analogue of "degree 0" turns out to be the condition that K_X is torsion – some multiple of K_X is linearly equivalent to the 0 divisor. When K_X is ample, torsion, or antiample we say that our variety has "pure type".

With these changes, the trichotomy we found for curves seems to hold up well in higher dimensions. However, many of statements are still only conjectural – these will be designated with a question mark in the table below.

K_X	antiample	$\sim_{\mathbb{Q}} 0$	ample
examples	$\mathbb{P}^n$ Fanos	abelian varieties hyperkählers Calabi-Yaus	high degree hypersurfaces in $\mathbb{P}^n$
fundamental group/$\mathbb{C}$	1	almost abelian	??
rational curves on $X/\mathbb{C}$	dense	contained in countable union of proper subsets	not dense?
rational points over # field	potentially dense?	??	not dense?

While this conceptual picture is very appealing, at first glance it seems to only address a very limited collection of varieties. The main conjecture of the MMP is that *any* variety admits a "decomposition" into these varieties of pure types: at least after passing to a birational model, we can find a fibration with pure type fibers.

CONJECTURE 2.3 (Guiding conjecture of the MMP). *Any smooth projective variety X admits either:*

(i) a birational model $\psi : X \dashrightarrow X'$ and a morphism $\pi : X' \to Z$ with connected fibers to a variety of smaller dimension such that the general fiber F of π has K_F antiample, or

(ii) a birational model $\psi : X \dashrightarrow X'$ and a morphism $\pi : X' \to Z$ with connected fibers to a variety of smaller dimension such that the general fiber F of π has K_F torsion, or

(iii) a birational model $\psi : X \dashrightarrow X'$ and a birational morphism $\pi : X' \to Z$ such that K_Z is ample.

We will refer to the outcomes respectively as cases (i), (ii), (iii). It is clear why Conjecture 2.3 is so powerful – it suggests that we can leverage results for pure type varieties to study any variety via a suitable fibration.

Historically, this conjecture has its roots in the Kodaira-Enriques classification of surfaces, which categorizes birational equivalence classes of surfaces exactly according to the ability to find a morphism with fibers of a given pure type. (The fact that the birational map ψ may not be defined everywhere is a new feature in higher dimensions.)

REMARK 2.4. Even when X is a smooth surface, the varieties predicted by Conjecture 2.3 may have certain "mild" singularities. In this section we will ignore singularities completely, but the reader should remember they are there.

Implicit in the statement of Conjecture 2.3 is that the three cases have a hierarchy ordered by negativity: we look for an antiample fibration, then (failing to find any) a torsion fibration, then (failing that too) we expect to be in case (iii).

The justification is that it is quite easy to construct subvarieties with ample canonical divisor – for example, take complete intersections of sufficiently positive very ample divisors. The "most special" subvarieties are those with antiample canonical divisor, and we should look for these subvarieties first. (As we will see soon, this hierarchy is more properly motivated by the birational properties of K_X.)

The apparent asymmetry of case (iii) is also justified by this logic. Every variety admits many rational maps with fibers of general type, and the existence of such maps tells us essentially nothing about the variety. In contrast, Conjecture 2.3 has real geometric consequences.

REMARK 2.5. Another common perspective on the MMP is that it identifies a "distinguished set" of representatives of each fixed birational equivalence class of varieties. In dimension 2, the Kodaira-Enriques classification identifies a unique smooth birational model of any surface and we obtain a birational classification of surfaces. In higher dimensions, the analogous constructions are not unique, and so this perspective is slightly less useful.

Conjecture 2.3 suggests the following questions:

(a) How can we identify the target Z, or equivalently, the rational map $\phi = \pi \circ \psi : X \dashrightarrow Z$?
(b) How can we identify the rational map ψ and the birational model X'? What properties of X' distinguish it as the "right" birational model?
(c) How can we predict the case (i), (ii), (iii) of X based on the geometry of K_X?

We will answer these questions in the following subsections.

2.1. Canonical models. We first turn to Question (a): how to identify the variety Z? In other words, how can we naturally choose a rational map $\phi : X \dashrightarrow Z$ which captures the essential geometric features of X? For now we will focus on cases (ii) and (iii); case (i) is somewhat different.

Rational maps are constructed from sections of line bundles on X. For arbitrary varieties we only really have access to one line bundle: the canonical bundle. Furthermore, the canonical bundle encodes fundamental information about the curvature of our variety X. Thus it is no surprise that our "canonical map" ϕ should be constructed from the canonical divisor K_X.

A fundamental principle of birational geometry is that the geometry of a divisor L is best reflected not by sections of L but by working with all multiples of L simultaneously. We obtain access to this richer structure by the following seminal theorem of [**BCHM10**].

THEOREM 2.6 ([**BCHM10**] Corollary 1.1.2). *Let X be a smooth projective variety. Then the pluricanonical ring of sections*

$$R(X, K_X) := \bigoplus_{m \geq 0} H^0(X, mK_X)$$

is finitely generated.

Suppose that some multiple of K_X has sections (which should happen – and can only happen – in cases (ii) and (iii)). Since the pluricanonical ring is finitely generated, we can take its Proj.

DEFINITION 2.7. Suppose that some multiple of K_X has sections. The canonical model of X is defined to be $\operatorname{Proj} R(X, K_X)$.

As suggested by the notation, we obtain a rational map $\phi : X \dashrightarrow \operatorname{Proj} R(X, K_X)$ which is canonically determined by X. It is expected, but not yet proved, that this rational map exactly coincides with the map $\phi : X \dashrightarrow Z$ predicted by Conjecture 2.3 in cases (ii) and (iii).

REMARK 2.8. It is worth mentioning briefly how case (i) fits into the picture. In this case, no multiple of K_X has sections and $\operatorname{Proj} R(X, K_X)$ is empty. Our solution is simply to add on a sufficiently large ample divisor A to ensure that multiples of $K_X + A$ have sections, and then to work with $\operatorname{Proj} R(X, K_X + A)$. Of course, there are now many choices of rational map corresponding to the possible choices of ample divisor A. This is reflected in the richer birational MMP structure for varieties falling into case (i).

There are two reasons why Theorem 2.6 does not solve Conjecture 2.3. First, it is unclear whether any multiples of K_X have sections; if not, we can not hope to obtain any geometry from the canonical map. Second, it is a priori unclear whether ϕ has the special structure predicted by the conjecture: does it factor as a birational map $\psi : X \dashrightarrow X'$ followed by a fibration with pure type fibers?

2.1.1. *Obstructions from negativity.* It is worth exploring this second point in more detail. Let us identify the obstruction to the existence of a factorization of rational maps $\phi = \pi \circ \psi$, where ψ is birational and π has pure type fibers.

Suppose first for simplicity that the rational map $\phi : X \dashrightarrow \operatorname{Proj} R(X, K_X)$ is defined everywhere. Due to the properties of the Proj construction, there is an ample divisor A on $\operatorname{Proj} R(X, K_X)$ and a positive integer m such that $mK_X = \phi^* A + E$, where E is an effective divisor contained in the base locus of mK_X. We consider two cases:

- $\dim \operatorname{Proj} R(X, K_X) < \dim X$. By adjunction, for a general fiber F we have $mK_F \sim E|_F$. Thus E represents the obstruction to the torsionness of K_F.
- $\dim \operatorname{Proj} R(X, K_X) = \dim X$. If the singularities of the canonical model are sufficiently mild, its canonical divisor is well defined. Then the canonical divisor of the canonical model necessarily coincides with the ample divisor A. The mildness of the singularities of the canonical model are controlled by E.

In either case we see that the divisor E represents an "obstruction" to the conclusion of Conjecture 2.3.

In the general case (when ϕ is only rational), a similar argument shows that the base locus of multiples of K_X prevents the fibers from having pure type. We are naturally led to try to "remove" the base locus of multiples of K_X via a birational transformation. (In fact, essentially the only way of proving that a divisor has finitely generated section ring is to find a birational model of X on which the strict transform of the divisor has no base locus.)

The problem of "removing" the base locus of multiples of K_X is traditionally divided into two steps. First, we find a birational model X' such that $K_{X'}$ has non-negative intersection against every curve. This condition on $K_{X'}$ is necessary, but not sufficient, for ensuring an empty base locus. We then hope to use this condition to prove that multiples of $K_{X'}$ have no base locus.

2.2. Running the MMP. We next turn to Question (b): how to find the birational model $\psi : X \dashrightarrow X'$ predicted by Conjecture 2.3? As discussed above, we would like to remove the base locus of K_X. This is accomplished by an inductive procedure known as "running the minimal model program." In brief, we would like to repeatedly contract curves which have negative intersection against the canonical divisor.

First, we need to know that if X carries K_X-negative curves – that is, curves with negative intersection against K_X – then there is a morphism $f : X \to Y$ with connected fibers to a normal projective Y which contracts some of these curves. This foundational result is known as the Cone Theorem, and can be applied whenever the singularities of X are sufficiently mild (the version we will use in this section is due to [**Kaw84a**]).

The MMP procedure starts when f is a birational map. There are two possibilities to consider:

- (Divisorial contraction) The morphism f could contract an irreducible divisor. In this case the variety Y only has mild singularities, and we can continue the MMP by applying the Cone Theorem to Y.
- (Flipping contraction) The morphism f could contract a locus of codimension ≥ 2. In this case the variety Y has harsh singularities: we cannot apply the Cone Theorem to Y. The key insight is that we should not work with Y, but with a related variety X^+: by [**HM10a**], [**BCHM10**] there is a diagram

$$\begin{array}{ccc} X & \dashrightarrow & X^+ \\ & {\scriptstyle f}\searrow \quad \swarrow{\scriptstyle f^+} & \\ & Y & \end{array}$$

  where X^+ has mild singularities, X and X^+ are isomorphic in codimension 1, and the contracted curves for f are K_X-negative but the contracted curves for f^+ are K_{X^+}-positive. Thus in passing from X to X^+ we have eliminated some "negativity" of the canonical divisor. Since X^+ again only has mild singularities, we can continue by applying the MMP by applying the Cone Theorem to X^+. This diagram is known as a flip diagram, and the rational map $X \dashrightarrow X^+$ is known as a flip.

The main conjecture of the MMP is that we can do only a finite sequence of such birational transformations. It is not hard to see that there can be only finitely many steps in the MMP which contract a divisor: each such step drops the Picard rank, while flips preserve the Picard rank. However, it is much more subtle to determine whether there can be an infinite sequence of flips:

CONJECTURE 2.9 (Termination of flips). *There is no infinite sequence of flips.*

2.2.1. *End result of the MMP.* Assuming that we can only do a finite sequence of birational steps as above, the process will end with a birational model $\psi : X \dashrightarrow X'$. This X' must satisfy one of the following two conditions.

The first possibility is that X' contains $K_{X'}$-negative curves, but the corresponding contraction $f : X' \to Z$ maps to a variety of smaller dimension. This f achieves the desired conclusion for case (i): the fibers of f have antiample canonical divisor, and the birational map ψ is the map we are looking for.

The second possibility is that $K_{X'}$ is not negative against any curve in X'. Conjecturally $\psi : X \dashrightarrow X'$ is exactly the birational map predicted by cases (ii) and (iii) (see the next section for more details). At the very least, we have successfully eliminated the "intersection-theoretic" contributions to the base locus of K_X. This condition on X' is useful enough to merit its own terminology:

DEFINITION 2.10. A variety X' is called a minimal model if $K_{X'}$ has non-negative intersection against every curve. We say X' is a minimal model for X if it is a minimal model achieved by running the MMP for X (but a more precise definition will be given in Definition 4.2).

Note that this usage is slightly different than the classical usage for surfaces involving (-1)-curves. For smooth surfaces, the non-existence of (-1)-curves is necessary but not sufficient to be a minimal model in our sense. An important feature which only appears in dimension > 2 is that X might admit many different minimal models if we make different choices of morphism while applying the Cone Theorem.

The termination of flips conjecture would imply:

CONJECTURE 2.11 (Existence of minimal models). *Suppose that X is not uniruled. Then X admits a minimal model.*

Let us briefly contrast the notion of minimal and canonical models:

(1) Minimal models are expected to exist whenever X is not uniruled, and canonical models exist when some multiple of K_X has sections. These two conditions on X are expected to be equivalent by the Abundance Conjecture below, and should correspond to cases (ii) and (iii).
(2) A minimal model for X is always birational to X, but a canonical model for X may have smaller dimension (in case (ii)).
(3) There can be many minimal models for X, but there is only one canonical model for X.
(4) There should be a morphism (defined everywhere) from any minimal model for X to the canonical model for X, as predicted by the Abundance Conjecture below.
(5) If X' is a minimal model for X then $K_{X'}$ may have vanishing intersection against some curves, but (conjecturally) all such curves should be contracted by the morphism to the canonical model.

2.3. Abundance Conjecture. Finally, suppose that by running the MMP we have found a minimal model X' whose canonical divisor $K_{X'}$ has non-negative degree against every curve. We hope that $K_{X'}$ has no base locus. This is the content of the following conjecture:

CONJECTURE 2.12 (Abundance Conjecture). *Suppose that $K_{X'}$ has non-negative degree against every curve. Then there is a morphism $\pi : X' \to Z$ and an ample divisor A on Z such that some multiple of $K_{X'}$ is linearly equivalent to $\pi^* A$.*

Note that if the Abundance Conjecture is true, then Z will be the canonical model:

$$\begin{aligned} Z &= \operatorname{Proj} \bigoplus_{m=0}^{\infty} H^0(X', mK_{X'}) \\ &= \operatorname{Proj} \bigoplus_{m=0}^{\infty} H^0(X, mK_X) \end{aligned}$$

due to the birational invariance of plurigenera.

As suggested by the notation, the morphism $\pi : X' \to Z$ in the Abundance Conjecture should be the same as the morphism π in Conjecture 2.3. Let us analyze the connection more carefully. Assuming the Abundance Conjecture:

- In case (ii), we will necessarily have $\dim(Z) < \dim(X)$. Using adjunction, we see that some multiple satisfies $mK_X|_F \sim 0$ for a general fiber F. So F has torsion pure type as desired.
- In case (iii), Z will be necessarily be birational to X. Since canonical divisors relate well over birational maps, the identification $K_X \sim_{\mathbb{Q}} \pi^*A$ actually allows us (after a more careful setup and argument) to identify $A = K_Z$. So X will be birational to a variety Z with ample canonical divisor.

According to the Abundance Conjecture, a numerical property for K_X – having non-negative intersection against every curve – implies a section property – some multiple has no base locus. Such implications are quite rare: usually one can not deduce holomorphic information from intersection theory. In fact, this is an important theme in the study of the canonical bundle which applies in much more generality:

PRINCIPLE 2.13. The behavior of sections of multiples of K_X is governed by intersection theoretic properties of ω_X.

We finally answer Question (c): how can we determine which case (i), (ii), (iii) X falls into? The case depends on the birational positivity of K_X. According to Principle 2.13, we can use either numerical or sectional forms of positivity. Although we have not yet seen the relevant definitions, the *conjectural* picture is summarized below:

	Case (i)	Case (ii)	Case (iii)
Sectional property	$\kappa(X) = -\infty$	$0 \leq \kappa(X) < \dim X$, $\kappa(X) = \dim Z$	$\kappa(X) = \dim X$
Numerical properties	$K_X \notin \overline{\mathrm{Eff}}^1(X)$, equivalently $\nu(K_X) = -\infty$	K_X on the boundary of $\overline{\mathrm{Eff}}^1(X)$, $\nu(K_X) = \dim Z$	$K_X \in \overline{\mathrm{Eff}}^1(X)^\circ$, equivalently $\nu(K_X) = \dim X$
Curve properties	uniruled, equivalently dominated by K_X-negative curves	dominated by K_X-trivial curves (but not negative ones)	neither of the previous conditions

3. Background

A pair (X, Δ) is a normal variety X and an effective $\mathbb{R}$-Weil divisor Δ on X such that $K_X + \Delta$ is $\mathbb{R}$-Cartier.

We refer to any standard reference for a summary of the various singularity types: terminal, canonical, Kawamata log terminal (henceforth klt), log canonical (henceforth lc), etc.

3.1. Birational geometry of divisors. The Neron-Severi space $N^1(X)$ is the vector space of $\mathbb{R}$-Cartier divisors up to numerical equivalence. The dual space is $N_1(X)$. We define the pseudo-effective and nef cones:

- the pseudo-effective cone of divisors $\overline{\mathrm{Eff}}^1(X)$ is the closure of the cone in $N^1(X)$ generated by the classes of effective Cartier divisors;
- the pseudo-effective cone of curves $\overline{\mathrm{Eff}}_1(X)$ is the closure of the cone in $N_1(X)$ generated by the classes of effective curves
- the nef cone of divisors $\mathrm{Nef}^1(X)$ is the dual of $\overline{\mathrm{Eff}}_1(X)$;
- the nef cone of curves $\mathrm{Nef}_1(X)$ is the dual of $\overline{\mathrm{Eff}}^1(X)$.

The nef cone of divisors can also be interpreted as the closure of the cone generated by all ample Cartier divisors (see [**Kle66**]). The nef cone of curves can also be interpreted as the closure of the cone generated by all irreducible curves which deform to cover X (see [**BDPP13**]).

We say that an $\mathbb{R}$-Cartier divisor D is

- pseudo-effective, if its numerical class is contained in $\overline{\mathrm{Eff}}^1(X)$.
- big, if its numerical class is contained in $\overline{\mathrm{Eff}}^1(X)^\circ$.
- nef, if its numerical class is contained in $\mathrm{Nef}^1(X)$.
- ample, if its numerical class is contained in $\mathrm{Nef}^1(X)^\circ$.

The movable cone of divisors $\mathrm{Mov}^1(X)$ is the closure of the cone generated by all divisors whose base locus has codimension ≥ 2. Equivalently, it is the closure of the cone generated by all divisors such that every irreducible component deforms to cover X.

Suppose X has dimension n. Given an $\mathbb{R}$-Cartier divisor D, its Iitaka dimension is

$$\kappa(D) = \max\left\{ k \in \mathbb{Z}_{\geq 0} \,\middle|\, \limsup_{m\to\infty} \frac{\dim H^0(X, \mathcal{O}_X(\lfloor mD \rfloor))}{m^k} > 0 \right\}.$$

unless every $H^0(X, \lfloor mD \rfloor) = 0$, in which case we formally set $\kappa(D) = -\infty$. It is well-known that

- D is big if and only if $\kappa(D) = n$.
- if D is not pseudo-effective then $\kappa(D) = -\infty$;

however, the converse of the latter statement is false. By combining the seminal results of [**MM86**] and [**BDPP13**], a smooth variety X is uniruled if and only if K_X is not pseudo-effective.

We define the numerical dimension of an $\mathbb{R}$-Cartier divisor in a similar way. Choose any sufficiently ample Cartier divisor A. Then

$$\nu(D) = \max\left\{ k \in \mathbb{Z}_{\geq 0} \,\middle|\, \limsup_{m\to\infty} \frac{\dim H^0(X, \mathcal{O}_X(\lfloor mD \rfloor) + A)}{m^k} > 0 \right\}.$$

unless D is not pseudo-effective, in which case we formally set $\nu(D) = -\infty$. It is well-known that D is big if and only if $\nu(D) = n$. On the other extreme, divisors satisfying $\nu(D) = 0$ are very "rigid": they are numerically equivalent to a unique effective divisor E, and even after perturbing by a small ample divisor, we can not deform E away from its support. The prototypical example of a divisor D satisfying $\nu(D) = 0$ is an exceptional divisor of a blow-up.

There is always an inequality $\kappa(D) \leq \nu(D)$. A divisor D is said to be abundant if $\kappa(D) = \nu(D)$. When D is pseudo-effective, this condition turns out to be equivalent to a very natural geometric statement: after a birational modification, the positive part in the Nakayama-Zariski decomposition of D is the pullback of a big divisor on a variety of dimension $\kappa(D) = \nu(D)$.

We refer to [**Nak04**], [**Leh12**], and [**Eck15**] for more details on these invariants.

3.2. Birational contractions. A birational contraction is a birational map $\phi : X \dashrightarrow Y$ which does not extract any divisor (or equivalently, the inverse map ϕ^{-1} does not contract any divisor to a smaller dimensional locus). This implies that the divisor theory on Y is "controlled" by the divisor theory on X.

One reason why birational contractions are useful is that any birational map $\psi : X \dashrightarrow X'$ constructed via a sequence of flips and divisorial contractions is a birational contraction. In fact, there is an additional property which (more-or-less) identifies the outcomes of the MMP amongst all birational contractions.

Definition 3.1. Let (X, Δ) be a pair. A birational contraction $\phi : X \dashrightarrow Y$ is called a $(K_X + \Delta)$-negative birational contraction if $(Y, \phi_*\Delta)$ is a pair and for some (equivalently any) resolution

$$\begin{array}{ccc} & W & \\ {\scriptstyle g_X} \swarrow & & \searrow {\scriptstyle g_Y} \\ X & \overset{\phi}{\dashrightarrow} & Y \end{array}$$

we have that $g_X^*(K_X + \Delta) = g_Y^*(K_Y + \phi_*\Delta) + \sum a_i E_i$ where each a_i is positive and E_i varies over all the g_Y-exceptional divisors.

If we allow some $a_i = 0$, the contraction is called $(K_X + \Delta)$-non-positive.

4. Main conjectures

We now state precisely the main conjectures of the MMP. We focus almost exclusively on establishing when the loosely phrased Principle 2.2 and Conjecture 2.3 hold for a variety X and on related issues.

Remark 4.1. I will precisely state the theorems proved in the literature, with the somewhat annoying consequence of frequently switching back and forth between singularity types. I will also indicate when log canonical pairs are known to be irredeemably worse behaved.

While the extensions to worse singularities (semi log canonical pairs) are very important, they introduce an additional layer of technicality not in keeping with the spirit of this paper and will not be discussed. We will also not work in the relative setting, for the same reason, despite the important additional theoretical flexibility it provides. Essentially all of the results stated below go through unchanged.

We split the main conjectures into three parts: existence of a (good) minimal model, termination of flips, and various flavors of the abundance conjecture. Note the existence of a minimal model usually comes down to the termination of a "special" sequence of flips, which distinguishes the problem from the termination of arbitrary sequences of flips.

4.0.1. *Existence of a good minimal model.* The definition of a minimal model is supposed to encode the end result of the MMP, without keeping track of the steps taken. (Recall that running the MMP is more-or-less the same as identifying a $(K_X + \Delta)$-negative contraction). The definition of a *good* minimal model furthermore encodes the expected existence of the "pure type map" given by the canonical model. These are exactly the structures predicted by Conjecture 2.3 in cases (ii) and (iii).

DEFINITION 4.2. Let (X, Δ) be a lc pair. A minimal model of (X, Δ) is a $(K_X + \Delta)$-negative birational contraction $\psi : X \dashrightarrow X'$ such that $(X', \psi_*\Delta)$ is a lc pair and $K_{X'} + \psi_*\Delta$ is nef.

A good minimal model of (X, Δ) is a minimal model $\psi : X \dashrightarrow X'$ such that $K_{X'} + \psi_*\Delta$ is semiample – that is, $\mathbb{R}$-linearly equivalent to the pullback of an ample divisor under some morphism $\pi : X' \to Z$.

In dimension ≥ 3, the following conjectures have their roots in [**Mor82**].

CONJECTURE 4.3. *(Existence of minimal models) Let (X, Δ) be a lc pair. If $K_X + \Delta$ is pseudo-effective, then (X, Δ) admits a minimal model.*

CONJECTURE 4.4. *(Existence of good minimal models) Let (X, Δ) be a lc pair. If $K_X + \Delta$ is pseudo-effective, then (X, Δ) admits a good minimal model.*

A minimal model can only exist if $K_X + \Delta$ is pseudo-effective. To understand the non-pseudo-effective case, by analogy with the notion of a minimal model, we must identify the expected result in case (i) of Conjecture 2.3:

DEFINITION 4.5. Let (X, Δ) be a lc pair. A birational Mori fiber space structure for (X, Δ) is a $(K_X + \Delta)$-negative birational contraction $\psi : X \dashrightarrow X'$ and a morphism $\pi : X' \to Z$ with connected fibers such that $\dim(Z) < \dim(X)$ and $(K_{X'} + \psi_*\Delta)|_F$ is antiample for a general fiber F of π. (It is also common to insist that π have relative Picard rank 1, in which case we may insist that $K_{X'} + \psi_*\Delta$ is antiample along every fiber. We will however use the more general version.)

There is no need for a "existence of Mori fiber spaces conjecture" in the klt case since (as discussed below) the existence has already been proved by [**BCHM10**].

4.0.2. *Termination of flips.*

DEFINITION 4.6. A flip consists of a lc pair (X, Δ) and a diagram of birational maps

$$\begin{array}{ccc} X & \overset{\phi}{\dashrightarrow} & X^+ \\ & {\scriptstyle f}\searrow \quad \swarrow{\scriptstyle f^+} & \\ & Y & \end{array}$$

such that f and f^+ have exceptional locus of codimension at least 2, $(X^+, \phi_*\Delta)$ is a lc pair, $K_X + \Delta$ is f-antiample, and $K_{X^+} + \phi_*\Delta$ is f^+-ample.

Flips are known to exist in the lc case by [**Bir12**, Corollary 1.2], [**HX12**, Corollary 1.8]. The termination of flips conjecture predicts that there is no infinite sequence of flips.

WARNING 4.7. A flop is the similar diagram where both maps are crepant. While klt flops exist, lc flops need not exist (see [**Kol08**, Exercise96]). Furthermore, even for klt flops there can be a non-terminating sequence by [**Rei83**] or [**Ogu14**].

4.0.3. *Abundance conjecture.* The abundance conjecture predicts that the asymptotic sectional properties of the canonical divisor are controlled by its numerical properties. Results of this type were first proved for varieties of general type by [**Ben83**] and [**Kaw84c**] in dimension 3 and by [**Sho85**] in general. They were also first proved for varieties of intermediate Kodaira dimensions by [**Kaw85b**]. Two common versions of the abundance conjecture are:

CONJECTURE 4.8 (Semi-ample abundance conjecture). *If (X, Δ) is a lc pair such that $K_X + \Delta$ is nef, then $K_X + \Delta$ is $\mathbb{R}$-semi-ample.*

CONJECTURE 4.9 (Non-vanishing conjecture). *If (X, Δ) is a lc pair such that $K_X + \Delta$ is pseudo-effective, then $\kappa(K_X + \Delta) \geq 0$.*

These phrasings are useful but slightly unsatisfactory, since they implicitly rely on the existence of a minimal model for use in applications. My preference is for the following version, which again predicts that the sectional properties are controlled by numerical properties but in a cleaner "birational" sense.

CONJECTURE 4.10 (Abundance conjecture). *Let (X, Δ) be a lc pair. Then $K_X + \Delta$ is abundant: $\kappa(K_X + \Delta) = \nu(K_X + \Delta)$.*

It is common to write C_n as a shorthand to denote "the statement of Conjecture C for varieties of dimension at most n." The easy relationships between the conjectures are:

- Termination of $\text{flips}_n \implies$ Existence of minimal models_n.
- Existence of minimal models_n + Semi-ample $\text{abundance}_n \iff$ Existence of good minimal models_n.
- Existence of good minimal $\text{models}_n \implies \text{Non-vanishing}_n$ + Semi-ample abundance_n.

As we will see below, there are other (quite difficult) implications between the various statements in the literature.

5. Unconditional results

We now discuss what is known about the main conjectures (and also mention a few useful technical corollaries). We divide these results into "unconditional" and "conditional" results. In this section we discuss "unconditional" results, meaning theorems that can directly be verified on a single variety. In the following section we discuss "conditional" results, which usually means results which "follow from induction on dimension, given some assumptions."

5.1. Existence of good minimal models. The most important unconditional advances in the MMP are due to [**BCHM10**]. One of the key technical advances in [**BCHM10**] is a special case of the termination of flips. Recall that while running the MMP, we allowed ourselves to choose any $(K_X+\Delta)$-negative face of $\overline{\mathrm{Eff}}_1(X)$ at each step. [**BCHM10**] shows that if we instead limit our choice by "scaling an ample divisor", then under some conditions the sequence of flips must terminate.

The first result establishes Conjecture 2.3 in the case when $K_X + \Delta$ is not pseudo-effective (thus establishing that "case (i)" holds whenever it is possible).

THEOREM 5.1 ([**BCHM10**], Corollary 1.3.2). *Let (X, Δ) be a dlt pair. Suppose that $K_X + \Delta$ is not pseudo-effective. Then (X, Δ) admits a birational Mori fiber space structure.*

The next result establishes Conjecture 2.3 in the case when $K_X + \Delta$ is big (thus establishing that "case (iii)" holds whenever it possible).

THEOREM 5.2 ([**BCHM10**], Theorem 1.2). *Let (X, Δ) be a klt pair. Suppose that $K_X + \Delta$ is big. Then (X, Δ) admits a good minimal model.*

In fact, something more is true: the techniques of [**BCHM10**] can be applied so long as Δ itself is big. Thus there is one additional case: when K_X itself is not big, but upon adding a big Δ it lands on the boundary of the pseudo-effective cone. This additional case is applicable to uniruled varieties.

THEOREM 5.3 ([**BCHM10**], Theorem 1.2). *Let (X, Δ) be a dlt pair. Suppose that $K_X + \Delta$ is pseudo-effective and Δ is big $\mathbb{R}$-Cartier. Then (X, Δ) admits a good minimal model.*

Since we will refer to these results later, we state:

CONDITION 5.4. (X, Δ) is a pair such that either Δ is big $\mathbb{R}$-Cartier or $K_X + \Delta$ is big.

Note that the only unsettled cases of the existence of good minimal models for klt pairs are all "case (ii)": situations where $K_X + \Delta$ lies on the boundary of the pseudo-effective cone. These are the situations where $0 \leq \kappa(K_X + \Delta) \leq \nu(K_X + \Delta) < n$.

Perhaps the next most natural case to consider is when $K_X + \Delta \equiv 0$; is it then true that $K_X + \Delta \sim_{\mathbb{Q}} 0$ as predicted by the Abundance Conjecture? The answer is yes for any log canonical (or even semi-log canonical) pair as proved by [**Gon13**]. Other results in this direction have been proved in [**Kaw85a**], [**Nak04**], [**Gon11**], [**CKP12**], [**Kaw13b**]. In fact, a stronger result for dlt pairs was proved by [**Nak04**], [**Dru11**], [**Gon11**].

THEOREM 5.5 ([**Nak04**] V.4.8 Theorem, [**Dru11**] Corollaire 3.4, [**Gon11**] Theorem 1.2). *Let (X, Δ) be a dlt pair. Suppose that $\nu(K_X + \Delta) = 0$. Then (X, Δ) admits a good minimal model.*

In other words, if $K_X + \Delta$ is numerically rigid, then it is actually $\mathbb{R}$-linearly equivalent to an exceptional divisor for a birational map. It is interesting that there are complete results on both extremes – both for the most positive divisors ($\nu = \dim X$) and the most rigid divisors ($\nu = 0$).

Note that the condition $\nu = 0$ implies the condition $\kappa = 0$, but not conversely; indeed, if one could establish the $\kappa = 0$ case then the existence of good minimal models would follow for varieties of positive Kodaira dimension.

It turns out that the numerical dimension is a very useful tool for working with minimal models. The most general statement, which subsumes most of the statements we have made already, is due to [**Lai10**]. (The statement cited is for terminal varieties but the argument works as well for klt varieties.)

THEOREM 5.6 ([**Lai10**] Theorem 4.4). *Let (X, Δ) be a klt pair. Suppose that $K_X + \Delta$ is pseudo-effective and abundant: $\kappa(K_X + \Delta) = \nu(K_X + \Delta)$. Then (X, Δ) admits a good minimal model.*

Note that no induction assumption is necessary in the statement. This theorem is most useful in situations where abundance is known automatically: if $K_X + \Delta$ is big, if $\nu(K_X + \Delta) = 0$, or if $\kappa(K_X + \Delta) = n - 1$. Recently an additional step has been taken:

THEOREM 5.7 ([**LP16**] Theorem B). *Let X be a terminal normal variety with K_X nef. If $\nu(K_X) = 1$ and $\chi(X, \mathcal{O}_X) \neq 0$ then $\kappa(K_X) \geq 0$.*

Also, after much hard work the main conjectures of the MMP are known for all varieties of small dimension:

THEOREM 5.8.

- *Termination of flips is known in dimension ≤ 3. See* [**Mor88**], [**Kol89**], [**Sho93**], [**Kaw92c**].
- *All forms of the Abundance Conjecture (and hence existence of good minimal models) are known in dimension ≤ 3. See* [**Kaw92a**], [**Miy88b**], [**Miy88a**], [**KMM94**].
- *The existence of minimal models (via a special case of termination of flips) is known in dimension ≤ 4. See* [**AHK07**], [**Sho09**], [**Bir09b**].
- *The existence of minimal models is known for klt pairs (X, Δ) with $\kappa(K_X + \Delta) \geq 0$ in dimension ≤ 5. See* [**Bir10a**].

5.2. Structure of minimal models. Once we have established the existence of a good minimal model for (X, Δ), it is natural to ask if the set of all good minimal models has any kind of structure. It is important to distinguish between two options:

- We can consider all minimal models of (X, Δ) as abstract varieties, up to isomorphism.
- We can consider possible ways to construct a minimal model for (X, Δ) by running the MMP $\psi : X \dashrightarrow X'$. We identify $\psi : X \dashrightarrow X'$ and $\phi : X \dashrightarrow X''$ only if the rational map $\psi^{-1} \circ \phi$ extends to an isomorphism. This is the same as counting the number of distinct subcones $\psi^{-1}\mathrm{Amp}^1(X') \subset \overline{\mathrm{Eff}}^1(X)$ defined by outcomes of the MMP.

For surfaces, we only have a finite number of minimal models using either of the counting methods. (Note that this is not true for the "classical" definition of minimal model, where we only insist that there are no (-1)-curves. The blow-up of $\mathbb{P}^2$ in 9 points gives a counter-example.) But in higher dimensions the situation is more subtle.

5.2.1. *Structure of minimal models.* We first simply consider minimal models as abstract varieties. The strongest known results are for varieties of general type by [**BCHM10**] (which proves something stronger as we will soon see).

THEOREM 5.9 ([**BCHM10**], Corollary 1.1.5). *Let (X, Δ) be a klt pair with $K_X + \Delta$ big. Then there are only finitely many minimal models for (X, Δ) as abstract varieties, up to isomorphism.*

For intermediate Kodaira dimensions, the situation is much more subtle. In particular, the following question of Kawamata is open:

QUESTION 5.10. Does every variety X only admit finitely many minimal models as abstract varieties, up to isomorphism?

Certain cases of this question are known.

THEOREM 5.11 ([**Kaw97b**] Theorem 4.5). *Let X be a smooth projective variety of dimension 3 with $\kappa(X) > 0$. Then X only has finitely many minimal models as abstract varieties, up to isomorphism.*

REMARK 5.12. One can also phrase a stronger question: are there only finitely many minimal models in a fixed birational equivalence class? (That is, are there only finitely many terminal $\mathbb{Q}$-factorial normal varieties X' birational to X with $K_{X'}$ nef?) Both Kawamata's original question and the results of [**Kaw97b**] are phrased in this generality.

5.2.2. *Set of MMP outcomes.* In this section we discuss possible outcomes of the MMP $\phi : X \dashrightarrow X'$, and not just the abstract varieties X'. As discussed already, the "first step" of the MMP is to choose a K_X-negative extremal ray or face of $\overline{\mathrm{Eff}}_1(X)$. A key insight of Mori is to interpret a step of the MMP via the curve classes it contracts in $\overline{\mathrm{Eff}}_1(X)$. The precise statement is known as the Cone Theorem. Building on [**Kaw84c**, Theorem 4.5] and [**Kol84**, Theorem 1], we have:

THEOREM 5.13 ([**Fuj11**] Theorem 1.4). *Let (X, Δ) be a lc pair. There are countably many $(K_X + \Delta)$-negative rational curves C_i such that $0 < -(K_X + \Delta) \cdot C_i < 2 \dim X$ and*

$$\overline{\mathrm{Eff}}_1(X) = \overline{\mathrm{Eff}}_1(X)_{K_X+\Delta\geq 0} + \sum \mathbb{R}_{\geq 0}[C_i].$$

The rays $\mathbb{R}_{\geq 0}[C_i]$ only accumulate along the hyperplane $(K_X + \Delta)^{\perp}$.

The $K_X + \Delta$-negative faces of $\overline{\mathrm{Eff}}_1(X)$ are in bijection with $K_X + \Delta$-negative contractions.

Note that if (X, Δ) is klt and $K_X + \Delta$ is big, then a perturbation argument shows that there are only finitely many $K_X + \Delta$-negative minimal rays.

As mentioned before, analyzing outcomes of the MMP (up to isomorphism) is essentially the same as counting regions in the pseudo-effective cone corresponding to the pullback of the nef cone on the various results of the program. Thus it will still be useful to phrase results in terms of the structure of cones.

We now discuss the final outcomes of the MMP in the three cases (i), (ii), (iii). Often one can interpret the set of outcomes via a "cone theorem" describing the structure of the cone of curves. When $K_X + \Delta$ is big, the finiteness noted above persists through the entire MMP process:

THEOREM 5.14 ([**BCHM10**], Corollary 1.1.5). *Let (X, Δ) be a klt pair with $K_X + \Delta$ big. Let T be a subcone of $\overline{\mathrm{Eff}}^1(X)$ over a compact set. Suppose that every ray in T is generated by a pair (X, Δ) satisfying Condition 5.4. Then there are only finitely many birational contractions defined by running the MMP for the corresponding divisors.*

In other words, the region in $\overline{\mathrm{Eff}}^1(X)$ consisting of divisors which are proportional to a pair as in Condition 5.4 looks "Mori Dream Space-like", in the sense that it admits a chamber decomposition satisfying the same properties as a Mori Dream Space. (See **[HK00]** or **[CL13]** for a discussion of this viewpoint.) This is particularly useful for log Fano varieties, since it shows that they are Mori Dream Spaces.

If we pass to the situation when $K_X + \Delta$ lies on the pseudo-effective boundary the picture becomes much more complicated. **[Les15]** gives an example of a non-uniruled terminal threefold with infinitely many K_X-negative extremal rays; any resolution of this variety will admit an infinite set of minimal model outcomes. Nevertheless the set of MMP outcomes conjecturally has a rich structure. Given an effective divisor Δ on X, we denote by

- $\mathrm{Aut}(X, \Delta)$ the automorphisms of X which preserve Δ,
- $\mathrm{PsAut}(X, \Delta)$ the birational maps $\psi : X \dashrightarrow X$ which are an isomorphism in codimension 1 and which preserve Δ.

We will only phrase the conjecture in the case when $K_X + \Delta$ is numerically trivial; for the essentially identical relative statement which addresses all case (ii) morphisms, see **[Tot08]**.

CONJECTURE 5.15 (Kawamata-Morrison Cone Conjectures). *Let (X, Δ) be a klt pair such that $K_X + \Delta$ is numerically trivial.*

(1) There is a finite rational polyhedral cone Π which is a fundamental domain for the action of $\mathrm{Aut}(X, \Delta)$ on the effective nef cone (that is, the intersection of $\mathrm{Nef}^1(X)$ with the cone generated by all effective divisors).
(2) There is a finite rational polyhedral cone Π' which is a fundamental domain for the action of $\mathrm{PsAut}(X, \Delta)$ on the effective movable cone (that is, the intersection of $\mathrm{Mov}^1(X)$ with the cone generated by all effective divisors).

These conjectures also predict that there are only finitely many equivalence classes under the group action of faces of the cones corresponding to actual morphisms or marked small $\mathbb{Q}$-factorial modifications.

WARNING 5.16. These conjectures are false for log canonical pairs. **[Tot08]** gives the example where X is $\mathbb{P}^2$ blown up at 9 very general points and Δ is the strict transform of the elliptic curve through the points. In this case the cone is not rational polyhedral but the automorphism group of X is trivial.

The Kawamata-Morrison Cone Conjectures are known for abelian varieties (**[PS12b]**) and (essentially) for hyperkähler manifolds (**[AV14]** and **[MY14]** for 1, **[Mar11]** for 2). They are also known completely in dimension ≤ 2 (**[Ste85]**, **[Nam85]**, **[Tot10]**) and the relative versions are known in dimension 3 over a positive dimensional base (**[Kaw97b]**). Many additional special cases have been proved in **[PS12a]**, **[CPS14]**, **[Ogu01]**, **[Ogu14]**, **[CO11]**, **[LP13]**, **[Bor91]**, **[Wil94]**, **[Ueh04]**, **[Sze99]**, **[Zha14]**, etc.

Finally, in the case when $K_X+\Delta$ is not pseudo-effective, we should instead look for all possible birational Mori fiber space structures. Again, this can be interpreted as a structure theorem for a suitable cone of curves. Building on **[Bat92]**, **[Ara10]** we have:

THEOREM 5.17 (**[Leh12]**). *Let (X, Δ) be a dlt pair. There are countably many $(K_X + \Delta)$-negative movable curves C_i such that*

$$\overline{\mathrm{Eff}}_1(X)_{K_X+\Delta\geq 0} + \mathrm{Nef}_1(X) = \overline{\mathrm{Eff}}_1(X)_{K_X+\Delta\geq 0} + \overline{\sum \mathbb{R}_{\geq 0}[C_i]}.$$

The rays $\mathbb{R}_{\geq 0}[C_i]$ only accumulate along hyperplanes that support both $\mathrm{Nef}_1(X)$ *and* $\overline{\mathrm{Eff}}_1(X)_{K_X+\Delta\geq 0}$.

The birational equivalence classes of birational Mori fiber space structures are in bijection with the faces of this cone which admit a supporting hyperplane not intersecting $\overline{\mathrm{Eff}}_1(X)_{K_X+\Delta\geq 0}$.

The presence of the term $\overline{\mathrm{Eff}}_1(X)_{K_X+\Delta\geq 0}$ on both sides has the effect of "rounding out the cone" and can not be removed. However, Batyrev conjectures a stronger statement: the accumulation of rays only occurs along the hyperplane $(K_X+\Delta)^{\perp}$.

REMARK 5.18. Again, one can pose a harder question: if we fix a smooth projective variety X, what is the set of all minimal models X' birationally equivalent to X equipped with a birational contraction $X \dashrightarrow X'$? We identify $\psi : X \dashrightarrow X'$ and $\phi : X \dashrightarrow X''$ only if the rational map $\psi^{-1} \circ \phi$ extends to an isomorphism. This is the same as counting the number of distinct subcones $\psi^{-1}\mathrm{Amp}^1(X') \subset \overline{\mathrm{Eff}}^1(X)$ defined by birational contractions to minimal models.

This set will usually be larger than the set of runs of the MMP, since we now allow flops as well as K_X-negative contractions. An important example of **[Rei83]** shows that the number of minimal models marked with a rational map can be infinite (and a minimal model can admit an infinite sequence of flops). Note however that the Kawamata-Morrison Cone Conjectures also predict a nice structure for this set.

5.2.3. *Structure of MMP outcomes.* Given two outcomes of the $(K_X+\Delta)$-MMP

$$\phi_1 : X \dashrightarrow Y_1 \qquad \text{and} \qquad \phi_2 : X \dashrightarrow Y_2,$$

it is natural to ask for a relationship between Y_1 and Y_2. Since both are constructed from X by a sequence of simple steps, it is natural to wonder whether the induced rational map $Y_1 \dashrightarrow Y_2$ can also be factored as a sequence of simple steps. As always we will need to discuss the various cases separately.

First suppose we are in case (ii) or (iii), so the final outcome of the MMP is a minimal model. Given two minimal models $\phi_1 : X \dashrightarrow Y_1$ and $\phi_2 : X \dashrightarrow Y_2$, we would like to factor the induced map $\psi : Y_1 \dashrightarrow Y_2$ into a series of simple steps.

THEOREM 5.19 (**[Kaw08]** Theorem 1). *Let (X, Δ) and (X', Δ') be two terminal $\mathbb{Q}$-factorial pairs where Δ, Δ' are $\mathbb{Q}$-divisors such that there is a birational map $\phi : X \dashrightarrow X'$ sending $\phi_*\Delta = \Delta'$. Suppose that $K_X + \Delta$ and $K_{X'} + \Delta'$ are nef. Then ϕ decomposes into a sequence of flops.*

Now suppose that $K_X + \Delta$ is not pseudo-effective and that we are given two Mori fiber space outcomes of the $(K_X + \Delta)$-MMP:

$$\phi_1 : X \dashrightarrow Y_1 \quad \text{with contraction} \quad \pi_1 : Y_1 \to Z_1$$
$$\phi_2 : X \dashrightarrow Y_2 \quad \text{with contraction} \quad \pi_2 : Y_2 \to Z_2$$

As discussed above, one would like to factor the rational map $\psi : Y_1 \dashrightarrow Y_2$ into a number of "basic steps." We should of course also keep track of the Mori fibrations π_1, π_2. This picture is modeled on dimension 2, where any two minimal ruled surfaces over a curve are connected by a series of elementary transformations. In this case, the study of these "basic steps" is known as the Sarkisov program. The main goal of the program was accomplished in [**HM13**], which decomposes any two outcomes of the MMP as above into a finite sequence of Sarkisov links. These links come in four types, where each step is characterized by outcomes of the two-ray game. We refer to [**HM13**] for the precise statement. See also [**Kal13**] for a discussion of the relations in the Sariskov program.

5.3. Relating the geometry of (X, Δ) to the canonical model. Given a canonical model Z of (X, Δ), it is natural to ask how the geometry of (X, Δ) is related to the geometry of Z. Even when $\Delta = 0$ and X and Z are smooth, it is too much to hope for the positivity of K_X and K_Z to be directly related due to the presence of singularities of the canonical map. Instead, one must account for the discriminant locus of Z by including a boundary divisor Δ_Z.

For simplicity, we will assume right away that $\pi : X \to Z$ is a morphism with connected fibers such that $K_X + \Delta$ is trivial along the fibers (so the canonical model map is a special example). To obtain a clean statement one first must pass to a birational model of π to ensure that all the singularities of π are detected in codimension 1 on the base.

The following statement builds on Kodaira's canonical bundle formula for elliptic fibrations and is due to [**Kaw97b**], [**Kaw98**], [**FM00**], [**Amb04**], [**Amb05**], [**FG14b**], [**Fuj15**]. See [**Fuj15**, Section 3] for an extensive discussion of this result and the inputs of various authors. The log canonical version is due to [**FG14b**] and generalizes results of Ambro in the klt case.

THEOREM 5.20 ([**FG14b**]). *Suppose that (X, Δ) is an lc pair and that $\pi : X \to Z$ is a morphism with connected fibers such that $(K_X + \Delta)|_F \sim_{\mathbb{Q}} 0$ for a general fiber F of π. Then there exists:*

- *a log smooth model (X', Δ') of (X, Δ), where we have $\mu : X' \to X$ birational and $K_{X'} + \Delta' = \mu^*(K_X + \Delta) + E$ for an effective μ-exceptional divisor E,*
- *a smooth variety Z' and an effective divisor $\Delta_{Z'}$*
- *a morphism $\pi' : X' \to Z'$ birationally equivalent to π,*
- *a $\mathbb{Q}$-Cartier divisor B on X' which we express as the difference $B = B^+ - B^-$ of effective divisors with no common components*

such that

(1) *$K_{X'} + \Delta' = \pi'^*(K_{Z'} + \Delta_{Z'}) + B$,*
(2) *there is a positive integer b (clearing denominators) such that*

$$H^0(X', mb(K_{X'} + \Delta')) = H^0(Z', mb(K_{Z'} + \Delta_{Z'}))$$

for any positive integer m,

(3) B^- is f'-exceptional and μ-exceptional,
(4) $\pi'_* \mathcal{O}_{X'}(\lfloor lB^+ \rfloor) = \mathcal{O}_{Z'}$ for every positive integer l.

Furthermore, the divisor $\Delta_{Z'}$ admits a decomposition $\Delta_{Z'} \sim_{\mathbb{Q}} D + M$, where

- *D is the "discriminant part": it is effective and is explicitly determined as in* [**FM00**] *via the log canonical thresholds of $K_{X'} + \Delta'$ over prime divisors in Z'. The pair (Z', D) is lc, and is klt if (X, Δ) is klt.*
- *M is the "moduli part": it is nef.*

If for a general fiber F of π the pair $(F, \Delta|_F)$ has a good minimal model, then by choosing Z' appropriately we may ensure that M is nef and abundant. In this case, if (X, Δ) is klt then we may ensure $(Z', \Delta_{Z'})$ is klt (by choosing $\Delta_{Z'}$ appropriately in its $\mathbb{Q}$-linear equivalence class).

Some remarks are in order. There are two ways in which a divisor can be "trivial" with respect to a map: it can be contracted to a locus of codimension ≥ 2, or the pushforward of the corresponding sheaf can be trivial. In our situation the negative B^- satisfies the stronger first property and can thus essentially be ignored when comparing sections of divisors on X' and Z'. The positive part B^+ satisfies the weaker second property and thus does not affect the comparison of sections. So the conclusion (1) expresses a very tight relationship between the pair upstairs and the pair downstairs.

The main point of the theorem is to understand the positivity of M. In Kodaira's original formula, the moduli part of the elliptic fibration was pulled back from the moduli space of pointed elliptic curves. Morally speaking, in general the moduli part M on the base Z' should be the pullback of an ample divisor over a rational morphism $g : Z' \dashrightarrow \mathcal{H}$ to a "universal parameter space" $\mathcal{H}$ for the fibers of the map. So in fact, one expects the moduli part to be birationally semiample instead of just nef (or nef and abundant). It is an important problem to clarify this potential link to the geometry of the map.

A closely related approach is given by Campana's theory of orbifolds; see [**Cam11**] and the references therein.

6. Conditional results

6.1. Termination of flips. [**HMX14**] establishes an inductive procedure for establishing termination of flips (and hence the existence of minimal models). Building on [**Bir07**], [**dFEM10**] [**dFEM11**], the paper proves:

THEOREM 6.1 ([**HMX14**] Corollary 1.2). *Assume termination of flips in dimension $\leq n-1$. Then termination of flips holds for any klt pair (X, Δ) of dimension n such that $K_X + \Delta$ is numerically equivalent to an effective divisor.*

Many similar but weaker statements have been proven by [**Bir10a**], [**Bir11**], [**Bir12b**]; for example, [**Bir11**, Corollary 1.7] proves the analogous statement with "termination of flips" replaced by "existence of minimal models".

Note that the condition on $K_X + \Delta$ in Theorem 6.1 would follow from various flavors of the Abundance Conjecture. (In particular Klt Termination of flips$_{n-1}$ + Klt Non-vanishing$_n$ $\implies$ Klt Termination of flips$_n$.) Thus Abundance is in some sense the only missing piece of the minimal model program, and we focus on this conjecture henceforth.

6.2. Abundance conjecture. One way to try to prove the main conjectures of the MMP by induction is to choose a divisor on X and to try to "lift sections" from this divisor to all of X using vanishing theorems. This approach has been very successful under certain conditions. Since the big case is reasonably well-understood, we only mention statements which are potentially applicable to the remaining case (ii). These results are far too varied to summarize here, so we only give a couple easily-stated results which illustrate recent developments. The first is the key to Siu's proof of invariance of plurigenera.

THEOREM 6.2 ([**Pău07**] Theorem 1). *Let $f : X \to Y$ be a smooth morphism with connected fibers. Let L be a divisor on X and let h_L be a singular hermitian metric on $\mathcal{O}_X(L)$ with positive curvature. Suppose that the restriction of h_L to a fiber X_0 is well-defined. Then for any positive integer m, the restriction map*

$$H^0(X, \mathcal{O}_X(mK_X + L)) \to H^0(X_0, \mathcal{O}_{X_0}(mK_X + L))$$

surjects onto sections of $\mathcal{O}_{X_0}(mK_X + L) \otimes \mathcal{I}(h_L|_{X_0})$.

THEOREM 6.3 ([**DHP13**] Corollary 1.8). *Let $(X, S+B)$ be a plt pair such that $K_X + S + B$ is nef and is $\mathbb{Q}$-linearly equivalent to an effective divisor D satisfying $S \subset \operatorname{Supp}(D) \subset \operatorname{Supp}(S + B)$. Then the restriction map*

$$H^0(X, \mathcal{O}_X(m(K_X + S + B))) \to H^0(S, \mathcal{O}_S(m(K_X + S + B)))$$

is surjective for all sufficiently divisible integers m.

Another way to prove the main conjectures by induction is to try to "lift positivity" from the base of a morphism. There are several such theorems in the literature. Usually, one requires that the fibers of the morphism be trivial in some way.

The first such statement concerns the Iitaka fibration of $K_X + \Delta$. For this map the restriction of $K_X + \Delta$ to the fibers is "sectionally trivial." While the following theorem of [**Lai10**] is only stated in the terminal case, it is also true for klt pairs.

THEOREM 6.4 ([**Lai10**] Theorem 4.4). *Let (X, Δ) be a klt pair. Suppose that $\kappa(X, \Delta) \geq 0$ and (for simplicity) the Iitaka fibration for $K_X + \Delta$ is a morphism $f : X \to Y$. If $(F, \Delta|_F)$ has a good minimal model for the general fiber of f, then (X, Δ) has a good minimal model.*

Another version assumes that the restriction of $K_X + \Delta$ to the fibers of the morphism are "numerically trivial." Statements of this kind appear in [**Amb04**], [**Amb05**]. Building on these, we have:

THEOREM 6.5 ([**GL13**] Theorem 1.3). *Let (X, Δ) be a klt pair. Suppose that $f : X \to Y$ is a surjective morphism with connected fibers such that $\nu((K_X+\Delta)|_F) = 0$ for a general fiber F. Then there is a birational model Y' of Y and a klt pair $(Y', \Delta_{Y'})$ such that (X, Δ) has a good minimal model if and only if $(Y', \Delta_{Y'})$ has a good minimal model.*

An important advantage is that there is no assumption on the existence of log pluricanonical sections or on termination of flips.

7. Additional results

In this subsection we briefly outline results which do not "fit into the narrative" but are absolutely essential for understanding the modern MMP. I will focus on directions of current research.

7.1. Boundedness results. There are many kinds of boundedness statements arising from the minimal model program. Such statements are closely related to the boundedness of the moduli functor for stable pairs. We refer to [**HM10b**] for an informative discussion of such results and the recent results of Birkar, Hacon, McKernan, and Xu for the precise statements.

7.2. Moduli of stable pairs. The modern approach to constructing moduli of stable pairs relies heavily on the minimal model program. We refer to the excellent survey paper [**Kov09**] and to the upcoming book of Kollár for more technical details.

7.3. Foliations. Surprisingly, foliations seem to exhibit a beautiful structure analogous to the structure for varieties provided by MMP. There are two kinds of results in this direction.

First, Brunella, McQuillan, and Mendes have proved some MMP-type structure results for foliations on surfaces. The next cases are under active investigation and there are many interesting open questions. We refer to [**Bru15**] for an overview of this area.

Second, one can deduce the existence of rational curves from the presence of "negative" foliations. We refer to [**BM01**] and [**KSCT07**] and the subsequent literature for this important circle of ideas.

7.4. Characteristic p. It is interesting to develop the MMP over arbitrary fields. Recent work has focused on algebraically closed fields of positive characteristic.

The first step is to understand how to generalize the technical results underlying the MMP – namely, vanishing theorems. As is well-known, even the Kodaira vanishing theorem can fail in characteristic p. Nevertheless, there is a large body of work which successfully establishes some good analogues in positive characteristic. This area is too broad to summarize here; we instead refer to the recent survey paper [**ST12**] and the references therein.

The second step is to apply these results to obtain a MMP theory. We refer to recent work of Birkar, Hacon, Tanaka, Xu, and their collaborators for such applications.

7.5. Existence of rational curves. Mori's celebrated bend-and-break theorem relates the negativity of K_X against curves with the existence of rational curves.

THEOREM 7.1 ([**MM86**] Theorem 5). *Let X be a smooth variety and let A be an ample Cartier divisor on X. Suppose that C is an irreducible curve on X such that $K_X \cdot C < 0$. Then through every point of C there is a rational curve T on X satisfying $A \cdot T \leq 2\dim(X)\frac{A \cdot C}{-K_X \cdot C}$.*

It is important to understand the behavior of rational curves on singular varieties as well. In this direction, we have:

THEOREM 7.2. [**Kaw91**, Theorem 1] *Let (X, Δ) be a klt pair. Every $K_X + \Delta$-negative extremal ray of $\overline{\mathrm{Eff}}_1(X)$ is spanned by the class of a rational curve C satisfying $0 < -(K_X + \Delta) \cdot C \leq 2\dim X$.*

Since many birational maps can be interpreted using the minimal model program, this shows the existence of rational curves in a wide variety of situations. There are related results due to [**HM07**], [**Tak08**], [**BBP13**].

An important open problem is:

QUESTION 7.3. Suppose that (X, Δ) is a klt pair such that $K_X + \Delta$ is not pseudo-effective. Is there a rational curve C through a very general point of X such that $(K_X + \Delta) \cdot C < 0$?

By running the MMP to obtain a variety X' with a Mori fiber space structure, it is clear that X' is covered by rational curves C satisfying $(K_{X'} + \Delta') \cdot C < 0$. However, it is not at all clear whether the preimage of these curves on X satisfy the desired condition. Even the surface case, which was settled by [**KM99**], is quite difficult and requires many new techniques.

7.6. MMP for quasi-projective varieties. Suppose that U is a quasi-projective variety. By Hironaka's resolution of singularities, U admits a projective completion X such that the complement of U is a simple normal crossing divisor. An important idea of Iitaka ("Iitaka's philosophy") is that we can understand the geometry of U by studying the log pair (X, Δ). Often (but not always!) a theorem for projective varieties has an analogue for quasi-projective varieties: we replace projective invariants – plurigenera, rational curves, the cotangent bundle – by their log versions – log plurigenera, log rational curves, the log cotangent bundle.

Recently, there has been new progress towards establishing Iitaka's philosophy. We will mention only the most recent developments. For understanding log rational curves, log degenerations have been a key tool: see [**KM99**], [**CZ14**], [**CZ15**]. Another approach is to study the log cotangent bundle: see Campana's theory of orbifolds (for example [**Cam11**]) and [**Zhu15**]. Finally, [**GKKP11**] and [**GKP14**] discuss when forms extend from an open subset to the compactification in the klt setting.

7.7. Running the MMP for moduli spaces. Suppose that X is a moduli space (for example, the moduli space of stable curves or a Hilbert scheme of points on a surface). By running the MMP, we obtain a special sequence of birational models of X connected by divisorial contractions and flips. If we construct X using GIT, one can obtain a sequence of birational models by varying the linearization. Surprisingly, these birational models often seem to admit interpretations as moduli spaces as well. An interesting and exciting field, initiated by Hasset and Keel for $\overline{M}_{g,n}$, is to explicitly construct the moduli problems for the birational models constructed abstractly by the MMP. Another important setting is moduli spaces of sheaves arising from stability conditions in the derived category.

The literature is far too vast to survey here; we direct readers to the survey paper [**FS13**] for moduli spaces of curves and [**ABCH13**] and [**BM14**] for the use of stability conditions in studying Hilbert schemes of points on a surface.

7.8. MMP and derived categories. An interesting idea originating from the work of Bondal and Orlov ([**BO**]) is that the birational structure of the MMP should be reflected on the level of derived categories. More precisely, the contractions constructed by the cone theorem should naturally yield semi-orthogonal decompositions of the derived category, and flops should yield a derived equivalence of some kind. Furthermore, the different models should correspond to the variation of a stability condition. [**Kaw09**] gives a good introduction to the questions and technical difficulties of the area.

In addition to a number of special examples, progress has been made for surfaces (see for example [**Tod13**], [**Tod14**]), threefolds (see [**Bri02**] and the many subsequent generalizations, and [**Tod13**]) and for toric varieties (see [**Kaw06**], [**Kaw13a**]). There are also interesting connections to non-commutative algebras; see for example [**IW14**].

7.9. Singularities, the dual complex, and Berkovich spaces. Suppose that $0 \in X$ is a singularity and that $\phi : Y \to X$ is a birational map such that the preimage of 0 is a simple normal crossing divisor E. One can associate to E its dual complex $D(E)$ encoding how the components of E intersect. While the dual complex depends on the choice of resolution, certain topological features of the dual complex (such as the homotopy type) turn out to be independent of the resolution chosen (see [**Pay13**]). Interestingly, certain algebraic features of the singularity are captured by the topological properties of $D(E)$. This relationship has been studied for a long time for the dual graphs of resolutions of surface singularities.

New advances in the MMP have opened up the study of higher dimension varieties. By using the MMP to systematically contract components of E, [**dFKX12**] (following up on [**Kol13a**], [**Kol14**], [**KK14**]) identifies an "optimal" dual complex which is well-defined up to piecewise linear homomorphism. Building on this viewpoint, [**NX16a**] uses the MMP to study Berkovich spaces (which exhibit structure similar to a limit of dual complexes). This circle of ideas has found further applications in [**NX16b**] and [**KX15**].

7.10. Rational points over number fields. Suppose that X is a smooth projective variety over a number field. As discussed in Section 2, conjecturally the behavior of rational points on X is constrained by the Kodaira dimension of X. Thus the minimal model program should be an essential tool for analyzing rational points. However, there are currently not many number-theoretic results using the full strength of the MMP, mainly due to the difficulty of the area. Even for surfaces (for which the MMP is comparatively easy), the behavior of rational points is still quite far from being understood.

Recently the MMP has found interesting applications to Manin's Conjecture. Suppose that X is a Fano variety and that $\mathcal{O}_X(L)$ is an adelically metrized ample line bundle inducing a height function H on the points of X. Manin's Conjecture predicts that (after a finite base change) the number of rational points on X of bounded H-height is controlled by certain geometric invariants associated to X and L. Building on [**BT98**] and [**HTT15**], [**LTT15**] uses the MMP to systematically analyze these geometric invariants. The results indicate that often one will need to remove a thin set of points, rather than the points in a closed subset, for Manin's Conjecture to hold.

References

[ABCH13] Daniele Arcara, Aaron Bertram, Izzet Coskun, and Jack Huizenga, *The minimal model program for the Hilbert scheme of points on $\mathbb{P}^2$ and Bridgeland stability*, Adv. Math. **235** (2013), 580–626. MR3010070

[AHK07] Valery Alexeev, Christopher Hacon, and Yujiro Kawamata, *Termination of (many) 4-dimensional log flips*, Invent. Math. **168** (2007), no. 2, 433–448. MR2289869

[Ale94] Valery Alexeev, *Boundedness and K^2 for log surfaces*, Internat. J. Math. **5** (1994), no. 6, 779–810. MR1298994

[Amb04] Florin Ambro, *Nef dimension of minimal models*, Math. Ann. **330** (2004), no. 2, 309–322. MR2089428

[Amb05] Florin Ambro, *The moduli b-divisor of an lc-trivial fibration*, Compos. Math. **141** (2005), no. 2, 385–403. MR2134273

[And85] Tetsuya Ando, *On extremal rays of the higher-dimensional varieties*, Invent. Math. **81** (1985), no. 2, 347–357. MR799271

[Ara10] Carolina Araujo, *The cone of pseudo-effective divisors of log varieties after Batyrev*, Math. Z. **264** (2010), no. 1, 179–193. MR2564937

[AS95] Urban Angehrn and Yum Tong Siu, *Effective freeness and point separation for adjoint bundles*, Invent. Math. **122** (1995), no. 2, 291–308. MR1358978

[AV14] Ekaterina Amerik and Misha Verbitsky, *Morrison-Kawamata cone conjecture for hyperkahler manifolds*, ArXiv14083892 Math (2014).

[AW98] Marco Andreatta and Jaroslaw A. Wiśniewski, *On contractions of smooth varieties*, J. Algebraic Geom. **7** (1998), no. 2, 253–312. MR1620110

[Bat92] Victor V. Batyrev, *The cone of effective divisors of threefolds*, Proceedings of the International Conference on Algebra, Part 3 (Novosibirsk, 1989), Contemp. Math., vol. 131, Amer. Math. Soc., Providence, RI, 1992, pp. 337–352. MR1175891

[BBP13] Sébastien Boucksom, Amaël Broustet, and Gianluca Pacienza, *Uniruledness of stable base loci of adjoint linear systems via Mori theory*, Math. Z. **275** (2013), no. 1-2, 499–507. MR3101817

[BC15] Caucher Birkar and Jungkai Alfred Chen, *Varieties fibred over abelian varieties with fibres of log general type*, Adv. Math. **270** (2015), 206–222. MR3286535

[BCHM10] Caucher Birkar, Paolo Cascini, Christopher D. Hacon, and James McKernan, *Existence of minimal models for varieties of log general type*, J. Amer. Math. Soc. **23** (2010), no. 2, 405–468. MR2601039

[BCZ04] Gavin Brown, Alessio Corti, and Francesco Zucconi, *Birational geometry of 3-fold Mori fibre spaces*, The Fano Conference, Univ. Torino, Turin, 2004, pp. 235–275. MR2112578

[BDPP13] Sébastien Boucksom, Jean-Pierre Demailly, Mihai Păun, and Thomas Peternell, *The pseudo-effective cone of a compact KÄhler manifold and varieties of negative Kodaira dimension*, J. Algebraic Geom. **22** (2013), no. 2, 201–248. MR3019449

[Ben83] X. Benveniste, *Sur l'anneau canonique de certaines variétés de dimension* 3 (French), Invent. Math. **73** (1983), no. 1, 157–164. MR707354

[Ben86] X. Benveniste, *Sur les applications pluricanoniques des variétés de type très général en dimension* 3 (French), Amer. J. Math. **108** (1986), no. 2, 433–449. MR833363

[Ber09] Bo Berndtsson, *Curvature of vector bundles associated to holomorphic fibrations*, Ann. of Math. (2) **169** (2009), no. 2, 531–560. MR2480611

[Bir07] Caucher Birkar, *Ascending chain condition for log canonical thresholds and termination of log flips*, Duke Math. J. **136** (2007), no. 1, 173–180. MR2271298

[Bir09a] Caucher Birkar, *The Iitaka conjecture $C_{n,m}$ in dimension six*, Compos. Math. **145** (2009), no. 6, 1442–1446. MR2575089

[Bir09b] Caucher Birkar, *Log minimal models according to Shokurov*, Algebra Number Theory **3** (2009), no. 8, 951–958. MR2587409

[Bir10a] Caucher Birkar, *On existence of log minimal models*, Compos. Math. **146** (2010), no. 4, 919–928. MR2660678

[Bir10b] Caucher Birkar, *On termination of log flips in dimension four*, Math. Ann. **346** (2010), no. 2, 251–257. MR2563688

[Bir11] Caucher Birkar, *On existence of log minimal models II*, J. Reine Angew. Math. **658** (2011), 99–113. MR2831514

[Bir12] Caucher Birkar, *Existence of log canonical flips and a special LMMP*, Publ. Math. Inst. Hautes Études Sci. **115** (2012), 325–368, DOI 10.1007/s10240-012-0039-5. MR2929730

[Bir12b] Caucher Birkar, *On existence of log minimal models and weak Zariski decompositions*, Math. Ann. **354** (2012), no. 2, 787–799. MR2965261

[BM97] Andrea Bruno and Kenji Matsuki, *Log Sarkisov program*, Internat. J. Math. **8** (1997), no. 4, 451–494. MR1460896

[BM01] Fedor Bogomolov and Michael McQuillan, *Rational curves on foliated varieties*, IHES Preprint (2001).

[BM14] Arend Bayer and Emanuele Macrì, *MMP for moduli of sheaves on K3s via wall-crossing: nef and movable cones, Lagrangian fibrations*, Invent. Math. **198** (2014), no. 3, 505–590. MR3279532

[BO] Alexey Bondal and Dmitri Orlov, *Semiorthogonal decomposition for algebraic varieties*, arXiv:9506012 [math.AG].

[Bor91] Ciprian Borcea, *On desingularized Horrocks-Mumford quintics*, J. Reine Angew. Math. **421** (1991), 23–41. MR1129573

[BP08] Bo Berndtsson and Mihai Păun, *Bergman kernels and the pseudoeffectivity of relative canonical bundles*, Duke Math. J. **145** (2008), no. 2, 341–378. MR2449950

[BP12] Bo Berndtsson and Mihai Păun, *Quantitative extensions of pluricanonical forms and closed positive currents*, Nagoya Math. J. **205** (2012), 25–65. MR2891164

[Bri02] Tom Bridgeland, *Flops and derived categories*, Invent. Math. **147** (2002), no. 3, 613–632. MR1893007

[Bru15] Marco Brunella, *Birational geometry of foliations*, IMPA Monographs, vol. 1, Springer, Cham, 2015. MR3328860

[BT98] Victor V. Batyrev and Yuri Tschinkel, *Tamagawa numbers of polarized algebraic varieties*, Astérisque **251** (1998), 299–340. Nombre et répartition de points de hauteur bornée (Paris, 1996). MR1679843

[Cam92] F. Campana, *Connexité rationnelle des variétés de Fano* (French), Ann. Sci. École Norm. Sup. (4) **25** (1992), no. 5, 539–545. MR1191735

[Cam11] F. Campana, *Special orbifolds and birational classification: a survey*, Classification of algebraic varieties, EMS Ser. Congr. Rep., Eur. Math. Soc., Zürich, 2011, pp. 123–170. MR2779470

[CC10a] Jungkai A. Chen and Meng Chen, *Explicit birational geometry of 3-folds of general type, II*, J. Differential Geom. **86** (2010), no. 2, 237–271. MR2772551

[CC10b] Jungkai A. Chen and Meng Chen, *Explicit birational geometry of threefolds of general type, I* (English, with English and French summaries), Ann. Sci. Éc. Norm. Supér. (4) **43** (2010), no. 3, 365–394. MR2667020

[CH11] Jungkai Alfred Chen and Christopher D. Hacon, *Kodaira dimension of irregular varieties*, Invent. Math. **186** (2011), no. 3, 481–500. MR2854084

[CK15] Alessio Corti and Anne-Sophie Kaloghiros, *The Sarkisov program for Mori fibred Calabi-Yau pairs*, Algebr. Geom. **3** (2016), no. 3, 370–384. MR3504536

[CKP12] F. Campana, Vincent Koziarz, and Mihai Păun, *Numerical character of the effectivity of adjoint line bundles* (English, with English and French summaries), Ann. Inst. Fourier (Grenoble) **62** (2012), no. 1, 107–119. MR2986267

[CL12] Paolo Cascini and Vladimir Lazić, *New outlook on the minimal model program, I*, Duke Math. J. **161** (2012), no. 12, 2415–2467. MR2972461

[CL13] Alessio Corti and Vladimir Lazić, *New outlook on the minimal model program, II*, Math. Ann. **356** (2013), no. 2, 617–633. MR3048609

[CL14] Paolo Cascini and Vladimir Lazić, *On the number of minimal models of a log smooth threefold* (English, with English and French summaries), J. Math. Pures Appl. (9) **102** (2014), no. 3, 597–616. MR3249723

[Cla07] Benoît Claudon, *Invariance for multiples of the twisted canonical bundle* (English, with English and French summaries), Ann. Inst. Fourier (Grenoble) **57** (2007), no. 1, 289–300. MR2316240

[CO11] Serge Cantat and Keiji Oguiso, *Birational automorphism groups and the movable cone theorem for Calabi-Yau manifolds of Wehler type via universal Coxeter groups*, Amer. J. Math. **137** (2015), no. 4, 1013–1044. MR3372313

[Cor95] Alessio Corti, *Factoring birational maps of threefolds after Sarkisov*, J. Algebraic Geom. **4** (1995), no. 2, 223–254. MR1311348

[Cor07] Alessio Corti (ed.), *Flips for 3-folds and 4-folds*, Oxford Lecture Series in Mathematics and its Applications, vol. 35, Oxford University Press, Oxford, 2007. MR2352762

[CPS14] Izzet Coskun and Artie Prendergast-Smith, *Fano manifolds of index $n-1$ and the cone conjecture*, Int. Math. Res. Not. IMRN **9** (2014), 2401–2439. MR3207372

[CZ14] Qile Chen and Yi Zhu, *Very free curves on Fano complete intersections*, Algebr. Geom. **1** (2014), no. 5, 558–572. MR3296805

[CZ15] Qile Chen and Yi Zhu, $\mathbb{A}^1$*-connected varieties of rank one over nonclosed fields*, Math. Ann. **364** (2016), no. 3-4, 1505–1515, DOI 10.1007/s00208-015-1257-1. MR3466876

[DB12] Lorenzo Di Biagio, *Pluricanonical systems for 3-folds and 4-folds of general type*, Math. Proc. Cambridge Philos. Soc. **152** (2012), no. 1, 9–34. MR2860415

[dFEM10] Tommaso de Fernex, Lawrence Ein, and Mircea Mustaţă, *Shokurov's ACC conjecture for log canonical thresholds on smooth varieties*, Duke Math. J. **152** (2010), no. 1, 93–114. MR2643057

[dFEM11] Tommaso de Fernex, Lawrence Ein, and Mircea Mustaţă, *Log canonical thresholds on varieties with bounded singularities*, Classification of algebraic varieties, EMS Ser. Congr. Rep., Eur. Math. Soc., Zürich, 2011, pp. 221–257.

[dFKX12] Tommaso de Fernex, János Kollár, and Chenyang Xu, *The dual complex of singularities*, 2012, arXiv:1212.1675 [math.AG], to appear in "Proceedings of the conference in honor of Yujiro Kawamatas 60th birthday, Advanced Studies in Pure Mathematics".

[DHP13] Jean-Pierre Demailly, Christopher D. Hacon, and Mihai Păun, *Extension theorems, non-vanishing and the existence of good minimal models*, Acta Math. **210** (2013), no. 2, 203–259. MR3070567

[Dru11] Stéphane Druel, *Quelques remarques sur la décomposition de Zariski divisorielle sur les variétés dont la première classe de Chern est nulle* (French), Math. Z. **267** (2011), no. 1-2, 413–423. MR2772258

[Eck15] Thomas Eckl, *Numerical analogues of the Kodaira dimension and the abundance conjecture*, Manuscripta Math. **150** (2016), no. 3-4, 337–356. MR3514733

[EP12] Lawrence Ein and Mihnea Popa, *Extension of sections via adjoint ideals*, Math. Ann. **352** (2012), no. 2, 373–408. MR2874961

[FG12] Osamu Fujino and Yoshinori Gongyo, *On canonical bundle formulas and subadjunctions*, Michigan Math. J. **61** (2012), no. 2, 255–264. MR2944479

[FG14a] Osamu Fujino and Yoshinori Gongyo, *Log pluricanonical representations and the abundance conjecture*, Compos. Math. **150** (2014), no. 4, 593–620. MR3200670

[FG14b] Osamu Fujino and Yoshinori Gongyo, *On the moduli b-divisors of lc-trivial fibrations* (English, with English and French summaries), Ann. Inst. Fourier (Grenoble) **64** (2014), no. 4, 1721–1735. MR3329677

[FM00] Osamu Fujino and Shigefumi Mori, *A canonical bundle formula*, J. Differential Geom. **56** (2000), no. 1, 167–188. MR1863025

[FS13] Maksym Fedorchuk and David Ishii Smyth, *Alternate compactifications of moduli spaces of curves*, Handbook of moduli. Vol. I, Adv. Lect. Math. (ALM), vol. 24, Int. Press, Somerville, MA, 2013, pp. 331–413. MR3184168

[Fuj78] Takao Fujita, *The sheaf of relative canonical forms of a Kähler fiber space over a curve*, Proc. Japan Acad. Ser. A Math. Sci. **54** (1978), no. 7, 183–184. MR510945

[Fuj86] Takao Fujita, *Zariski decomposition and canonical rings of elliptic threefolds*, J. Math. Soc. Japan **38** (1986), no. 1, 19–37. MR816221

[Fuj94] Takao Fujita, *Approximating Zariski decomposition of big line bundles*, Kodai Math. J. **17** (1994), no. 1, 1–3. MR1262949

[Fuj00] Osamu Fujino, *Abundance theorem for semi log canonical threefolds*, Duke Math. J. **102** (2000), no. 3, 513–532. MR1756108

[Fuj03] Osamu Fujino, *Algebraic fiber spaces whose general fibers are of maximal Albanese dimension*, Nagoya Math. J. **172** (2003), 111–127. MR2019522

[Fuj05] Osamu Fujino, *On termination of 4-fold semi-stable log flips*, Publ. Res. Inst. Math. Sci. **41** (2005), no. 2, 281–294. MR2138024

[Fuj10] Osamu Fujino, *Finite generation of the log canonical ring in dimension four*, Kyoto J. Math. **50** (2010), no. 4, 671–684. MR2740690

[Fuj11] Osamu Fujino, *Non-vanishing theorem for log canonical pairs*, J. Algebraic Geom. **20** (2011), no. 4, 771–783. MR2819675

[Fuj12] Osamu Fujino, *A transcendental approach to Kollár's injectivity theorem*, Osaka J. Math. **49** (2012), no. 3, 833–852. MR2993068

[Fuj13a] Osamu Fujino, *On maximal Albanese dimensional varieties*, Proc. Japan Acad. Ser. A Math. Sci. **89** (2013), no. 8, 92–95. MR3127923

[Fuj13b] Osamu Fujino, *A transcendental approach to Kollár's injectivity theorem II*, J. Reine Angew. Math. **681** (2013), 149–174. MR3181493

[Fuj15] Osamu Fujino, *Some remarks on the minimal model program for log canonical pairs*, J. Math. Sci. Univ. Tokyo **22** (2015), no. 1, 149–192. MR3329193

[GKKP11] Daniel Greb, Stefan Kebekus, Sándor J. Kovács, and Thomas Peternell, *Differential forms on log canonical spaces*, Publ. Math. Inst. Hautes Études Sci. **114** (2011), 87–169. MR2854859

[GKP14] Daniel Greb, Stefan Kebekus, and Thomas Peternell, *Reflexive differential forms on singular spaces. Geometry and cohomology*, J. Reine Angew. Math. **697** (2014), 57–89. MR3281652

[GL13] Yoshinori Gongyo and Brian Lehmann, *Reduction maps and minimal model theory*, Compos. Math. **149** (2013), no. 2, 295–308. MR3020310

[GM93] Antonella Grassi and David R. Morrison, *Automorphisms and the Kähler cone of certain Calabi-Yau manifolds*, Duke Math. J. **71** (1993), no. 3, 831–838. MR1240604

[Gon11] Yoshinori Gongyo, *On the minimal model theory for dlt pairs of numerical log Kodaira dimension zero*, Math. Res. Lett. **18** (2011), no. 5, 991–1000. MR2875871

[Gon13] Yoshinori Gongyo, *Abundance theorem for numerically trivial log canonical divisors of semi-log canonical pairs*, J. Algebraic Geom. **22** (2013), no. 3, 549–564. MR3048544

[Hac14] Christopher D. Hacon, *On the log canonical inversion of adjunction*, Proc. Edinb. Math. Soc. (2) **57** (2014), no. 1, 139–143. MR3165017

[HK00] Yi Hu and Sean Keel, *Mori dream spaces and GIT*, Michigan Math. J. **48** (2000), 331–348. Dedicated to William Fulton on the occasion of his 60th birthday. MR1786494

[HK10] Christopher D. Hacon and Sándor J. Kovács, *Classification of higher dimensional algebraic varieties*, Oberwolfach Seminars, vol. 41, Birkhäuser Verlag, Basel, 2010. MR2675555

[HM06] Christopher D. Hacon and James McKernan, *Boundedness of pluricanonical maps of varieties of general type*, Invent. Math. **166** (2006), no. 1, 1–25. MR2242631

[HM07] Christopher D. Hacon and James Mckernan, *On Shokurov's rational connectedness conjecture*, Duke Math. J. **138** (2007), no. 1, 119–136. MR2309156

[HM10a] Christopher D. Hacon and James McKernan, *Existence of minimal models for varieties of log general type. II*, J. Amer. Math. Soc. **23** (2010), no. 2, 469–490. MR2601040

[HM10b] Christopher D. Hacon and James McKernan, *Boundedness results in birational geometry*, Proceedings of the International Congress of Mathematicians. Volume II, Hindustan Book Agency, New Delhi, 2010, pp. 427–449. MR2827803

[HM13] Christopher D. Hacon and James McKernan, *The Sarkisov program*, J. Algebraic Geom. **22** (2013), no. 2, 389–405. MR3019454

[HMX13] Christopher D. Hacon, James McKernan, and Chenyang Xu, *On the birational automorphisms of varieties of general type*, Ann. of Math. (2) **177** (2013), no. 3, 1077–1111. MR3034294

[HMX14] Christopher D. Hacon, James McKernan, and Chenyang Xu, *ACC for log canonical thresholds*, Ann. of Math. (2) **180** (2014), no. 2, 523–571. MR3224718

[HP02] Christopher D. Hacon and Rita Pardini, *On the birational geometry of varieties of maximal Albanese dimension*, J. Reine Angew. Math. **546** (2002), 177–199. MR1900998

[HTT15] Brendan Hassett, Sho Tanimoto, and Yuri Tschinkel, *Balanced line bundles and equivariant compactifications of homogeneous spaces*, Int. Math. Res. Not. **15** (2015), 6375–6410. MR3384482

[HX12] Christopher D. Hacon and Chenyang Xu, *Existence of log canonical closures*, Invent. Math. **192** (2013), no. 1, 161–195. MR3032329

[HX15] Christopher D. Hacon and Chenyang Xu, *On the three dimensional minimal model program in positive characteristic*, J. Amer. Math. Soc. **28** (2015), no. 3, 711–744. MR3327534

[Iit77] S. Iitaka, *Complex analysis and algebraic geometry* (1977), 175–189. MR0569688

[IW14] Osamu Iyama and Michael Wemyss, *Singular derived categories of $\mathbb{Q}$-factorial terminalizations and maximal modification algebras*, Adv. Math. **261** (2014), 85–121. MR3213296

[Jia13] Xiaodong Jiang, *On the pluricanonical maps of varieties of intermediate Kodaira dimension*, Math. Ann. **356** (2013), no. 3, 979–1004. MR3063904

[JLT13] Zhi Jiang, Martí Lahoz, and Sofia Tirabassi, *On the Iitaka fibration of varieties of maximal Albanese dimension*, Int. Math. Res. Not. IMRN **13** (2013), 2984–3005. MR3072998

[Kal13] Anne-Sophie Kaloghiros, *Relations in the Sarkisov program*, Compos. Math. **149** (2013), no. 10, 1685–1709. MR3123306

[Kaw82] Yujiro Kawamata, *Kodaira dimension of algebraic fiber spaces over curves*, Invent. Math. **66** (1982), no. 1, 57–71. MR652646

[Kaw84a] Yujiro Kawamata, *The cone of curves of algebraic varieties*, Ann. of Math. (2) **119** (1984), no. 3, 603–633. MR744865

[Kaw84b] Yujiro Kawamata, *Elementary contractions of algebraic 3-folds*, Ann. of Math. (2) **119** (1984), no. 1, 95–110. MR736561

[Kaw84c] Yujiro Kawamata, *On the finiteness of generators of a pluricanonical ring for a 3-fold of general type*, Amer. J. Math. **106** (1984), no. 6, 1503–1512. MR765589

[Kaw85a] Y. Kawamata, *Pluricanonical systems on minimal algebraic varieties*, Invent. Math. **79** (1985), no. 3, 567–588. MR782236

[Kaw85b] Yujiro Kawamata, *Minimal models and the Kodaira dimension of algebraic fiber spaces*, J. Reine Angew. Math. **363** (1985), 1–46. MR814013

[Kaw87] Yujiro Kawamata, *The Zariski decomposition of log-canonical divisors*, Algebraic geometry, Bowdoin, 1985 (Brunswick, Maine, 1985), Proc. Sympos. Pure Math., vol. 46, Amer. Math. Soc., Providence, RI, 1987, pp. 425–433.

[Kaw91] Yujiro Kawamata, *On the length of an extremal rational curve*, Invent. Math. **105** (1991), no. 3, 609–611. MR1117153

[Kaw92a] Yujiro Kawamata, *Abundance theorem for minimal threefolds*, Invent. Math. **108** (1992), no. 2, 229–246. MR1161091

[Kaw92b] Yujiro Kawamata, *Boundedness of Q-Fano threefolds*, Proceedings of the International Conference on Algebra, Part 3 (Novosibirsk, 1989), Contemp. Math., vol. 131, Amer. Math. Soc., Providence, RI, 1992, pp. 439–445.

[Kaw92c] Yujiro Kawamata, *Termination of log flips for algebraic 3-folds*, Internat. J. Math. **3** (1992), no. 5, 653–659. MR1189678

[Kaw97a] Yujiro Kawamata, *On the cone of divisors of Calabi-Yau fiber spaces*, Internat. J. Math. **8** (1997), no. 5, 665–687. MR1468356

[Kaw97b] Yujiro Kawamata, *Subadjunction of log canonical divisors for a subvariety of codimension* 2, Birational algebraic geometry (Baltimore, MD, 1996), Contemp. Math., vol. 207, Amer. Math. Soc., Providence, RI, 1997, pp. 79–88.

[Kaw98] Yujiro Kawamata, *Subadjunction of log canonical divisors. II*, Amer. J. Math. **120** (1998), no. 5, 893–899. MR1646046

[Kaw06] Yujiro Kawamata, *Derived categories of toric varieties*, Michigan Math. J. **54** (2006), no. 3, 517–535. MR2280493

[Kaw07] Masayuki Kawakita, *Inversion of adjunction on log canonicity*, Invent. Math. **167** (2007), no. 1, 129–133. MR2264806

[Kaw08] Yujiro Kawamata, *Flops connect minimal models*, Publ. Res. Inst. Math. Sci. **44** (2008), no. 2, 419–423. MR2426353

[Kaw09] Yujiro Kawamata, *Derived categories and birational geometry*, Algebraic geometry—Seattle 2005. Part 2, Proc. Sympos. Pure Math., vol. 80, Amer. Math. Soc., Providence, RI, 2009, pp. 655–665. MR2483950 (2009m:14018)

[Kaw13a] Yujiro Kawamata, *Derived categories of toric varieties II*, Michigan Math. J. **62** (2013), no. 2, 353–363. MR3079267

[Kaw13b] Yujiro Kawamata, *On the abundance theorem in the case of numerical Kodaira dimension zero*, Amer. J. Math. **135** (2013), no. 1, 115–124. MR3022959

[KK10] János Kollár and Sándor J. Kovács, *Log canonical singularities are Du Bois*, J. Amer. Math. Soc. **23** (2010), no. 3, 791–813. MR2629988

[KK14] Michael Kapovich and János Kollár, *Fundamental groups of links of isolated singularities*, J. Amer. Math. Soc. **27** (2014), no. 4, 929–952. MR3230815

[Kle66] Steven L. Kleiman, *Toward a numerical theory of ampleness*, Ann. of Math. (2) **84** (1966), 293–344. MR0206009

[KM87] Yujiro Kawamata and Kenji Matsuki, *The number of the minimal models for a 3-fold of general type is finite*, Math. Ann. **276** (1987), no. 4, 595–598. MR879538

[KM98] János Kollár and Shigefumi Mori, *Birational geometry of algebraic varieties*, Cambridge Tracts in Mathematics, vol. 134, Cambridge University Press, Cambridge, 1998. With the collaboration of C. H. Clemens and A. Corti; Translated from the 1998 Japanese original. MR1658959

[KM99] Seán Keel and James McKernan, *Rational curves on quasi-projective surfaces*, Mem. Amer. Math. Soc. **140** (1999), no. 669, viii+153. MR1610249

[KMM92a] János Kollár, Yoichi Miyaoka, and Shigefumi Mori, *Rational connectedness and boundedness of Fano manifolds*, J. Differential Geom. **36** (1992), no. 3, 765–779. MR1189503

[KMM92b] János Kollár, Yoichi Miyaoka, and Shigefumi Mori, *Rationally connected varieties*, J. Algebraic Geom. **1** (1992), no. 3, 429–448. MR1158625

[KMM94] Sean Keel, Kenji Matsuki, and James McKernan, *Log abundance theorem for threefolds*, Duke Math. J. **75** (1994), no. 1, 99–119. MR1284817

[KMMT00] János Kollár, Yoichi Miyaoka, Shigefumi Mori, and Hiromichi Takagi, *Boundedness of canonical* $\mathbf{Q}$*-Fano 3-folds*, Proc. Japan Acad. Ser. A Math. Sci. **76** (2000), no. 5, 73–77. MR1771144

[Kol84] János Kollár, *The cone theorem. Note to a paper: "The cone of curves of algebraic varieties" [Ann. of Math. (2)* **119** *(1984), no. 3, 603–633; MR0744865 (86c:14013b)] by Y. Kawamata*, Ann. of Math. (2) **120** (1984), no. 1, 1–5. MR750714

[Kol86a] János Kollár, *Higher direct images of dualizing sheaves. I*, Ann. of Math. (2) **123** (1986), no. 1, 11–42. MR825838

[Kol86b] János Kollár, *Higher direct images of dualizing sheaves. II*, Ann. of Math. (2) **124** (1986), no. 1, 171–202. MR847955

[Kol89] János Kollár, *Flops*, Nagoya Math. J. **113** (1989), 15–36. MR986434

[Kol91] János Kollár, *Extremal rays on smooth threefolds*, Ann. Sci. École Norm. Sup. (4) **24** (1991), no. 3, 339–361. MR1100994

[Kol08] János Kollár, *Exercises in the birational geometry of algebraic varieties*, ArXiv08092579 Math (2008).

[Kol13a] János Kollár, *Links of complex analytic singularities*, Surveys in differential geometry. Geometry and topology, Surv. Differ. Geom., vol. 18, Int. Press, Somerville, MA, 2013, pp. 157–193. MR3087919

[Kol13b] János Kollár, *Singularities of the minimal model program*, Cambridge Tracts in Mathematics, vol. 200, Cambridge University Press, Cambridge, 2013, With a collaboration of Sándor Kovács.

[Kol14] János Kollár, *Simple normal crossing varieties with prescribed dual complex*, Algebr. Geom. **1** (2014), no. 1, 57–68. MR3234114

[Kov09] Sándor J. Kovács, *Young person's guide to moduli of higher dimensional varieties*, Algebraic geometry—Seattle 2005. Part 2, Proc. Sympos. Pure Math., vol. 80, Amer. Math. Soc., Providence, RI, 2009, pp. 711–743.

[KSCT07] Stefan Kebekus, Luis Solá Conde, and Matei Toma, *Rationally connected foliations after Bogomolov and McQuillan*, J. Algebraic Geom. **16** (2007), no. 1, 65–81. MR2257320

[KX15] János Kollár and Chenyang Xu, *The dual complex of Calabi-Yau pairs*, Invent. Math. **205** (2016), no. no. 3, 527–557. MR3539921

[Lai10] Ching-Jui Lai, *Varieties fibered by good minimal models*, Math. Ann. **350** (2011), no. 3, 533–547. MR2805635

[Leh12] Brian Lehmann, *A cone theorem for nef curves*, J. Algebraic Geom. **21** (2012), no. 3, 473–493. MR2914801

[Leh13] Brian Lehmann, *Comparing numerical dimensions*, Algebra Number Theory **7** (2013), no. 5, 1065–1100. MR3101072

[Les15] John Lesieutre, *Some constraints on positive entropy automorphisms of smooth threefolds*, 2015, arXiv:1503.07834 [math.AG].

[LP13] Vladimir Lazić and Thomas Peternell, *On the cone conjecture for Calabi-Yau manifolds with Picard number two*, Math. Res. Lett. **20** (2013), no. 6, 1103–1113. MR3228624

[LP16] Vladimir Lazić and Thomas Peternell, *Abundance for varieties with many differential forms*, 2016, arXiv:1601.01602 [math.AG].

[LTT15] Brian Lehmann, Sho Tanimoto, and Yuri Tschinkel, *Balanced line bundles on Fano varieties*, 2015, arXiv:1409.5901 [math.AG], to appear in J. Reine Agnew. Math.

[Mar11] Eyal Markman, *A survey of Torelli and monodromy results for holomorphic-symplectic varieties*, Complex and differential geometry, Springer Proc. Math., vol. 8, Springer, Heidelberg, 2011, pp. 257–322. MR2964480

[Mat90] Kenji Matsuki, *An approach to the abundance conjecture for 3-folds*, Duke Math. J. **61** (1990), no. 1, 207–220. MR1068386

[Mat91] Kenji Matsuki, *Termination of flops for 4-folds*, Amer. J. Math. **113** (1991), no. 5, 835–859. MR1129294

[Mat02] Kenji Matsuki, *Introduction to the Mori program*, Universitext, Springer-Verlag, New York, 2002.

[Miy77] Yoichi Miyaoka, *On the Chern numbers of surfaces of general type*, Invent. Math. **42** (1977), 225–237. MR0460343

[Miy88a] Yoichi Miyaoka, *Abundance conjecture for 3-folds: case $\nu = 1$*, Compositio Math. **68** (1988), no. 2, 203–220. MR966580

[Miy88b] Yoichi Miyaoka, *On the Kodaira dimension of minimal threefolds*, Math. Ann. **281** (1988), no. 2, 325–332. MR949837

[MM86] Yoichi Miyaoka and Shigefumi Mori, *A numerical criterion for uniruledness*, Ann. of Math. (2) **124** (1986), no. 1, 65–69. MR847952

[Mor79] Shigefumi Mori, *Projective manifolds with ample tangent bundles*, Ann. of Math. (2) **110** (1979), no. 3, 593–606. MR554387

[Mor82] Shigefumi Mori, *Threefolds whose canonical bundles are not numerically effective*, Ann. of Math. (2) **116** (1982), no. 1, 133–176. MR662120

[Mor85] Shigefumi Mori, *On 3-dimensional terminal singularities*, Nagoya Math. J. **98** (1985), 43–66. MR792770

[Mor88] Shigefumi Mori, *Flip theorem and the existence of minimal models for 3-folds*, J. Amer. Math. Soc. **1** (1988), no. 1, 117–253. MR924704

[MY14] Eyal Markman and Kota Yoshioka, *A proof of the Kawamata-Morrison cone conjecture for holomorphic symplectic varieties of $K3^{[n]}$ or generalized Kummer deformation type*, Int. Math. Res. Not. IMRN **24** (2015), 13563–13574. MR3436156

[Nak86] Noboru Nakayama, *Invariance of the plurigenera of algebraic varieties under minimal model conjectures*, Topology **25** (1986), no. 2, 237–251. MR837624

[Nak04] Noboru Nakayama, *Zariski-decomposition and abundance*, MSJ Memoirs, vol. 14, Mathematical Society of Japan, Tokyo, 2004.

[Nam85] Yukihiko Namikawa, *Periods of Enriques surfaces*, Math. Ann. **270** (1985), no. 2, 201–222. MR771979

[NX16a] Johannes Nicaise and Chenyang Xu, *The essential skeleton of a degeneration of algebraic varieties*, Amer. J. Math. **138** (2016), no. 6, 1645–1667, DOI 10.1353/ajm.2016.0049. MR3595497

[NX16b] Johannes Nicaise and Chenyang Xu, *Poles of maximal order of motivic zeta functions*, Duke Math. J. **165** (2016), no. 2, 217–243, DOI 10.1215/00127094-3165648. MR3457672

[Ogu01] Keiji Oguiso, *On the finiteness of fiber-space structures on a Calabi-Yau 3-fold*, J. Math. Sci. (New York) **106** (2001), no. 5, 3320–3335. Algebraic geometry, 11. MR1878052

[Ogu14] Keiji Oguiso, *Automorphism groups of Calabi-Yau manifolds of Picard number 2*, J. Algebraic Geom. **23** (2014), no. 4, 775–795. MR3263669

[Pău07] Mihai Păun, *Siu's invariance of plurigenera: a one-tower proof*, J. Differential Geom. **76** (2007), no. 3, 485–493. MR2331528

[Pău12] Mihai Păun, *Relative critical exponents, non-vanishing and metrics with minimal singularities*, Invent. Math. **187** (2012), no. 1, 195–258. MR2874938

[Pay13] Sam Payne, *Boundary complexes and weight filtrations*, Michigan Math. J. **62** (2013), no. 2, 293–322. MR3079265

[PS12a] Arthur Prendergast-Smith, *The cone conjecture for some rational elliptic threefolds*, Math. Z. **272** (2012), no. 1-2, 589–605. MR2968244

[PS12b] Artie Prendergast-Smith, *The cone conjecture for abelian varieties*, J. Math. Sci. Univ. Tokyo **19** (2012), no. 2, 243–261. MR2987307

[Rei80] Miles Reid, *Canonical 3-folds*, Journées de Géometrie Algébrique d'Angers, Juillet 1979/Algebraic Geometry, Angers, 1979, Sijthoff & Noordhoff, Alphen aan den Rijn—Germantown, Md., 1980, pp. 273–310. MR605348

[Rei83] Miles Reid, *Minimal models of canonical 3-folds*, Algebraic varieties and analytic varieties (Tokyo, 1981), Adv. Stud. Pure Math., vol. 1, North-Holland, Amsterdam, 1983, pp. 131–180.

[Sho84] V. V. Shokurov, *The closed cone of curves of algebraic 3-folds* (Russian), Izv. Akad. Nauk SSSR Ser. Mat. **48** (1984), no. 1, 203–208. MR733364

[Sho85] V. V. Shokurov, *A nonvanishing theorem* (Russian), Izv. Akad. Nauk SSSR Ser. Mat. **49** (1985), no. 3, 635–651. MR794958

[Sho92] V. V. Shokurov, *Three-dimensional log perestroikas*, Izv. Ross. Akad. Nauk Ser. Mat. **56** (1992), 1, 105–203.

[Sho93] V. V. Shokurov, *Semi-stable 3-fold flips*, Izv. Ross. Akad. Nauk Ser. Mat. **57** (1993), 2, 165–222.

[Sho96] V. V. Shokurov, *3-fold log models*, J. Math. Sci. **81** (1996), no. 3, 2667–2699. Algebraic geometry, 4. MR1420223

[Sho03] V. V. Shokurov, *Prelimiting flips*, Tr. Mat. Inst. Steklova **240** (2003), no. Biratsion. Geom. Linein. Sist. Konechno Porozhdennye Algebry, 82–219.

[Sho04] V. V. Shokurov, *Letters of a bi-rationalist. V. Minimal log discrepancies and termination of log flips*, Tr. Mat. Inst. Steklova **246** (2004), no. Algebr. Geom. Metody, Svyazi i Prilozh., 328–351.

[Sho09] V. V. Shokurov, *Letters of a bi-rationalist. VII. Ordered termination* (Russian, with Russian summary), Tr. Mat. Inst. Steklova **264** (2009), no. Mnogomernaya Algebraicheskaya Geometriya, 184–208; English transl., Proc. Steklov Inst. Math. **264** (2009), no. 1, 178–200. MR2590847

[Siu98] Yum-Tong Siu, *Invariance of plurigenera*, Invent. Math. **134** (1998), no. 3, 661–673. MR1660941

[ST12] Karl Schwede and Kevin Tucker, *A survey of test ideals*, Progress in commutative algebra 2, Walter de Gruyter, Berlin, 2012, pp. 39–99. MR2932591

[Ste85] Hans Sterk, *Finiteness results for algebraic K3 surfaces*, Math. Z. **189** (1985), no. 4, 507–513. MR786280

[Sze99] BalÁzs SzendrÖi, *Some finiteness results for Calabi-Yau threefolds*, J. London Math. Soc. (2) **60** (1999), no. 3, 689–699. MR1753808

[Tak06] Shigeharu Takayama, *Pluricanonical systems on algebraic varieties of general type*, Invent. Math. **165** (2006), no. 3, 551–587. MR2242627

[Tak07] Shigeharu Takayama, *On the invariance and the lower semi-continuity of plurigenera of algebraic varieties*, J. Algebraic Geom. **16** (2007), no. 1, 1–18. MR2257317

[Tak08] Shigeharu Takayama, *On the uniruledness of stable base loci*, J. Differential Geom. **78** (2008), no. 3, 521–541. MR2396253

[Tir12] Sofia Tirabassi, *On the tetracanonical map of varieties of general type and maximal Albanese dimension*, Collect. Math. **63** (2012), no. 3, 345–349. MR2957974

[Tod13] Yukinobu Toda, *Stability conditions and extremal contractions*, Math. Ann. **357** (2013), no. 2, 631–685. MR3096520

[Tod14] Yukinobu Toda, *Stability conditions and birational geometry of projective surfaces*, Compos. Math. **150** (2014), no. 10, 1755–1788. MR3269467

[Tot08] Burt Totaro, *Hilbert's 14th problem over finite fields and a conjecture on the cone of curves*, Compos. Math. **144** (2008), no. 5, 1176–1198. MR2457523

[Tot10] Burt Totaro, *The cone conjecture for Calabi-Yau pairs in dimension 2*, Duke Math. J. **154** (2010), no. 2, 241–263. MR2682184

[TX09] Gueorgui Todorov and Chenyang Xu, *Effectiveness of the log Iitaka fibration for 3-folds and 4-folds*, Algebra Number Theory **3** (2009), no. 6, 697–710. MR2579391

[Ueh04] Hokuto Uehara, *Calabi-Yau threefolds with infinitely many divisorial contractions*, J. Math. Kyoto Univ. **44** (2004), no. 1, 99–118. MR2062709

[Uen83] Kenji Ueno, *Birational classification of algebraic varieties*, Recent progress of algebraic geometry in Japan, North-Holland Math. Stud., vol. 73, North-Holland, Amsterdam, 1983, pp. 82–105. MR722143

[Var08] Dror Varolin, *A Takayama-type extension theorem*, Compos. Math. **144** (2008), no. 2, 522–540. MR2406122

[Wil94] P. M. H. Wilson, *Minimal models of Calabi-Yau threefolds*, Classification of algebraic varieties (L'Aquila, 1992), Contemp. Math., vol. 162, Amer. Math. Soc., Providence, RI, 1994, pp. 403–410. MR1272711

[Xu12] Chenyang Xu, *Strong rational connectedness of surfaces*, J. Reine Angew. Math. **665** (2012), 189–205. MR2908743

[Xu14] Jinsong Xu, *The third and fourth pluricanonical maps of threefolds of general type*, Math. Proc. Cambridge Philos. Soc. **157** (2014), no. 2, 209–220. MR3254589

[Xu15] Chenyang Xu, *On the base-point-free theorem of 3-folds in positive characteristic*, J. Inst. Math. Jussieu **14** (2015), no. 3, 577–588. MR3352529

[Zha14] De-Qi Zhang, *Birational automorphism groups of projective varieties of Picard number two*, Automorphisms in birational and affine geometry, Springer Proc. Math. Stat., vol. 79, Springer, Cham, 2014, pp. 231–238. MR3229354

[Zhu15] Yi Zhu, *Log rationally connected surfaces*, Math. Res. Lett. **23** (2016), no. 5, 1527–1536. MR3601077

Department of Mathematics, Boston College, Chestnut Hill, Massachusetts 02467
E-mail address: `lehmannb@bc.edu`

Positive characteristic algebraic geometry

Zsolt Patakfalvi, Karl Schwede, and Kevin Tucker

ABSTRACT. These lecure notes contain a detailed introduction to the recent F-singularity theory methods in positive characteristic algebraic geometry. We carefully define the different local and global notions of F-singularity theory, show how they relate to other notions of algebraic geometry, as well as we include some applications.

CONTENTS

1. Introduction

The goal of these notes is to give a geometric introduction to recent methods utilizing the Frobenius morphism (see [**ST12a**] for more algebraic aspects) in higher dimensional algebraic geometry. These methods have been used extensively to move arguments from characteristic zero to positive characteristic. The main obstacle in doing so is usually some theorems of either analytic or p-adic origin that are known not to hold in positive characteristic: e.g., Kodaira-vanishing [**Ray78a**], Kawamata-Viehweg vanishing [**Xie10**], some semi-positivity statements [**MB81**, 3.2], etc. The

2010 *Mathematics Subject Classification.* Primary 14J17, 14J20, 14E30.

The current version of these notes were written for the The Boot Camp for the 2015 Algebraic Geometry Summer Research Institute, which was supported by NSF grant DMS #1500652. The first version of these notes arose from a "Positive Characteristic Algebraic Geometry Workshop" held at the University of Illinois at Chicago in March 2014, supported by NSF RTG grant #1246844 and UIC..

The second author was partially supported by NSF FRG Grant DMS #1265261/1501115, NSF CAREER Grant DMS #1252860/1501102 and a Sloan Fellowship.

The third author was partially supported by NSF grant DMS #1419448/1602070, and a Sloan fellowship.

method of Frobenius splitting gives a weak replacement to these theorems. We refer to the introduction of 5 for a list of global geometry results obtained in the past few years using the method of Frobenius split (including the existence of minimal models for threefolds).

As the goal of the notes is to serve as an introduction to the subject, we do not give the proofs of the above mentioned global results (except one example in 5). Instead, we aim for an introductory display of the techniques used in the field. In particular, the notes could serve the basis of a introductory course to the subject. Further, we list many open questions, which we hope will be attacked soon by the readers of this article. We would like to draw attention to the many exercises contained in the notes. We believe that the best approach to reading this article is to solve the exercises.

Throughout the notes we assume a basic knowledge of the language of $\mathbb{Q}$-divisors (in the form in which it is widely used in birational geometry, e.g., [**KM98**]), reflexive sheaves (c.f., [**Har94a, Har80**]), and Grothendieck duality for finite maps (c.f., [**Har77**] [**Har66**]). A big part of the assumed background knowledge is discussed in the appendices of [**ST12b**].

2. Frobenius Splittings

The main tool we have to study a scheme X with characteristic $p > 0$ is the Frobenius endomorphism $F : X \to X$ or p-th power map. Similarly, the e-iterated Frobenius $F^e : X \to X$ is always the identity on the underlying topological space of X determined by taking p^e-th powers on functions. It is often convenient to think of Frobenius as the morphism of sheaves of $\mathcal{O}_X$-modules $\mathcal{O}_X \to F^e_* \mathcal{O}_X$.

We say that X is *F-finite* if the Frobenius endomorphism is a finite morphism, or equivalently $F^e_* \mathcal{O}_X$ is coherent for some or all $e \in \mathbb{N}$. One checks easily that this is the case in essentially every geometric (or even arithmetic) situation of interest, and *F-finite will be a standing assumption throughout this survey.*

EXERCISE 2.1. Check that a quasi-projective variety over a perfect field is always F-finite.

EXERCISE 2.2. When $X = \operatorname{Spec}(R)$ is affine and M is a finitely generated R-module, justify the notation $F^e_* M$ for the module given by restriction of scalars for the p^e-th power map on R.

Note that F^e_* is always an exact functor, and for any coherent sheaf $\mathcal{M}$ on X we have $H^i(X, F^e_* \mathcal{M}) = H^i(X, \mathcal{M})$ (though it is perhaps more accurate to write $F^e_* H^i(X, \mathcal{M})$ to keep track of the linearity).

THEOREM 2.3 (Kunz). *X is regular if and only if Frobenius is flat, or equivalently (as X is assumed F-finite) $F^e_* \mathcal{O}_X$ is locally free for some or all $e \in \mathbb{N}$.*

Recall that the terms regular and smooth are interchangeable for varieties over perfect fields. In this case, we can think of the iterates of Frobenius as an infinite flag

$$\mathcal{O}_X \subseteq F_* \mathcal{O}_X \subseteq F^2_* \mathcal{O}_X \subseteq \cdots \subseteq F^e_* \mathcal{O}_X \subseteq \cdots$$

where $F^e_* \mathcal{O}_X$ is a vector bundle of rank $p^{e(\dim X)}$.

EXERCISE 2.4. Check this for $X = \mathbb{A}^n_k$ where k is a perfect field of positive characteristic. In particular, if $S = k[x_1, \ldots, x_n]$ we have that $F^e_* S$ is the free S-module of rank p^{en} on the basis of monomials where the exponents on each of the variables is strictly less than p^e.

EXERCISE 2.5. Assume $k = k^p$ once more. Recall that any vector bundle on $\mathbb{P}^1_k$ is isomorphic to a direct sum of line bundles $\mathcal{O}_{\mathbb{P}^1}(a)$ for $a \in \mathbb{Z}$. Identify $F^e_* \mathcal{O}_{\mathbb{P}^1}$ as a direct sum of line bundles.

Note that the key idea for the previous exercise is to make use of the projection formula: for a line bundle $\mathscr{L}$ and coherent sheaf $\mathscr{M}$ on X, we have

$$\mathscr{L} \otimes_{\mathcal{O}_X} F^e_* \mathscr{M} = F^e_* \left((F^{e*} \mathscr{L}) \otimes_{\mathcal{O}_X} \mathscr{M} \right) = F^e_* \left(\mathscr{L}^{p^e} \otimes_{\mathcal{O}_X} \mathscr{M} \right).$$

DEFINITION 2.6. We say that X is globally F-split (or simply F-split) if $\mathcal{O}_X \to F^e_* \mathcal{O}_X$ has a global splitting (in the category of $\mathcal{O}_X$-modules). We say that X is locally F-split (or simply F-pure) if $\mathcal{O}_X \to F^e_* \mathcal{O}_X$ splits in a sufficiently small affine neighborhood of any point $x \in X$.

EXAMPLE 2.7. A smooth variety is always locally F-split, but may or may not be globally F-split. Our next goal is to explore what happens for smooth projective curves, where the situation depends on the genus. Genus zero curves are always F-split, genus one curves are sometimes F-split, and higher genus curves are never F-split.

EXERCISE 2.8. Show that locally F-split and globally F-split are the same for affine varieties.

LEMMA 2.9. *Say that X is a smooth variety over $k = k^p$. Then*

$$\mathscr{H}om_{\mathcal{O}_X}(F^e_* \mathcal{O}_X, \mathcal{O}_X) = F^e_* \mathcal{O}_X((1 - p^e) K_X).$$

PROOF INGREDIENTS. You will need but two tricks. First, twisting and untwisting by ω_X, we have

$$\mathscr{H}om_{\mathcal{O}_X}(F^e_* \mathcal{O}_X, \mathcal{O}_X) = \mathscr{H}om_{\mathcal{O}_X}(F^e_* \mathcal{O}_X(p^e K_X), \mathcal{O}_X(K_X)).$$

The second trick is to use duality for a finite morphism, which can be thought of as simply saying that $F^e_*(\_\_)$ commutes with the Grothendieck-duality functor $\mathscr{H}om_{\mathcal{O}_X}(\_\_, \omega_X)$. In other words, for a coherent sheaf $\mathscr{M}$ on X, we have

$$\mathscr{H}om_{\mathcal{O}_X}(F^e_* \mathscr{M}, \omega_X) = F^e_* \mathscr{H}om_{\mathcal{O}_X}(\mathscr{M}, \omega_X).$$ □

EXERCISE 2.10. Prove the Lemma. Then explain how to do the same thing when X is only normal, instead of smooth, by restricting to the smooth locus and using that all of the sheaves in play are reflexive.

EXERCISE 2.11. If X is locally F-split, show it is reduced and weakly normal[1].

EXAMPLE 2.12. If X is F-split, then it follows from Lemma 2.9 that $-nK_X$ is effective for some $n \in \mathbb{N}$. In particular, if X has general type (*e.g.* a smooth projective curve of genus at least two), it is never globally F-split.

[1]A reduced ring of equal characteristic $p > 0$ is called *weakly normal* if for every $r \in K(R)$ the total ring of fractions of R such that $r^p \in R$ also satisfies $r \in R$.

LEMMA 2.13. *Suppose that X is a projective variety over $k = k^p$. Then X is F-split if and only if the section ring*

$$R(X, A) = \bigoplus_{m \geq 0} H^0(X, \mathcal{O}_X(mA))$$

with respect to some (or equivalently any) ample divisor A on X is F-split.

PROOF IDEAS. The main idea of the proof is straightforward enough – twisting a given F-splitting by $\mathcal{O}_X(mA)$ gives a splitting of $\mathcal{O}_X(mA) \to F^e_* \mathcal{O}_X(mp^e A)$ for any $m \geq 0$. Taking global sections and keeping track of the gradings as in the exercise below, it is then easy to see why $R(X, A)$ must be F-split. For the other direction, we need only take Proj of a graded F-splitting. Be careful though, if $M = \oplus_{n \in \mathbb{Z}} \Gamma(X, \mathcal{M} \otimes \mathcal{O}_X(nA))$ then $F_* M$ is not generally the same as $\oplus_{n \in \mathbb{Z}} \Gamma(X, (F_* \mathcal{M}) \otimes \mathcal{O}_X(nA))$. □

EXERCISE 2.14. Suppose that $X = \operatorname{Spec}(R)$, where $R = \bigoplus_{m \geq 0} R_m$ is a standard-graded k-algebra ($R_0 = k$ and $R = k[R_1]$). For a $\mathbb{Z}$-graded R-module M, explain how to think of $F^e_* M$ as a $\frac{1}{p^e}\mathbb{Z}$-graded R-module. In particular, show that we can break up $F^e_* M$ into a direct sum of $\mathbb{Z}$-graded R-modules

$$F^e_* M = \bigoplus_{0 \leq i < p^e} [F^e_* M]_{i/p^e \bmod \mathbb{Z}}$$

where $[F^e_* M]_{i/p^e \bmod \mathbb{Z}} = \bigoplus_{m \in \mathbb{Z}} M_{i+mp^e}$ as an abelian group, and the elements of R act via p^e-th powers. Finally, show that R is F-split if and only if it has a graded F-splitting, and if and only if $R_\mathfrak{m}$ is F-split where $\mathfrak{m} = R_+$ is the homogeneous maximal ideal.

EXERCISE 2.15. If $R \subseteq S$ is a split inclusion of rings and S is F-split, then conclude R is also F-split. In particular, conclude a Veronese subring of a graded F-split ring is always F-split.

PROPOSITION 2.16 (Fedder's Criterion). *Let $S = k[x_0, \ldots, x_n] \supseteq I$ homogeneous, $\mathfrak{m} = \langle x_0, \ldots, x_n \rangle$, and $R = S/I$. Then R is F-split if and only if $(I^{[p]} : I) \not\subseteq \mathfrak{m}^{[p]}$, where $J^{[p]} = \langle j^p | j \in J \rangle$ denotes the p-th bracket or Frobenius power of an ideal J.*

EXAMPLE 2.17. $R = k[x, y, z]/\langle f = x^3 + y^3 + z^3 \rangle$ is F-split if and only if p is congruent to 1 modulo 3. Indeed, one easily checks that these are precisely the characteristics for which $f^{p-1} \notin \langle x^p, y^p, z^p \rangle$.

EXAMPLE 2.18. Suppose that X is a smooth genus one curve over $k = \overline{k}$. Recall that X is said to be ordinary or have Hasse invariant one if the action of Frobenius on $H^1(X, \mathcal{O}_X)$, namely

$$F^* : H^1(X, \mathcal{O}_X) \to H^1(X, \mathcal{O}_X),$$

is injective. In this case, having chosen a base point, X has precisely p^e distinct p^e-torsion points for the group law on X.

In contrast, when the action of Frobenius on $H^1(X, \mathcal{O}_X)$ is not injective (and hence identically zero), we say that X is supersingular and has Hasse invariant zero. In this case, having chosen a base point, the identity element is the only p^e-torsion point for the group law.

In Theorem 4.21 of Hartshorne, it is shown that X is ordinary if and only if, when embedded as a cubic hypersurface $X = \mathbb{V}(f) \subseteq \mathbb{P}^2$, the term $(xyz)^{p-1}$ appears with a nonzero coefficient in f^{p-1}. By Fedder's criterion, it follows that X is F-split if and only if X is ordinary.

EXERCISE 2.19. Suppose X is a smooth projective genus one curve. First, recall Atiyah's classification of indecomposable vector bundles on X. For each $r > 0$, there exists a unique indecomposable vector bundle E_r of rank r and degree zero with $H^0(X, E_r) \neq 0$. Moreover, one can construct the E_r inductively by setting $E_1 = \mathcal{O}_X$ and letting E_{r+1} be the unique nontrivial extension

$$0 \to E_{r-1} \to E_r \to \mathcal{O}_X \to 0.$$

Any other indecomposable vector bundle of rank r and degree zero has the form $E_r \otimes \mathscr{L}$ for a uniquely determined line bundle $\mathscr{L}$ of degree zero.

Use this classification to determine the structure of $F^e_*\mathcal{O}_X$. If X is ordinary and $x_0 \in X$ is a fixed base point for the group law, let $x_0, \ldots, x_{p^e-1}$ be the p^e distinct p^e-torsion points. Show that $F^e_*\mathcal{O}_X = \bigoplus_{0\leq i<p^e} \mathcal{O}_X(x_i - x_0)$. In contrast, when X is supersingular, show that $F^e_*\mathcal{O}_X$ is the unique indecomposable vector bundle of rank p^e with degree zero having a nonzero global section.

FEDDER'S CRITERION PROOF IDEAS. We sketch the two main ideas of the proof, and leave the audience to flesh out the details.

First, one must show that every $\phi \in \operatorname{Hom}_R(F_*R, R)$ is a quotient of a $\tilde{\phi} \in \operatorname{Hom}_S(F_*S, S)$. Conversely, $\tilde{\psi} \in \operatorname{Hom}_S(F_*S, S)$ gives rise to a map $\psi \in \operatorname{Hom}_R(F_*R, R)$ whenever I is $\tilde{\psi}$-compatible (*i.e.* $\tilde{\psi}(F_*I) \subseteq I$). The key point here is that F_*S is a free S-module, which allows one to lift such (potential) splittings from R up to S. The second main idea of the proof is to once again make use of duality. We have that

$$\operatorname{Hom}_S(F_*\omega_S, \omega_S) = F_* \operatorname{Hom}_S(\omega_S, \omega_S) = F_*S.$$

In other words, identifying $S = \omega_S$, every element of $\operatorname{Hom}_S(F_*S, S)$ has the form $\Phi_S(F_*x \cdot \_\_)$ for a uniquely determined element $x \in S$. (One can take Φ_S to be the trace of Frobenius as discussed below.)

Putting these together, if we start with a splitting ϕ of R, we can lift it to $\tilde{\phi}$ on S. Then $\tilde{\phi}(\_\_) = \Phi_S(F_*c \cdot \_\_)$ for some $c \in S$. As $\tilde{\phi}(F_*I) \subseteq I$, one can conclude $c \in (I^{[p]} : I)$. As $\tilde{\phi}$ must be surjective since ϕ was a splitting, one can also conclude $c \notin \langle x_0^p, \ldots, x_n^p \rangle$. Conversely, if we can start by taking $c \in (I^{[p]} : I) \setminus \mathfrak{m}^{[p]}$, one can simply take $\Phi_S(F_*c \cdot \_\_)$ and reverse the process to get a surjective map in $\operatorname{Hom}_R(F_*R, R)$ – which is just as good as having an honest F-splitting. □

EXERCISE 2.20. Show that the dual of the Frobenius map $F^e : \mathcal{O}_X \to F^e_*\mathcal{O}_X$ of a normal variety takes the form $\operatorname{Tr}_{F^e} : F^e_*\omega_X = \mathscr{H}om_{\mathcal{O}_X}(F^e_*\mathcal{O}_X, \omega_X) \to \omega_X$ given by evaluation at one. We call this the *trace* of F^e. Show that Tr_{F^e} generates $\mathscr{H}om_{\mathcal{O}_X}(F^e_*\omega_X, \omega_X)$ as an $F^e_*\mathcal{O}_X$-module.

In case $X = \mathbb{A}^n = \operatorname{Spec}(S = k[x_1, \ldots, x_n])$, show also that one may take the projection onto the free summand generated by $(x_1 \cdots x_n)^{p-1}$ as a F^e_*S-generator of $\operatorname{Hom}_S(F^e_*S, S)$. Conclude this agrees with the trace of F^e up to (pre)multiplication by a unit.

EXERCISE 2.21. For X smooth, recall the Cartier isomorphism for the algebraic De Rham complex $\Omega^\bullet_X$, namely $\mathscr{H}^i(F^e_*\Omega^\bullet_X) \simeq \Omega^i_X$. Write this down explicitly for

$\mathbb{A}^n$ in coordinates, and check that the induced map

$$F^e_* \omega_X \to \mathscr{H}^n(F^e_* \Omega^\bullet_X) \simeq \omega_X$$

agrees (up to unit) with the trace of F^e.

EXERCISE 2.22. If A is a finitely generated $\mathbb{Z}$-algebra and $\mathfrak{m}$ is a maximal ideal of A, show that $A/\mathfrak{m}$ is a finite field.

DEFINITION 2.23. Suppose that X is a complex algebraic variety. One can form a finitely generated $\mathbb{Z}$-algebra domain A and an arithmetic family $\mathscr{X} \to \operatorname{Spec}(A)$ so that $\mathscr{X}_0 \otimes_{\operatorname{Frac}(A)} \mathbb{C} = X$. We call $\mathscr{X}$ a model of X over $\operatorname{Spec}(A)$. X is said to have dense (local or global, respectively) F-split type if $\mathscr{X}_{\mathfrak{m}}$ is (local or global, respectively) F-split for a dense set of closed points $\mathfrak{m} \in \operatorname{Spec}(A)$.

CONJECTURE 2.24. *If X is a complex Abelian variety (or more generally log Calabi-Yau), then X has dense F-split type.*

PROPOSITION 2.25. *Suppose that X is a globally F-split projective variety.*

(a) *If $\mathscr{L}$ is any line bundle so that $H^i(X, \mathscr{L}^{\otimes n}) = 0$ for $n \gg 0$, then $H^i(X, \mathscr{L}) = 0$.*
(b) *If A is an ample divisor on X, then $H^i(X, \mathcal{O}_X(A)) = 0$ for all $i > 0$.*
(c) *Suppose X is Cohen-Macaulay. If A is an ample divisor on X and $i < \dim X$, then $H^i(X, \mathcal{O}_X(-A)) = 0$.*
(d) *Suppose X is normal. If A is an ample divisor on X then $H^i(X, \omega_X(A)) = 0$ for $i > 0$.*

DEFINITION 2.26. If D is an effective Cartier divisor on a normal variety X, we say that X is e-F-split along D if $\mathcal{O}_X \to F^e_* \mathcal{O}_X \subseteq F^e_* \mathcal{O}_X(D)$ splits (globally).

LEMMA 2.27. *If X is e-F-split along an ample effective divisor A and $\mathscr{L}$ is nef, then $H^i(X, \mathscr{L}) = 0$ for all $i > 0$.*

EXERCISE 2.28. Suppose X is normal. If X is e-F-split along an ample effective divisor A, then $-K_X$ is big.

DEFINITION 2.29. We say that a normal variety X is globally F-regular if, for every effective Cartier divisor D, there is some $e > 0$ so that X is e-F-split along D. We say that X is locally F-regular (or strongly F-regular) if, for every effective Cartier divisor D, there is some $e > 0$ so that $\mathcal{O}_X \to F^e_* \mathcal{O}_X(D)$ is split in a neighborhood of any point.

EXERCISE 2.30. Suppose that $X = \operatorname{Spec}(R)$ is affine. Then X is globally F-regular if and only if X is locally F-regular if and only if, for all $0 \neq c \in R$, there exists $e > 0$ such that $R \to F^e_* R$ given by $1 \mapsto F^e_* c$ is split.

EXERCISE 2.31. Show that a regular ring is always strongly F-regular, and that a strongly F-regular ring is a Cohen-Macaulay normal domain.

EXERCISE 2.32. If X is a smooth projective curve, then X is globally F-regular if and only if X has genus zero.

EXERCISE 2.33. If X is globally F-regular and D is big and nef, then $H^i(X, \omega_X(D)) = 0$ for all $i > 0$.

3. Trace of Frobenius and Global Sections

Suppose X is a normal quasi-projective variety over a perfect field k of characteristic $p > 0$. We just read about Frobenius splittings $F^e_* \mathcal{O}_X \to \mathcal{O}_X$. We will generalize this notion. Fix $\mathscr{L}$ to be a line bundle and suppose that we have a non-zero $\mathcal{O}_X$-linear map $\phi : F^e_* \mathscr{L} \to \mathcal{O}_X$.

Twist ϕ by $\mathscr{L}$ and apply F^e_* and by the projection formula we get:

$$F^{2e}_*(\mathscr{L}^{1+p^e}) \to F^e_* \mathscr{L}$$

We compose this back with ϕ to obtain:

$$\phi^2 : F^{2e}_*(\mathscr{L}^{1+p^e}) \to \mathcal{O}_X.$$

This is an abuse of notation of course.

By repeating this operation we get

$$\phi^n : F^{ne}_*(\mathscr{L}^{1+p^e+\ldots+p^{(n-1)e}}) \to \mathcal{O}_X.$$

THEOREM 3.1 ([**HS77, Lyu97, Gab04**]). *With notation as above, there exists* n_0 *such that* $\operatorname{Image}(\phi^n) = \operatorname{Image}(\phi^{n+1})$ *as sheaves for all* $n \geq n_0$.

SKETCH OF A PROOF. Let U_n be the locus where $\operatorname{Image}(\phi^n)$ and $\operatorname{Image}(\phi^{n+1})$ coincide. It is easy to see that the U_n are ascending open sets and so they stabilize by Noetherian induction. By localizing at the generic point of $X \setminus U_n$ for $n \gg 0$ one can assume that X is the spectrum of a local ring with closed point x and that $U_n = X \setminus x$. The trick is then to show that $\mathfrak{m}_x^l \cdot \operatorname{Image}(\phi^m) \subseteq \operatorname{Image}(\phi^n)$ for some $l > 0$ and all $n \geq m \gg 0$. Finding this l is left as an exercise to the reader. □

These images measure the singularities of the pair (X, ϕ). We'll see the relevance of ϕ momentarily.

DEFINITION 3.2. We define $\sigma(X, \phi) = \operatorname{Image}(\phi^n)$ for $n \gg 0$. We say the (X, ϕ) is *F-pure* if $\sigma(X, \phi) = \mathcal{O}_X$.

EXERCISE 3.3. Suppose that $\phi : F^e_* \mathscr{L} \to \mathcal{O}_X$ is surjective. Prove that (X, ϕ) is F-pure.

Let us now describe the significance of ϕ in terms of divisor pairs that you might be more familiar with.

PROPOSITION 3.4. *There is a bijection of sets*

$$\left\{ \begin{matrix} \text{effective } \mathbb{Q}\text{-divisors } \Delta \text{ such that} \\ (p^e - 1)(K_X + \Delta) \text{ is Cartier} \end{matrix} \right\} \longleftrightarrow \left\{ \begin{matrix} \text{line bundles } \mathscr{L} \text{ and non-zero} \\ \mathcal{O}_X\text{-linear maps } \phi : F^e_* \mathscr{L} \to \mathcal{O}_X \end{matrix} \right\} \Big/ \sim$$

where the equivalence relation on the left declares two maps to be equivalent if they agree up to multiplication by a unit from $\Gamma(X, \mathcal{O}_X)$.

SKETCH. There is always a map $F^e_* \mathcal{O}_X(K_X) \to \mathcal{O}_X(K_X)$, hence by the projection formula (and reflexification) we get a map $F^e_* \mathcal{O}_X((1-p^e)K_X) \to \mathcal{O}_X$. If Δ is effective, we then get the composition

$$F^e_* \mathcal{O}_X((1-p^e)(K_X + \Delta)) \subseteq F^e_* \mathcal{O}_X((1-p^e)K_X) \to \mathcal{O}_X.$$

Thus we obtain the $\Rightarrow$ direction for $\mathscr{L} = \mathcal{O}_X((1-p^e)(K_X + \Delta))$. The reverse direction follows easily from the isomorphism $F^e_*(\mathscr{L}^{-1} \otimes \mathcal{O}_X((1-p^e)K_X)) \cong \mathscr{H}om_{\mathcal{O}_X}(F^e_* \mathscr{L}, \mathcal{O}_X)$ coming from Grothendieck duality for finite maps. Given this, a map ϕ determines a global section of $\mathscr{L}^{-1} \otimes \mathcal{O}_X((1-p^e)K_X)$ which determines an effective Weil divisor D_ϕ. Set $\Delta = \frac{1}{p^e-1} D_\phi$. $\square$

Given any effective $\mathbb{Q}$-divisor Δ such that $K_X + \Delta$ is $\mathbb{Q}$-Cartier with index not divisible by the characteristic p, one can find $e > 0$ such that $(p^e-1)(K_X+\Delta)$ is Cartier and hence find a corresponding $\phi_{e,\Delta}$.

EXERCISE 3.5. If $K_X + \Delta$ is $\mathbb{Q}$-Cartier with index not divisible by $p > 0$, show that there exists an $e > 0$ such that $(p^e - 1)(K_X + \Delta)$ is Cartier.

In particular, this is exactly how the above was presented in the mini-lecture. Given such a Δ one can form $\phi_{e,\Delta} : F^e_* \mathcal{O}_X\big((1-p^e)(K_X+\Delta)\big) \to \mathcal{O}_X$ for sufficiently divisible e. 3.1 says that $\sigma(X,\Delta)$ has stable image for $e \gg 0$ and sufficiently divisible. Translated this way consider the following exercise:

EXERCISE 3.6. With notation as above, suppose that $d > e$ are integers such that $(p^i - 1)(K_X + \Delta)$ is Cartier for $i = d, e$. Show that

$$\phi_{d,\Delta} : F^d_* \mathcal{O}_X\big((1-p^d)(K_X+\Delta)\big) \to \mathcal{O}_X$$

factors through

$$\phi_{e,\Delta} : F^e_* \mathcal{O}_X\big((1-p^e)(K_X+\Delta)\big) \to \mathcal{O}_X.$$

Further show that for any b, c satisfying the condition on i above, the composition

$$\begin{array}{rl} & F^{b+c}_* \mathcal{O}_X\big((1-p^b)(K_X+\Delta) + p^b(1-p^c)(K_X+\Delta)\big) \\ \xrightarrow{F^b_*\left(\phi_{b,\Delta} \otimes \mathcal{O}_X((1-p^c)(K_X+\Delta))\right)} & F^c_* \mathcal{O}_X\big((1-p^c)(K_X+\Delta)\big) \\ \xrightarrow{\phi_{c,\Delta}} & \mathcal{O}_X \end{array}$$

is simply $\phi_{b+c,\Delta}$.

DEFINITION 3.7. Given a $\mathbb{Q}$-divisor Δ such that $K_X + \Delta$ is $\mathbb{Q}$-Cartier with index not divisible by $p > 0$, we say that the pair (X, Δ) is F-pure if $(X, \phi_{e,\Delta})$ is F-pure for some (equivalently any) $\phi_{e,\Delta}$ corresponding to Δ as in 3.4.

EXERCISE 3.8. Suppose that X is an affine cone over a smooth hypersurface Y in $\mathbb{P}^n$. Further suppose that $H^0(Y, F^e_* \omega_Y(p^e n)) \to H^0(Y, \omega_Y(n))$ surjects for all $n \geq 0$ and all $e \geq 0$. Let $\pi : X' \to X$ be the blowup of X at the cone point (a resolution of singularities). Show that $\sigma(X, 0) = \pi_* \mathcal{O}_{X'}(K_{X'/X} + E)$ where $E \cong Y$ is the exceptional divisor.

Hint: Recall that $\omega_X = \oplus_{i \in \mathbb{Z}} H^0(Y, \omega_Y(i))$ which is just $\mathcal{O}_X$ with a shift in grading (since X is a hypersurface).

REMARK 3.9. Suppose we now consider a general situation of a pair (X, Δ) with $K_X + \Delta$ $\mathbb{Q}$-Cartier. Philosophically, one should expect that $\sigma(X, \Delta)$ to correspond with something like the following ideal sheaf $\pi_* \mathcal{O}_{X'}(\lceil K_{X'} - \pi^*(K_X + \Delta) + \varepsilon E \rceil)$ where $\pi : X' \to X$ is a log resolution and $E = \mathrm{Supp}(\pi^{-1}_* \Delta) \cup \mathrm{Exc}_\pi$. See [**FST11, Tak13, BST13**].

We can globalize this process.

DEFINITION 3.10. Suppose that X is a proper variety, $\Delta \geq 0$ is a $\mathbb{Q}$-divisor such that $K_X + \Delta$ is $\mathbb{Q}$-Cartier with index not divisible by p and M is a line bundle. We define

$$S^0(X, \sigma(X, \Delta) \otimes M)$$

to be the image

$$\mathrm{Image}\Big(H^0\big(X, F^e_*(\mathcal{O}_X((1-p^e)(K_X+\Delta)) \otimes M^{p^e})\big) \\ \to H^0(X, \sigma(X,\Delta) \otimes M\Big) \subseteq H^0(X, M)\Big)$$

for some $e \gg 0$ with $(p^e - 1)(K_X + \Delta)$ Cartier.

We make a similar definition even without the presence of Δ. We define $S^0(X, \omega_X \otimes M)$ to be the image

$$\mathrm{Image}\Big(H^0\big(X, F^e_*(\omega_X \otimes M^{p^e})\big) \to H^0(X, \omega_X \otimes M)\Big)$$

for some $e \gg 0$. Here the map is induced by $F^e_*\omega_X \cong F^e_*\mathcal{O}_X(K_X) \to \mathcal{O}_X(K_X) = \omega_X$ mentioned earlier.

EXERCISE 3.11. Suppose that X is Frobenius split by some map $\phi_{e,\Delta} : F^e_*\mathcal{O}_X \to \mathcal{O}_X$ which corresponds to a Weil divisor $(p^e-1)\Delta = D \sim (1-p^e)K_X$ as above. Show that $S^0(X, \sigma(X, \frac{1}{p^e-1}D) \otimes M) = H^0(X, M)$ for any line bundle M.

QUESTION 3.12. Find a way to compute $S^0(X, \omega_X \otimes M)$ for M that is not necessarily ample. In particular, how do you know when you have chosen $e > 0$ large enough?

Note that this object depends on the map used to construct $\sigma(X, \Delta)$ and not *simply* on the sheaf $\sigma(X, \Delta)$. Technically, we are relying on the structure of $\sigma(X, \Delta)$ as a Cartier module [**BB11, BS13**]. Likewise $S^0(X, \omega_X \otimes M)$ depends on the map $F^e_*\omega_X \to \omega_X$ (fortunately, that map is canonically determined).

REMARK 3.13. The letter S^0 should be thought of as a capitalized or globalized σ.

The sections $S^0(X, \sigma(X, \Delta))$ are important because they behave as though Kodaira (or Kawamata-Viehweg) vanishing were true. Perhaps this should not be surprising, indeed in order to construct counter-examples to Kodaira vanishing in characteristic $p > 0$ [**Ray78b**] one typically first finds a curve X and line bundle M such that $S^0(X, \omega_X \otimes M) \neq H^0(X, \omega_X \otimes M)$. This failure of equality is the key ingredient in proving that Kodaira vanishing fails.

Consider the following example.

EXAMPLE 3.14. Suppose that X is a smooth projective variety, M is a Cartier divisor, Δ is a $\mathbb{Q}$-divisor and D is a normal Cartier divisor such that D and Δ have no common components. We also assume that $M - K_X - \Delta - D$ is an ample $\mathbb{Q}$-divisor. Then $S^0(X, \sigma(X, \Delta + D) \otimes \mathcal{O}_X(M))$ surjects onto $S^0(D, \sigma(D, \Delta|_D) \otimes \mathcal{O}_D(M|_D))$.

This is actually quite easy. First consider the following commutative diagram:

$$\begin{array}{ccc}
F^e_*\mathcal{O}_X((1-p^e)(K_X+\Delta+D)-D) & \xrightarrow{\alpha} & \mathcal{O}_X(-D) \\
\downarrow & & \downarrow \\
F^e_*\mathcal{O}_X((1-p^e)(K_X+\Delta+D)) & \xrightarrow{\beta} & \mathcal{O}_X \\
\downarrow & & \downarrow \rho \\
F^e_*\mathcal{O}_D((1-p^e)(K_X+\Delta+D)|_D) & \xrightarrow{\gamma} & \mathcal{O}_D
\end{array}$$

Note that the lower left term is $F^e_*\mathcal{O}_D((1-p^e)(K_D+\Delta|_D))$ and it is not difficult[2] to see that the map γ is the one whose image on global sections is the one used to define $\sigma(D,\Delta|_D)$.

$$
\begin{array}{ccc}
H^0(X, F^e_*\mathcal{O}_X((1-p^e)(K_X+\Delta+D)-D+p^eM)) & \xrightarrow{\alpha} & H^0(X,\mathcal{O}_X(-D+M)) \\
\downarrow & & \downarrow \\
H^0(X, F^e_*\mathcal{O}_X((1-p^e)(K_X+\Delta+D)+p^eM)) & \xrightarrow{\beta} & H^0(X,\mathcal{O}_X(M)) \\
\delta\downarrow & & \downarrow\rho \\
F^e_*H^0(D,\mathcal{O}_D((1-p^e)(K_D+\Delta|_D)+p^eM|_D)) & \xrightarrow[\gamma]{} & H^0(D,\mathcal{O}_D(M))
\end{array} \tag{1}
$$

The map labeled ρ need not be surjective. However note that

$$
\begin{aligned}
&H^1(X,\mathcal{O}_X((1-p^e)(K_X+\Delta+D)-D+p^eM) \\
&\quad = H^1(X,\mathcal{O}_X(M-D)\otimes\mathcal{O}_X((p^e-1)(M-K_X-\Delta-D)))
\end{aligned}
$$

which vanishes due to Serre vanishing for $e \gg 0$. It follows that δ is surjective. The result follows from an easy diagram chase since the image of β is $S^0(X,\sigma(X,\Delta+D)\otimes\mathcal{O}_X(M))$ and the image of γ is $S^0(D,\sigma(D,\Delta|_D)\otimes\mathcal{O}_D(M|_D))$.

There are some details in the above example that I leave as an exercise.

EXERCISE 3.15. Show that diagram 3.14 commutes. To do this, start with the following diagram:

$$
\begin{array}{ccccccccc}
0 & \longrightarrow & F^e_*\mathcal{O}_X(-D) & \longrightarrow & F^e_*\mathcal{O}_X & \longrightarrow & F^e_*\mathcal{O}_D & \longrightarrow & 0 \\
 & & \uparrow & & & & & & \\
 & & F^e_*\mathcal{O}_X(-p^eD) & & \uparrow F^e & & \uparrow F^e & & \\
 & & \uparrow F^e & & & & & & \\
0 & \longrightarrow & \mathcal{O}_X(-D) & \longrightarrow & \mathcal{O}_X & \longrightarrow & \mathcal{O}_D & \longrightarrow & 0
\end{array}
$$

Apply $\mathscr{H}om_{\mathcal{O}_X}(\_\_,\omega_X)$ and then twist everything by $\mathcal{O}_X(-K_X-D)$. Additionally twist the F^e_*-row by $F^e_*\mathcal{O}_X((1-p^e)\Delta)$ and note we still have a map.

EXERCISE 3.16. In general (on a normal variety), a divisor Δ with $(p^e-1)(K_X+\Delta+D)$ Cartier induces a map $F^e_*\mathcal{O}_X((1-p^e)(K_X+\Delta+D)) \to \mathcal{O}_X$ which induces a map $F^e_*\mathcal{O}_D((1-p^e)(K_X+\Delta+D)|_D) \to \mathcal{O}_D$ by restriction. This map then induces a divisor Δ_D called the *F-different*. It was recently shown by O. Das that the F-different coincides with the usual notion of *different* in full generality, see **[Das15]**. Verify carefully in the case of the previous exercise[3] that the F-different is exactly $\Delta|_D$.

3.1. Examples and computations of S^0.

EXERCISE 3.17. Use a diagram similar to the one in the example above to show the following. Suppose that C is a curve and L is ample. Show that $S^0(C,\omega_C\otimes L)$ defines a base point free linear system if $\deg L \geq 2$ and that $S^0(C,\omega_C\otimes L)$ defines an embedding if $\deg L \geq 3$.

[2]Well, maybe it's a little difficult, it is an exercise below.

[3]or at least when D is Cartier and $(p^e-1)\Delta$ is Cartier

Hint: To show that it defines a base point free linear system, choose a point $Q \in C$ and consider $\omega_C \otimes L \to (\omega_C \otimes L)/(\omega_C \otimes L \otimes \mathcal{O}_C(-Q))$.

One can also ask when $S^0 \neq 0$, [**Tan15**] showed this when $\deg L \geq 1$ and $C \neq \mathbb{P}^1$.

QUESTION 3.18. Suppose X is a surface, or maybe a ruled surface to start, and L is a line bundle on X. What conditions on L imply that $S^0(X, \omega_X \otimes L)$

- is nonzero?
- defines a base point free linear system?
- induces an embedding $X \subseteq \mathbb{P}^n$?

What about other special varieties (smooth cubics, elliptic surfaces?)

EXERCISE 3.19. Suppose that (X, Δ) is as above and that M is a line bundle such that $M - K_X - \Delta$ is ample. Further suppose that N is an ample divisor such that $\mathcal{O}_X(N)$ is globally generated by S^0. Use Castelnuovo-Mumford regularity to prove that $\sigma(X, \Delta) \otimes \mathcal{O}_X(M + (\dim X)N)$ is globally generated.

Hint: Consider the map

$$F^e_* \mathcal{O}_X((1-p^e)(K_X+\Delta)+(p^e-1)M+M+p^e(\dim X)N) \to \mathcal{O}_X(M+(\dim X)N)$$

and show that the source is globally generated as an $\mathcal{O}_X$-module using Castelnuovo-Mumford regularity. See [**Kee08, Sch14**].

QUESTION 3.20 (Hard). With notation as above, find a constant $c(d)$, depending only on $d = \dim X$ such that if X is a smooth projective variety and L is ample and $M_{c(d)} = K_X + L$, then $S^0(X, \sigma(X, 0) \otimes M)$ yields a base point free linear system.

Later in 6, we will describe several other ways to produce sections in S^0.

4. F-singularities versus Singularities of the MMP

In this section we discuss the relationship between both global and local notions of F-singularity theory and of Minimal Model Program. The loose picture is summarized below:

globally F-regular	$\sim$	log-Fano
strongly F-regular	$\sim$	Kawamata log terminal
globally F-split	$\sim$	log-Calabi-Yau
sharply F-pure	$\sim$	log canonical
work in proofs		interesting geometry

Unfortunately there is a slight discrepancy between the left side, which works for proofs, and the right side, which is geometrically interesting. So, whenever one would like to prove something geometrically interesting has to deal with this discrepancy, the extent of which is the main topic of this section.

We work over a perfect base-field k of characteristic $p > 0$. Unless otherwise stated, a pair (X, Δ) denotes a normal, essentially finite type scheme X over k endowed with an effective $\mathbb{Q}$-divisor Δ.

First, recall the following definition.

DEFINITION 4.1. A pair (X, Δ) is globally F-split if the natural map $\mathcal{O}_X \to F^e_* \mathcal{O}_X(\lceil (p^e - 1)\Delta \rceil)$ admits a splitting for some integer $e > 0$. A pair (X, Δ) is sharply F-pure if it has an affine open cover by globally F-split pairs.

Note that $\lceil (p^e - 1)\Delta \rceil$ is a Weil divisor. As we observed before, there is an associated sheaf $\mathcal{O}_X(\lceil (p^e - 1)\Delta \rceil)$ defined with sections

$$\mathcal{O}_X(\lceil (p^e - 1)\Delta \rceil)(U) := \{ f \in K(X) | (f)|_U + \lceil (p^e - 1)\Delta \rceil|_U \geq 0 \}.$$

Further, this sheaf is S_2 or equivalently reflexive (see [**Har94a**, **Har80**] for generalities on reflexive sheaves). In particular its behavior is determined in codimension one.

Now we define the following related notions that, as we will see, are related to globally F-split and sharply F-pure as log canonical is related to Kawamata log terminal.

DEFINITION 4.2. A pair (X, Δ) is globally F-regular if for every effective $\mathbb{Z}$-Weil divisor D, the natural map $\mathcal{O}_X \to F^e_* \mathcal{O}_X(\lceil (p^e - 1)\Delta \rceil + D)$ admits a splitting for some integer $e > 0$. A pair (X, Δ) is strongly F-regular if it has an affine open cover by globally F-regular pairs.

REMARK 4.3. Strongly F-regular is a local condition, i.e., (X, Δ) is strongly F-regular, if and only if for each $x \in X$, $(\operatorname{Spec} \mathcal{O}_{X,x}, \Delta_x)$ is strongly F-regular [**ST12b**, Exercise 3.10]. Also, then every affine open set and every local ring of a strongly F-regular pair is strongly F-regular.

EXERCISE 4.4. Let $0 \leq a_1, a_2, a_3 \leq 1$ be rational numbers. Show that $(\mathbb{P}^1, a_1 P_1 + a_2 P_2 + a_3 P_3)$ is globally F-regular (resp. globally F-split) if and only if for some integer $e > 0$,

$$x^{\lceil (p^e-1)a_1 \rceil} y^{\lceil (p^e-1)a_2 \rceil} (x+y)^{\lceil (p^e-1)a_3 \rceil}$$

has a non-zero monomial $x^i y^j$ with $0 \leq i, j < p^e - 1$ (resp. $0 \leq i, j \leq p^e - 1$) for some $e > 0$.

Hint: Show that we can assume that $P_1 = 0$, $P_2 = 1$ and $P_3 = \infty$. Describe then explicitly the maps

$$\begin{aligned} & H^0\left(\mathbb{P}^1, \lfloor (1-p^e)(K_X + a_1 P_1 + a_2 P_2 + a_3 P_3) - D \rfloor\right) \\ & \quad \to H^0\left(\mathbb{P}^1, (1-p^e)K_X\right) \\ & \quad \to H^0\left(\mathbb{P}^1, \mathcal{O}_{\mathbb{P}^1}\right) \end{aligned} \tag{2}$$

(resp., $H^0\left(\mathbb{P}^1, \lfloor (1-p^e)(K_X + a_1 P_1 + a_2 P_2 + a_3 P_3) \rfloor\right) \to H^0\left(\mathbb{P}^1, (1-p^e)K_X\right) \to H^0\left(\mathbb{P}^1, \mathcal{O}_{\mathbb{P}^1}\right)$). Concluding the globally F-split case is then easy. In the globally F-regular case one has to deal with D, since if $\deg D > 1$ then the existence of a non-zero monomial $x^i y^j$ with $0 \leq i, j < p^e - 1$ does not guarantee that the composition of the maps in 2 is surjective. The idea here is to show that in this situation 2 becomes surjective if one replaces e by ne for any $n \gg 0$. For this, show that if $x^i y^j$ is a non-zero monomial of $x^{\lceil (p^e-1)a_1 \rceil} y^{\lceil (p^e-1)a_2 \rceil} (x+y)^{\lceil (p^e-1)a_3 \rceil}$ such that $0 \leq i, j < p^e - 1$ and i and j are minimal with these properties (or rather precisely i is minimal and j is minimal amongst having that i), then $x^{i \frac{p^{ne}-1}{p^e-1}} y^{j \frac{p^{ne}-1}{p^e-1}}$ is a non-zero monomial of $\left(x^{\lceil (p^e-1)a_1 \rceil} y^{\lceil (p^e-1)a_2 \rceil} (x+y)^{\lceil (p^e-1)a_3 \rceil}\right)^{\frac{p^{ne}-1}{p^e-1}}$. Now, using that $\lceil (p^{ne}-1)a_i \rceil \leq \lceil (p^e-1)a_i \rceil \frac{p^{ne}-1}{p^e-1}$ we obtain a non-zero coefficient $x^{i'}y^{j'}$ of $x^{\lceil (p^{ne}-1)a_1 \rceil} y^{\lceil (p^{ne}-1)a_2 \rceil} (x+y)^{\lceil (p^{ne}-1)a_3 \rceil}$, such that $i' + \frac{p^{ne}-1}{p^e-1}, j' + \frac{p^{ne}-1}{p^e-1} \leq p^{ne}-1$. Conclude then the surjectivity of 2 for e replaced by ne when $n \gg 0$.

EXERCISE 4.5. Show that $\left(\mathbb{P}^1, \frac{1}{2}P_1 + \frac{1}{2}P_2 + \frac{n-1}{n}P_3\right)$ is globally F-regular if and only if $p \neq 2$.

Hint: Apply the previous exercise. You can make your life easier by permuting the points.

EXERCISE 4.6. Show that if X is globally F-regular, then the splitting condition of the definition happens for all $e \gg 0$ (where the bound depends on D). That is, for each effective divisor D, there is an integer e_D, such that $\mathcal{O}_X \to F^e_* \mathcal{O}_X(\lceil (p^e-1)\Delta \rceil + D)$ admits a splitting for all integers $e \geq e_D$.

Hint: The splitting ϕ_e for e gives a splitting for $2e$ by restricting $\phi_e \circ F^e_*(\phi_e)$ from $F^{2e}_* \mathcal{O}_X \left(\lceil (p^e-1)\Delta \rceil + p^e \lceil (p^e-1)\Delta \rceil + \frac{p^{2e}-1}{p^e-1} D \right)$ to $F^{2e}_* \mathcal{O}_X(\lceil (p^{2e}-1)\Delta \rceil + D)$ (it has to be used somewhere that F_* of a reflexive sheaf is reflexive, which can be deduced from [**Har94a**, 1.12]). Then by induction one obtains by a similar method a splitting for every divisible enough e. To obtain it for every big enough e, replace D by $D' := \lceil \Delta \rceil + D$. The same argument then yields a splitting ψ_n of $\mathcal{O}_X \to F^{ne}_* \mathcal{O}_X \left(\lceil (p^{ne}-1)\Delta \rceil + \frac{p^{ne}-1}{p^e-1} D' \right)$. Finally note that by the choice of D', for every $r \geq e$, there is a natural inclusion $F^r_* \mathcal{O}_X \left(\lceil (p^r-1)\Delta \rceil + D \right)$ into $F^{me}_* \mathcal{O}_X \left(\lceil (p^{me}-1)\Delta \rceil + \frac{p^{me}-1}{p^e-1} D' \right)$, where $m = \left\lceil \frac{r}{e} \right\rceil$. Restricting ψ_m via this inclusion yields the desired splitting for every $r \geq e$.

EXERCISE 4.7. Show that in the definition of globally F-regular we could have assumed D to be an ample Cartier divisor as long as X is projective.

Hint: Use a similar argument to that in 4.6: the key is that by replacing e by ne we may replace D by $\frac{p^{ne}-1}{p^e-1} D$.

EXERCISE 4.8. Show that (X, D) is globally F-regular if and only if for every effective $\mathbb{Z}$-divisor D, the natural map $\mathcal{O}_X \to F^e_* \mathcal{O}_X(\lfloor p^e \Delta \rfloor + D)$ admits a splitting for some integer $e > 0$.

Hint: Since we have splitting for all D we may swallow into D any difference in the expression that is bounded as e grows. Now note that

$$-\Delta \leq \lceil (p^e-1)\Delta \rceil - \lfloor p^e \Delta \rfloor \leq 2\lceil \Delta \rceil - \Delta.$$

REMARK 4.9. The analogous statement to 4.8 does not hold for globally F-split pairs.

EXERCISE 4.10. Show that (X, Δ) is globally F-regular if and only if for all effective divisors D the natural map

$$H^0\Big(X, F^e_* \mathcal{O}_X(\lfloor (1-p^e)(K_X+\Delta) \rfloor - D)\Big) \to H^0(X, \mathcal{O}_X)$$

is surjective for some $e > 0$. Recall that the natural map $F^e_* \mathcal{O}_X(\lfloor (1-p^e)(K_X + \Delta) \rfloor - D) \to \mathcal{O}_X$ is constructed as the composition

$$F^e_* \mathcal{O}_X(\lfloor (1-p^e)(K_X+\Delta) \rfloor - D) \to F^e_* \mathcal{O}_X((1-p^e)K_X) \to \mathcal{O}_X,$$

where the latter map is the twist of the Grothendieck trace map by $\mathcal{O}_X(-K_X)$ using the projection formula (and that all the sheaves involved are S_2).

Similarly show that (X, Δ) is globally F-regular if and only if for all effective divisors D the natural map

$$H^0\Big(X, F^e_* \mathcal{O}_X(\lceil (1-p^e)K_X - p^e\Delta \rceil - D)\Big) \to H^0(X, \mathcal{O}_X)$$

is surjective for some $e > 0$.

Hint: use duality theory, that is, apply $R\,\mathrm{Hom}(\_, \omega_X)$ to $\mathcal{O}_X \to F^e_*\mathcal{O}_X(\lceil (p^e - 1)\Delta \rceil + D)$ and then apply also $\_ \otimes \mathcal{O}_X(-K_X)$. For the second claim use 4.8.

DEFINITION 4.11. A pair (X, Δ) is klt (Kawamata log terminal) if $K_X + \Delta$ is $\mathbb{Q}$-Cartier and for all birational morphisms from normal varieties $f : Y \to X$, $\lceil K_Y - f^*(K_X + \Delta) \rceil \geq 0$, where K_Y and K_X are chosen such that $f_* K_Y = K_X$.

DEFINITION 4.12. A pair (X, Δ) is lc (log canonical) if for all birational morphisms from normal varieties $f : Y \to X$, $K_Y - f^*(K_X + \Delta) \geq -E$ for some reduced divisor E, where K_Y and K_X are chosen such that $f_* K_Y = K_X$.

REMARK 4.13. If (X, Δ) admits a log-resolution $Y \to X$, then it is enough to require the above conditions only for one log-resolution $f : Y \to X$ [**KM98**, 2.32]. In particular, this applies to dimensions at most three by [**Cut09, CP08, CP09**].

EXERCISE 4.14. Show that for a sharply F-pure X (that is, $(X, 0)$ is sharply F-pure) with ω_X a line bundle, X is log canonical.

Hint: Given a birational morphism $f : Y \to X$ of normal schemes, show first that there is commutative diagram

$$\begin{array}{ccc} f_* F^e_*(\omega_Y(p^e E)) = F^e_* f_*(\omega_Y(p^e E)) & \longrightarrow & f_*(\omega_Y(E)) \\ \downarrow & & \downarrow \\ F^e_* \omega_X & \longrightarrow & \omega_X \end{array}$$

for every integer $e > 0$ and any exceptional $\mathbb{Z}$-divisor E. Twist then this diagram with ω_X^{-1} and conclude the exercise.

EXERCISE 4.15. Show that if (X, Δ) is globally F-split, then there is a $\Delta' \geq \Delta$, such that (X, Δ') is globally F-split, $(p^e - 1)(K_X + \Delta')$ is Cartier. Furthermore, show that Δ' can be chosen so that (X, Δ') is log Calabi-Yau, that is, $K_X + \Delta' \sim_{\mathbb{Q}} 0$.

Hint: Let Γ be the divisor given by the section of $\mathcal{O}_X(\lfloor (1-p^e)(K_X + \Delta) \rfloor)$ for an $e > 0$ for which there is a splitting. Using the dual picture (4.10) show that this induces a splitting of $\mathcal{O}_X \to F^e_* \mathcal{O}_X(D + \lceil (p^e - 1)\Delta \rceil)$. Define then $\Delta' := \frac{D + \lceil (p^e - 1)\Delta \rceil}{p^e - 1}$.

(Note: There is a corresponding statement with globally F-regular and log-Fano, however the proof is more tedious. See [**SS10**, Theorem 4.3].)

EXERCISE 4.16. Show that if (X, Δ) is a sharply F-pure pair such that $K_X + \Delta$ is $\mathbb{Q}$-Cartier then (X, Δ) is log canonical.

Hint: Soup up the argument of 4.14. First using 4.15 assume that $(p^e - 1)(K_X + \Delta)$ is $\mathbb{Q}$-Cartier. Then construct a map

$$\mathscr{F}_e := F^e_* \mathcal{O}_Y(\lceil K_Y - f^* p^e (K_X + \Delta) + \varepsilon p^e E \rceil) \to \mathscr{F}_0 := \mathcal{O}_Y(\lceil K_Y - f^*(K_X + \Delta) + \varepsilon E \rceil),$$

where E is the reduced exceptional divisor. Then note, that

$$f_* \mathscr{F}_e \cong F^e_* \Big(\mathscr{O}_X \big((1-p^e)(K_X + \Delta) \big) \otimes f_* \big(\mathscr{O}_Y (\lceil K_Y - f^*(K_X + \Delta) + \varepsilon p^e E \rceil) \big) \Big).$$

Show then that there is a commutative diagram as follows using the commutativity of the diagram on the regular locus.

$$\begin{array}{ccc} f_* \mathscr{F}^e & \longrightarrow & f_* \mathscr{F}_0 \\ \downarrow & & \downarrow \\ F^e_* \mathscr{O}_X((1-p^e)(K_X + \Delta)) & \longrightarrow & \mathscr{O}_X \end{array}.$$

EXERCISE 4.17. Show that (X, Δ) is globally F-regular if and only if for every effective Cartier divisor A, $(X, \Delta + \varepsilon A)$ is globally F-split for some rational $\varepsilon > 0$. Further, show that (X, Δ) is klt if for every effective Cartier divisor A, $(X, \Delta + \varepsilon A)$ is log canonical for some $\mathbb{Q} \ni \varepsilon > 0$, and the other direction also holds if (X, Δ) admits a log-resolution (which is known up to dimension 3). Conclude that if (X, Δ) is a strongly F-regular pair with $K_X + \Delta$ being $\mathbb{Q}$-Cartier, then (X, Δ) is klt.

Hint: For the first statement again use the trick that showed up earlier that by replacing e by ne we may replace D by $\frac{p^{ne}-1}{p^e-1} D$. For the second one given a birational morphism $f : Y \to X$, choose an A that contains the images of the exceptional divisors of f on X. For the backwards direction of the second statement use [**KM98**, Prop. 2.36.(2)]. For the conclusion use 4.16.

Having shown 4.17, the question is how far are the two singularity classes. One answer to this is as follows.

THEOREM 4.18 ([**Har98**, **MS97**, **Har01**, **Smi00**]). *If (X, Δ) is a klt pair over a field k_1 of characteristic zero, then for every model $(X_A, \Delta_A) \to \operatorname{Spec} A$ of $(X, \Delta) \to \operatorname{Spec} k_1$ over a finitely generated $\mathbb{Z}$-algebra A, the set*

$$\{ p \in \operatorname{Spec} A | (X_p, \Delta_p) \text{ is strongly } F\text{-regular} \}$$

is open and dense.

Since the primary focus of these notes is the fixed characteristic situation, instead of going into the direction of 4.18, we would like to understand the difference between klt and strongly F-regular in positive characteristic. This is mostly known only for surfaces [**Har98**], which we will discuss thoroughly below. The higher dimensional case, as well as some surface questions on slc singularities are posed as research questions below.

So, take a klt surface singularity X. In any characteristic X admits a minimal resolution $f : Y \to X$. The exceptional divisor of f is a normal crossing curve, with all components being smooth rational curves and the dual graph of which is star shaped. A list of the possible dual graphs can be found on [**Har98**, page 50-51]. It is presumably easier to work on Y instead of X, since it is smooth with well-described normal crossing divisors. The next exercise is the first step in this direction.

EXERCISE 4.19. Let $f : (Y, \Gamma) \to (X, \Delta)$ be a proper birational morphism of pairs of dimension at least two, that is, we require $f_*\Gamma = \Delta$ (e.g., Y is a resolution either of a variety or of a local ring of a variety X). Show the following statements.

(a) Let $H \geq 0$ be a Cartier divisor on X and $\Gamma \geq 0$ a divisor for which $f_*\Gamma = \Delta$. Assume further that
$$H^0(Y, F^e_*\mathcal{O}_Y(\lfloor(1-p^e)(K_Y+\Gamma)\rfloor - f^*H)) \to H^0(Y, \mathcal{O}_Y)$$
is surjective for some integer $e > 0$. Then for the same value of e,
$$H^0(X, F^e_*\mathcal{O}_X(\lfloor(1-p^e)(K_X+\Delta)\rfloor - H)) \to H^0(X, \mathcal{O}_X)$$
is surjective as well.

(b) If (Y, Γ) is globally F-split then so is (X, Δ).

(c) If (Y, Γ) is globally F-regular then so is (X, Δ).

(d) Give a counterexample to the converse of the previous statement

Hint: Fermat cubic surface.

PROPOSITION 4.20. *Consider the following situation:*

(a) *(Y, D) a pair such that Y is regular and D is normal crossing (e.g., Y is smooth over k),*

(b) *$G \geq 0$ a divisor on Y,*

(c) *E is a prime divisor on Y, such that* $\operatorname{coeff}_E D = 1$,

(d) *$(E, (D-E)|_E)$ is globally F-regular,*

(e) *for every integer $n \geq 0$, $H^1(Y, \lceil(1-p^e)K_Y - p^eD\rceil - G + nE) = 0$ for every $e \gg 0$.*

Then, the image of $H^0(Y, F^e_*\mathcal{O}_Y(\lceil(1-p^e)K_Y - p^eD\rceil - G + (n+1)E)) \to H^0(Y, \mathcal{O}_Y)$ maps surjectively onto $H^0(E, \mathcal{O}_E)$ for all $e \gg 0$, where n is the biggest integer such that $nE \leq G$.

PROOF. By c, E is what we will call in Section 6.2 an F-pure center of (X, D); all that is needed here is the commutativity of certain diagrams similar to those in Exercises 3.14 and 3.15. Let n be the biggest integers such that $nE \leq G$. Consider the following commutative diagram (note that $\operatorname{coeff}_E(\lfloor p^eD\rfloor - (p^e-1)E) = 1$, hence the appearance of $n+1$ instead of n):

$$\begin{array}{ccc}
H^0(Y, F^e_*\mathcal{O}_Y(\lceil(1-p^e)K_Y - p^eD\rceil - G + (n+1)E)) & \xrightarrow{\ \beta\ } & H^0(Y, F^e_*\mathcal{O}_E(\lceil(1-p^e)K_E - p^e(D-E)_E\rceil - (G-nE)_E)) \\
\downarrow & & \downarrow \\
H^0(Y, F^e_*\mathcal{O}_Y((1-p^e)(K_Y+E))) & \twoheadrightarrow & H^0(Y, F^e_*\mathcal{O}_E((1-p^e)K_E)) \\
\downarrow & & \downarrow \\
H^0(Y, \mathcal{O}_Y) & \longrightarrow & H^0(E, \mathcal{O}_E)
\end{array}$$
$$\alpha : H^0(Y, F^e_*\mathcal{O}_E(\lceil(1-p^e)K_E - p^e(D-E)_E\rceil - (G-nE)_E)) \to H^0(E, \mathcal{O}_E)$$

We need to prove that both α and β are surjective. The surjectivity of α follows from the global F-regularity assumption and the surjectivity of β from the long exact sequence of cohomology and assumption e. $\square$

COROLLARY 4.21. *Let Y be a smooth, projective variety over the algebraically closed ground field k. Assume further that $-(K_Y + D)$ is ample and E is a codimension one F-pure center such that $(E, D - E)|_E)$ is globally F-regular. Then (Y, D) and consequently Y is globally F-regular.*

EXERCISE 4.22. Show that $\mathbb{P}^n$ is globally F-regular.

Hint: use induction starting with $n = 0$, apply 4.21 in the inductional step.

APPLICATION 4.23. We apply 4.20 to the situation when $X = \operatorname{Spec} A$ is the local ring of a normal singularity, $f : Y \to X$ is a log-resolution of singularities and $E \subseteq f^{-1}(P)$ is an adequately chosen prime divisor, where P is the closed point of X. Assume also that k is algebraically closed. Since X is local, it is strongly F-regular if and only if it is globally F-regular. By 4.19, we should prove for every effective Cartier divisor H on X we are supposed to prove that the map

$$H^0(Y, F^e_* \mathcal{O}_Y((1 - p^e)K_Y - f^*H)) \to H^0(Y, \mathcal{O}_Y) \cong A$$

is surjective. For this it is enough to choose an adequate boundary D on Y (supported on the exceptional divisor), and show that

$$H^0(Y, F^e_* \mathcal{O}_Y(\lceil (1 - p^e)K_Y - p^e D \rceil - f^*H)) \to H^0(Y, \mathcal{O}_Y) \cong A.$$

At this point we would like to apply 4.20. However, there we see that instead of f^*H we would need to have $f^*H - nE$. The solution is to find such D and E, such that $\lceil -p^e D \rceil + nE \leq 0$ and prove then that

$$\psi : H^0(Y, F^e_* \mathcal{O}_Y(\lceil (1 - p^e)K_Y - p^e D) \rceil - f^*H + nE)) \to H^0(Y, \mathcal{O}_Y) \cong A.$$

is surjective. Note that the condition $\lceil -p^e D \rceil + nE \leq 0$ is automatically satisfied for every $e \gg 0$, since we have chosen $\operatorname{coeff}_E D = 1$. Further, note that the image of this map is an ideal I in A. Since A is local, then it is enough to prove that the image of I in A/m (where $m \subseteq A$ is the maximal ideal) is non-zero. That is, we want to prove that the image of ψ maps surjectively onto $H^0(E, \mathcal{O}_E)$. So, we just need to check that the conditions of 4.20 apply for adequately chosen D and E.

Out of the conditions of 4.20, (a) and (b) are automatically satisfied. For condition (c) we just need to choose E to be one of the exceptional prime divisors in D that has coefficient 1. The other two conditions are harder, so we just summarize our findings:

X is strongly F-regular, if for every effective Cartier divisor H on X we can find an exceptional effective divisor D, a component E of D with coefficient 1 such that

(a) $(E, (D - E)|_E)$ is globally F-regular,
(b) for every integer $n \geq 0$, $H^1(Y, \lceil (1 - p^e)K_Y - p^e D) \rceil - f^*H + nE) = 0$ for every $e \gg 0$.

We show how 4.23 works for a two dimensional klt singularity. So, let X be such a singularity and $f : Y \to X$ its minimal resolution. We want to use the following theorem.

THEOREM 4.24 ([**KK**, 2.2.1]). *If $f : Y \to X$ is a log resolution of a two dimensional normal affine scheme essentially of finite type over k, and L is an f-nef exceptional $\mathbb{Q}$-divisor, then $H^1(Y, K_Y + \lceil L \rceil) = 0$.*

First we work through one concrete example, the canonical D_5 surface singularity, and leave the general case for exercises. Consider a canonical D_5 surface singularity. So, the dual graph is

(3)

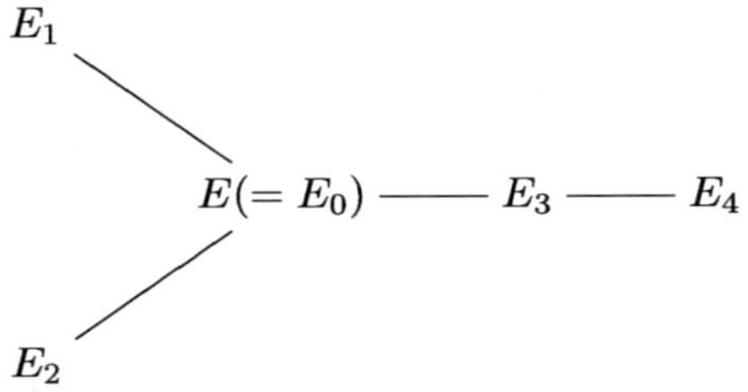

where all the self-intersections are -2 and all the exceptional curves are smooth rational curves. Let E be the prime divisor corresponding to the center of this graph as shown on 3. According to 4.24 we want

$$(1-p^e)(K_Y+D)-f^*H+nE$$

to be nef for $n \geq 0$ and every $e \gg 0$ (in particular, K_Y+D to be anti-nef). Further we need $\operatorname{coeff}_E D = 1$ and $(E, (D-E)|_E)$ to be globally F-regular. It turns out that there is a good choice:

$$D := \frac{1}{2}E_1 + \frac{1}{2}E_2 + \frac{2}{3}E_3 + \frac{1}{3}E_4 + E$$

which can be pictured as

(4)

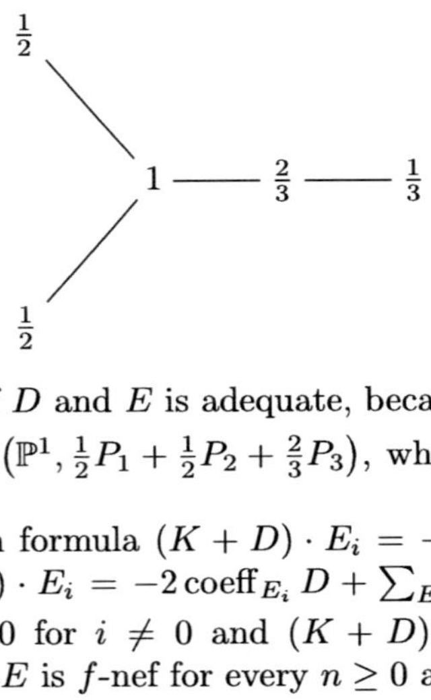

Indeed, the above choice of D and E is adequate, because:

(a) $(E, (D-E)_E) \cong \left(\mathbb{P}^1, \frac{1}{2}P_1 + \frac{1}{2}P_2 + \frac{2}{3}P_3\right)$, which is globally F-regular by 4.5 if $p \neq 2$.

(b) By the adjunction formula $(K+D)\cdot E_i = -2 - E_i^2 + D \cdot E_i = D \cdot E_i$, therefore $(K+D)\cdot E_i = -2\operatorname{coeff}_{E_i} D + \sum_{E_i \cap E_j \neq \emptyset} \operatorname{coeff}_{E_j} D$ and hence $(K+D)\cdot E_i = 0$ for $i \neq 0$ and $(K+D)\cdot E = -\frac{1}{3}$. In particular, $-p^e(K_Y+D)+nE$ is f-nef for every $n \geq 0$ and $e \gg 0$. Further,

$$\lceil (1-p^e)K_Y - p^e D \rceil - f^*H + nE = K_Y + \lceil -p^e(K_Y+D) - f^*H + nE \rceil,$$

hence by 4.24, $H^1(Y, \lceil (1-p^e)(K_Y+D) \rceil - G + nE) = 0$ for every $n \geq 0$ and $e \gg 0$.

So, we have just proven that:

THEOREM 4.25 (**[Har98]**). *A klt surface singularity of type D_5 is strongly F-regular if $p \neq 2$.*

The general case (the singularities other than canonical D_5) is left for the following few exercises. So, let $X = \operatorname{Spec} A$ be a general klt surface singularity, where A is a local ring. As mentioned above the dual graph of the exceptional curve is star shaped. The star can have either two or three arms. In the three arm case we group the graphs according to the absolute value of the determinants of the intersection matrices of the arms. Then the possible cases are $(2,2,d)$, $(2,3,3)$, $(2,3,4)$ and $(2,3,5)$. We set $E = E_0$ to be the vertex of the star in general, and we define D by the properties

$$\operatorname{coeff}_E D = 1 \ ; \ (K_Y + D) \cdot E_i = 0 \ (i \neq 0).$$

Let E_1, E_2 (and possibly E_3) be the curves in the arms that are adjacent to $E = E_0$.

EXERCISE 4.26. Show that

$$D = E_0 + \sum_i \frac{d_i - 1}{d_i} E_i + \dots$$

where (d_1, d_2, d_3) is the type of the graph. Conclude then that $(K + D) \cdot E = \sum_i \frac{d_i - 1}{d_i} - 1 < 0$.

Hint: Consider the intersection matrix of the i-th arm:

$$\begin{pmatrix} e_1 & 1 & 0 & \dots & & & \\ 1 & e_2 & 1 & 0 & \dots & & \\ 0 & 1 & e_3 & 1 & 0 & \dots & \\ \vdots & \vdots & \vdots & & & & \end{pmatrix}$$

Using the adjunction formula, write up a linear equation for computing the coefficients of D in the i-th arm involving the above matrix. Then use Cramer's rule and that the above matrix has determinant d_i.

EXERCISE 4.27. Show that the followings are globally F-regular.

(a) $\left(\mathbb{P}^1, \frac{1}{2}P_1 + \frac{1}{2}P_2\right)$
(b) $\left(\mathbb{P}^1, \frac{1}{2}P_1 + \frac{2}{3}P_2 + \frac{2}{3}P_3\right)$for $p \neq 2,3$
(c) $\left(\mathbb{P}^1, \frac{1}{2}P_1 + \frac{2}{3}P_2 + \frac{3}{4}P_3\right)$for $p \neq 2,3$
(d) $\left(\mathbb{P}^1, \frac{1}{2}P_1 + \frac{2}{3}P_2 + \frac{4}{5}P_3\right)$for $p \neq 2,3,5$

Hint: Use 4.4.

So, we have proved the theorem:

THEOREM 4.28 (**[Har98]**). *Every klt surface singularity is strongly F-regular if $p > 5$.*

4.1. Questions.

QUESTION 4.29. DEPTH OF SHARPLY F-PURE SINGULARITIES. Let (X, Δ) be a log canonical pair (allowing Δ to be an $\mathbb{R}$-divisor) and let $x \in X$ be a point which is not a log canonical center of (X, Δ). Suppose that $0 \leq \Delta' \leq \Delta$ is another $\mathbb{Q}$-divisor and D a $\mathbb{Z}$-divisor, such that $D \sim_{\mathbb{Q},\mathrm{loc}} \Delta'$. Show then that

$$\operatorname{depth}_x \mathcal{O}_X(-D) \geq \min\{3, \operatorname{codim}_X x\}.$$

or give a counterexample. Note that in characteristic zero this was shown in **[Kol11]** and in positive characteristic $p > 0$ in **[PS13]** with the extra assumption that the indices are not divisible by p and the singularities are sharply F-pure. So, the interesting case is when either p divides some of the appearing indices or the singularities are log canonical but not sharply F-pure. In two or three dimensions this could be in particular approachable by an argument similar to that of [**Kol11**], since then some of the required vanishings can still hold (because the dimension is small). Further, the above question is also interesting for slc singularities (see 4.1.2) as stated in [**Kol11**, Thm 3].

Other related question is that in the above situation if Z is the reduced subscheme supported on the union of some log canonical centers, then show or disprove

$$\operatorname{depth}_x \mathscr{I}_Z \geq \min\{3, \operatorname{codim}_X x + 1\}.$$

The known and unknown cases for this question are completely analogous to the previous one (see [**Kol11**, Thm 3] and [**PS13**, Thm 3.6]).

A third related question is to show or disprove that if (X, Δ) has dlt singularities and $0 \leq D$ a $\mathbb{Z}$-divisor such that $\Delta \sim_{\mathbb{Q},\mathrm{loc}} D$, then $\mathcal{O}_X(-D)$ is Cohen-Macaulay. The situation here is also analogous (see [**Kol11**, Thm 2] and [**PS13**, Thm 3.1]) .

QUESTION 4.30. FROBENIUS STABLE GRAUERT-RIEMENSCHNEIDER VANISHING. If $f : Y \to X$ is a resolution of a normal variety, then is the natural Frobenius map $F^e_* R^i f_* \omega_Y \to R^i f_* \omega_Y$ zero for any $i > 0$ and $e \gg 0$?

EXERCISE 4.31. Check 4.30 for X being a cone over a smooth projective variety.

4.1.1. *Three dimensional singularities.* The main question, which is probably hard in general, but should be kept in mind as a guiding principle, is the following:

QUESTION 4.32. In a fixed dimension n is there a prime p_0, such that for all $p \geq p_0$, klt is equivalent to strongly F-regular for pairs with zero boundaries? What if we allow boundaries with coefficients in a finite set?

Most likely, there is either an easy counterexample to 4.32 or it is extremely hard. So, first we should set our sights to dimension 3:

QUESTION 4.33. Answer 4.32 for $n = 3$. One can also restrict further to terminal singularities. That is, is there a prime p_0, such that for all $p \geq p_0$, every terminal singularity X is strongly F-regular? The latter question seems to be more approachable, since in characteristic zero there is a good classification of three dimensional terminal singularities [**Mor85**]. So, first the question would be how much of that holds in positive characteristic. However, even without solving this question, it would be interesting to take examples of log-resolutions in characteristic zero that show up in any characteristic and find the primes for which they are strongly F-regular. One can use that in characteristic zero we know nice resolutions of terminal singularities (c.f., [**Che13**]).

4.1.2. *Classify slc surface singularities.* Slc (semi-log canonical) singularities are the natural non-normal versions of log canonical singularities. Their main use, as well as the motivation for their introduction [**KSB88**], is in higher dimensional moduli theory. As in dimension one needs to add nodal curves to smooth curves to obtain a projective moduli space of curves, in higher dimensions one has to add slc singularities to log canonical singularities to obtain a (partially proven and partially conjectural) moduli space of canonical models. Note that such moduli space parameterizes birational equivalence classes of varieties of general type (which is shown in characteristic zero, for surfaces in positive characteristic and for threefolds in characteristic $p \geq 7$).

First, we define slc singularities. Let k be a field.

DEFINITION 4.34. A scheme X locally of finite type over k is *demi-normal* if it is equidimensional, S_2 and its codimension one points are nodes [**Kol13**, 5.1]. Here $x \in X$ is a *node*, if $\mathcal{O}_{X,x}$ is isomorphic to $R/(f)$ for some two dimensional regular local ring (R, m), $f \in m^2$ and f is not a square in m^2/m^3 [**Kol13**, 1.41].

PROPOSITION 4.35 ([**Kol13**, 1.41.1]). *Let (A, n) be a 1-dimensional local ring with residue field k and normalization $\overline{A}$. Let $\overline{n}$ be the intersection of the maximal ideals of $\overline{A}$. Assume that (A, n) is a quotient of a regular local ring. Then (A, n) is nodal if and only if* $\dim_k(\overline{A}/\overline{n}) = 2$ *and* $\overline{n} \subseteq A$.

EXERCISE 4.36. Show that the pinch point $\operatorname{Spec} \frac{k[x,y,t]}{(x^2-ty^2)}$ is demi-normal, even in characteristic 2.

DEFINITION 4.37 ([**Kol13**, 5.2]). If X is a reduced scheme, and $\pi : \overline{X} \to X$ its normalization, then the conductor ideal is the largest ideal sheaf $\mathscr{I} \subseteq \mathcal{O}_X$ that is also an ideal sheaf of $\pi_* \mathcal{O}_{\overline{X}}$. The conductor subschemes are defined as $D := \operatorname{Spec}(\mathcal{O}_X/\mathscr{I})$ and $\overline{D} := \operatorname{Spec}(\mathcal{O}_{\overline{X}}/\mathscr{I})$.

FACT 4.38 ([**Kol13**, 5.2]). *For a demi-normal scheme X, the conductor subschemes are the reduced divisors supported on the closure of the nodal locus, and on the preimage of that, respectively.*

DEFINITION 4.39 ([**Kol13**, 5.10]). Let X be a demi-normal scheme with normalization $\pi : \overline{X} \to X$ with conductor divisors $D \subseteq X$ and $\overline{D} \subseteq \overline{X}$, respectively. Let Δ be a $\mathbb{Q}$ (Weil-)divisor on X not containing common components with D, and let $\overline{\Delta} := \pi^* \Delta$ (defined as the divisorial part of $\pi^{-1}(\Delta)$). Then the pair (X, Δ) is *semi-log canonical*, or shortly just *slc*, if $K_X + \Delta$ is $\mathbb{Q}$-Cartier, and $(\overline{X}, \overline{D} + \overline{\Delta})$ is log canonical.

Now, we would like to see the F-singularities aspects of slc singularities. Though we defined F-purity in the normal setting, it works without modification in the G_1 and S_2 setting (c.f., [**PS13**]).

EXERCISE 4.40. Show that if $p = 2$, then the pinch-point $(\{x_1^2 + x_2^2 x_3 = 0\} \subseteq \mathbb{A}^n_{x_1,\dots,x_n}, n \geq 3)$ is not F-pure along the entire double line (i.e., along $x_1 = x_2 = 0$).

Hint: Use 2.16.

A more general wording of the previous exercise is as follows.

EXERCISE 4.41. Let X be an affine variety (where we do not assume that X is irreducible) and let x be a codimension one point of the conductor given by the conductor ideal $\mathscr{I} \subseteq \mathcal{O}_X$.

(a) Show that for every $\phi \in \operatorname{Hom}_{\mathcal{O}_X}(F^e_* \mathcal{O}_X, \mathcal{O}_X)$, $\phi(F^e_* \mathscr{I}) \subseteq \mathscr{I}$.
(b) Show that every $\phi \in \operatorname{Hom}_{\mathcal{O}_X}(F^e_* \mathcal{O}_X, \mathcal{O}_X)$ extends to

$$\phi^N \in \operatorname{Hom}_{\mathcal{O}_Y}(F^e_* \mathcal{O}_Y, \mathcal{O}_Y),$$

where Y is the normalization of X.
(c) Show that the conductor of the normalization is inseparable at x over the conductor of X, then X is not F-pure at x.

Hint: For point a extend ϕ first to the total ring of fractions K of $R := \Gamma(X, \mathcal{O}_X)$, then using this extension show that $\phi(s)t \in I := \Gamma(X, \mathscr{I})$ for every $s \in I$ and $t \in R^N := \Gamma(Y, \mathcal{O}_Y)$.

For point b show that if ϕ' is the above extension of ϕ to K, then $\phi'(F^e_* R^N) \subseteq R^n$. For that first show that we may assume that R is integral. Then show that $(\phi'(x))^m I \subseteq I$ for every $x \in F^e_* R^N$ and integer $m > 0$. Then deduce that $\phi'(x)$ is integral over R.

For point c assume that X is F-pure. Take then a splitting ϕ of $\mathcal{O}_X \to F_* \mathcal{O}_X$ guaranteed by the F-purity. Let ϕ^N be the splitting given by point b. Let B be the conductor on X and C the conductor on Y. Show then using point a that this yields a splitting of $\mathcal{O}_B \to F_* \mathcal{O}_B$ that extends to a splitting of $\mathcal{O}_C \to F_* \mathcal{O}_C$. The latter yields then a contradiction with the inseparability.

The following are the questions posed about slc singularities.

QUESTION 4.42. Classify slc surface singularities in characteristic $p > 0$. There are more approaches to the characteristic zero classification. Historically, one first passes to the canonical covers and then classifies only the Gorenstein slc surface singularities [**KSB88**, 4.21, 4.22, 4.23, 4.24]. The second goes by classifying the normalization (c.f., [**Kol13**, Section 3.3]) and then using Kollár's gluing theory [**Kol13**, 5.12, 5.13] to see when one can glue the normalization into a demi-normal singularity. Unfortunately, there are issues with both in positive characteristic. For the first one, the canonical cover can be inseparable in characteristic p, and for the second one, aspects of the gluing theory work only in characteristic zero [**Kol13**, 5.12, 5.13].

QUESTION 4.43. Find the F-pure ones out of the list of slc singularities found as an answer to the previous question. For the method see [**Har98**, **MS91**] and [**MS12**]. Note that the two articles together solve the question if the conductor of the normalization maps separably to the conductor of the singularity (which is always true if $p \neq 2$) and further the index is not divisible by p. So, the interesting cases are the ones left out.

5. Global Applications

In this section we present one application to global geometry of the theory of F-singularities and then we list a few problems. The known applications are somewhat different in nature with little connection between some of them (a non-complete list is: [**Sch14**, **Hac15**, **Mus13**, **MS14**, **CHMtaS14**, **HX15**, **CTX15**, **Zha14**, **Pat14**, **Pat13**, **HP16**]). In particular, the chosen application is admittedly somewhat random.

Throughout the section the base field k is algebraically closed, and of characteristic $p > 0$. The algebraically closed assumption is not necessary in 5.1 but it is important in 5.3.1 among other places.

5.1. Semi-positivity of pushforwards. The application we discuss in more details here is the semi-positivity of sheaves of the form $f_* \left(\omega^m_{X/Y} \right)$. We use the following notations. This is very far from the most general situation where the results work (see [**Pat14**] for a more general situation).

NOTATIONS 5.1. X is a Gorenstein, projective variety, $f : X \to Y$ is a surjective, morphism to a smooth projective curve with normal (and connected) geometric generic fiber, and hence the fibers are normal (and connected) over an open set of Y. Fix also a closed point $y_0 \in Y$ such that $X_0 := X_{y_0}$ is normal.

We prove the following theorem. Recall that a vector bundle $\mathscr{E}$ on a projective scheme Z is nef, if for all finite maps $\tau : C \to Z$ from smooth curves, and all line bundle quotients $\tau^* \mathscr{E} \twoheadrightarrow \mathscr{L}$, $\deg \mathscr{L} \geq 0$.

THEOREM 5.2 ([**Pat14**]). *In the situation of 5.1, if $\omega_{X/Y}$ is f-ample and X_0 is sharply F-pure, then $f_*(\omega_{X/Y}^m)$ is a nef vector bundle for every $m \gg 0$.*

Note that the corresponding theorem in characteristic zero was proven using the consequence of Hodge theory that $f_* \omega_{X/Y}$ is semi-positive. The corresponding statement in characteristic p is false, so the proof necessarily eludes such considerations.

Recall that a coherent sheaf $\mathscr{F}$ on a scheme X is *generically globally generated*, if there is homomorphism $\mathcal{O}_X^{\oplus m} \to \mathscr{F}$ which is surjective over a dense open set. Further, if Z is a normal variety and $\mathscr{L}$ a line bundle on Z, then we use the notation $S^0(Z, \mathscr{L})$ for $S^0(Z, \sigma(X,0) \otimes \mathscr{L})$.

PROPOSITION 5.3. *In the situation of 5.1, choose a Cartier divisor N and set $\mathscr{N} := \mathcal{O}_X(N)$. Assume that $N - K_{X/Y}$ is f-ample and nef, and $H^0(X_0, \mathscr{N}|_{X_0}) = S^0(X_0, \mathscr{N}|_{X_0})$. Then $f_* \mathscr{N} \otimes \omega_Y(2y_0)$ is generically globally generated.*

PROOF. Set $M := N + f^* K_Y + 2X_0$ and $\mathscr{M} := \mathcal{O}_X(M)$. Consider the commutative diagram below.

$$\begin{array}{ccccc} f_* \mathscr{M} & \longrightarrow & (f_* \mathscr{M}) \otimes k(y_0) & \hookrightarrow & H^0(X_0, \mathscr{M}|_{X_0}) , \\ \uparrow & & & & \| \\ S^0(X, \sigma(X, X_0) \otimes \mathscr{M}) \otimes \mathcal{O}_Y & & \longrightarrow & & S^0(X_0, \mathscr{M}|_{X_0}) \end{array} \tag{5}$$

where $H^0(X_0, \mathscr{M}|_{X_0}) = S^0(X_0, \mathscr{M}|_{X_0})$, because

$$H^0(X_0, \mathscr{M}|_{X_0}) \cong H^0(X_0, \mathscr{N}|_{X_0}) = S^0(X_0, \mathscr{N}|_{X_0}) \cong S^0(X_0, \mathscr{M}|_{X_0}).$$

Note that

(6) $M - K_X - X_0 = N + f^* K_Y + 2X_0 - K_{X/Y} - f^* K_Y - X_0 = N - K_{X/Y} + X_0.$

Note also that $N - K_{X/Y}$ is nef and f-ample by assumption. Furthermore, X_0 is the pullback of an ample divisor from Y. Hence, $N - K_{X/Y} + X_0$ is ample and then by 6 so is $M - K_X - X_0$. Hence, 1 implies that the bottom horizontal arrow in 5 is surjective. This finishes our proof. □

EXERCISE 5.4. If $\mathscr{F}$ is a vector bundle on a smooth curve Y and $\mathscr{L}$ is a line bundle such that for every $m > 0$, $(\bigotimes_{i=1}^m \mathscr{F}) \otimes \mathscr{L}$ is generically globally generated, then $\mathscr{F}$ is nef.

NOTATIONS 5.5. For a morphism $f : X \to Y$ of schemes, define

$$X_Y^{(m)} := \underbrace{X \times_Y X \times_Y \cdots \times_Y X}_{m \text{ times}},$$

and let $f_Y^{(m)} : X_Y^{(m)} \to Y$ be the natural induced map. If $\mathscr{F}$ is a sheaf of $\mathcal{O}_X$-modules, then

$$\mathscr{F}_Y^{(m)} := \bigotimes_{i=1}^{m} p_i^* \mathscr{F},$$

where p_i is the i-th projection $X_Y^{(m)} \to X$. In most cases, we omit Y from our notation. I.e., we use $X^{(m)}$, $f^{(m)}$ and $\mathscr{F}^{(m)}$ instead of $X_Y^{(m)}$, $f_Y^{(m)}$ and $\mathscr{F}_Y^{(m)}$, respectively.

EXERCISE 5.6. If $\mathscr{N}$ is a line bundle on a normal Gorenstein variety X, then

$$S^0\left(X^{(m)}, \mathscr{N}^{(m)}\right) \cong S^0(X, \mathscr{N})^{\otimes m}.$$

(Here $X^{(m)}$ and $\mathscr{N}^{(m)}$ are taken over $\operatorname{Spec} k$.)

Hint: the main issue is showing that the trace map is the box product of the trace maps, it is easier to show it on the regular locus and then just extend globally since the complement has large codimension.

EXERCISE 5.7. Show that in the situation of 5.5, $f_*^{(m)} \mathscr{N}^{(m)} \cong \bigotimes_{i=1}^{m} f_* \mathscr{N}$ for any line bundle $\mathscr{N}$ on X and integer $m > 0$

Hint: do induction, for one induction step use the projection formula and flat base-change.

PROPOSITION 5.8. *In the situation of Notation 5.1, choose a Cartier divisor N and set $\mathscr{N} := \mathcal{O}_X(N)$. Assume that $N - K_{X/Y}$ is nef and f-ample, and $H^0(X_0, \mathscr{N}|_{X_0}) = S^0(X_0, \mathscr{N}|_{X_0})$. Then $f_* \mathscr{N}$ is a nef vector bundle.*

PROOF. First we claim that $X^{(n)}$ is a variety, that is, it is integral for every integer $n > 0$. Indeed, since X is Gorenstein and Y is smooth, X is relatively Gorenstein over Y, and hence so is $X^{(n)}$. Therefore, $X^{(n)}$ is also absolutely (so not only relatively over Y) Gorenstein. In particular, X is S_2 and then we can check reducedness only at generic points. However, $X^{(n)}$ is flat over Y, so all generic points lie over the generic point of Y. Further over the generic point of Y, X^n is reduced and irreducible since the geometric generic fiber of f is assumed to be normal (and connected). This concludes our claim.

Hence the assumptions of 5.1 are satisfied for $f^{(n)} : X^{(n)} \to Y$, $N^{(n)} - K_{X^{(n)}/Y} = (N - K_{X/Y})^{(n)}$ is nef and $f^{(n)}$-ample, and further by 5.6 and the Künneth formula,

$$\begin{aligned} & H^0\left(X_0^{(n)}, \mathscr{N}^{(n)}|_{X_0^{(n)}}\right) \\ & = H^0\left(X_0, \mathscr{N}|_{X_0}\right)^{\otimes n} \\ & \cong S^0\left(X_0, \mathscr{N}|_{X_0}\right)^{\otimes n} \\ & \cong S^0\left(X_0^{(n)}, \mathscr{N}^{(n)}|_{X_0^{(n)}}\right). \end{aligned}$$

Hence Proposition 5.3 applies to $X^{(n)}$ and $N^{(n)}$, and consequently, $f_*^{(n)}(\mathscr{N}^{(n)}) \otimes \omega_Y(2y_0)$ generically globally generated for every $n > 0$.

By 5.7, $f_*^{(n)}\left(\mathscr{N}^{(n)}\right) \cong \bigotimes_{i=1}^{n} f_* \mathscr{N}$. Therefore, $f_* \mathscr{N}$ is a vector bundle, such that $(\bigotimes_{i=1}^{n} f_* \mathscr{N}) \otimes \omega_Y(2y_0)$ is generically globally generated for every $n > 0$. Hence, by 5.4, $f_* \mathscr{N}$ is a nef vector bundle. This concludes our proof.

□

PROPOSITION 5.9. *In the situation of Proposition* 5.8, *if furthermore $\mathscr{N}_y$ globally generated for all $y \in Y$, then $\mathscr{N}$ is nef.*

PROOF. Consider the following commutative diagram for every $y \in Y$.

$$
\begin{array}{ccc}
f^* f_* \mathscr{N} & \longrightarrow & \mathscr{N} \\
\downarrow & & \downarrow \\
H^0(X_y, \mathscr{N}) \otimes \mathcal{O}_{X_y} & \longrightarrow & \mathscr{N}_y
\end{array}
\tag{7}
$$

The left vertical arrow is an isomorphism for all but finitely many y by cohomology and base-change. The bottom horizontal arrow is surjective for all $y \in Y$ by assumption. Hence $f^* f_* \mathscr{N} \to \mathscr{N}$ is surjective except possibly at points lying over finitely many points of $y \in Y$. To show that $\mathscr{N}$ is nef, we have to show that $\deg(\mathscr{N}|_C) \geq 0$ for every smooth projective curve C mapping finitely to X. By assumption this follows if C is vertical (globally generated implies nef). So, we may assume that C maps surjectively onto Y. However, then $(f^* f_* \mathscr{N})|_C \to \mathscr{N}|_C$ is generically surjective. Since $f_* \mathscr{N}$ is nef by Proposition 5.8, so is $f^* f_* \mathscr{N}$ and hence $\deg(\mathscr{N}|_C) \geq 0$ has to hold. □

The next exercises tell us how to satisfy the condition $H^0 = S^0$ from the previous proposition.

EXERCISE 5.10. Let (X, Δ) be a pair such that $(p^e - 1)(K_X + \Delta)$ is Cartier for some $e > 0$. Show then that the restriction of the natural map $F_*^{ne} \mathcal{O}_X((1 - p^{ne})(K_X + \Delta)) \to \mathcal{O}_X$ to $\sigma(X, \Delta) \otimes \mathcal{O}_X((1 - p^{ne})(K_X + \Delta))$ yields for every $n \gg 0$ a surjective homomorphism

$$F_*^{ne}(\sigma(X, \Delta) \otimes \mathcal{O}_X((1 - p^{ne})(K_X + \Delta))) \to \sigma(X, \Delta).$$

Hint: Since the image of the original homomorphism is $\sigma(X, \Delta)$ for every $n \gg 0$, it is clear that the image of the restricted homomorphism is contained in $\sigma(X, \Delta)$. To prove surjectivity, take an n', such that

$$F_*^{n'e}(\mathcal{O}_X((1 - p^{n'e})(K_X + \Delta))) \to \sigma(X, \Delta).$$

is surjective and show that the composition of

$$
\begin{aligned}
F_*^{ne}(F_*^{n'e}(\mathcal{O}_X((1 - p^{n'e})(K_X + \Delta))) \otimes \mathcal{O}_X((1 - p^{ne})(K_X + \Delta))) \to \\
\to F_*^{ne}(\sigma(X, \Delta) \otimes \mathcal{O}_X((1 - p^{ne})(K_X + \Delta))) \to \sigma(X, \Delta)
\end{aligned}
$$

is also surjective.

EXERCISE 5.11. Show that if L is a line bundle on a projective pair (X, Δ) such that $(p^e - 1)(K_X + \Delta)$ is Cartier for some $e > 0$, then for each $n \gg 0$, $S^0(X, \sigma(X, \Delta) \otimes L)$ is equal to the image of the homomorphism

$$H^0(X, L \otimes F_*^{ne}(\sigma(X, \Delta) \otimes \mathcal{O}_X((1 - p^{ne})(K_X + \Delta))) \to H^0(X, L \otimes \sigma(X, \Delta)).$$

Hint: Let V_n be the image of the above homomorphism. It is immediate that $V_n \subseteq S^0(X, \sigma(X, \Delta) \otimes L)$. For the other containment, use a trick similar to the previous exercise: take an integer $n' > 0$, such that

$$F_*^{n'e}(\mathcal{O}_X((1 - p^{n'e})(K_X + \Delta))) \to \sigma(X, \Delta).$$

is surjective, and consider the composition of

$$H^0(X, L \otimes F_*^{ne}(F_*^{n'e}(\mathcal{O}_X((1-p^{n'e})(K_X+\Delta))) \otimes \mathcal{O}_X((1-p^{ne})(K_X+\Delta)))) \to$$
$$\to H^0(X, L \otimes F_*^{ne}(\sigma(X,\Delta) \otimes \mathcal{O}_X((1-p^{ne})(K_X+\Delta)))) \to H^0(X, L \otimes \sigma(X,\Delta)).$$

Show that the image of the above homomorphism contains V_n, and also that this image equals $S^0(X, \sigma(X,\Delta) \otimes L)$ for n big enough.

EXERCISE 5.12. Show that if L is an ample line bundle on a projective pair (X, Δ) such that $(p^e - 1)(K_X + \Delta)$ is Cartier for some $e > 0$, then there is an integer $n > 0$, such that for every nef line bundle N, $S^0(X, \sigma(X,\Delta) \otimes L^n \otimes N) = H^0(X, \sigma(X,\Delta) \otimes L^n \otimes N)$.

Hint: Let $e > 0$ be an integer such that $(p^e - 1)(K_X + \Delta)$ is Cartier. According to 5.11, it is enough to show that

$$H^0\left(X, L^n \otimes N \otimes F_*^{(i+1)e}\left(\sigma(X,\Delta) \otimes \mathcal{O}_X((1-p^{(i+1)e})(K_X+\Delta))\right)\right)$$
$$\to H^0\left(X, L^n \otimes N \otimes F_*^{ie}\left(\sigma(X,\Delta) \otimes \mathcal{O}_X((1-p^{ie})(K_X+\Delta))\right)\right)$$

is surjective for every i. Show then that the above maps are induced from the exact sequence

$$0 \longrightarrow \mathscr{B} \longrightarrow F_*^e(\sigma(X,\Delta) \otimes \mathcal{O}_X((1-p^e)(K_X+\Delta))) \longrightarrow \sigma(X,\Delta) \longrightarrow 0.$$

Use then Fujita vanishing [**Fuj83**] to show surjectivity.

THEOREM 5.13. *In the situation of Notation* 5.1, *if X_0 is sharply F-pure and $K_{X/Y}$ is f-nef, then $K_{X/Y}$ is nef.*

PROOF. Using relative Fujita vanishing [**Kee08**] and 5.12, there is an ample enough line bundle $\mathscr{Q}$ on X, such that for all $i > 0$ and f-nef line bundle $\mathscr{K}$,

$$H^0(X_0, \mathscr{Q} \otimes \mathscr{K}|_{X_0}) = S^0(X_0, (\mathscr{Q} \otimes \mathscr{K})|_{X_0}) \tag{8}$$

and

$$\mathscr{Q} \otimes \mathscr{K}|_{X_y} \text{ is globally generated for all } y \in Y. \tag{9}$$

Let Q be a divisor of $\mathscr{Q}$. We prove by induction that $qK_{X/Y} + Q$ is nef for all $q \geq 0$. For $q = 0$ the statement is true by the choice of Q. Hence, we may assume that we $(q-1)K_{X/Y} + Q$ is nef. Now, we verify that the conditions of Proposition 5.9 hold for $N := qK_{X/Y} + Q$ and $\mathscr{N} := \mathcal{O}_X(N)$. Indeed:

- the divisor
$$N - K_{X/Y} = (q-1)K_{X/Y} + Q$$
is not only f-ample, but also nef by the inductional hypothesis,
- using the f-nefness of $K_{X/Y}$ and (8),
$$H^0(X_0, \mathscr{N}|_{X_0}) = S^0(X_0, \mathscr{N}|_{X_0}),$$
- since all the summands of N are f-nef, so is N,
- for every $y \in Y$, $N|_{X_y}$ is globally generated by (9).

Hence Proposition 5.9 implies that N is nef. This finishes our inductional step, and hence the proof of the nefness of $qK_{X/Y} + Q$ for every $q \geq 0$. However, then $K_{X/Y}$ has to be nef as well. This concludes our proof. □

PROOF OF 5.2. By 5.13, we know that $K_{X/Y}$ is not only f-ample, but also nef. Further for every $m \gg 0$,

$$H^{C}(X_0, \omega_{X/Y}^m|_{X_0}) = \underbrace{H^0(X_0, \omega_{X_0}^m)}_{X \text{ is Gorenstein over } Y} = \underbrace{S^0(X_0, \omega_{X_0}^m)}_{\text{by } 5.12} = \underbrace{S^0(X_0, \omega_{X/Y}^m|_{X_0})}_{X \text{ is Gorenstein over } Y}.$$

Then 5.8 applies to $N = \omega_{X/Y}^m$ for each $m \gg 0$, which concludes our proof. □

5.2. Miscellaneous exercises.

EXERCISE 5.14. Show that a Del-Pezzo surface X over an algebraically closed field k of characteristic $p > 0$ is globally F-regular if

(a) $K_X^2 \geq 4$,
(b) $K_X^2 = 3$, and $p > 2$,
(c) $K_X^2 = 2$, and $p > 3$ and
(d) $K_X^2 = 1$, and $p > 5$.

Further show that in the missing cases there are examples of both globally F-regular and not globally F-regular del-Pezzos.

Hint: Use 3.14.

EXERCISE 5.15. Let X be a smooth projective variety over k. Show that Frobenius stable canonical ring $R_S(X) := \bigoplus_{n \geq 0} S^0(X, \omega_X^m)$ is an ideal of the canonical ring $R(X) := \bigoplus_{n\geq 0} H^0(X, \omega_X^m)$. Further, show that $R_S(X)$ is a birational invariant (of smooth projective varieties over k). Then show that it does not change during a run of the MMP, where for singular varieties $R_S(X) := \bigoplus_{n\geq 0} S^0\left(X, \omega_X^{[m]}\right)$ ($\mathscr{F}^{[m]}$ denotes the reflexive power, that is, the double dual $(\mathscr{F}^{\otimes m})^{\vee\vee}$ of the tensor power, see **[Har94b]** for the theory of reflexive sheaves). Deduce then that if the index of the canonical model X_{can} is coprime to p, then for m divisible enough, $S^0(X, \omega_X^m) = H^0\left(X_{\text{can}}, \sigma(X_{\text{can}}, 0) \otimes \omega_{X_{\text{can}}}^{[m]}\right)$. Give then an example of an X for which $R_S(X)$ is not finitely generated as a ring.

Hint: To show that $R_S(X)$ is a birational invariant, for two birational varieties Z and Y take a normal variety W that maps with a birational morphism to both Z and Y. For example one can take W to be the normalization of the closure of the graph of the birational equivalence. Finally show that $R_S(Z) = R_S(W) = R_S(Y)$ by showing that $H^0(Z, \omega_Z^m) \cong H^0\left(W, \omega_W^{[m]}\right)$ and $H^0\left(Z, \omega_Z^{1+(m-1)p^e}\right) \cong H^0\left(W, \omega_W^{[1+(m-1)p^e]}\right)$. There is one more subtlety: the trace maps also have to be identified, however that is not hard to do because of the open sets where the maps $W \to Z$ and $W \to Y$ are isomorphisms.

Showing that MMP does not change $R_S(X)$ is similar, using that for each step the discrepancies do not decrease. For the final conclusion use 5.12.

DEFINITION 5.16. We define the Frobenius stable Kodaira-dimension $\kappa_S(X)$ of a smooth projective variety X over k to be the growth rate of $\dim_k S^0(X, \omega_X^m)$. That is, it is the integer d, such that there are positive real numbers a and b for which

$$am^d < \dim_k S^0(X, \omega_X^m) < bm^d,$$

for every divisible enough m. We say $d = -\infty$ if $S^0(X, \omega_X^m) = 0$ for every $m > 0$.

EXERCISE 5.17. Show that if $\kappa_S(X) \geq 0$ or X is of general type then $\kappa_S(X) = \kappa(X)$. Find examples for which $\kappa_S(X) = -\infty$ and $\kappa(X)$ is any number between 0 and $\dim X - 1$.

(Note: solutions can be found in [**HP16**, Section 4]).

EXERCISE 5.18. Let X be an irreducible hypersurface of degree d in $\mathbb{P}^n$ $(n \geq 2)$ defined by $f(x_0, \dots, x_n) = 0$ such that $X \cap D(x_0) \neq \emptyset$ (i.e., f is not a polynomial of x_0). Let $\tilde{f}(x_1, \dots, x_n) := f(1, x_1, \dots, x_n)$ and let P_l be the space of polynomials of degree at most l in the variables $x_1, \dots, x_n$. Further, let Φ_e be the k-linear map $k[x_1, \dots, x_n] \to k[x_1, \dots, x_n]$, for which

$$\Phi_e\left(\prod_{i=1}^{n} x_i^{j_i}\right) = \begin{cases} \prod_{i=1}^{n} x_i^{\frac{j_i - p^e + 1}{p^e}} & \text{if } p^e | j_i - p^e + 1 \text{ for all } i \\ 0 & \text{otherwise} \end{cases}.$$

Consider then

$$V_e := \Phi_e\left(\tilde{f}^{p^e-1} \cdot P_{(d-n-1)(1+(m-1)p^e)}\right).$$

That is, V_e is the image via Φ_e of the space containing all the polynomials that are obtained by multiplying a degree at most $(d-n-1)(1+(m-1)p^e)$ polynomial $p^e - 1$ times with $\tilde{f}$. Let $W_e \subseteq V_e$ be the subspace of V_e containing the polynomials divisible by $\tilde{f}$.

Show that then the image of

$$H^0(X, \omega_X^{m-1} \otimes F_*^e \omega_X) \to H^0(X, \omega_X^m)$$

can be identified with the quotient V_e/W_e. In particular, $S^0(X, \omega_X^m)$ can be identified with V_e/W_e for $e \gg 0$.

Hint: Show that there is a commutative diagram as follows.

$$\begin{array}{ccc} H^0(\mathbb{P}^n, \mathcal{O}_{\mathbb{P}^n}((1+(m-1)p^e)(d-n-1))) & \longrightarrow & H^0\left(X, \omega_X^{1+(m-1)p^e}\right) \\ \downarrow {\scriptstyle \cdot f^{p^e-1}} & & \\ H^0(\mathbb{P}^n, \mathcal{O}_{\mathbb{P}^n}((1+(m-1)p^e)(d-n-1) + (p^e-1)d)) & & \downarrow {\scriptstyle \mathrm{tr}} \\ \downarrow {\scriptstyle \mathrm{tr}} & & \\ H^0(\mathbb{P}^n, \mathcal{O}_{\mathbb{P}^n}(m(d-n-1))) & \longrightarrow & H^0(X, \omega_X^m) \end{array}$$

Then show that the horizontal arrows are surjective and restrict everything to $D(x_0)$.

EXERCISE 5.19. Let X be the surface in $\mathbb{P}^3$ defined by $x^5+y^5+z^5+v^5$. Compute $S^0(\omega_X)$ for each prime p. Further, for $p = 2$, show that $S^0(\omega_X) = S^0(\omega_X^2) = 0$, but $S^0(\omega_X^3) \neq 0$.

Hint: use 5.18.

5.3. Problems.

QUESTION 5.20. Fujita's conjecture states that if X is a smooth projective variety and L an ample line bundle on it, then $K_X + (\dim X + 1)L$ is free and $K_X + (\dim X + 2)L$ is very ample. This is not known in positive characteristic even for surfaces. For surfaces there are results when L is special [**Eke88, SB91b,**

Ter99, DCF15]. However, the full conjecture is not known even in the surfaces case.

QUESTION 5.21. What is the lowest m such that the semi-positivity of 5.2 holds? The question is already interesting if one fixes the dimension, so for example for families of surfaces. Note that for $m = 1$ the semi-positivity is known not to hold even for families of curves by [**MB81**, 3.2].

QUESTION 5.22. Compute the semi-stable rank of the Hasse-Witt matrix or equivalently the dimension of $S^0(X, \omega_X)$ of general elements in the components of the moduli space of surfaces. By a conjecture of Grothendieck this is non-zero, so finding a component where this is zero would be interesting. A few related articles: [**Lie08a, Lie09, Lie08b, Mir84, Hir99**].

5.3.1. *Log-Fano Varieties.*

QUESTION 5.23. If $I \subseteq (0,1) \cap \mathbb{Q}$ is a finite subset, then is there a p_0 depending only on I, such that for all log-Del Pezzos (X, Δ) for which the coefficients of Δ are in

$$D(I) := \left\{ \frac{m + \sum_{j=1}^{l} a_j i_j}{m+1} \,\middle|\, m, l \in \mathbb{N}, a_j \in \mathbb{N}, i_j \in I \right\} \cap [0,1]$$

(X, Δ) is strongly F-regular? Note that this is the two dimensional version of [**CGS16**, 4.1], and it might be hard in full generality. However, the question is already interesting for $I = \{1\}$ or for $D(I)$ replaced by any smaller set, for example by $\{1\}$. Other fixed I's are also interesting.

The main interest in the question stems from applying it to threefold singularities, as in [**CGS16**], or to threefold fibrations the geometric generic fibers of which are log-Del Pezzos (in this case it would yield semi-positivity statements using 5.2). The question is interesting in any higher dimension as well, so for log-Del Pezzo replaced by log-Fano.

QUESTION 5.24. Classify globally F-regular smooth Fano threefolds of Picard number 1. According to [**SB97**], the classification in any characteristic agrees with the characteristic zero classification [**IP99**], which is a finite list of deformation equivalence classes. Hence it should be possible to determine a list of globally F-regular ones as in 5.14.

QUESTION 5.25. For the cases when not all the Del-Pezzos are globally F-regular (see 5.14), describe the locus of the non globally F-regular ones in the moduli space. That is, describe its dimension. Is it closed?

QUESTION 5.26. In [**MS03**] it is hinted to conclude the classification of smooth Fano threefolds along the line of the characteristic zero classification [**IP99, MM82, MM03**]. The Picard number one and two cases are done in [**SB97**] and [**Sai03**]. Finish the higher Picard number cases.

5.3.2. *MMP.*

QUESTION 5.27. Classify the singularities that the geometric general fiber of the Iitaka fibration of 3-folds can have. Recall that the Iitaka fibration is a fibration $f : X \to Y$ such that $\omega_X \cong f^*\mathscr{L}$ for some big line bundle $\mathscr{L}$. Here you can assume X to be smooth for the first. In general, it would be interesting to have an answer for X with terminal singularities, which is most likely hard. Note that in dimension two only cusps appear and only in characteristics 2 and 3 (see [**Mum69**, **BM77**, **BM76**, **SS10**] and [**CD89**, Chapter V]).

QUESTION 5.28. Can one run MMP for threefolds of characteristic 2, 3 or 5? The situation for characteristic $p > 5$ has been recently mostly cleared out in a series of papers [**HX15**, **CTX15**, **Bir16**]. If $p \leq 5$, then though some generalized extremal contractions are known to exist [**Kee99**, 0.5], it is not known whether flips exist.

QUESTION 5.29. Prove full cone theorem, and full contraction theorems for threefolds (at least in characteristic $p > 5$). The existing statements, though are fantastic achievements, have a few unnatural hypotheses: line bundles are not semi-ample, only endowed with a map [**Kee99**, 0.5], there are finitely many extremal rays that are not known to contain rational curves [**Kee99**, 0.6] (see also [**Kee99**, 5.5.4]), if $K_X + B$ is not pseudo-effective then only a weak cone theorem is known [**CTX15**, 1.7]. The main question is whether these hypotheses can be removed.

QUESTION 5.30. Abundance for threefolds?

QUESTION 5.31. MMP for 4-folds?

5.3.3. *Surface and threefold inequalities.* For minimal surfaces of general type a few inequalities govern the possible values of standard numerical invariants (e.g., $K_X^2 \geq 1$, $\chi(\mathcal{O}_X) \geq 1$, $K_X^2 \geq 2p_g - 4$ (Noether inequality), $K_X^2 \leq 9\chi(\mathcal{O}_X)$, c.f., [**BCP06**, Section 1.2]). Many of these are known either to hold, or there is a good understanding of when they fail in positive characteristic [**Lie13**, Sections 8.2, 8.3, 8.4]. There are a few questions concerning surfaces left and very little is known for threefolds.

QUESTION 5.32. Is there a smooth surface of general type with $\chi(\mathcal{O}_X) \leq 0$? Note that the possibilities are quite restricted (see [**SB91a**, Theorem 8]).

QUESTION 5.33. Minimal Gorenstein threefolds of general type are known to have $\chi(\mathcal{O}_X) > 0$ in characteristic zero. Is this true in positive characteristic? If not, what are the exceptions? The expectations are statements as [**Lie13**, 8.4,8.5].

QUESTION 5.34. If X is a minimal threefold of general type then is there a lower bound for K_X^3 in terms of a (linear) function of the geometric genus $\rho_g(X)$? Note that in characteristic zero this has been shown in [**Kob92**], and the analogous statement is known for surfaces of positive characteristic [**Lie08a**].

6. Seshadri constants, F-pure centers and test ideals

In 3 we introduced S^0 and showed how global sections can be extended from them via adjunction along divisors. At the start of this section we discuss other ways to produce sections in S^0. First we recall Seshadri constants and F-pure centers and how to use them to produce sections in S^0. Then, building upon and generalizing the definition of F-pure centers we introduce test ideals and explore a number of open questions about them.

We begin with Seshadri constants.

6.1. Seshadri constants. Recall the following definition originally found in [**Dem93**].

DEFINITION 6.1 (Seshadri constants). [**Laz04a**] Suppose X is a projective variety and L is an ample (or big and nef) line bundle. Choose $z \in X$ a smooth closed point. let $\pi : Y \to X$ be the blowup of z with exceptional divisor E. We define the Seshadri constant of L at z as

$$\varepsilon(L, z) = \sup\{t > 0 \mid \pi^* L - tE \text{ is nef}\}.$$

The Seshadri constant is a local measure of positivity of L at z. Its usefulness comes from the following theorem (and variants).

THEOREM 6.2. *If X is a projective variety over $\mathbb{C}$ and if $\varepsilon(L, z) > \dim X$ for some smooth closed point $z \in X$, then $\omega_X \otimes L$ is globally generated at z.*

PROOF. Let $n = \dim X$ so that $\pi^* L - nE$ is big and nef. Consider the following diagram:

$$\begin{array}{ccc}
0 & & 0 \\
\downarrow & & \downarrow \\
H^0(Y, \omega_Y(\pi^* L - nE)) & \longrightarrow & H^0(X, \mathfrak{m}_z \otimes \omega_X \otimes L) \\
\downarrow & & \downarrow \\
H^0(Y, \omega_Y(\pi^* L - (n-1)E)) & \xrightarrow{\sim} & H^0(X, \omega_X \otimes L) \\
\downarrow & & \downarrow{\scriptstyle \alpha} \\
H^0(E, \omega_E(\pi^* L - nE)) & \xrightarrow{\sim} & H^0(X, \omega_X \otimes L/\mathfrak{m}_z) \\
\downarrow & & \downarrow \\
H^1(Y, \omega_Y(\pi^* L - nE)) & \longrightarrow & H^1(X, \mathfrak{m}_z \otimes \omega_X \otimes L)
\end{array}$$

It is an exercise left to the reader to verify the isomorphisms indicated. By Kawamata-Viehweg vanishing $H^1(Y, \omega_Y(\pi^* L - nE)) = 0$ and hence α surjects proving that $\omega_X \otimes L$ is globally generated at z. □

EXERCISE 6.3. Work in characteristic $p > 0$ and assume that L is ample. Show that in fact that if $\varepsilon(L, z) > \dim X$ then $S^0(X, \omega_X \otimes L)$ has a section which does not vanish at z.

Hint: Add a further left column to the diagram with horizontal right-pointing arrows induced by Frobenius. Replace then Kawamata-Viehweg vanishing with Serre vanishing.

QUESTION 6.4 (Perhaps hard). Are there lower bounds for Seshadri constants at very general points in characteristic $p > 0$ ala [**EKL95**]?

There is another version of Seshadri constants which has recently been studied in characteristic $p > 0$ [**MS14**]. This definition is inspired by the characterization of ordinary Seshadri constants described by separation of jets [**Laz04a**, Chapter 5].

Given a positive integer e, we say that a line bundle L on X *separates p^e-Frobenius jets at z* if the restriction map

$$H^0(X, L) \to H^0(X, L \otimes \mathcal{O}_X/\mathfrak{m}_z^{[p^e]}) \tag{10}$$

is surjective (here $\mathfrak{m}_z^{[p^e]}$ is the ideal generated by the p^eth powers of the elements of $\mathfrak{m}_z$).

Let $s_F(L^m, z)$ be the largest $e \geq 1$ such that L^m separates p^e-Frobenius jets at z (if there is no such e, then we put $s_F(L^m, Z) = 0$).

DEFINITION 6.5 (F-Seshadri constants). [**MS14**] Let L be ample, the *Frobenius-Seshadri constant of L at z* is

$$\epsilon_F(L, z) := \sup_{m \geq 1} \frac{p^{s_F(L^m;z)} - 1}{m}.$$

It is not difficult to show that $\frac{\epsilon(L,z)}{n} \leq \epsilon_F(L, z) \leq \epsilon(L, z)$, see [**MS14**, Proposition 2.12]. Hence 6.4 is equivalent to:

QUESTION 6.6. Are there lower bounds for F-Seshadri constants at very general points in characteristic $p > 0$?

We also have

THEOREM 6.7. *If $\epsilon_F(L, z) > 1$ then $S^0(X, \sigma(X, 0) \otimes \mathcal{O}_X(L + K_X))$ globally generates $\omega_X \otimes L$ at z.*

EXERCISE 6.8. Use the above theorem to give another proof of the fact that if $\epsilon(L, z) > \dim X$ then $S^0(X, \sigma(X, 0) \otimes \mathcal{O}_X(L + K_X))$ globally generates $\omega_X \otimes L$ at z.

Perhaps a more approachable question (inspired by the behavior of the usual Seshadri constant) is:

QUESTION 6.9. If L_1 and L_2 are ample divisors and $z \in X$ is a smooth closed point then show that:

$$\epsilon_F(L_1 \otimes L_2; z) \geq \epsilon_F(L_1; z) + \epsilon_F(L_2; z).$$

Since the Frobenius Seshadri constant is distinct from the ordinary Seshadri constant, another interesting question is:

QUESTION 6.10. What does the F-Seshadri constant correspond to in characteristic zero? Is there a geometric characteristic zero definition that behaves in the same way philosophically? (The toric case for torus invariant points is handled at the end of [**MS14**]).

6.2. F-pure centers. Now we move on to F-pure centers. If the reader is introduced to S^0, one exercise to keep in mind is 6.20 which generalizes 3.14 from the case of divisors to higher codimension F-pure centers.

DEFINITION 6.11. Suppose (X, Δ) is a pair with $\Delta \geq 0$ an effective $\mathbb{Q}$-divisor such that $(p^e - 1)(K_X + \Delta)$ is Cartier. Let $\phi_{e,\Delta} : F^e_* \mathscr{L} \to \mathcal{O}_X$ be the corresponding map. We say that a subvariety $Z \subseteq X$ is *an F-pure center of* (X, Δ) if

- $\phi_{e,\Delta}(F^e_*(I_Z \cdot \mathscr{L})) \subseteq I_Z$ and
- (X, Δ) is F-pure (ie $\phi_{e,\Delta}$ is surjective) at the generic point of Z.

REMARK 6.12. In the case that $\phi_{e,\Delta} : F^e_* \mathcal{O}_X \to \mathcal{O}_X$ is a splitting of Frobenius[4], the F-pure centers are called the *compatibly split subvarieties of* $\phi_{e,\Delta}$.

EXERCISE 6.13. With (X, Δ) as above, show that Z is an F-pure center if and only if

- for every Cartier divisor H containing Z and every $\varepsilon > 0$, $(X, \Delta + \varepsilon H)$ is not F-pure and
- (X, Δ) is F-pure at the generic point of Z.

If you know the definition of a log canonical center, prove that the analogous characterization can be used to define log canonical centers.

Let's do an example. First we state (but don't prove) a variant of Fedder's criterion for F-pure centers.

LEMMA 6.14. *Suppose that $S = k[x_1, \ldots, x_n]$ and $R = S/I$ and $X = \operatorname{Spec} R$. If K_X is $\mathbb{Q}$-Cartier with index not divisible by $p > 0$, then, after localizing further if necessary, $I^{[p^e]} : I = \langle g_e \rangle + I^{[p^e]}$ for some single polynomial g_e (depending on e). Then $V(Q) = Z \subseteq X$ is an F-pure center of $(X, 0)$ if and only if $g_e \in Q^{[p^e]} : Q$ and $g_e \notin Q^{[p^e]}$.*

EXERCISE 6.15. Suppose that $X = \operatorname{Spec} k[x_1, \ldots, x_n]/\langle x_1 x_2 \cdots x_n \rangle$. Identify the F-pure centers of $(X, 0)$.

Let's prove inversion of F-adjunction.

EXERCISE 6.16 (Inversion of F-adjunction). Suppose (X, Δ) is a pair and $Z \subseteq X$ is an F-pure center. For simplicity assume that Z is normal. Show that the map $\phi_{e,\Delta} : F^e_* \mathscr{L} \to \mathcal{O}_X$ induces a map $\phi_Z : F^e_* \mathscr{L}|_Z \to \mathcal{O}_Z$ and hence an effective $\mathbb{Q}$-divisor Δ_Z (this wasn't covered in the lecture but see the notes above). Show that $(K_X + \Delta)|_Z \sim_{\mathbb{Q}} K_Z + \Delta_Z$. Further show that (X, Δ) if F-pure in a neighborhood of Z if and only if (Z, Δ_Z) is F-pure.

It is not hard to show that $Z \subseteq X$ is a log canonical center of (X, Δ) then Z is also an F-pure center as long as (X, Z) is F-pure at the generic point of Z (using the fact that F-pure singularities are log canonical [**HW02**]).

EXERCISE 6.17. Prove the above assertion.

Based on recent work [**MS11, Mus12**] consider the following conjecture.

CONJECTURE 6.18 (Weak ordinarity). *Suppose that X is a smooth projective variety over $\mathbb{C}$, $A = \operatorname{Spec} R$ where R is a finitely generated $\mathbb{Z}$-algebra containing $\mathbb{Z}$ and $X_A \to A$ is a spreading out of X over A (so that $X_A \times_A \mathbb{C} \cong X$), in other words a family of characteristic $p > 0$ models of X. Then there exists a Zariski dense set of closed points $U \subseteq A$ such that for every point $p \in U$ we have that if $X_p = X_A \times_A k(p)$ then the Frobenius morphism:*

$$H^i(X, \mathcal{O}_X) \to H^i(X, F^e_* \mathcal{O}_X)$$

is bijective.

---

[4]Where again $\Delta \geq 0$ is necessarily an effective $\mathbb{Q}$-divisor such that $(p^e - 1)(K_X + \Delta)$ is Cartier, as above.

In [**Tak13**], it was shown that this conjecture implies that if (X, Δ) is log canonical and (X_A, Δ_A) is a family of characteristic $p > 0$ models of (X, Δ) as above, then (X_p, Δ_p) is F-pure for a Zariski dense set of $p \in A$. Hence we are inspired to ask:

QUESTION 6.19. Assume the weak ordinarity conjecture and suppose that $\{Z_i\}$ is the (finite) set of log canonical centers of a log canonical pair (X, Δ). Is it true that for any $(X_A, \Delta_A) \to A$ and $\{(Z_i)_A\} \to A$ as above, there exists a Zariski-dense set of closed points $p \in A$ such that $\{(Z_i)_p\}$ is exactly the set of F-pure centers of (X_p, Δ_p)?

Comment: It is easy to see that the $(Z_i)_p$ are F-pure centers (exercise), but it may be more difficult to see that this is all of them.

Finally, we mention that F-pure centers can be used to construct global sections.

EXERCISE 6.20. Formulate and prove a generalization of 3.14. In particular, show that $S^0(X, \sigma(X, \Delta) \otimes \mathcal{O}_X(M))$ surjects onto $S^0(Z, \sigma(Z, \Delta_Z) \otimes \mathcal{O}_X(M))$ under appropriate hypotheses (here Z will be an F-pure center of (X, Δ)).

The question is then how to construct F-pure centers. The common techniques involving sections of high multiplicity (*cf.* for example [**Laz04b**, Chapter 10]) seem to not work since the F-pure threshold can have a p in the denominator (this actually gets quite subtle, simple perturbation techniques seem not to work). For some ways around this issue see the recent work of [**CTX15**].

6.3. Test ideals. In this subsection we introduce test ideals, a characteristic $p > 0$ analog of multiplier ideals.

Suppose X is integral and normal and Δ is a $\mathbb{Q}$-divisor. As before, we have observed that if the index of $K_X + \Delta$ is not divisible by p, then we obtain a map $\phi_\Delta : F^e_* \mathscr{L} \to \mathcal{O}_X$. In this setting we define:

DEFINITION 6.21. With notation as above, we define the *test ideal* $\tau(X, \Delta)$ to be the smallest non-zero ideal sheaf J such that $\phi_\Delta(F^e_*(J \cdot \mathscr{L})) \subseteq J$.

We give a definition in the local setting $X = \operatorname{Spec} R$ but when $K_X + \Delta$ has no $\mathbb{Q}$-Cartier requirements.

DEFINITION 6.22. Assume $X = \operatorname{Spec} R$ is integral and normal and $\Delta \geq 0$ is a $\mathbb{R}$-divisor. Then we define the *test ideal* $\tau(X, \Delta)$ to be the smallest non-zero ideal sheaf J such that for all $e \geq 0$ and all $\phi \in \operatorname{Hom}_R(F^e_* \mathcal{O}_X(\lceil (p^e - 1)\Delta \rceil), \mathcal{O}_X)$ we have $\phi(F^e_* J) \subseteq J$.

It is probably worth remarking that $\operatorname{Hom}_R(F^e_* \mathcal{O}_X(\lceil (p^e - 1)\Delta \rceil), \mathcal{O}_X)$ consists precisely of those ϕ' corresponding to divisors Δ' such that $(p^e - 1)(K_X + \Delta') \sim 0$ and such that $\Delta' \geq \Delta$.

Since in either definition, $\tau(X, \Delta)$ is the *smallest* such ideal, it is not obvious that the test ideal exists. To show it exists (locally), you first find a $c \in R$ such that for every $\mathfrak{a} \subseteq R$ an ideal, we have $c \in \phi(F^e_* \mathfrak{a})$ for some $\phi \in \operatorname{Hom}_R(F^e_* \mathcal{O}_X(\lceil (p^e - 1)\Delta \rceil), \mathcal{O}_X)$. This is tricky, and such elements are called *test elements*. However, once the c is found constructing $\tau(X, \Delta)$ is simply

$$\tau(X, \Delta) = \sum_{e \geq 0} \sum_{\phi} \phi(F^e_*(c \cdot \mathcal{O}_X))$$

in the second case (which subsumes the first) where ϕ varies over

$$\mathrm{Hom}_R(F^e_* \mathcal{O}_X(\lceil (p^e - 1)\Delta \rceil), \mathcal{O}_X).$$

This is clearly the smallest ideal satisfying the condition of 6.22 and containing c. But since any ideal satisfying 6.22 obviously contains J, we have constructed $\tau(X, \Delta)$.

EXERCISE 6.23. Show that the formation $\tau(X, \Delta)$ of commutes with localization and so our local definition glues to a global one.

REMARK 6.24. It is natural to think of τ as a robust version of σ in the same way that the augmented base locus is a robust version of the stable base locus. In fact, there is a global version of S^0 which builds in a "test element". Assume that (X, Δ) is a pair and that $(p^n - 1)(K_X + \Delta)$ is Cartier (where n is the smallest positive integer with that property), set $\mathscr{L}_e = \mathcal{O}_X((1 - p^e)(K_X + \Delta))$ so that we have a map $\phi^e_\Delta : F^e_* \mathscr{L}_e \to \mathcal{O}_X$ for each e divisible by n. For any Cartier divisor M define $P^0(X, \tau(X, \Delta) \otimes \mathcal{O}_X(M))$ to be

$$\bigcap_{0 \leq D \subseteq X} \bigcap_{e_0 \geq 0} \left(\sum_{e \geq e_0} \mathrm{Tr}_{F^e} \left(H^0(X, \left(F^e_* \mathcal{O}_X(\lceil K_X - p^e(K_X + \Delta) + p^e M - D \rceil) \right)\right) \right)$$

where the intersection runs over all effective Weil divisors ≥ 0 on X.

EXERCISE 6.25. For simplicity, assume that $(p-1)(K_X + \Delta)$ so that there are no roundings. Then show that for $P^0 \subseteq S^0$ and further that the image of the maps used to define P^0 at the sheaf level is simply $\tau(X, \Delta) \otimes M$

One of the main reasons people have been interested in test ideals is that if $(X_{\mathbb{C}}, \Delta_{\mathbb{C}})$ is a log $\mathbb{Q}$-Gorenstein pair defined over $\mathbb{C}$ and $X_A \to A$ is a family of characteristic p models, then $\tau(X_p, \Delta_p) = \left(\mathcal{J}(X_{\mathbb{C}}, \Delta_{\mathbb{C}})\right)_p$ for all closed points p in an open dense set of A. In other words, the multiplier ideal reduces to the test ideal for $p \gg 0$ (see [**Tak04**]). Note that if the coefficients of Δ vary, then this is not true. It is conjectured then that for H an effective Cartier divisor then $\tau(X_p, \Delta_p + tH) = \left(\mathcal{J}(X_{\mathbb{C}}, \Delta_{\mathbb{C}}) + tH\right)_p$ for all t and for all p is a Zariski dense set of points of A but this is related to some very hard conjectures, see [**MS11, Mus12, BST13**] for discussion.

It would be natural to try to generalize the results on the behavior of test ideals to the non-$\mathbb{Q}$-Gorenstein setting.

QUESTION 6.26. Suppose that $(X_{\mathbb{C}}, \Delta_{\mathbb{C}})$ is a pair but that $K_{X_{\mathbb{C}}} + \Delta_{\mathbb{C}}$ is not assumed to be $\mathbb{Q}$-Cartier. Is it true that $\tau(X_p, \Delta_p) = \left(\mathcal{J}(X_{\mathbb{C}}, \Delta_{\mathbb{C}})\right)_p$ for all closed points p in an open and Zariski-dense set of A (where $X_A \to A$ is as above)?

6.4. Jumping numbers. Suppose that (X, Δ) is as above and that H is a Cartier divisor. Consider the behavior of $\tau(X, \Delta + tH)$ as $t \in \mathbb{R}_{\geq 0}$ varies. Since we have asserted that test ideals are characteristic $p > 0$ analogs of multiplier ideals, one should expect that $\tau(X, \Delta + tH)$ jumps (as t varies) at a set of rational $t \in \mathbb{Q}_{\geq 0}$ without limit points, at least assuming that $K_X + \Delta$ is Cartier. Indeed, this is the case (in the log-$\mathbb{Q}$-Gorenstein setting).

First an easy exercise (you can assume the existence of test elements):

EXERCISE 6.27. If $t \geq t'$ show that $\tau(X, \Delta + tH) \subseteq \tau(X, \Delta + t'H)$.

DEFINITION 6.28. An *F-jumping number of* (X, Δ) *with respect to* H is a real number $t \geq 0$ such that $\tau(X, \Delta + tH) \neq \tau(X, \Delta + (t - \varepsilon)H)$ for any $\varepsilon > 0$.

First you shall prove that this notion is sensible.

EXERCISE 6.29. Suppose that $t \geq 0$ is a rational number. Prove that there exists an $\varepsilon > 0$ such that $\tau(X, \Delta + (t + \varepsilon)H) = \tau(X, \Delta + tH)$.

Hint: Absorb the bigger t into the test element somehow and use the fact that the test ideal can be written as a finite sum.

QUESTION 6.30. Assume now that (X, Δ) is a pair and that $K_X + \Delta$ is not necessarily $\mathbb{Q}$-Cartier. Does the set of F-jumping numbers of (X, Δ) with respect to H have any limit points? Are the F-jumping numbers rational?

One would assume that the F-jumping numbers are not necessarily rational based upon the situation for the multiplier ideal [**Urb12**]. However, one might hope that the F-jumping numbers do not have limit points (as is also hoped in characteristic zero). For some partial progress, see [**TT08, Bli13, KSSZ14**]. A first target might be to try to consider the case of an isolated singularity with $\Delta = 0$. Some relevant methods might be found for instance in [**LS01**].

6.5. Test ideals of non-$\mathbb{Q}$-Gorenstein rings.

QUESTION 6.31. Suppose that (X, Δ) is a pair and that $K_X + \Delta$ is not assumed to be $\mathbb{Q}$-Gorenstein. Then does there exist a $\Delta' \geq \Delta$ such that $K_X + \Delta'$ is $\mathbb{Q}$-Cartier (with index not divisible by p hopefully) such that $\tau(X, \Delta) = \tau(X, \Delta')$.

The answer to this is known to be yes if $\tau(X, \Delta) = \mathcal{O}_X$ at least in the affine setting, see [**SS10, Sch11**]. More generally, it is also known that there exist finitely many $\Delta_i \geq \Delta$ such that $\tau(X, \Delta) = \sum_i \tau(X, \Delta_i)$ and such that each $(K_X + \Delta_i)$ is $\mathbb{Q}$-Cartier with index not divisible by $p > 0$, see [**Sch11**]. As a general strategy on this problem, one would hope that some general choice of Δ' would work but more work is required. In characteristic zero the analog is true if one defines $\mathcal{J}(X, \Delta)$ is the non-log-$\mathbb{Q}$-Gorenstein setting as in [**DH09**].

EXERCISE 6.32. Suppose that R is a local ring and that $X = \operatorname{Spec} R$. Suppose that we know that $\tau(X, \Delta) = \sum_i \tau(X, \Delta_i)$ with $K_X + \Delta_i$ $\mathbb{Q}$-Cartier as before and also that $\tau(X, \Delta) = \mathcal{O}_X$. Prove that $\tau(X, \Delta_i) = \tau(X, \Delta)$ for some i.

Here $\mathcal{J}(X_{\mathbb{C}}, \Delta_{\mathbb{C}})$ is defined as in [**DH09**]. Some recent work on this question was done in [**dFDTT15**]. It is easily seen to be true in the toric case [**Bli04**]. Perhaps isolated or even isolated graded singularities would be a natural place to start (see [**LS99**] for some ideas using commutative algebra language).

EXERCISE 6.33. Show that the containment $\left(\mathcal{J}(X_{\mathbb{C}}, \Delta_{\mathbb{C}})\right)_p \subseteq \tau(X_p, \Delta_p)$ holds.

Hint: Use the fact that we know it holds in the $\mathbb{Q}$-Gorenstein setting.

7. Numerical Invariants

DEFINITION 7.1. Assume $X = \operatorname{Spec} R$ is integral and normal and $\Delta \geq 0$ is an $\mathbb{R}$-divisor. Suppose that (X, Δ) is sharply F-pure and that H is a Cartier divisor. The F-pure threshold of (X, Δ) along H is $\operatorname{fpt}((X, \Delta), H) = \sup\{t \in \mathbb{R} \mid (X, \Delta + tH) \text{ is sharply } F\text{-pure}\}$. When $\Delta = 0$, then we write $\operatorname{fpt}(X, H)$ instead of $\operatorname{fpt}((X, 0), H)$.

Of course, one should compare the definition about with that of the Log Canonical Threshold (LCT) in characteristic zero (which simply replaces sharply F-pure with Log Canonical throughout).

EXERCISE 7.2. When (X, Δ) is strongly F-regular, show that $\mathrm{fpt}((X, \Delta), H)$ is the smallest jumping number of (X, Δ) with respect to H.

It is known that if $K_X + \Delta$ is $\mathbb{Q}$-Cartier, then $\mathrm{fpt}((X, \Delta), H) \in \mathbb{Q}$ (indeed all jumping numbers are rational). But in general, it is an open question, compare with [**Urb12**].

PROBLEM 7.3. Find (X, Δ) and H so that $\mathrm{fpt}((X, \Delta), H) \notin \mathbb{Q}$ (or show this can't happen).

It follows easily from 4.16 that if $K_X + \Delta$ is $\mathbb{Q}$-Cartier and (X, Δ) is sharply F-pure, then (X, Δ) is log canonical. Hence in a fixed characteristic $\mathrm{fpt}((X, \Delta), H) \geq \mathrm{lct}((X, \Delta), H)$. As we vary the characteristic however we do get some more subtle behavior.

EXERCISE 7.4. Working in characteristic $p > 0$, let $X = \mathbb{A}^2$ and $H = V(y^2 - x^3)$. Show that the following formula for $\mathrm{fpt}(X, H)$ holds.

$$\mathrm{fpt}(X, H) = \begin{cases} \frac{1}{2} & \text{if } p = 2 \\ \frac{2}{3} & \text{if } p = 3 \\ \frac{5}{6} & \text{if } p \equiv 1 \mod 6 \\ \frac{5}{6} - \frac{1}{6p} & \text{if } p \equiv 5 \mod 6 \end{cases}$$

EXERCISE 7.5. If $X = \mathbb{A}^2$ and $H = V(xy(x + y))$. Find the formula for $\mathrm{fpt}(X, H)$ in terms of the characteristic.

THEOREM 7.6. *Suppose $(X_{\mathbb{C}}, \Delta_{\mathbb{C}})$ is KLT and $H_{\mathbb{C}}$ is a Cartier divisor all defined over $\mathbb{C}$. Take a model $\mathscr{X} \to \mathrm{Spec}(A)$ of X (together with Δ and H) over a finitely generated $\mathbb{Z}$-algebra domain A (so that $\mathscr{X}_0 \otimes_{\mathrm{Frac}(A)} \mathbb{C} = X_{\mathbb{C}}$). Then for every $\varepsilon > 0$, there exists a dense open set $U \subseteq \mathrm{Spec}\, A$, so that*

$$\mathrm{lct}((X_{\mathbb{C}}, \Delta_{\mathbb{C}}), H_{\mathbb{C}}) - \varepsilon < \mathrm{fpt}((X_{\mathfrak{m}}, \Delta_{\mathfrak{m}}), H_{\mathfrak{m}})$$

for all maximal ideals $\mathfrak{m} \in \mathrm{Spec}\, A$. Informally this says that

$$\lim_{p \to \infty} \mathrm{fpt}((X_p, \Delta_p), H_p) = \mathrm{lct}((X_{\mathbb{C}}, \Delta_{\mathfrak{c}}), H_{\mathbb{C}}).$$

EXERCISE 7.7. Use 4.18 to prove 7.6.

While we do have that limit, in many examples it appears the two invariants frequently agree on the nose, in other words that

$$\mathrm{fpt}((X_{\mathfrak{m}}, \Delta_{\mathfrak{m}}), H_{\mathfrak{m}}) = \mathrm{lct}((X_{\mathbb{C}}, \Delta_{\mathbb{C}}), H_{\mathbb{C}})$$

for a Zariski dense set of closed points $\mathfrak{m} \in \mathrm{Spec}\, A$. Hence we have the following very important conjecture.

CONJECTURE 7.8. *Suppose that $X_{\mathbb{C}}$ is a normal affine algebraic variety over $\mathbb{C}$, and $\Delta \geq 0$ is an $\mathbb{R}$-divisor, and H is a Cartier divisor. Take a model $\mathscr{X} \to \mathrm{Spec}(A)$ of X (together with Δ and H) over a finitely generated $\mathbb{Z}$-algebra domain A (so that*

$\mathscr{X}_0 \otimes_{\mathrm{Frac}(A)} \mathbb{C} = X_{\mathbb{C}}$). *Then for a dense set of maximal ideals* $\mathfrak{m} \in \operatorname{Spec} A$, *we have that* $\operatorname{fpt}((\mathscr{X}_{\mathfrak{m}}, \Delta_{\mathfrak{m}}), H_{\mathfrak{m}})$ *equals* $\operatorname{lct}((X_{\mathbb{C}}, \Delta), H)$.

EXERCISE 7.9. If (X, Δ) is strongly F-regular and H is a Cartier divisor so that $c = \operatorname{fpt}((X, \Delta), H) \in \mathbb{Q}$ so that $K_X + \Delta + cH$ has index relatively prime to p, show that $(X, \Delta + cH)$ is sharply F-pure and $\tau(X, \Delta + cH)$ is a radical ideal.

With notation as above, if $K_X + \Delta + cH$ has index divisible by p, then it can happen that $\tau(X, \Delta + cH)$ is not radical [**MY09**].

QUESTION 7.10. Does the FPT satisfy analogous ACC properties to the LCT?

EXERCISE 7.11. Suppose X is normal and K_X is Cartier. Let H be a Cartier divisor. If ξ is a jumping number of the test ideals $\tau(X, tH)$ for $t \in \mathbb{R}_{\geq 0}$, show also that $p\xi$ is a jumping number.

Hint: consider the image of $\tau(X, ptH)$ under the trace map. Also, using the ideas of the following section, do this again and show that $p^e\xi$ is a jumping number of the ideals $\tau(X, \Delta + tH)$ whenever $(p^e - 1)(K_X + \Delta)$ is Cartier.

DEFINITION 7.12. Suppose that $X = \operatorname{Spec} R$ where $R = S/I$ is a quotient of a polynomial ring $S = k[x_1, \ldots, x_n]$. Let R_d be the image the polynomials of degree less than or equal to d in S. Define a function $\delta : R \to \mathbb{Z}_{\geq 0} \cup \{-\infty\}$ by $\delta(f) = d$ whenever $f \in R_d \setminus R_{d-1}$. We say that δ is a gauge for R. It is easy to see that the following properties hold.

(a) $\delta(f) = -\infty$ if and only if $f = 0$.
(b) Each R_d is a finite dimensional k vector space.
(c) $\bigcup_d R_d = R$.
(d) $\delta(f + g) \leq \max\{\delta(f), \delta(g)\}$.
(e) $\delta(fg) \leq \delta(f) + \delta(g)$.

In the exercises below, we will make use of the notation from the definition.

EXERCISE 7.13. Suppose that $\phi : F^e_* R \to R$ is an R-linear map. Show that there exists a constant K so that

$$\delta(\phi(F^e_* f)) \leq \delta(f)/p^e + K/p^e$$

for all $f \in R$. If ϕ corresponds to a divisor Δ on $X = \operatorname{Spec} R$, show how to use this to get a bound on the degrees of generators for the test ideals $\tau(X, \Delta + t\operatorname{div}(g))$ when $g \in R$ in terms of $\delta(g)$. See the survey [**BS13**] for more details.

EXERCISE 7.14. Use the previous exercise to conclude that the set of F-jumping numbers of $\tau(X, \Delta + t\operatorname{div}(g))$ are a discrete set of rational numbers.

8. More on test ideals and F-Singularities in Families

8.1. Bertini theorems for test ideals. In 6.5 we thought about choosing general boundaries Δ_i. The choice of a general Δ_i also seems related to the following other problem.

QUESTION 8.1 (Bertini's theorem for test ideals). Suppose that X is a quasi-projective variety and that (X, Δ) is a pair with $K_X + \Delta$ even $\mathbb{Q}$-Cartier with index not divisible by $p > 0$. Suppose that H is a general hyperplane section of X. Is it true that $\tau(X, \Delta) \cdot \mathcal{O}_H = \tau(H, \Delta|_H)$?

Note that the containment $\supseteq$ was shown in [**Tak13**, Proposition 2.12(1)] under some assumptions. The other direction seems to be hard. The main approach seems to be to show the following.

QUESTION 8.2. If $f : U \to V$ is a flat family with regular (but not necessarily smooth) fibers then is $\tau(V, \Delta) \cdot \mathcal{O}_U \subseteq \tau(U, f^*\Delta)$? This is known in the case that $\tau(V, \Delta) = \mathcal{O}_V$, in other words where (V, Δ) is strongly F-regular, see [**SZ13**] and [**HH94**, Theorem 7.3]. It would be an interesting problem to study on its own.

We need one other property.

QUESTION 8.3. suppose $\phi : Y \to S$ is a morphism of finite type between schemes essentially of finite type over a field $k = \overline{k}$ and that $\Delta \geq 0$ is a $\mathbb{Q}$-divisor on Y such that $K_Y + \Delta$ is $\mathbb{Q}$-Cartier with index not divisible by p. If $\overline{g} \in \mathcal{O}_{Y_s}$ is such that $\overline{g} \in \tau(Y_s, \Delta_s)$ for some geometric point $s \in S$, is it true that $\mathrm{Image}(g) \in \tau(Y_t, \Delta_t)$ for some $g \in \mathcal{O}_{Y \times_S T}$ (some base change) restricting to $\overline{g}$ and all (geometric) t in a neighborhood of s?

Perhaps something like this result can be obtained using the method of [**PSZ13**]. We sketch the main idea for how to use this to get Bertini's theorem. Suppose that solutions to both these interim problems were positive. Now consider the following general strategy taken from [**CGM86**].

Let Z be the reduced closed subscheme of $\mathbb{P}^n_k \times_k (\mathbb{P}^n_k)^\vee$ obtained by taking the closure of the set

$$\big\{(x, H) \in \,\big|\, x \in H\big\}.$$

We claim that the projection map $\beta : Z \to \mathbb{P}^n_k$ is flat. Indeed, it is clearly generically flat, and since it is finite type it is flat at some closed point. But for any $z \in Z$, $\mathcal{O}_{Z,z}$ is certainly isomorphic to that of any other z. Furthermore, the inclusion of rings $\mathcal{O}_{\mathbb{P}^n_k, \beta(z)} \to \mathcal{O}_{Z,z}$ is also independent of the choice of z, up to isomorphism. Thus β is flat in general.

We form a commutative diagram:

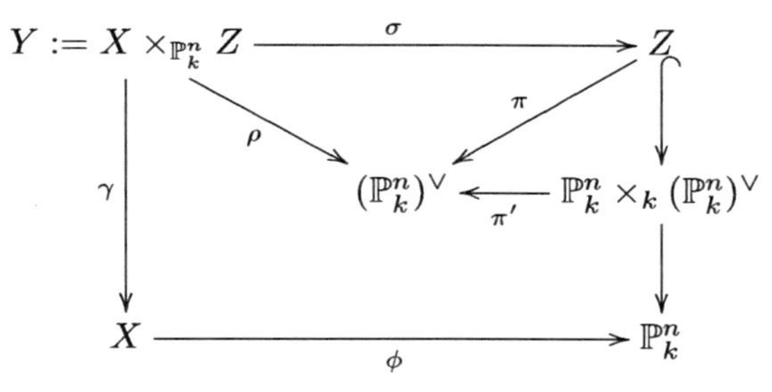

We describe each map appearing above.

- σ is the projection.
- π' is the projection and thus so is π.
- $\rho = \pi \circ \sigma$.
- γ is the projection. Note that γ is flat since it is a base change of the projection $Z \to \mathbb{P}^n_k$.

EXERCISE 8.4. Assuming what you need to from above, show how the above diagram can be used to prove Bertini's theorem for test ideals.

Hint: The fibers of ρ are hyperplane sections of X, and so one would like to show that the geometric fiber over the generic point of $(\mathbb{P}^n_k)^\vee$ behaves in a way controllable by 8.3. One should try to use 8.2 to prove this, for more details see [**CGM83**].

8.2. Test ideals by finite covers and alterations. Suppose that (X, Δ) is a pair such that $K_X + \Delta$ is $\mathbb{Q}$-Cartier (of any index). We have the following theorem

THEOREM 8.5 ([**BST15**]). *There exists a finite surjective separable map from a normal variety $f : Y \to X$ such that the trace map $\mathrm{Tr}_{Y/X} : F_* \mathcal{O}_Y \to \mathcal{O}_X$ sends $\mathrm{Tr}_{Y/X}(f_* \mathcal{O}_Y(\lceil K_Y - \pi^*(K_X + \Delta) \rceil)) = \tau(X, \Delta)$.*

Indeed, one can also replace $f : Y \to X$ to be a (sufficiently large) regular alteration if you'd like Y to be smooth.

Let us do a couple exercises in order to prove this result. First we need a lemma of [**HL07**] and generalized in [**SS12**] which we will take as given.

LEMMA 8.6. *Suppose that $(R, \mathfrak{m})$ is a local ring such that the Frobenius map is finite. Suppose we are given an element $z \in H^i_{\mathfrak{m}}(R)$ such that the submodule of $H^i_{\mathfrak{m}}(R)$ generated by $\{z, z^p, z^{p^2}, \ldots\}$ is finitely generated. Then there exists a finite (separable) ring extension $R \subseteq S$ such that $H^i_{\mathfrak{m}}(R) \to H^i_{\mathfrak{m}S}(S)$ sends z to zero.*

EXERCISE 8.7 (Hard). Suppose that R is a domain, let $N \subseteq H^{\dim R}(R)$ be the largest proper submodule of $H^{\dim R}(R)$ such that $F(N) \subseteq N$. Show that there exists a finite extension of rings $R \subseteq S$ such that N goes to 0 in $H^{\dim R}_{\mathfrak{m}}(R) \to H^{\dim R}_{\mathfrak{m}S}(S)$.

Hint: First show that $\ker(H^{\dim R}_{\mathfrak{m}}(R) \to H^{\dim R}_{\mathfrak{m}S}(S))$ is a submodule of N. Then, using local duality, observe that $N^\vee = \omega_R / \tau(\omega_R)$ where $\tau(\omega_R)$ is the smallest non-zero submodule of ω_R such that $T^e(F^e_* \tau(\omega_R)) \subseteq \tau(\omega_R)$.

Suppose that $R \subseteq S$ is a finite extension of domains and $J_S = \mathrm{Image}(\omega_S \to \omega_R)$. Show that there is a finite extension of domains $S \subseteq S'$ such that the support of $J_{S'}/\tau(\omega_R)$ is strictly smaller than support of $J_S/\tau(\omega_R)$. Now proceed by Noetherian induction.

By local duality, the above result proves the theorem in the case that $\omega_R \cong R$ and $\Delta = 0$.

To obtain the more general theorem, one has to understand the behavior of test ideals under finite maps. We'll talk about this shortly, but first we highlight a more general question.

QUESTION 8.8. Suppose (X, Δ) is a pair but do *not* assume that $K_X + \Delta$ is $\mathbb{Q}$-Cartier. Does there exist a finite separable map (or alteration) such that $\mathrm{Tr}_{Y/X}(f_* \mathcal{O}_Y(\lceil K_Y - \pi^*(K_X + \Delta) \rceil) = \tau(X, \Delta)$?

This is a generalization of the biggest open question in tight closure theory (characteristic $p > 0$ commutative algebra). Let us state this another version of this question.

EXERCISE 8.9. Suppose an affirmative answer to the above question. Suppose that $X = \mathrm{Spec}\, R$ and $\tau(X, 0) = \mathcal{O}_X$ (in other words, that X is F-regular). Show that for any finite extension of $R \subseteq S$ we have that $S \cong R \oplus M$ as R-modules. In other words that $R \hookrightarrow S$ splits as a map of R-modules. This was proven for $\mathbb{Q}$-Gorenstein rings first in [**Sin99a**].

Let us now introduce the other machinery needed to prove 8.5.

THEOREM 8.10 ([**ST14**]). *Suppose that $R \subseteq S$ is a finite separable extension of normal F-finite*[5] *domain. Set $X = \operatorname{Spec} R$ and $y = \operatorname{Spec} S$. Then for any $\mathbb{Q}$-divisor Δ on X we have*

$$\operatorname{Tr}(f_*\tau(Y, f^*\Delta - \operatorname{Ram}_{Y/X})) = \tau(X, \Delta).$$

We prove the theorem in a special case.

EXERCISE 8.11. Prove the above theorem in the case that $\Delta \geq 0$, $(p^e - 1)(K_X + \Delta) \sim 0$ and that $f^*\Delta - \operatorname{Ram}_{Y/X}$ is effective.

Hint: There is a diagram

$$\begin{array}{ccc} f_*F^e_*\omega_Y & \longrightarrow & f_*\omega_Y \\ \downarrow & & \downarrow \\ F^e_*\omega_X & \longrightarrow & \omega_X \end{array}$$

twist it appropriately and recall that $\operatorname{Ram}_{Y/X} = K_Y - f^*K_X$.

EXERCISE 8.12. Prove 8.5 using the above methods.

Hint: You can use the fact that given an $\mathbb{Q}$-Cartier divisor Γ on X, there exists a finite separable cover $f : Y \to X$ such that $f^*\Gamma$ is Cartier. You can also use the fact that if D is Cartier, then $\tau(X, D) = \tau(X) \otimes \mathcal{O}_X(-D)$.

8.3. Other types of singularities and deformation thereof. We introduce other types of singularities.

Note that for any variety X, with dualizing complex $\omega_X^\bullet$ we have $T : F_*\omega_X^\bullet \to \omega_X^\bullet$ dual to $\mathcal{O}_X \to F_*\mathcal{O}_X$.

DEFINITION 8.13 (F-injective and F-rational singularities). Suppose that X is a variety with dualizing complex $\omega_X^\bullet$. We say that X is *F-injective* if the canonical maps $\mathbf{h}^i(F_*\omega_X^\bullet) \to \mathbf{h}^i\omega_X^\bullet$ surject[6] for all i.

We say that X is *F-rational* if X is Cohen-Macaulay and if, locally, for every effective Cartier divisor D we have that $F^e_*(\omega_X \otimes \mathcal{O}_X(-D)) \to \omega_X$ surjects for some $e > 0$.

EXERCISE 8.14. Show that X is F-rational if and only if X is Cohen-Macaulay and if the only nonzero submodule $J \subseteq \omega_X$ such that $T(F_*J) \subseteq J$ is ω_X.

DEFINITION 8.15. Recall that X is *pseudo-rational* if and only if it is Cohen-Macaulay and for any proper birational map $\pi : Y \to X$ we have that $\pi_*\omega_Y = \omega_X$.

EXERCISE 8.16. Prove that F-rational singularities are pseudo-rational.

Hint: Use the previous exercise and show that $T(F_*\pi_*\omega_Y) \subseteq \pi_*\omega_Y$.

THEOREM 8.17. *Suppose $(R, \mathfrak{m})$ is a local ring and $f \in R$ is a non-zero divisor. Then if R/f is F-rational, so is R. Likewise if R/f is F-injective and Cohen-Macaulay, so is R.*

---

[5] Meaning the Frobenius map is finite.

[6] By local duality, this is the same as requiring that $H^i_x(\mathcal{O}_{X,x}) \to H^i_x(F_*\mathcal{O}_{X,x})$ inject for all i.

EXERCISE 8.18. Prove the F-injective half (the second statement) of the above theorem.

Hint: Consider the short exact sequence $0 \to \omega_R \xrightarrow{\cdot f} \omega_R \to \omega_{R/f} \to 0$ and determine how it is effected by Frobenius (or rather T). Then use Nakayama's lemma.

CONJECTURE 8.19. *With notation as above, if R/f is F-injective, then R is F-injective.*

Some recent progress on this conjecture was made by Jun Horiuchi, Lance Edward Miller and Kazuma Shimomoto [**HMS14**]. F-injective singularities seem to correspond closely to Du Bois singularities in characteristic zero [**BST13**] and it was recently shown that Du Bois singularities satisfy the condition of 8.17, see [**KS11**].

One can also ask the same types of questions for F-pure singularities. The analog 8.17 for F-pure singularities is known to hold if R is $\mathbb{Q}$-Gorenstein with index of K_R not divisible by p. It is also known to fail without the $\mathbb{Q}$-Gorenstein hypothesis [**Fed83, Sin99b**]. This leaves one open target

QUESTION 8.20. Suppose that $(R, \mathfrak{m})$ is a local $\mathbb{Q}$-Gorenstein ring and that $f \in R$ is a non-zero divisor such that R/f is F-pure. Is it true that R is also F-pure? One can ask also for a definition of σ for non $\mathbb{Q}$-Gorenstein rings that behaves well under restriction.

8.4. Bertini theorems for F-rational and F-injective singularities. We have already explored Bertini theorems for test ideals. There is a somewhat less ambitious target which also may be reasonable.

QUESTION 8.21. Suppose that $X \subseteq \mathbb{P}^n$ is a normal and F-injective (respectively F-rational) quasi-projective variety. Then is $X \cap H$ also F-injective (respectively F-rational) for all general hyperplanes $H \subseteq \mathbb{P}^n$?

The problem below, analogous to 8.2 seems to be the missing piece needed to prove that F-injective and F-rational singularities satisfy Bertini-type theorems.

QUESTION 8.22. If $f : U \to V$ is a flat family with regular (but not necessarily smooth) fibers and V is F-rational (respectively normal and F-injective), is it true that U is also F-rational (respectively F-injective)?

It turns out that this is not true for F-injectivity without the normality hypothesis [**Ene09**, Section 4]. Even worse, 8.21 has a negative answer for F-injectivity without the normality hypothesis. Let us sketch the background necessary in order to understand this failure.

DEFINITION 8.23. Suppose that $X = \operatorname{Spec} R$ is a variety over a field of characteristic $p > 0$. We say that X is *weakly normal* if $z \in K(R)$ and $z^p \in R$ implies that $z \in R$. We say that X is WN1 if it is weakly normal and the normalization morphism $\operatorname{Spec} S = X^{\mathrm{N}} \to X = \operatorname{Spec} R$ is unramified in codimension one.[7]

In [**CGM83**] the authors showed that if weakly normal surfaces that are not WN1 have general hyperplanes that are *not* weakly normal. If you combine this with

[7]Recall that an extension of local rings $(R, \mathfrak{m}) \subseteq (S, \mathfrak{n})$ is *unramified* if $\mathfrak{m} \cdot S = \mathfrak{n}$ and $R/\mathfrak{m} \subseteq S/\mathfrak{n}$ is separable.

the fact that F-injective singularities are weakly normal [**Sch09**], all we must do in order to construct an F-injective quasi-projective variety whose general hyperplane is not F-injective is to find an F-injective but not WN1 surface.

EXERCISE 8.24. Suppose k is an algebraically closed field of characteristic $p > 0$. Consider $R = k[x, y]$ and let $I = \langle y(y-1) \rangle$ so that $V(I)$ is two lines. Let $Y = \operatorname{Spec} k[t]$ and let $V(I) \to Y$ be the map which is the identity on one component but Frobenius on the other. Let S to be the pullback of the diagram of rings $\{R \to R/I \leftarrow k[t]\}$ with maps as described above. In other words, $\operatorname{Spec} S$ is the pushout of $\{X \leftarrow V(I) \to Y\}$.

Show that $\operatorname{Spec} S$ is F-injective but not WN1.

Hint: To show that $\operatorname{Spec} S$ is not WN1 note that X is its normalization. To show that it is F-injective, analyze the short exact sequence $0 \to S \to R \oplus k[t] \xrightarrow{-} R/I \to 0$.[8]

Acknowledgments. The authors would like to thank Andrew Bydlon, Eric Canton, Alberto Chiecchio, Hong R. Zong, Sheng Meng, and the anonymous referee for pointing out typos in these notes.

References

[BCP06] Ingrid C. Bauer, Fabrizio Catanese, and Roberto Pignatelli, *Complex surfaces of general type: some recent progress*, Global aspects of complex geometry, Springer, Berlin, 2006, pp. 1–58, DOI 10.1007/3-540-35480-8_1. MR2264106

[BST13] B. BHATT, K. SCHWEDE, AND S. TAKAGI: *The weak ordinarity conjecture and F-singularities*, arXiv:1307.3763, to appear in a volume of Advanced Studies in Pure Mathematics.

[Bir16] Caucher Birkar, *Existence of flips and minimal models for 3-folds in char p* (English, with English and French summaries), Ann. Sci. Éc. Norm. Supér. (4) **49** (2016), no. 1, 169–212. MR3465979

[Bli04] Manuel Blickle, *Multiplier ideals and modules on toric varieties*, Math. Z. **248** (2004), no. 1, 113–121, DOI 10.1007/s00209-004-0655-y. MR2092724

[Bli13] Manuel Blickle, *Test ideals via algebras of p^{-e}-linear maps*, J. Algebraic Geom. **22** (2013), no. 1, 49–83, DOI 10.1090/S1056-3911-2012-00576-1. MR2993047

[BB11] Manuel Blickle and Gebhard Böckle, *Cartier modules: finiteness results*, J. Reine Angew. Math. **661** (2011), 85–123, DOI 10.1515/CRELLE.2011.087. MR2863904

[BS13] Manuel Blickle and Karl Schwede, *p^{-1}-linear maps in algebra and geometry*, Commutative algebra, Springer, New York, 2013, pp. 123–205, DOI 10.1007/978-1-4614-5292-8_5. MR3051373

[BST15] Manuel Blickle, Karl Schwede, and Kevin Tucker, *F-singularities via alterations*, Amer. J. Math. **137** (2015), no. 1, 61–109, DOI 10.1353/ajm.2015.0000. MR3318087

[BM76] E. Bombieri and D. Mumford, *Enriques' classification of surfaces in char. p. III*, Invent. Math. **35** (1976), 197–232, DOI 10.1007/BF01390138. MR0491720

[BM77] E. Bombieri and D. Mumford, *Enriques' classification of surfaces in char. p. II*, Complex analysis and algebraic geometry, Iwanami Shoten, Tokyo, 1977, pp. 23–42. MR0491719

[CGS16] Paolo Cascini, Yoshinori Gongyo, and Karl Schwede, *Uniform bounds for strongly F-regular surfaces*, Trans. Amer. Math. Soc. **368** (2016), no. 8, 5547–5563, DOI 10.1090/tran/6515. MR3458390

[CHMtaS14] Paolo Cascini, Christopher Hacon, Mircea Mustaţă, and Karl Schwede, *On the numerical dimension of pseudo-effective divisors in positive characteristic*, Amer. J. Math. **136** (2014), no. 6, 1609–1628, DOI 10.1353/ajm.2014.0047. MR3282982

[8] Here, the map denoted $-$ is the difference of the maps on each of the two components.

[CTX15] Paolo Cascini, Hiromu Tanaka, and Chenyang Xu, *On base point freeness in positive characteristic* (English, with English and French summaries), Ann. Sci. Éc. Norm. Supér. (4) **48** (2015), no. 5, 1239–1272. MR3429479

[Che13] J. A. Chen: *Explicit resolution of three dimensional terminal singularities*, 2013.

[CP08] Vincent Cossart and Olivier Piltant, *Resolution of singularities of threefolds in positive characteristic. I. Reduction to local uniformization on Artin-Schreier and purely inseparable coverings*, J. Algebra **320** (2008), no. 3, 1051–1082, DOI 10.1016/j.jalgebra.2008.03.032. MR2427629

[CP09] Vincent Cossart and Olivier Piltant, *Resolution of singularities of threefolds in positive characteristic. II*, J. Algebra **321** (2009), no. 7, 1836–1976, DOI 10.1016/j.jalgebra.2008.11.030. MR2494751

[CD89] François R. Cossec and Igor V. Dolgachev, *Enriques surfaces. I*, Progress in Mathematics, vol. 76, Birkhäuser Boston, Inc., Boston, MA, 1989. MR986969

[CGM83] Caterina Cumino, Silvio Greco, and Mirella Manaresi, *Bertini theorems for weak normality*, Compositio Math. **48** (1983), no. 3, 351–362. MR700745

[CGM86] Caterina Cumino, Silvio Greco, and Mirella Manaresi, *An axiomatic approach to the second theorem of Bertini*, J. Algebra **98** (1986), no. 1, 171–182, DOI 10.1016/0021-8693(86)90020-7. MR825140

[Cut09] Steven Dale Cutkosky, *Resolution of singularities for 3-folds in positive characteristic*, Amer. J. Math. **131** (2009), no. 1, 59–127, DOI 10.1353/ajm.0.0036. MR2488485

[Das15] Omprokash Das, *On strongly F-regular inversion of adjunction*, J. Algebra **434** (2015), 207–226, DOI 10.1016/j.jalgebra.2015.03.025. MR3342393

[dFDTT15] Tommaso de Fernex, Roi Docampo, Shunsuke Takagi, and Kevin Tucker, *Comparing multiplier ideals to test ideals on numerically $\mathbb{Q}$-Gorenstein varieties*, Bull. Lond. Math. Soc. **47** (2015), no. 2, 359–369, DOI 10.1112/blms/bdv006. MR3335128

[DH09] Tommaso de Fernex and Christopher D. Hacon, *Singularities on normal varieties*, Compos. Math. **145** (2009), no. 2, 393–414, DOI 10.1112/S0010437X09003996. MR2501423

[Dem93] Jean-Pierre Demailly, *A numerical criterion for very ample line bundles*, J. Differential Geom. **37** (1993), no. 2, 323–374. MR1205448

[DCF15] Gabriele Di Cerbo and Andrea Fanelli, *Effective Matsusaka's theorem for surfaces in characteristic p*, Algebra Number Theory **9** (2015), no. 6, 1453–1475, DOI 10.2140/ant.2015.9.1453. MR3397408

[EKL95] Lawrence Ein, Oliver Küchle, and Robert Lazarsfeld, *Local positivity of ample line bundles*, J. Differential Geom. **42** (1995), no. 2, 193–219. MR1366545

[Eke88] Torsten Ekedahl, *Canonical models of surfaces of general type in positive characteristic*, Inst. Hautes Études Sci. Publ. Math. **67** (1988), 97–144. MR972344

[Ene09] Florian Enescu, *Local cohomology and F-stability*, J. Algebra **322** (2009), no. 9, 3063–3077, DOI 10.1016/j.jalgebra.2009.04.025. MR2567410

[Fed83] Richard Fedder, *F-purity and rational singularity*, Trans. Amer. Math. Soc. **278** (1983), no. 2, 461–480, DOI 10.2307/1999165. MR701505

[FST11] Osamu Fujino, Karl Schwede, and Shunsuke Takagi, *Supplements to non-lc ideal sheaves*, Higher dimensional algebraic geometry, RIMS Kôkyûroku Bessatsu, B24, Res. Inst. Math. Sci. (RIMS), Kyoto, 2011, pp. 1–46. MR2809647

[Fuj83] Takao Fujita, *Vanishing theorems for semipositive line bundles*, Algebraic geometry (Tokyo/Kyoto, 1982), Lecture Notes in Math., vol. 1016, Springer, Berlin, 1983, pp. 519–528, DOI 10.1007/BFb0099977. MR726440

[Gab04] Ofer Gabber, *Notes on some t-structures*, Geometric aspects of Dwork theory. Vol. I, II, Walter de Gruyter, Berlin, 2004, pp. 711–734. MR2099084

[Hac15] Christopher D. Hacon, *Singularities of pluri-theta divisors in Char $p > 0$*, Algebraic geometry in east Asia—Taipei 2011, Adv. Stud. Pure Math., vol. 65, Math. Soc. Japan, Tokyo, 2015, pp. 117–122. MR3380778

[HP16] Christopher D. Hacon and Zsolt Patakfalvi, *Generic vanishing in characteristic $p > 0$ and the characterization of ordinary abelian varieties*, Amer. J. Math. **138** (2016), no. 4, 963–998, DOI 10.1353/ajm.2016.0031. MR3538148

[HX15] Christopher D. Hacon and Chenyang Xu, *On the three dimensional minimal model program in positive characteristic*, J. Amer. Math. Soc. **28** (2015), no. 3, 711–744, DOI 10.1090/S0894-0347-2014-00809-2. MR3327534

[Har98] Nobuo Hara, *Classification of two-dimensional F-regular and F-pure singularities*, Adv. Math. **133** (1998), no. 1, 33–53, DOI 10.1006/aima.1997.1682. MR1492785

[Har01] Nobuo Hara, *Geometric interpretation of tight closure and test ideals*, Trans. Amer. Math. Soc. **353** (2001), no. 5, 1885–1906, DOI 10.1090/S0002-9947-01-02695-2. MR1813597

[HW02] Nobuo Hara and Kei-Ichi Watanabe, *F-regular and F-pure rings vs. log terminal and log canonical singularities*, J. Algebraic Geom. **11** (2002), no. 2, 363–392, DOI 10.1090/S1056-3911-01-00306-X. MR1874118

[Har66] Robin Hartshorne, *Residues and duality*, Lecture notes of a seminar on the work of A. Grothendieck, given at Harvard 1963/64. With an appendix by P. Deligne. Lecture Notes in Mathematics, No. 20, Springer-Verlag, Berlin-New York, 1966. MR0222093

[Har77] Robin Hartshorne, *Algebraic geometry*, Springer-Verlag, New York-Heidelberg, 1977. Graduate Texts in Mathematics, No. 52. MR0463157

[Har80] Robin Hartshorne, *Stable reflexive sheaves*, Math. Ann. **254** (1980), no. 2, 121–176, DOI 10.1007/BF01467074. MR597077

[Har94a] Robin Hartshorne, *Generalized divisors on Gorenstein schemes*, Proceedings of Conference on Algebraic Geometry and Ring Theory in honor of Michael Artin, Part III (Antwerp, 1992), K-Theory **8** (1994), no. 3, 287–339, DOI 10.1007/BF00960866. MR1291023

[Har94b] Robin Hartshorne, *Generalized divisors on Gorenstein schemes*, Proceedings of Conference on Algebraic Geometry and Ring Theory in honor of Michael Artin, Part III (Antwerp, 1992), K-Theory **8** (1994), no. 3, 287–339, DOI 10.1007/BF00960866. MR1291023

[HS77] Robin Hartshorne and Robert Speiser, *Local cohomological dimension in characteristic p*, Ann. of Math. (2) **105** (1977), no. 1, 45–79, DOI 10.2307/1971025. MR0441962

[Hir99] Masayuki Hirokado, *Singularities of multiplicative p-closed vector fields and global 1-forms of Zariski surfaces*, J. Math. Kyoto Univ. **39** (1999), no. 3, 455–468. MR1718718

[HH94] Melvin Hochster and Craig Huneke, *F-regularity, test elements, and smooth base change*, Trans. Amer. Math. Soc. **346** (1994), no. 1, 1–62, DOI 10.2307/2154942. MR1273534

[HMS14] Jun Horiuchi, Lance Edward Miller, and Kazuma Shimomoto, *Deformation of F-injectivity and local cohomology*, Indiana Univ. Math. J. **63** (2014), no. 4, 1139–1157, DOI 10.1512/iumj.2014.63.5313. MR3263925

[HL07] Craig Huneke and Gennady Lyubeznik, *Absolute integral closure in positive characteristic*, Adv. Math. **210** (2007), no. 2, 498–504, DOI 10.1016/j.aim.2006.07.001. MR2303230

[IP99] V. A. Iskovskikh and Yu. G. Prokhorov, *Fano varieties*, Algebraic geometry, V, Encyclopaedia Math. Sci., vol. 47, Springer, Berlin, 1999, pp. 1–247. MR1668579

[KSSZ14] Mordechai Katzman, Karl Schwede, Anurag K. Singh, and Wenliang Zhang, *Rings of Frobenius operators*, Math. Proc. Cambridge Philos. Soc. **157** (2014), no. 1, 151–167, DOI 10.1017/S0305004114000176. MR3211813

[Kee99] Seán Keel, *Basepoint freeness for nef and big line bundles in positive characteristic*, Ann. of Math. (2) **149** (1999), no. 1, 253–286, DOI 10.2307/121025. MR1680559

[Kee08] Dennis S. Keeler, *Fujita's conjecture and Frobenius amplitude*, Amer. J. Math. **130** (2008), no. 5, 1327–1336, DOI 10.1353/ajm.0.0015. MR2450210

[Kob92] Masanori Kobayashi, *On Noether's inequality for threefolds*, J. Math. Soc. Japan **44** (1992), no. 1, 145–156, DOI 10.2969/jmsj/04410145. MR1139663

[KSB88] J. Kollár and N. I. Shepherd-Barron, *Threefolds and deformations of surface singularities*, Invent. Math. **91** (1988), no. 2, 299–338, DOI 10.1007/BF01389370. MR922803

[Kol11] János Kollár, *A local version of the Kawamata-Viehweg vanishing theorem*, Pure Appl. Math. Q. **7** (2011), no. 4, Special Issue: In memory of Eckart Viehweg, 1477–1494, DOI 10.4310/PAMQ.2011.v7.n4.a18. MR2918170

[Kol13] János Kollár, *Singularities of the minimal model program*, Cambridge Tracts in Mathematics, vol. 200, Cambridge University Press, Cambridge, 2013. With a collaboration of Sándor Kovács. MR3057950

[KK] J. Kollár and S. J. Kovács: *Birational geometry of log surfaces*, https://web.math.princeton.edu/~kollar/FromMyHomePage/BiratLogSurf.ps.

[KM98] János Kollár and Shigefumi Mori, *Birational geometry of algebraic varieties*, Cambridge Tracts in Mathematics, vol. 134, Cambridge University Press, Cambridge, 1998. With the collaboration of C. H. Clemens and A. Corti; Translated from the 1998 Japanese original. MR1658959

[KS11] S. J. Kovács and K. Schwede: *Du Bois singularities deform*, arXiv:1107.2349, to appear in Advanced Studies in Pure Mathematics.

[Laz04a] Robert Lazarsfeld, *Positivity in algebraic geometry. II*, Ergebnisse der Mathematik und ihrer Grenzgebiete. 3. Folge. A Series of Modern Surveys in Mathematics [Results in Mathematics and Related Areas. 3rd Series. A Series of Modern Surveys in Mathematics], vol. 49, Springer-Verlag, Berlin, 2004. Positivity for vector bundles, and multiplier ideals. MR2095472

[Laz04b] Robert Lazarsfeld, *Positivity in algebraic geometry. II*, Ergebnisse der Mathematik und ihrer Grenzgebiete. 3. Folge. A Series of Modern Surveys in Mathematics [Results in Mathematics and Related Areas. 3rd Series. A Series of Modern Surveys in Mathematics], vol. 49, Springer-Verlag, Berlin, 2004. Positivity for vector bundles, and multiplier ideals. MR2095472

[Lie08a] Christian Liedtke, *Algebraic surfaces of general type with small c_1^2 in positive characteristic*, Nagoya Math. J. **191** (2008), 111–134. MR2451222

[Lie08b] Christian Liedtke, *Uniruled surfaces of general type*, Math. Z. **259** (2008), no. 4, 775–797, DOI 10.1007/s00209-007-0248-7. MR2403741

[Lie09] Christian Liedtke, *Non-classical Godeaux surfaces*, Math. Ann. **343** (2009), no. 3, 623–637, DOI 10.1007/s00208-008-0284-6. MR2480705

[Lie13] Christian Liedtke, *Algebraic surfaces in positive characteristic*, Birational geometry, rational curves, and arithmetic, Springer, New York, 2013, pp. 229–292, DOI 10.1007/978-1-4614-6482-2_11. MR3114931

[Lyu97] Gennady Lyubeznik, *F-modules: applications to local cohomology and D-modules in characteristic $p > 0$*, J. Reine Angew. Math. **491** (1997), 65–130, DOI 10.1515/crll.1997.491.65. MR1476089

[LS99] Gennady Lyubeznik and Karen E. Smith, *Strong and weak F-regularity are equivalent for graded rings*, Amer. J. Math. **121** (1999), no. 6, 1279–1290. MR1719806

[LS01] Gennady Lyubeznik and Karen E. Smith, *On the commutation of the test ideal with localization and completion*, Trans. Amer. Math. Soc. **353** (2001), no. 8, 3149–3180, DOI 10.1090/S0002-9947-01-02643-5. MR1828602

[MS91] V. B. Mehta and V. Srinivas, *Normal F-pure surface singularities*, J. Algebra **143** (1991), no. 1, 130–143, DOI 10.1016/0021-8693(91)90255-7. MR1128650

[MS97] V. B. Mehta and V. Srinivas, *A characterization of rational singularities*, Asian J. Math. **1** (1997), no. 2, 249–271, DOI 10.4310/AJM.1997.v1.n2.a4. MR1491985

[MS12] Lance Edward Miller and Karl Schwede, *Semi-log canonical vs F-pure singularities*, J. Algebra **349** (2012), 150–164, DOI 10.1016/j.jalgebra.2011.08.035. MR2853631

[Mir84] Rick Miranda, *Nonclassical Godeaux surfaces in characteristic five*, Proc. Amer. Math. Soc. **91** (1984), no. 1, 9–11, DOI 10.2307/2045258. MR735553

[MB81] L. Moret-Bailly: *Familles de courbes et de variétés abéliennes sur $\mathbb{P}^1$ (II. Exemples)*, Astérisque (1981), no. 86, 125–140.

[Mor85] Shigefumi Mori, *On 3-dimensional terminal singularities*, Nagoya Math. J. **98** (1985), 43–66. MR792770

[MM82] Shigefumi Mori and Shigeru Mukai, *Classification of Fano 3-folds with $B_2 \geq 2$*, Manuscripta Math. **36** (1981/82), no. 2, 147–162, DOI 10.1007/BF01170131. MR641971

[MM03] Shigefumi Mori and Shigeru Mukai, *Classification of Fano 3-folds with $B_2 \geq 2$*, Manuscripta Math. **36** (1981/82), no. 2, 147–162, DOI 10.1007/BF01170131. MR641971

[MS03] Shigefumi Mori and Natsuo Saito, *Fano threefolds with wild conic bundle structures*, Proc. Japan Acad. Ser. A Math. Sci. **79** (2003), no. 6, 111–114. MR1992032

[Mum69] David Mumford, *Enriques' classification of surfaces in* char p. *I*, Global Analysis (Papers in Honor of K. Kodaira), Univ. Tokyo Press, Tokyo, 1969, pp. 325–339. MR0254053

[Mus12] Mircea Mustaţă, *Ordinary varieties and the comparison between multiplier ideals and test ideals II*, Proc. Amer. Math. Soc. **140** (2012), no. 3, 805–810, DOI 10.1090/S0002-9939-2011-11240-1. MR2869065

[Mus13] Mircea Mustaţă, *The non-nef locus in positive characteristic*, A celebration of algebraic geometry, Clay Math. Proc., vol. 18, Amer. Math. Soc., Providence, RI, 2013, pp. 535–551. MR3114955

[MS14] Mircea Mustaţă and Karl Schwede, *A Frobenius variant of Seshadri constants*, Math. Ann. **358** (2014), no. 3-4, 861–878, DOI 10.1007/s00208-013-0976-4. MR3175143

[MS11] Mircea Mustaţă and Vasudevan Srinivas, *Ordinary varieties and the comparison between multiplier ideals and test ideals*, Nagoya Math. J. **204** (2011), 125–157. MR2863367

[MY09] Mircea Mustaţă and Ken-Ichi Yoshida, *Test ideals vs. multiplier ideals*, Nagoya Math. J. **193** (2009), 111–128. MR2502910

[Pat13] Zs. Patakfalvi: *On subadditivity of Kodaira dimension in positive characteristic*, http://arxiv.org/abs/1308.5371 (2013).

[Pat14] Zsolt Patakfalvi, *Semi-positivity in positive characteristics* (English, with English and French summaries), Ann. Sci. Éc. Norm. Supér. (4) **47** (2014), no. 5, 991–1025. MR3294622

[PS13] Zsolt Patakfalvi and Karl Schwede, *Depth of F-singularities and base change of relative canonical sheaves*, J. Inst. Math. Jussieu **13** (2014), no. 1, 43–63, DOI 10.1017/S1474748013000066. MR3134015

[PSZ13] Zs. Patakfalvi, K. Schwede, and W. Zhang: *F-singularities in families*, http://arxiv.org/abs/1305.1646 (2013).

[Ray78a] M. Raynaud, *Contre-exemple au "vanishing theorem" en caractéristique $p > 0$* (French), C. P. Ramanujam—a tribute, Tata Inst. Fund. Res. Studies in Math., vol. 8, Springer, Berlin-New York, 1978, pp. 273–278. MR541027

[Ray78b] M. Raynaud, *Contre-exemple au "vanishing theorem" en caractéristique $p > 0$* (French), C. P. Ramanujam—a tribute, Tata Inst. Fund. Res. Studies in Math., vol. 8, Springer, Berlin-New York, 1978, pp. 273–278. MR541027

[Sai03] Natsuo Saito, *Fano threefolds with Picard number 2 in positive characteristic*, Kodai Math. J. **26** (2003), no. 2, 147–166, DOI 10.2996/kmj/1061901059. MR1993671

[SS12] Akiyoshi Sannai and Anurag K. Singh, *Galois extensions, plus closure, and maps on local cohomology*, Adv. Math. **229** (2012), no. 3, 1847–1861, DOI 10.1016/j.aim.2011.12.021. MR2871158

[SS10] Matthias Schütt and Tetsuji Shioda, *Elliptic surfaces*, Algebraic geometry in East Asia—Seoul 2008, Adv. Stud. Pure Math., vol. 60, Math. Soc. Japan, Tokyo, 2010, pp. 51–160. MR2732092

[Sch09] Karl Schwede, *F-injective singularities are Du Bois*, Amer. J. Math. **131** (2009), no. 2, 445–473, DOI 10.1353/ajm.0.0049. MR2503989

[Sch11] Karl Schwede, *Test ideals in non-$\mathbb{Q}$-Gorenstein rings*, Trans. Amer. Math. Soc. **363** (2011), no. 11, 5925–5941, DOI 10.1090/S0002-9947-2011-05297-9. MR2817415

[Sch14] Karl Schwede, *A canonical linear system associated to adjoint divisors in characteristic $p > 0$*, J. Reine Angew. Math. **696** (2014), 69–87, DOI 10.1515/crelle-2012-0087. MR3276163

[SS10] Karl Schwede and Karen E. Smith, *Globally F-regular and log Fano varieties*, Adv. Math. **224** (2010), no. 3, 863–894, DOI 10.1016/j.aim.2009.12.020. MR2628797

[ST12a] Christopher Francisco, Lee Klingler, Sean Sather-Wagstaff, and Janet C. Vassilev (eds.), *Progress in commutative algebra 2*, Walter de Gruyter GmbH & Co. KG, Berlin, 2012. Closures, finiteness and factorization. MR2920512

[ST12b] Karl Schwede and Kevin Tucker, *A survey of test ideals*, Progress in commutative algebra 2, Walter de Gruyter, Berlin, 2012, pp. 39–99. MR2932591

[ST14] Karl Schwede and Kevin Tucker, *On the behavior of test ideals under finite morphisms*, J. Algebraic Geom. **23** (2014), no. 3, 399–443, DOI 10.1090/S1056-3911-2013-00610-4. MR3205587

[SZ13] Karl Schwede and Wenliang Zhang, *Bertini theorems for F-singularities*, Proc. Lond. Math. Soc. (3) **107** (2013), no. 4, 851–874, DOI 10.1112/plms/pdt007. MR3108833

[SB91a] N. I. Shepherd-Barron, *Geography for surfaces of general type in positive characteristic*, Invent. Math. **106** (1991), no. 2, 263–274, DOI 10.1007/BF01243913. MR1128215

[SB91b] N. I. Shepherd-Barron, *Unstable vector bundles and linear systems on surfaces in characteristic p*, Invent. Math. **106** (1991), no. 2, 243–262, DOI 10.1007/BF01243912. MR1128214

[SB97] N. I. Shepherd-Barron, *Fano threefolds in positive characteristic*, Compositio Math. **105** (1997), no. 3, 237–265, DOI 10.1023/A:1000158618674. MR1440723

[Sin99a] Anurag K. Singh, **Q**-*Gorenstein splinter rings of characteristic p are F-regular*, Math. Proc. Cambridge Philos. Soc. **127** (1999), no. 2, 201–205, DOI 10.1017/S0305004199003710. MR1735920

[Sin99b] Anurag K. Singh, *Deformation of F-purity and F-regularity*, J. Pure Appl. Algebra **140** (1999), no. 2, 137–148, DOI 10.1016/S0022-4049(98)00014-0. MR1693967

[Smi00] Karen E. Smith, *The multiplier ideal is a universal test ideal*, Comm. Algebra **28** (2000), no. 12, 5915–5929, DOI 10.1080/00927870008827196. Special issue in honor of Robin Hartshorne. MR1808611

[Tak04] Shunsuke Takagi, *An interpretation of multiplier ideals via tight closure*, J. Algebraic Geom. **13** (2004), no. 2, 393–415, DOI 10.1090/S1056-3911-03-00366-7. MR2047704

[Tak13] Shunsuke Takagi, *Adjoint ideals and a correspondence between log canonicity and F-purity*, Algebra Number Theory **7** (2013), no. 4, 917–942, DOI 10.2140/ant.2013.7.917. MR3095231

[TT08] Shunsuke Takagi and Ryo Takahashi, *D-modules over rings with finite F-representation type*, Math. Res. Lett. **15** (2008), no. 3, 563–581, DOI 10.4310/MRL.2008.v15.n3.a15. MR2407232

[Tan15] Hiromu Tanaka, *The trace map of Frobenius and extending sections for threefolds*, Michigan Math. J. **64** (2015), no. 2, 227–261, DOI 10.1307/mmj/1434731922. MR3359024

[Ter99] Hiroyuki Terakawa, *The d-very ampleness on a projective surface in positive characteristic*, Pacific J. Math. **187** (1999), no. 1, 187–199, DOI 10.2140/pjm.1999.187.187. MR1674325

[Urb12] Stefano Urbinati, *Discrepancies of non-$\mathbb{Q}$-Gorenstein varieties*, Michigan Math. J. **61** (2012), no. 2, 265–277, DOI 10.1307/mmj/1339011527. MR2944480

[Xie10] Qihong Xie, *Counterexamples to the Kawamata-Viehweg vanishing on ruled surfaces in positive characteristic*, J. Algebra **324** (2010), no. 12, 3494–3506, DOI 10.1016/j.jalgebra.2010.09.009. MR2735396

[Zha14] Yuchen Zhang, *Pluri-canonical maps of varieties of maximal Albanese dimension in positive characteristic*, J. Algebra **409** (2014), 11–25, DOI 10.1016/j.jalgebra.2014.03.019. MR3198833

EPFL SB MATHGEOM CAG, MA B3 444 (Bâtiment MA), Station 8, CH-1015 Lausanne, Switzerland

E-mail address: zsolt.patakfalvi@epfl.ch

Department of Mathematics, The University of Utah, 155 S 1400 E Room 233, Salt Lake City, Utah 84112

E-mail address: schwede@math.utah.edu

Department of Mathematics, Statistics and Computer Science, University of Illinois at Chicago, 322 Science and Engineering Offices (M/C 249), 851 S. Morgan Street, Chicago, Illinois 60607-7045

E-mail address: kftucker@uic.edu

The geometry of the moduli space of curves and abelian varieties

Orsola Tommasi

ABSTRACT. Both algebraic curves and abelian varieties are basic objects in algebraic geometry and the study of their moduli spaces is a classical topic, with connections to number theory and other areas of geometry. In these expository notes, we discuss and compare the main geometric properties of moduli spaces of curves and abelian varieties, also in view of the relationships between them arising from Torelli and Prym-type maps.

1. The main players

The aim of the present paper is to review and compare the geometry of the following two moduli spaces:

(1) the moduli space $\mathcal{M}_{g,n}$ of smooth projective curves of genus g with n distinct marked points.
(2) the moduli space $\mathcal{A}_g$ of principally polarized abelian varieties.

The objects of $\mathcal{M}_{g,n}$ are $n+1$-tuples $(C, p_1, \dots, p_n)$ where C is a smooth irreducible projective variety of dimension 1 and $p_1, \dots, p_n$ are distinct points on C. If the Euler characteristic of $C \setminus \{p_1, \dots, p_n\}$ is negative, that is, if the integers $g, n \geq 0$ satisfy the condition $2g - 2 + n > 0$, then the automorphism group of any such n-pointed curve is finite and $\mathcal{M}_{g,n}$ exists as a smooth Deligne–Mumford stack of dimension $3g-3+n$. The coarse moduli space $M_{g,n}$ is a quasi-projective variety of the same dimension $3g-3+n$ with locally quotient singularities corresponding to the curves with non-trivial isomorphism group.

An abelian variety is simply a smooth projective variety A with a group structure on its set of points. In this review, we focus on the case of *principally polarized abelian varieties* (we will recall the definition in §5). This choice is motivated by the fact that most abelian varieties arising from geometrical constructions are principally polarized. This is the case, for instance, for Jacobian of curves and for intermediate Jacobians of threefolds. We will denote the moduli space of principally polarized abelian varieties of dimension g by $\mathcal{A}_g$. It is again a Deligne–Mumford stack, of dimension $g(g+1)/2$, and its coarse moduli space A_g is a quasi-projective variety with locally quotient singularities.

2010 *Mathematics Subject Classification.* Primary 14H10, 14K10, 14H40, 14-02.

Although the stacks $\mathcal{M}_{g,n}$ and $\mathcal{A}_g$ are defined over $\mathbb{Z}$ (see [**DM69**, **FC90**]) and have fascinating arithmetic properties, in this review we will restrict ourselves to the field $\mathbb{C}$ of complex numbers unless specified otherwise.

The study of the geometry of $\mathcal{M}_{g,n}$ and $\mathcal{A}_g$ is a very broad and rich subject. For this reason, this paper has no pretense of being a complete review. Its goal is rather to give something of the flavor of the techniques and kind of questions that arise. For reviews on different sets of topics on moduli spaces of curves and abelian varieties, we refer the reader to [**Oor81**, **Loo00**, **FP13**, **Har13**] for curves and [**vdGO99**, **HS02**, **Gru09**, **vdG13**] for principally polarized abelian varieties.

The first two parts of this survey paper are devoted to the discussion of the geometry of $\mathcal{M}_{g,n}$ and $\mathcal{A}_g$, respectively. For both spaces, we will focus on two geometrical problems: the choice of an appropriate compactification and the birational geometry of the space. In the case of $\mathcal{M}_{g,n}$, we present the moduli space of Deligne–Mumford stable curves as the most natural compactification. For $\mathcal{A}_g$, we introduce the Baily–Borel–Satake compactification and the toroidal compactifications constructed in [**AMRT75**]. We will discuss the geometrical features of three toroidal compactifications: the perfect cone or first Voronoi compactification $\mathcal{A}_g^{\mathrm{perf}}$, the second Voronoi compactification $\mathcal{A}_g^{\mathrm{Vor}}$ and the matroidal partial compactification $\mathcal{A}_g^{\mathrm{matr}}$. This illustrates the principle that the appropriate choice of a compactification for $\mathcal{A}_g$ depends very much on the geometric properties one is interested in. As for birational geometry, we will mainly limit ourselves to a review of the results on the Kodaira dimension of both $\mathcal{M}_g$ and $\mathcal{A}_g$ and the related concept of *slope* of divisors. To get a more complete picture, we also give an overview of the progress in understanding the birational geometry of $\mathcal{M}_{0,n}$.

In the final part of the paper, we will tackle the relationship between moduli spaces of curves and abelian varieties that arise from the classical theory of Torelli and Prym maps, and describe the most important recent advance in the study of curve models for abelian varieties, i.e. the work [**ADF⁺15**] on the Prym–Tyurin–Kanev map to $\mathcal{A}_6$.

The paper is structured as follows. The first part is devoted to the moduli space of curves. In §2 and §3 we describe its Deligne–Mumford compactification and introduce the tautological ring. The discussion of topics in the birational geometry of $\mathcal{M}_{g,n}$ is done in §4.

The second part of the paper starts in §5 with a review of the analytic construction of $\mathcal{A}_g$ as a quotient of Siegel space by the symplectic group. The Baily–Borel–Satake compactification of $\mathcal{A}_g$ and the Kodaira dimension of $\mathcal{A}_g$ are discussed in §6 and §7, respectively. In §8, we introduce the toroidal compactifications of $\mathcal{A}_g$ and in §§8.1–8.3 we outline their properties in the case of $\mathcal{A}_g^{\mathrm{perf}}$, $\mathcal{A}_g^{\mathrm{Vor}}$ and $\mathcal{A}_g^{\mathrm{matr}}$.

The final part describes how to control the geometry of $\mathcal{A}_g$ by means of the geometry of $\mathcal{M}_{g,n}$ and its covers. In §9 we review the Torelli and Prym constructions and in §10 we explain how they lead to curve models for abelian varieties of small dimension. Finally, in §11 we discuss the results from [**ADF⁺15**] for abelian 6-folds.

Acknowledgements. This survey is based on my lectures at the Graduate Students Bootcamp for the Algebraic Geometry Summer Research Institute in Salt Lake City, Utah, July 6–10, 2015. I would like to thank the organizers İzzet Coşkun, Tommaso de Fernex, Angela Gibney and Max Lieblich for the invitation. Furthermore, I would like to thank the referee for helpful remarks on a previous version of this manuscript.

2. Deligne–Mumford stable curves

With the obvious exception of $\mathcal{M}_{0,3}$, the moduli space $\mathcal{M}_{g,n}$ is never complete. However, there is a natural way to compactify it by considering moduli of *Deligne–Mumford stable n-pointed curves* [**DM69**].

A stable curve of genus g with n marked point is an $n+1$-tuple $(C, p_1, \dots, p_n)$ where

- the curve C is a connected nodal curve of arithmetic genus g
- the points $p_1, \dots, p_n$ are distinct points in the non-singular locus of C
- every irreducible component T of C satisfies the stability condition

$$\chi(T \setminus (C_{\mathrm{sing}} \cup \{p_1, \dots, p_n\})) < 0.$$

The stability condition above is equivalent to requiring that each component of geometric genus 0 contains at least 3 special points, which can be either marked points or singular points of C.

The moduli space $\overline{\mathcal{M}}_{g,n}$ is then a proper Deligne–Mumford stack of dimension $3g-3+n$, defined over $\mathbb{Z}$. Its coarse moduli space $\overline{M}_{g,n}$ is a projective variety [**Knu83**].

An important feature of $\overline{\mathcal{M}}_{g,n}$ is its stratification by topological type. Namely, with each stable curve C with n marked points we can associate a labeled graph $\Gamma = \Gamma_C$ with n half-edges, called the *dual graph* of C. The dual graph Γ encodes all combinatorial information on the genera of the components of C and the position of the singular points, as follows:

- the *vertices* $v(C')$ of Γ are in bijection with the irreducible components C' of C;
- each vertex $v(C')$ is *labeled* by the genus $g(C')$ of the corresponding components;
- the edges joining $v(C')$ and $v(C'')$ correspond to the singular points that C' and C'' have in common;
- there are n numbered half-edges (or leaves) $l_1, \dots, l_n$; the half-edge l_j starts at the vertex $v(C')$ if and only if the jth marked point lies on the component C'.

In particular, two n-marked complex stable curves C and D have the same topological type if and only if their dual graphs Γ_C and Γ_D are isomorphic. Let us observe that it is easy to reconstruct the (arithmetic) genus of C from its dual graph: it is equal to the genus of the graph plus the sum of the labels on its vertices. For instance, the stable curve in Figure 1 has genus 6, because the genus of the underlying graph is 2 and the labels on the vertices add up to 4.

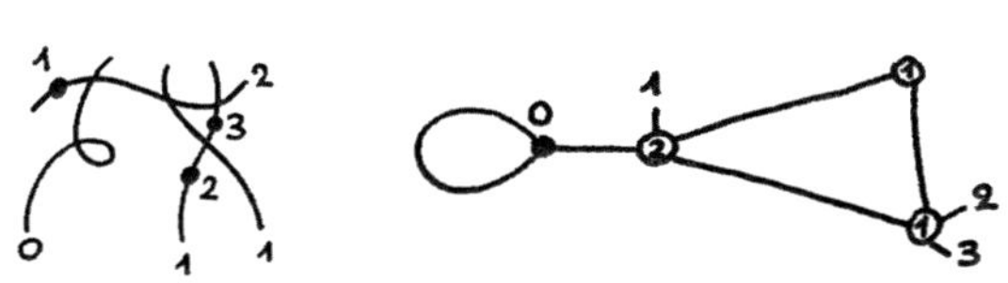

FIGURE 1. A stable curve of genus 6 with 3 marked points and the corresponding stable graph

If Γ is a stable graph of genus g with n half-edges, then the associated *gluing morphism* is

$$\mathrm{gluing}_\Gamma : \prod_{v\in V} \overline{\mathcal{M}}_{g(v),e(v)+l(v)} \to \overline{\mathcal{M}}_{g,n},$$

where for every vertex v of Γ we denoted by $e(v)$ and $l(v)$ the number of edges and the number of leaves, respectively, starting at v. Then the image of gluing_Γ is the closure in $\overline{\mathcal{M}}_{g,n}$ of the locus of curves with dual graph Γ.

For instance, for a graph Γ with two vertices joined by one edge, we get the gluing map

$$\overline{\mathcal{M}}_{g_1,n_1+1} \times \overline{\mathcal{M}}_{g_2,n_2+1} \to \overline{\mathcal{M}}_{g_1+g_2,n_1+n_2}$$

whose image is a boundary divisor of $\overline{\mathcal{M}}_{g,n}$ with $g = g_1 + g_2$ and $n = n_1 + n_2$.

In this way we can describe explicitly all *boundary components* of $\overline{\mathcal{M}}_g$ and $\overline{\mathcal{M}}_{g,n}$. In the case of $\overline{\mathcal{M}}_g$, they are

- the divisor Δ_0 whose general element is an irreducible nodal curve of geometric genus $g-1$;
- for all i, $1 \leq i \leq [\frac{g}{2}]$, the divisor Δ_i whose elements are the union of a curve of arithmetic genus i and a curve of arithmetic genus $g-i$ intersecting at a point.

The description in the case of $\overline{\mathcal{M}}_{g,n}$ with $n \geq 0$ is analogous, but requires to keep track of how the marked points are distributed on the components. Furthermore, if there are at least 2 marked points, also curves with a smooth genus 0 component and a genus g component occur as general elements of divisors in $\mathcal{M}_{g,n}$.

3. Tautological classes on the moduli space of curves

The study of the Chow (and the cohomology) ring of $\overline{\mathcal{M}}_{g,n}$ was initiated by Mumford in [**Mum83b**]. Mumford was interested in studying the rational Chow ring of $\overline{M}_{g,n}$. As $\overline{\mathcal{M}}_{g,n}$ is a stack, its coarse moduli space $\overline{M}_{g,n}$ is singular in general, so Mumford had to prove first that the singularities of $\overline{M}_{g,n}$ are sufficiently mild that it makes sense to consider the Chow ring with rational coefficients. By work of Looijenga [**Loo94**] and Boggi–Pikaart [**BP00**], it is now known that $\overline{\mathcal{M}}_{g,n}$ is the quotient stack of a smooth projective scheme by a finite group. This makes the construction of rational Chow groups and cohomology groups easier, because it reduces it to the construction of classes on the smooth cover that are invariant under the action of a finite group.

The definition of $\overline{\mathcal{M}}_{g,n}$ as a moduli space can be used to construct classes in its Chow ring. Natural examples include:

- The ψ-classes $\psi_1, \ldots, \psi_n$ obtained as the first Chern class of the cotangent line bundle at the ith marked point. For $i = 1, \ldots, n$, let us denote by s_i the section of the universal family corresponding to the ith marked point:

$$\mathcal{C}_{g,n} \underset{s_i}{\overset{\pi}{\rightleftarrows}} B$$

  then the class ψ_i is defined by $\psi_i := c_1(s_i^*\omega_\pi) \in A^1(\overline{\mathcal{M}}_{g,n})$.
- the λ-classes, defined as the Chern classes $\lambda_i = c_i(\mathbf{E})$ of the Hodge bundle $\mathbf{E} = \pi_*\omega_\pi$, whose fiber over C is $H^0(C, \omega_C)$.
- the κ-classes, also known as Mumford–Morita–Miller-classes, defined as $\kappa_j := p_*(\psi_{n+1}^{j+1}) \in A^j(\overline{\mathcal{M}}_{g,n})$ for the forgetful map $p : \overline{\mathcal{M}}_{g,n+1} \to \overline{\mathcal{M}}_{g,n}$.

These classes are not independent. For instance, Mumford showed that the λ-classes can always be expressed as linear combinations of polynomials in the κ-classes. This can be proved by applying Grothendieck–Riemann–Roch to the universal curve π and to ω_π.

The classes above make sense for every value of g and n and can be considered both in the Chow and in the cohomology ring of $\overline{\mathcal{M}}_{g,n}$. They are special case of the so-called *tautological classes.*

The Chow ring of $\overline{\mathcal{M}}_{g,n}$ becomes quickly very complicated when g and n increase, so that a complete study of it cannot be realistically expected for positive genus. Instead, for many geometric applications it is enough to know the *tautological ring*, which is generated by the tautological classes introduced above, the fundamental classes of the closure of the strata of $\overline{\mathcal{M}}_{g,n}$ in the stratification by topological type, and other classes that can be obtained from the ψ- and κ-classes by considering the natural gluing morphisms associated to any stable graph Γ of genus g with n half-edges. Let us recall that in the genus 0 case the Chow ring coincides with the tautological ring and is completely known by work of Keel [**Kee92**].

4. Birational geometry of $\mathcal{M}_{g,n}$

The key ingredients in the birational geometry of $\overline{\mathcal{M}}_{g,n}$ are the Hodge class $\lambda = \lambda_1$, the cotangent classes $\psi_1, \dots, \psi_n$ and the boundary divisors. The main underlying question is how to describe as explicitly as possible the *pseudo-effective cone* and the *nef cone.* The former is the convex cone inside $N^1(\overline{M}_{g,n}) := \mathrm{Pic}(\overline{M}_{g,n}) \otimes \mathbb{Q}/\equiv$ spanned by the effective divisors; the latter is defined analogously, but by taking the cone generated by nef divisors, i.e. by divisors with non-negative intersection with all curves in $\overline{M}_{g,n}$.

These questions have a completely different flavor depending on whether g is positive or equal to zero. For genus 0, the nice geometric properties of $\overline{\mathcal{M}}_{0,n}$ allow to reformulate birational geometry questions in a very combinatorial fashion. We will discuss this later in §4.2. When the genus g of the curves considered is positive and $n = 0$, a primary topic in the birational geometry is the study of the *slope* of divisors, which is directly connected to the Kodaira dimension of M_g. We will focus on this in §4.1. For a more complete review of results concerning the slope, we refer the reader to [**CFM13**].

As a complete discussion of the progress on the birational geometry of $\mathcal{M}_{g,n}$ is not within the scope of this review, we will not include other important topics on the birational geometry of $\mathcal{M}_{g,n}$, such as Mori theory and the Hassett–Keel log minimal Mori program. We refer the interested reader to the survey [**FS13**] and the references therein. For the most recent progress in the Hassett–Keel program, see also [**AFSvdW16**].

We would like to mention that for $g > 0$, the *contractions* of $\overline{M}_{g,n}$, i.e. the non-constant morphisms with connected fibers from $\overline{M}_{g,n}$ to projective varieties, are known. Specifically, Gibney, Keel and Morrison [**GKM02**] proved that each contraction of $\overline{M}_g$ is a birational morphism with target a compactification of M_g. This extends to the case of $\overline{M}_{g,n}$ with $g, n > 0$. There, contractions always factor through a forgetful map $\overline{M}_{g,n} \to \overline{M}_{g,n'}$ with $n' < n$.

4.1. Birational geometry of $\overline{M}_g$ with $g > 0$. In this subsection, we will assume $g \geq 2$ and (for simplicity) $n = 0$. Then every irreducible effective divisor D

not contained in the boundary of $\overline{M}_g$ can be written (up to numerical equivalence) as a linear combination of $\lambda := \lambda_1$ and the classes $\delta_0, \dots, \delta_{[\frac{g}{2}]}$ of the boundary divisors as

$$D \equiv a\lambda - \sum_{i=0}^{[\frac{g}{2}]} b_i \delta_i$$

with $a, b_i \in \mathbb{Q}_{>0}$.

DEFINITION 1. The *slope* of D is $s(D) := \frac{a}{\min_i b_i}$.

The slope of the canonical divisor is known. Namely, the class of the canonical divisor of the stack $\overline{\mathcal{M}}_g$ is known to be

$$K_{\overline{\mathcal{M}}_g} = 13\lambda - 2\sum_0^{[g/2]} \delta_i.$$

As the natural map $\overline{\mathcal{M}}_g \to \overline{M}_g$ to the coarse moduli spaces is ramified along the divisor Δ_1 of stable curves containing a subcurve of arithmetic genus 1, one has

$$K_{\overline{M}_g} = 13\lambda - 2\delta_0 - 3\delta_1 - 2\sum_{i\geq 1} \delta_i.$$

In particular, if D is an effective divisor with slope $s(D) < \frac{13}{2}$, we have

$$K_{\overline{M}_g} = D + a\lambda + \sum_{i\geq 0} b_i\delta_i \quad \text{with } a, b_i \in \mathbb{Q}_{>0},$$

i.e. $K_{\overline{M}_g}$ is the sum of the *effective* divisors D and $\sum_{i\geq 0} b_i\delta_i$ and the *big* divisor $a\lambda$. Hence $K_{\overline{M}_g}$ is ample and $\overline{M}_g$ is of general type. Let us remark that this kind of approach relies on the fact that $\overline{M}_g$ has canonical singularities for $g \geq 2$ [**HM82**], so that one can always extend the sections of $K_{\overline{M}_g}$ and its tensor powers defined over the smooth locus of $\overline{M}_g$ to desingularizations of $\overline{M}_g$.

Divisors of small slope play a fundamental role in the proof that $\overline{M}_g$ is of general type for $g \geq 24$ [**HM82, Har84, EH87**]. For instance, for curves of odd genus [**EH87**] considers the Brill–Noether divisor $\overline{M}^r_{g,d}$ in $\overline{M}_g$ of curves admitting a g^r_d with $g - r(r+1)(g+r-d) = -1$. By Brill–Noether theory, the general curve of genus g does not admit a linear series of this dimension and degree, and one can prove that $\overline{M}^r_{g,d}$ is indeed a divisor in $\overline{M}_g$. By using the degeneration technique of *limit linear series*, Harris and Mumford could prove that the slope of $\overline{M}^r_{g,d}$ is smaller than $13/2$.

However, the Brill–Noether divisors in general are not the effective divisors with the smallest possible slope. In particular, Farkas [**Far09b**] constructed an infinite sequence of divisors of slope $< 6 + \frac{12}{g+1}$ using the theory of syzygies. Moreover (see [**Far09a**]), in the case $g = 22$ the closure of the locus of curves with a linear series L of degree 25 and at least 7 linearly independent independent sections such that the natural map

$$\mathrm{Sym}^2(H^0(L)) \to H^0(L^{\otimes 2})$$

is *not* injective is a divisor D_{22} with slope $s(D_{22}) = \frac{17121}{2636} < \frac{13}{2}$. This implies that $\overline{M}_{22}$ is of general type. For $\overline{M}_{23}$, instead, it is known [**Far00**] that the Kodaira dimension is at least 2.

The approach for the case in which g is small is completely different and is based on the possibility of finding explicit geometric models for a general curve of a fixed low genus. For instance, the unirationality of $\overline{M}_g$ for $2 \leq g \leq 10$ is a classical result, whose proof is due to Severi and completed in modern mathematical language by Arbarello and Sernesi [**AS79**]. By work of several authors, most notably Chang and Ran, it is now known that $\overline{M}_g$ is unirational for $g \leq 14$ and has negative Kodaira dimension for $g = 15, 16$. The problem of determining the Kodaira dimension in the intermediate cases $17 \leq g \leq 21$ is to my knowledge still open.

4.2. Birational geometry of $\overline{M}_{0,n}$. The situation is very different for the case $g = 0$. First of all, in this case $\overline{M}_{0,n}$ is always a fine moduli space, so that one does not need to distinguish between the stack and its coarse moduli space. Moreover, the spaces $\overline{M}_{0,n}$ are smooth and can be described very explicitly as a blow-up of $\mathbb{P}^{n-3}$ using Kapranov's construction [**Kap93**]. Also the stratification by topological type is simpler to describe. Since all irreducible components of a rational stable curve are rational, a stable graph of genus 0 is simply a tree with numbered leaves, such that each vertex has valence at least 3.

For $g = 0$ the space $N^1(\overline{M}_{0,n})$ is generated by the classes of the boundary divisors [**Kee92**]. Hence, the question of describing the effective cone of $\overline{M}_{0,n}$ can be translated into the combinatorics of stable trees — and can be restated as asking which linear combination of boundary divisors are effective. By work of Keel and Vermeire [**Ver02**], it is definitely known that for $n \geq 6$ there are effective divisors that are *not* an effective sum of boundary divisors. Hence, the boundary divisors are not the only divisors generating extremal rays in the effective cone of $\overline{M}_{0,n}$.

However, it is not clear how to construct the additional generators. A breakthrough in this direction was given by Castravet and Tevelev's construction [**CT13**] of extremal divisors associated with certain combinatorial structures known as *hypertrees*. As the number of such hypertree divisors increases very rapidly with n, it was natural to expect that hypertree divisors and their pullbacks could provide a unified description of all missing extremal rays of the effective cone in $N^1(\overline{M}_{0,n})$. This conjectural expectation has been disproved by Opie [**Opi16**]. In her work, she considered a family of rigid, effective divisors in $\overline{M}_{1,n}$ constructed by Chen and Coskun in [**CC14**] and their pull-back to $\overline{M}_{0,n+2}$ under the gluing map identifying the $n+1$st and $n+2$nd marked points. Such divisors can be constructed as follows. Let $a = (a_1, \ldots, a_n) \in \mathbb{Z}^n$ be a sequence of integers with $a_1 + \cdots + a_n = 0$ and $\operatorname{ggd}(a_1, \ldots, a_n) = 1$. For any $(C, p_1, \ldots, p_{n+2}) \in \overline{M}_{0,n+2}$, let us denote by $\tilde{C}$ the curve obtained by identifying the last two marked points. Then the associated divisor D_a is the closure of the locus in $\overline{M}_{0,n+2}$ of rational n-pointed curves $(C, p_1, \ldots, p_{n+2})$ such that the linear combination $a_1[p_1] + \cdots + a_n[p_n] = 0$ vanishes in $\operatorname{Pic}^0(\tilde{C})$. In the special case $a = (n, 1, -1, \ldots, -1)$, one obtains in this way an extremal divisor which is not a hypertree divisor or the pull-back of a hypertree divisor.

The fact that the structure of $\overline{M}_{0,n}$ can be described very explicitly raises the hope that $\overline{M}_{0,n}$ shares most of the properties of a toric variety. For instance, in [**HK00**, Question 3.2] Hu and Keel pose the question of whether $\overline{M}_{0,n}$ is a Mori dream space, a class of varieties X with mild singularities (specifically, $\mathbb{Q}$-factorial) with $\operatorname{Pic}(X) \otimes \mathbb{Q} = N^1(X)$ and whose Cox ring is finitely generated. In particular, the nef cone of X should be generated by finitely many semiample divisors. (A semiample divisor is simply a divisor D such that $|mD|$ is base-point free for some $m > 0$).

Recently, Castravet and Tevelev [**CT15**] were able to answer this question in a negative sense: even in characteristic 0, the coarse moduli space $\overline{M}_{0,n}$ is *not* a Mori dream space for $n \geq 134$. Instead of working directly with $\overline{M}_{0,n}$, Castravet and Tevelev consider a different compactification of $M_{0,n}$, the Losev–Manin space $\overline{LM}_n$ introduced in [**LM00**], which is a smooth toric variety of dimension $n-3$. By a direct construction, they provide surjective morphisms from a certain projective modification of $\overline{LM}_{n+1}$ to $\overline{M}_{0,n}$ and from $\overline{M}_{0,n}$ to the blow-up of $\overline{LM}_n$ at the identity element e of the $n-3$-dimensional torus. In this way, on the one hand one obtains that if the blow-up of $\overline{LM}_n$ is a Mori dream space for some value of n, then all $\overline{M}_{0,m}$ with $m < n$ are Mori dream spaces as well. On the other hand, if $\mathrm{Bl}_e \overline{LM}_n$ is not a Mori dream space, also $\overline{M}_{0,n}$ is not a Mori dream space. Using a construction in toric geometry, the authors relate the birational geometry of the blow-up of $\overline{LM}_n$ to that of the blow-up of the weighted projective plane $\mathbb{P}(a,b,c)$ whenever the integer n is of the form $n = a+b+c+8$ with $a,b,c > 0$ coprime integers. This allows them to apply a result of Goto, Nishida and Watanabe that ensures that over a field of characteristic 0 the blow-up $\mathrm{Bl}_2 \mathbb{P}(a,b,c)$ is not a Mori dream space when a, b and c are integers of a certain form.

Hu and Keel's original motivation in studying Mori dream spaces (see [**HK00**, §3]) was to get a better understanding of the cone of effective divisors in $\overline{M}_{0,n}$. There is a long standing conjecture about this, which generalizes to $g \geq 0$ as well.

CONJECTURE 2 (F-conjecture). *A divisor D on $\overline{M}_{g,n}$ is nef if and only if $D \cdot C \geq 0$ holds for all irreducible curves C in the stratification by topological type.*

The 1-dimensional strata in the stratification by topological type are also known as *F-curves*. Basically, there are two types of F-curves. The first type of curve is parametrized by $\overline{\mathcal{M}}_{1,1}$. Let us call a stable curve of type *maximally degenerate* if its dual graph is a trivalent graph with only genus 0 vertices. Such maximally degenerate curves correspond to the dimension 0 strata in the stratification by topological type. Let us fix a stable maximally degenerate stable curve of genus $g-1$ and $n+1$ marked points and attach to it an elliptic curve E at the $n+1$ st marked point. Letting E move in $\overline{\mathcal{M}}_{1,1}$ gives a one-dimensional stratum of $\overline{M}_{g,n}$.

The other type of F-curves is parametrized by $\overline{\mathcal{M}}_{0,4}$. One starts by fixing two maximally degenerate one-pointed stable curves C_1 and C_2 of genus i and $g-1-i$, respectively. Then for every $(C, p_1, p_2, p_3, p_4) \in \overline{\mathcal{M}}_{0,4}$ one considers the curve obtained by attaching C_1 to C at p_1 and C_2 to C at p_2, and identifying p_3 and p_4 to obtain a nodal curve. One can also start by choosing a maximally degenerate two-pointed stable curve (C', q_1, q_2). Then one attaches C to C' by identifying p_i with q_i for $i = 1, 2$ and p_3 with p_4 to create a node.

The F-conjecture is still open. However, to settle it, it would be sufficient to prove it in the case of genus 0, where so many combinatorial tools are available:

THEOREM 3 ([**GKM02**]). *If the F-conjecture holds for $\overline{M}_{0,n}$ for* all $n \geq 3$, *then it holds for $\overline{M}_{g,n}$ for all values of g and n.*

5. Construction of $\mathcal{A}_g$

Abelian varieties are smooth projective varieties A with a group structure (defined by regular morphisms) on their set of points. In moduli theory, one always fixes an ample line bundle L on A, i.e. a *polarization.* This rigidifies the moduli problem, because the automorphism group of the pair (A, L) is always finite, even when A has a positive-dimensional automorphism group. In this review we will concentrate on the case of principal polarizations.

DEFINITION 4. A *principally polarized abelian variety* (for short, ppav) is a pair (A, L) where A is an abelian variety and L is a *principal polarization*, i.e. an ample line bundle on A with one-dimensional space of sections, considered up to translation on A.

Using complex analysis, it is easy to construct a parameter space for principally polarized abelian varieties, at least, if one works over the complex numbers. An attractive feature of this classical construction is that it works in the same way for all values of g.

The starting point is that an abelian variety over $\mathbb{C}$ is the same as a complex torus together with a polarization. In particular, every g-dimensional abelian variety is the quotient of $\mathbb{C}^g$ by a lattice and without loss of generality one can restrict to lattices of the form

$$\Lambda_\tau := \mathbb{Z}^g + \mathbb{Z}\tau_1 + \cdots + \mathbb{Z}\tau_g$$

with $\tau_1, \ldots, \tau_g \in \mathbb{C}^g$. By Riemann's bilinear relations, to obtain an algebraic torus the $g \times g$ matrix $\tau = (\tau_1, \ldots, \tau_g)$ must be an element of *Siegel space*

$$\mathbb{H}_g := \{\text{complex symmetric } g \times g\text{-matrices with positive definite imaginary part}\}.$$

For matrices $\tau \in \mathbb{H}_g$ the theta function

$$\theta(\tau, z) = \sum_{m \in \mathbb{Z}^g} \exp(\pi i m^t \tau m + 2\pi i n^t z)$$

is well-defined. Its zero locus $\Theta_\tau \subset A_\tau := \mathbb{C}^g/\Lambda_\tau$ is the divisor defining a principal polarization on A_τ. Two principally polarized abelian varieties A_τ and $A_{\tau'}$ with $\tau, \tau' \in \mathbb{H}_g$ are isomorphic if and only if

$$\tau' = (A\tau + B)(C\tau + D)^{-1}$$

for a symplectic matrix $M = \begin{pmatrix} A & B \\ C & D \end{pmatrix}$, i.e. an element of the symplectic group

$$\mathrm{Sp}(2g, \mathbb{Z}) := \{M \in \mathrm{Mat}(2g \times 2g, \mathbb{Z}) | MJM^t = J\},$$

where $J = \begin{pmatrix} 0 & I_g \\ -I_g & 0 \end{pmatrix}$ is the matrix defining the standard symplectic form on $\mathbb{Z}^{2g}$.

One can use this construction to identify $\mathcal{A}_g$ with the quotient $\mathrm{Sp}(2g, \mathbb{Z})\backslash\mathbb{H}_g$. This description shows that $\mathcal{A}_g$ is intimately related to the space of Siegel modular forms, i.e. of holomorphic maps on $\mathbb{H}_g$ that behave well with respect to the $\mathrm{Sp}(2g, \mathbb{Z})$-action.

There is also a natural construction that yields a finite cover of $\mathcal{A}_g$ which is a smooth scheme. Namely, one can rigidify the moduli problem by looking at moduli of ppav with a level n structure.

DEFINITION 5. A (principal) level n structure on an abelian variety A is an isomorphism $\alpha : (\mathbb{Z}/g\mathbb{Z})^{2g} \to A[n]$, where $A[n]$ denotes the subgroup of A consisting of the points whose order divides n.

Then one can realize the moduli space of ppav with a level n structure as the quotient

$$\mathcal{A}_{g,n} = \Gamma_g(n)\backslash\mathbb{H}_g,$$

of Siegel space by the *level n subgroup* $\Gamma_g(n) := \{M \in \mathrm{Sp}(2g,\mathbb{Z}) | M \equiv I_{2g} \mod n\}$.

Serre [**Ser62**] proved that a ppav A with a level n structure α has no non-trivial automorphisms if $n \geq 3$. In particular, the natural map $\mathcal{A}_{g,n} \to \mathcal{A}_g$ realizes $\mathcal{A}_g$ as the quotient of the smooth scheme $\mathcal{A}_{g,n}$ by the action of a finite group.

6. The Hodge classes and the Baily–Borel–Satake compactification

The existence of a universal family

$$\mathcal{X}_g \underset{s}{\overset{\pi}{\rightleftarrows}} \mathcal{A}_g$$

over the stack $\mathcal{A}_g$ allows to define a natural vector bundle, called the *Hodge bundle*

$$\mathbb{E} := \pi_*(\Omega^1_{\mathcal{X}_g/\mathcal{A}_g}) = s^*(\Omega^1_{\mathcal{X}_g/\mathcal{A}_g}).$$

The Chern classes $\lambda_i := c_i(\mathbb{E}) \in A^i(\mathcal{A}_g)$ are called the Hodge classes. They generate a subring of $A^\bullet(\mathcal{A}_g)$ known as the tautological subring. This is very well understood (see [**vdG99**]).

First, the relations between λ-classes are given by

$$(1 + \lambda_1 + \cdots + \lambda_g)(1 - \lambda_1 + \cdots + (-1)^g\lambda_g) = 1,$$

and

$$\lambda_g = 0.$$

Both relations can be proved using Grothendieck–Riemann–Roch, for the normalized universal theta divisor on $\mathcal{X}_g$ in the former case and for the structure sheaf of $\mathcal{X}_g$ in the latter case, respectively.

Furthermore, the class λ_1 — or, equivalently, the determinant of the Hodge bundle $\mathbb{E}$ — is ample on $\mathcal{A}_g$. Over $\mathbb{C}$ one can use the description of $\mathcal{A}_g$ as the quotient $\mathrm{Sp}(2g,\mathbb{Z})\backslash\mathbb{H}_g$ to see that the pull-back of the sections of $(\det\mathbb{E})^k$ to $\mathbb{H}_g$ give weight k Siegel modular forms, i.e. functions $f : \mathbb{H}_g \to \mathbb{C}$ satisfying

$$f\left((A\tau + B)(C\tau + D)^{-1}\right) = \det(C\tau + D)^k f(\tau)$$

for all symplectic matrices $\begin{pmatrix} A & B \\ C & D \end{pmatrix} \in \mathrm{Sp}(2g,\mathbb{Z})$. Baily and Borel proved in [**BB66**] that modular forms of sufficiently high degree define an embedding of $\mathcal{A}_g$ into projective space. Its image is the minimal compactification of $\mathcal{A}_g$, known as the *Baily–Borel–Satake compactification* $\mathcal{A}_g^{\mathrm{S}}$. The same compactification can be constructed by considering the structure of $\mathcal{A}_g$ as a locally symmetric space, and indeed it is the first case in which a Baily–Borel compactification was constructed.

7. Kodaira dimension of $\mathcal{A}_g$ and Mumford's partial compactification

Set-theoretically, we can stratify $\mathcal{A}_g^{\mathrm{S}}$ as the disjoint union of moduli spaces of ppav of dimension $\leq g$:

$$\mathcal{A}_g^{\mathrm{S}} = \mathcal{A}_g \sqcup \mathcal{A}_{g-1} \sqcup \cdots \sqcup \mathcal{A}_0.$$

This stratification is directly related to the structure of the Borel subgroups of the symplectic group, but also has an algebro-geometric counterpart in the semistable reduction theorem [**FC90**, I.2.6], asserting that 1-dimensional families of abelian varieties can always be completed (up to finite base change) to a family of *semiabelian varieties*, i.e. smooth group schemes obtained as extensions of the form

$$0 \to (\mathbb{G}_m)^r \to S \to A \to 0$$

where $\mathbb{G}_m$ denotes the multiplicative group and A is an abelian variety of dimension $g - r$. The integer r is called the *torus rank* of S.

In particular, if we consider a curve $B \subset \mathcal{A}_g$, the intersection of $\overline{B} \subset \mathcal{A}_g^{\mathrm{S}}$ with the boundary of $\mathcal{A}_g^{\mathrm{S}}$ is determined by its abelian part A. However, all information about the extension disappears in $\mathcal{A}_g^{\mathrm{S}}$, so that $\mathcal{A}_g^{\mathrm{S}}$ is not a moduli space of degenerations of abelian varieties. For instance, if one wants to encode the geometry of rank 1 degenerations of abelian varieties, the additional data needed is a description of the line bundle over the $g-1$-dimensional abelian variety A obtained by partially compactifying S. As line bundles over A are the points of the dual abelian variety $\hat{A}$, which is canonically isomorphic to A by the principal polarization, rank 1 semiabelian varieties of dimension g are parametrized by a finite quotient of the universal abelian variety $\mathcal{X}_{g-1}$ over $\mathcal{A}_{g-1}$, which has indeed dimension $g(g+1)/2-1$. More specifically, a line bundle L and its inverse L^{-1} define the same semiabelian variety, so that rank 1 degenerations are parametrized by $\mathcal{X}_{g-1}/\pm 1$. This suggests that the boundary of a modular compactification of $\mathcal{A}_g$ should have codimension 1. However, the boundary of $\mathcal{A}_g^{\mathrm{S}}$ has codimension g. Furthermore, although $\mathcal{A}_g^{\mathrm{S}}$ is known to be normal, its boundary is highly singular.

The partial compactification $\mathcal{A}'_g$ obtained by adding the divisor $\mathcal{X}_{g-1}/\pm 1$ to $\mathcal{A}_g$ is studied in [**Mum83a**]. There, it is used to gain information on the Kodaira dimension of $\mathcal{A}_g$. The strategy is similar to the one outlined in the case of the moduli space of curves. The rational Picard group of the coarse moduli space A'_g of $\mathcal{A}'_g$ is generated by $\lambda := \lambda_1$ and the class δ of the boundary. This allows to define the *slope* of a divisor $a\lambda - b\delta$ in A'_g as the ratio a/b. One can compute directly that the class of the canonical divisor is $(g+1)\lambda - \delta$.

Hence, if one can find an effective divisor of the form $a\lambda - b\delta$ with $a/b > g+1$, one automatically obtains that A'_g, and hence A_g, is of general type. In [**Mum83a**], Mumford considers the Andreotti–Mayer locus N_0, i.e. the closure in A'_g of the locus of abelian varieties whose Θ-divisor is singular, and proves that its slope is $6 + 12/(g+1)$. This implies that A_g is of general type for $g \geq 7$. This refines previous results by Tai in [**Tai82**], who had already proved that A_g was of general type for $g \geq 9$ by studying explicitly the spaces of sections of $K_{A_g}^{\otimes N}$.

8. Toroidal compactifications of $\mathcal{A}_g$

Ideally one would like to be able to construct a compactification of $\mathcal{A}_g$ which has properties similar to those of $\overline{\mathcal{M}}_{g,n}$: a compactification with a modular interpretation and which is smooth as an algebraic stack. In pursuit of a compactification with better properties than $\mathcal{A}_g^{\mathrm{S}}$, Ash, Mumford, Rapoport and Tai [**AMRT75**] introduced the toroidal compactifications of $\mathcal{A}_g$. The intuitive idea behind this is that degenerations of elements τ of Siegel space have imaginary part which is *positive semidefinite*. Indeed, the combinatorial gadget needed to define a toroidal compactification of $\mathcal{A}_g$ is a decomposition into convex polyhedral cones of the space $\mathrm{Sym}^2_{\geq 0}(\mathbb{R}^g)$ of positive semidefinite quadratic forms with rational radical. This cone decomposition Σ should satisfy certain admissibility properties. In particular, the cone decomposition Σ should define a stratification of $\mathrm{Sym}^2_{\geq 0}(\mathbb{R}^g)$ into open convex polyhedral cones, it should be $\mathrm{GL}(g,\mathbb{Z})$-invariant and contain only finitely many $\mathrm{GL}(g,\mathbb{Z})$-isomorphism classes of cones. Then, toroidal geometry provides a recipe to construct a toroidal compactification $\mathcal{A}_g^{\Sigma}$ of $\mathcal{A}_g$. This compactification is not necessarily a moduli space; however, it possesses a universal family of semiabelian varieties.

The compactification $\mathcal{A}_g^{\Sigma}$ has a natural stratification. Namely, it can be written as the disjoint union of locally closed strata $\beta(\sigma)$ corresponding to the $\mathrm{GL}(g,\mathbb{Z})$-orbits of cones $\sigma \in \Sigma$. If we denote the maximal rank of a quadratic form in Σ and the real dimension of the cone σ by r and d, respectively, then $\beta(\sigma)$ has complex codimension d in $\mathcal{A}_g^{\Sigma}$ is the finite quotient of a torus bundle over the fiber product over $\mathcal{A}_g$ of r copies of $\mathcal{X}_{g-r}$. For instance, rank 0 degenerations are just abelian varieties of dimension g and we have $\beta(0) = \mathcal{A}_g$. As already described, rank 1 degenerations give a divisor in $\mathcal{A}_g^{\Sigma}$, namely $\beta(\mathbb{R}_{>0}x_1^2) \cong \mathcal{X}_{g-1}/\pm 1$. The situation becomes more complicated for rank 2 degenerations. In this case, up to $\mathrm{GL}(2,\mathbb{Z})$-equivalence there are two cones that one needs to consider, namely the two-dimensional cone $\sigma_{1+1} := \mathbb{R}_{>0}x_1^2 + \mathbb{R}_{>0}x_2^2$ and the three-dimensional cone $\sigma_{C_3} := \mathbb{R}_{>0}x_1^2 + \mathbb{R}_{>0}x_2^2 + \mathbb{R}_{>0}(x_1 - x_2)^2$. They correspond to different types of degenerations of abelian varieties.

Among the explicitly known families of admissible cone decompositions (see e.g. [**Nam80**]), the most useful for geometric applications are the *perfect cone decomposition* Σ^{perf} (also known as *first Voronoi decomposition*), the *second Voronoi decomposition* Σ^{Vor} and the *central cone decomposition* Σ^{centr}. They give rise to the perfect cone compactification $\mathcal{A}_g^{\mathrm{perf}}$, the second Voronoi compactification $\mathcal{A}_g^{\mathrm{Vor}}$ and the central cone compactification $\mathcal{A}_g^{\mathrm{centr}}$, respectively. Each of these compactifications can be seen as the "most natural one" from a different geometrical point of view. For instance, Igusa [**Igu67**] proved that the central cone compactification can be characterized as the normalization of the blow-up of $\mathcal{A}_g^{\mathrm{S}}$ along the boundary. The other two compactifications will be considered in the next sections.

It is important to remark that all of these compactifications are singular if g is sufficiently large. Their existence can be detected using toric geometry. Let us recall that a convex cone $\sigma \subset \mathrm{Sym}^2(\mathbb{R}^g)$ is called *basic* if the extremal rays of σ are a $\mathbb{Z}$-basis of $\mathrm{Sym}^2(\mathbb{Z}^g)$. If the extremal rays of σ form a $\mathbb{Q}$-basis of $\mathrm{Sym}^2(\mathbb{Q}^g)$, the cone is called *simplicial*. In toric geometry, non-basic cones give rise to singular toric varieties and simplicial cones correspond to the existence of locally quotient singularities. This description of the singularities extends to the toroidal compactifications $\mathcal{A}_g^{\Sigma}$. For g sufficiently large, all cone decompositions mentioned above contain non-simplicial cones. Therefore, they give rise to algebraic stacks $\mathcal{A}_g^{\Sigma}$ with singularities which are not locally quotient ones. However, all cone decompositions above can be refined to obtain smooth toroidal compactifications. The existence of these smooth toroidal compactification is very important from a theoretical point of view and is one of the fundamental ingredients of [**FC90**].

8.1. Perfect cone compactification. The perfect cone decomposition Σ^{perf} is easy to define. Let $Q \in \mathrm{Sym}^2_{>0}(\mathbb{R}^g)$ be a positive definite quadratic form. Let us denote by $M(Q)$ the finite and non-empty set of non-trivial elements of the lattice $\mathbb{Z}^g$ on which Q attains its minimal value:

$$M(Q) = \left\{ \xi_0 \in \mathbb{Z}^g |\ Q(\xi_0) = \min_{\xi \in \mathbb{Z}^g \setminus \{0\}} Q(\xi) \right\}.$$

Then Σ^{perf} is the set of all cones of the form

$$\sigma(Q) := \sum_{\xi_0 \in M(Q)} \mathbb{R}_{>0}\xi_0^2$$

with $Q \in \Sigma_{>0}(\mathbb{R}^g)$. In particular, cones in Σ^{perf} are generated by quadratic forms of rank 1, so that up to the $\mathrm{GL}(g,\mathbb{Z})$-action there is only one 1-dimensional cone in Σ^{perf}, namely, the one spanned by x_1^2. As a consequence, the boundary of $\mathcal{A}_g^{\mathrm{perf}}$ is irreducible. This makes the birational geometry of its coarse moduli space A_g^{perf} a natural object of study, because the rational Picard group of A_g^{perf} then coincides with that of the partial compactification $\mathcal{A}'_g$, generated by λ and δ.

In **[SB06]**, Shepherd–Barron uses this property to prove that A_g^{perf} is the canonical model of $\mathcal{A}_g$ for $g \geq 12$. Let us recall that a canonical model of a quasi-projective variety Y is a normal complex projective variety birational to Y, with canonical singularities and ample canonical class. Shepherd–Barron's result relies on first characterizing the cone of ample divisors on A_g^{perf} and then proving that A_g^{perf} has canonical singularities for $g \geq 5$. Unfortunately, the proof that A_g^{perf} has canonical singularities has a gap; the interested reader can now refer to the preprint **[ASB16]** for how to complete the proof.

8.2. Second Voronoi compactification. The second Voronoi decomposition is combinatorially more involved than the perfect cone decomposition. Its definition relies on the fact that each positive semidefinite form Q has an associated Deloné decomposition, which subdivides the points in $\mathbb{R}^g$ according to which lattice point is nearest according to the metric associated to Q.

More precisely, let $\alpha \in \mathbb{R}^g \setminus \{0\}$ be arbitrary and let us denote by $m_Q(\alpha)$ the minimal value of $Q(\xi - \alpha)$ for $\xi \in \mathbb{Z}^g \setminus \{0\}$. Then a subset of $\mathbb{R}^g$ is called a *Deloné cell* if it is the convex element of the set of elements $a \in \mathbb{Z}^g \setminus \{0\}$ such that $Q(a - \alpha) = m_Q(\alpha)$ holds. The Deloné cells form a polyhedral subdivision of $\mathbb{R}^g$, invariant under translation by elements of the lattice $\mathbb{Z}^g$: this is called the *Deloné decomposition* corresponding to Q. If D is a fixed Deloné decomposition, then the set of all positive semidefinite quadratic forms with Deloné decomposition equal to D forms a convex polyhedral cone σ_D. Then one can define the second Voronoi decomposition as

$$\Sigma^{\mathrm{Vor}} = \{\sigma_D |\ D \text{ is a Deloné decomposition}\}.$$

The reason why the second Voronoi compactification $\mathcal{A}_g^{\mathrm{Vor}}$ is important geometrically is that it is the only compactification of $\mathcal{A}_g$ for which an interpretation as a moduli space is known. Namely, by work of Alexeev **[Ale02]**, it can be described as (the normalization of) a component of the moduli space of semiabelic pairs. A more precise description of this component using the language of logarithmic geometry was given by Olsson in **[Ols08]**.

8.3. Matroidal cones. As mentioned before, the combinatorial structure of $\mathcal{A}_g^{\mathrm{perf}}$ and $\mathcal{A}_g^{\mathrm{Vor}}$ is very complicated. A complete classification for $\mathrm{GL}(g,\mathbb{Z})$-orbits of cones in Σ^{Vor} is known only for $g \leq 5$; for $g = 6$ the number of orbits is known to be larger than 250,000 (see **[Val03**, Ch. 4]). For perfect cones the situation is better and a classification is known for $g \leq 7$ (see **[EVGS13]** and references therein). Hence, for none of these compactifications the combinatorics can be expressed as explicitly as the combinatorics of stable graphs in the case of moduli of curves.

Nevertheless, there is a partial compactification which is particularly nice from this point of view.

A $g \times n$ matrix with integer coefficients is called *totally unimodular* if each every square submatrix has determinant -1, 0 or 1. A cone in $\mathrm{Sym}^2_{\geq 0}(\mathbb{R}^g)$ is called *matroidal* if up to the $\mathrm{GL}(g, \mathbb{Z})$-action it is spanned by the rank 1 forms given by the columns of a totally unimodular matrix. The set of all matroidal cones gives rise to a polyhedral cone decomposition Σ^{matr} whose union is strictly contained in the set of positive semidefinite quadratic forms with rational radical. As such, it gives rise to a partial compactification $\mathcal{A}_g^{\mathrm{matr}}$ of $\mathcal{A}_g$. Furthermore, matroidal cones and totally unimodular matrices can be reinterpreted in combinatorial terms as regular matroids [**Oxl11**]. One can even prove that matroidal cones are always simplicial, so that the stack $\mathcal{A}_g^{\mathrm{matr}}$ has only locally quotient singularities.

Matroidal cones can also be characterized in a completely different way. Indeed, building on Alexeev and Brunyate's paper [**AB11**], Melo and Viviani [**MV12**] proved that Σ^{matr} equals the intersection of Σ^{perf} and Σ^{Vor}. Geometrically, this means that $\mathcal{A}_g^{\mathrm{matr}}$ is the largest open subset that can be naturally embedded in both the perfect cone and the second Voronoi compactification.

9. Torelli and Prym map

So far, we have seen an overview of the geometry of $\mathcal{A}_g$ and $\mathcal{M}_{g,n}$, with special attention to their birational geometry and the choice of a compactification. Of the two spaces, the construction of $\mathcal{A}_g$ over the complex numbers was particularly explicit and had the advantage of providing a direct relationship between forms and functions on $\mathcal{A}_g$ and modular forms for the symplectic group. Other constructions, like the choice of a compactification, are more straightforward in the case of the moduli space of curves. On the other hand, the study of the birational geometry of $\mathcal{A}_g$ and $\mathcal{M}_{g,n}$ uses techniques that are similar. In the case of $\mathcal{M}_{g,n}$, one profits from the wealth of explicit constructions coming from the geometry of curves. In the case of $\mathcal{A}_g$, the relation with Siegel space and the symplectic group can be useful in allowing an analytic approach to the problem.

However, it is important to keep in mind how intimately related the two spaces are. First and foremost, for every non-singular curve C the degree zero Picard group $JC := \mathrm{Pic}^0(C)$ is an abelian variety, with a natural choice of principal polarization given by taking the theta divisor Θ_C, defined as the image of $\mathrm{Sym}^{g-1} C \subset \mathrm{Pic}^{g-1}(C)$ under any translation map $\mathrm{Pic}^{g-1}(C) \xrightarrow{\cong} \mathrm{Pic}^0(C)$. This gives rise to the classical *Torelli map*

$$\begin{array}{ccc} \mathcal{M}_g & \longrightarrow & \mathcal{A}_g \\ C & \longmapsto & (JC, \Theta_C). \end{array}$$

By the Torelli theorem (see e.g. [**And58**]), this map is injective on coarse moduli spaces, in the sense that the Jacobian determines the curve C up to isomorphism. If one takes the stack structure into account, one notices that the general abelian variety has an automorphism $x \leftrightarrow -x$ of order 2, whereas the automorphism group of a general curve of genus 3 is trivial. So the Torelli map is a degree 2 map onto its image for $g \geq 3$. In genus 2 all curves have an involution, namely, the hyperelliptic involution, and the Torelli map is injective as a map of stack.

Of course, one would like to be able to extend the Torelli map to a map from the Deligne–Mumford compactification $\overline{\mathcal{M}}_g$ to some toroidal compactification of $\mathcal{A}_g$. Indeed, it has been known for a long time that the Torelli map extends to a map $\overline{\mathcal{M}}_g \to \mathcal{A}_g^{\mathrm{Vor}}$ [**Nam76a, Nam76b**], which is known to be compatible with the modular interpretation of the two moduli spaces [**Ale04**]. The combinatorial conditions necessary to ensure that the Torelli map extends to a given stable curve were studied by Alexeev and Brunyate in [**AB11**]. Using their more refined criterion, they could prove that the Torelli map does not extend to a regular map to $\mathcal{A}_g^{\mathrm{centr}}$ for $g \geq 9$ and that the image of $\mathcal{M}_g \to \mathcal{A}_g^{\mathrm{Vor}}$ is contained in a common open subset of $\mathcal{A}_g^{\mathrm{Vor}}$ and $\mathcal{A}_g^{\mathrm{perf}}$, i.e. to the matroidal locus, so that the extended Torelli map can also be viewed as a map $\overline{\mathcal{M}}_g \to \mathcal{A}_g^{\mathrm{Vor}}$.

The relationship between curves and abelian varieties can be pushed further by working with Prym varieties of covers of curves. Here, the classical case is that of étale double covers.

DEFINITION 6. An *admissible double cover* is a finite surjective morphism $\pi : \tilde{C} \to C$ such that

- C is a stable curve;
- the genus of $\tilde{C}$ satisfies $p_a(\tilde{C}) = 2p_a(C) - 1$;
- the restriction $\pi : \pi^{-1}(D) \to D$ to the preimage of any irreducible component $D \subset C$ has degree 2;
- the fixed points of the involution $i : \tilde{C} \to \tilde{C}$ given by sheet interchange are nodes of $\tilde{C}$, and i does not interchange the two branches of a node.

The moduli stack $\overline{\mathcal{R}}_g$ of admissible double covers of a genus g curve is a proper Deligne–Mumford stack. The natural map $\overline{\mathcal{R}}_g \to \overline{\mathcal{M}}_g$ that realizes it as a finite cover of degree $2^{2g} - 1$ of the moduli space of curves, branched over the divisor $\Delta_0 \subset \overline{\mathcal{M}}_g$ whose general element is an irreducible nodal curve. The open substack in which the curve C is smooth is denoted by $\mathcal{R}_g$.

A smooth admissible double cover $\pi : \tilde{C} \to C$ with $p_a(C) = g + 1$ has an associated principally polarized abelian variety of dimension g, known as the *Prym variety* $\mathrm{Prym}(\tilde{C}/C)$ of π. Following Mumford's work [**Mum74**], one can construct it by looking at the norm map

$$\begin{array}{rccc} \mathrm{Nm}: & J(\tilde{C}) & \longrightarrow & J(C) \\ & [D] & \longmapsto & [\pi_*(D)] \end{array}$$

and setting $\mathrm{Prym}(\tilde{C}/C)$ to be the connected component of the identity of the kernel of Nm. After translation, the Prym variety can also be thought of as a component of the preimage of the canonical divisor K_C under the norm map $\mathrm{Pic}^{2g-2}(\tilde{C}) \to \mathrm{Pic}^{2g-2}(C)$, namely the component $\mathrm{Nm}^{-1}(K_C)^{\mathrm{even}}$ corresponding to line bundles whose space of sections has even dimension. Then the locus

$$\Xi_C = \{L \in \mathrm{Nm}^{-1}(K_C)^{\mathrm{even}} |\ h^0(\tilde{C}, L) \geq 2\}$$

is the divisor defining the desired principal polarization. In this way one obtains the *Prym map* $\mathcal{R}_{g+1} \to \mathcal{A}_g$.

Let us observe that this construction makes sense also when C is a stable curve. In this case, one has to consider JC and $J\tilde{C}$ as the connected component of the identity in the Picard group. If the curve C only has separating nodes, i.e. if the dual graph of C is a tree, then JC is an abelian variety, but in general JC is just a semiabelian variety, namely the extension of the Jacobian of its normalization by $H^0(\Gamma(C), \mathbb{Z}) \otimes_{\mathbb{Z}} \mathbb{C}^*$. The construction of the norm map carries on in this settings and one can define $P(\tilde{C}/C)$ exactly as in the smooth case.

Analogously to the case of the Torelli map, one would like to be able to extend the Prym map to a map $\overline{\mathcal{R}}_{g+1} \to \mathcal{A}_g^\Sigma$ for some toroidal compactification $\mathcal{A}_g^\Sigma$ of $\mathcal{A}_g$. However, it turns out that the Prym map cannot be extended to any of the best known toroidal compactifications. The conditions for the existence of a regular extension of the Prym map to an admissible double cover $\tilde{C} \to C$ were studied in [**ABH02**] for the second Voronoi compactification and in [**CMLGH15**] for the perfect cone and the central cone decomposition.

10. Curve models of abelian varieties of dimension $g \leq 5$

The moduli space of principally polarized abelian varieties is in most instances very well behaved. Nevertheless, the richness of the geometry of the moduli space of curves makes it easier to study. The best situation occurs when there is a dominant map from a moduli space of curves to $\mathcal{A}_g$. Then one can combine the geometry of curves and abelian varieties to obtain very useful additional information.

The classical example of this is the Torelli map. In view of the Torelli theorem, the Torelli map is birational on the coarse moduli spaces whenever $\dim \mathcal{M}_g = \dim \mathcal{A}_g$ holds, that is, for $g \leq 3$. In particular, one can deduce the rationality of A_2 and A_3 directly from the rationality of M_2 and M_3. However, for $g \geq 4$ the general ppav of dimension g is no longer a Jacobian.

The situation for the Prym map is more complicated. For instance, the Prym map is known to be *generically injective* on the coarse moduli spaces for $g \geq 7$ by [**FS82**]. However, it is shown in [**Don92**] that the Prym map is never injective. For instance, its restriction to the locus of hyperelliptic curves has positive-dimensional fibers.

If one looks at the dimension of the spaces, one has $\dim(\mathcal{R}_{g+1}) \geq \dim(\mathcal{A}_g)$ for $g \leq 5$. Therefore, one expects it to be a dominant map in this range, and indeed a general ppav of dimension $g \leq 5$ is a Prym variety, as already proved by Wirtinger in [**Wir95**]. For instance, this implies that in this range the coarse moduli space R_{g+1} is unirational and this can be used to prove that A_g is unirational as well. This approach works for R_6 and A_5 [**Don84**] as well as for R_5 and A_4 [**Cle83, ILGS09**]. As we already explained, A_g is of general type for $g \geq 7$. The birational type of A_6 is an open problem.

11. Curve models of abelian 6-folds

The locus of Prym varieties has codimension 3 in $\mathcal{A}_6$, so that the Prym map cannot be used to control the geometry of $\mathcal{A}_6$ and answer questions about its birational type. However, there is now a construction available that allows to describe the general abelian 6-fold in terms of curve geometry. Namely, in [**ADF$^+$15**], it is proved that the generalized Prym–Tyurin–Kanev map from the space of E_6-covers of the projective line to $\mathcal{A}_6$ is dominant. The E_6-covers are a special class of degree 27 covers of $\mathbb{P}^1$ with monodromy group equal to the Weyl group of the E_6 lattice and 24 branch points. In particular, the Hurwitz space parametrizing them has dimension equal to $\dim \overline{\mathcal{M}}_{0,24} = 21$. Special examples of E_6-covers arise quite naturally if one considers a general pencil of cubic surfaces. Specifically, let $X \subset \mathbb{P}^4$ be a smooth cubic threefold and $\{S_t |\, t \in \mathbb{P}^1\}$ a general pencil of hyperplane sections of X. Such a pencil contains 24 nodal surfaces. Then

$$C = \{(L,t) \in \mathbb{G}(1,\mathbb{P}^4) \times \mathbb{P}^1 |\, L \subset S_t\} \longrightarrow \mathbb{P}^1$$

is a degree 27 cover ramified at 24 points. From the ramification pattern one deduces that C is a curve of genus 46. However, the expected dimension of the space of pencil of cubic surfaces is $16 < 21$, so that it is clear that the general E_6-cover does not come from a pencil of cubics.

Nevertheless, most features of pencil of cubics extend to general E_6-covers. Let us recall that $W(E_6)$ is the monodromy group for family of lines on smooth cubic surfaces. Hence, requiring that the monodromy group of the cover $C \to \mathbb{P}^1$ is $W(E_6)$ implies that one can identify consistently the fiber of $C \to \mathbb{P}^1$ with the 27 lines on a cubic surface. This allows to define a correspondence $D \subset C \times C$ by

$$D = \{((L_1, t), (L_2, t)) \in C \times C | \text{ the lines } L_1 \text{ and } L_2 \text{ are incident to each other}\}.$$

The correspondence D defines a map $JC \to JC$, simply by mapping the class of a point $(L, t) \in C$ to the sum of the classes of all $(M, t) \in C$ with M incident to L. Then Kanev [**Kan87**] proved that the locus in JC where D acts as multiplication by -5 is a ppav of dimension 6. The main result of [**ADF$^+$15**] is that the map from the Hurwitz scheme parametrizing E_6-coves to A_6 defined by this construction is dominant. As this Hurwitz scheme is a finite cover of $\overline{\mathcal{M}}_{0,24}/S_{24}$, it has dimension 21, hence the Prym–Tyurin–Kanev map is also generically injective.

The idea of the proof is again based on techniques from toroidal geometry. Namely, one can consider the extension of the Prym–Tyurin–Kanev map to E_6-covers $C \to R$ obtained by letting the pair $(\mathbb{P}^1, B)$ consisting of $\mathbb{P}^1$ and the branch locus B of the cover degenerate to a maximally degenerate 24-pointed stable rational curve (R, B') in $\overline{\mathcal{M}}_{0,24}/S_{24}$. Locally, the extension of the Prym–Tyurin–Kanev map to such a degenerate cover gives rise to a toroidal map. In particular, this allows to compute explicitly the rank of the map by analyzing the appropriate fan.

References

[AB11] Valery Alexeev and Adrian Brunyate, *Extending the Torelli map to toroidal compactifications of Siegel space*, Invent. Math. **188** (2012), no. 1, 175–196, DOI 10.1007/s00222-011-0347-2. MR2897696

[ABH02] V. Alexeev, Ch. Birkenhake, and K. Hulek, *Degenerations of Prym varieties*, J. Reine Angew. Math. **553** (2002), 73–116, DOI 10.1515/crll.2002.103. MR1944808

[ADF$^+$15] V. Alexeev, R. Donagi, G. Farkas, E. Izadi, and A. Ortega, *The uniformization of the moduli space of principally polarized abelian 6-folds*, Preprint arxiv:1507.05710.

[AFSvdW16] J. Alper, M. Fedorchuk, D. I. Smyth, and F. van der Wyck, *Second flip in the Hassett–Keel program: A local description*, Compos. Math., to appear.

[Ale02] Valery Alexeev, *Complete moduli in the presence of semiabelian group action*, Ann. of Math. (2) **155** (2002), no. 3, 611–708, DOI 10.2307/3062130. MR1923963

[Ale04] Valery Alexeev, *Compactified Jacobians and Torelli map*, Publ. Res. Inst. Math. Sci. **40** (2004), no. 4, 1241–1265. MR2105707

[AMRT75] A. Ash, D. Mumford, M. Rapoport, and Y. Tai, *Smooth compactification of locally symmetric varieties*, Math. Sci. Press, Brookline, Mass., 1975. Lie Groups: History, Frontiers and Applications, Vol. IV. MR0457437

[And58] Aldo Andreotti, *On a theorem of Torelli*, Amer. J. Math. **80** (1958), 801–828, DOI 10.2307/2372835. MR0102518

[AS79] Enrico Arbarello and Edoardo Sernesi, *The equation of a plane curve*, Duke Math. J. **46** (1979), no. 2, 469–485. MR534063

[ASB16] J. Armstrong and N. I. Shepherd-Barron, *The singularities of A_g^P*, Preprint arxiv:1604.05943.

[BB66] W. L. Baily Jr. and A. Borel, *Compactification of arithmetic quotients of bounded symmetric domains*, Ann. of Math. (2) **84** (1966), 442–528, DOI 10.2307/1970457. MR0216035

[BP00] M. Boggi and M. Pikaart, *Galois covers of moduli of curves*, Compositio Math. **120** (2000), no. 2, 171–191, DOI 10.1023/A:1001731524036. MR1739177

[CC14] Dawei Chen and Izzet Coskun, *Extremal effective divisors on* $\overline{\mathcal{M}}_{1,n}$, Math. Ann. **359** (2014), no. 3-4, 891–908, DOI 10.1007/s00208-014-1027-5. MR3231020

[CFM13] Dawei Chen, Gavril Farkas, and Ian Morrison, *Effective divisors on moduli spaces of curves and abelian varieties*, A celebration of algebraic geometry, Clay Math. Proc., vol. 18, Amer. Math. Soc., Providence, RI, 2013, pp. 131–169. MR3114939

[Cle83] C. Herbert Clemens, *Double solids*, Adv. in Math. **47** (1983), no. 2, 107–230, DOI 10.1016/0001-8708(83)90025-7. MR690465

[CMLGH15] S. Casalaina-Martin, R. Laza, S. Grushevsky, and K. Hulek, *Extending the Prym map to toroidal compactifications of the moduli space of abelian varieties*, With an appendix by M. Dutour Sikirić. Preprint arXiv:1403.1938v2; J. Eur. Math. Soc. (JEMS), to appear.

[CT13] Ana-Maria Castravet and Jenia Tevelev, *Hypertrees, projections, and moduli of stable rational curves*, J. Reine Angew. Math. **675** (2013), 121–180. MR3021449

[CT15] Ana-Maria Castravet and Jenia Tevelev, $\overline{M}_{0,n}$ *is not a Mori dream space*, Duke Math. J. **164** (2015), no. 8, 1641–1667, DOI 10.1215/00127094-3119846. MR3352043

[DM69] P. Deligne and D. Mumford, *The irreducibility of the space of curves of given genus*, Inst. Hautes Études Sci. Publ. Math. **36** (1969), 75–109. MR0262240

[Don84] Ron Donagi, *The unirationality of* $\mathcal{A}_5$, Ann. of Math. (2) **119** (1984), no. 2, 269–307, DOI 10.2307/2007041. MR740895

[Don92] Ron Donagi, *The fibers of the Prym map*, Curves, Jacobians, and abelian varieties (Amherst, MA, 1990), Contemp. Math., vol. 136, Amer. Math. Soc., Providence, RI, 1992, pp. 55–125, DOI 10.1090/conm/136/1188194. MR1188194

[EH87] David Eisenbud and Joe Harris, *The Kodaira dimension of the moduli space of curves of genus* $\geq$ 23, Invent. Math. **90** (1987), no. 2, 359–387, DOI 10.1007/BF01388710. MR910206

[EVGS13] Philippe Elbaz-Vincent, Herbert Gangl, and Christophe Soulé, *Perfect forms, K-theory and the cohomology of modular groups*, Adv. Math. **245** (2013), 587–624, DOI 10.1016/j.aim.2013.06.014. MR3084439

[Far00] Gavril Farkas, *The geometry of the moduli space of curves of genus 23*, Math. Ann. **318** (2000), no. 1, 43–65, DOI 10.1007/s002080000108. MR1785575

[Far09a] Gavril Farkas, *Birational aspects of the geometry of* $\overline{\mathcal{M}}_g$, Surveys in differential geometry. Vol. XIV. Geometry of Riemann surfaces and their moduli spaces, Surv. Differ. Geom., vol. 14, Int. Press, Somerville, MA, 2009, pp. 57–110, DOI 10.4310/SDG.2009.v14.n1.a3. MR2655323

[Far09b] Gavril Farkas, *Koszul divisors on moduli spaces of curves*, Amer. J. Math. **131** (2009), no. 3, 819–867, DOI 10.1353/ajm.0.0053. MR2530855

[FC90] Gerd Faltings and Ching-Li Chai, *Degeneration of abelian varieties*, Ergebnisse der Mathematik und ihrer Grenzgebiete (3) [Results in Mathematics and Related Areas (3)], vol. 22, Springer-Verlag, Berlin, 1990. With an appendix by David Mumford. MR1083353

[FP13] C. Faber and R. Pandharipande, *Tautological and non-tautological cohomology of the moduli space of curves*, Handbook of moduli. Vol. I, Adv. Lect. Math. (ALM), vol. 24, Int. Press, Somerville, MA, 2013, pp. 293–330. MR3184167

[FS82] Robert Friedman and Roy Smith, *The generic Torelli theorem for the Prym map*, Invent. Math. **67** (1982), no. 3, 473–490, DOI 10.1007/BF01398932. MR664116

[FS13] Maksym Fedorchuk and David Ishii Smyth, *Alternate compactifications of moduli spaces of curves*, Handbook of moduli. Vol. I, Adv. Lect. Math. (ALM), vol. 24, Int. Press, Somerville, MA, 2013, pp. 331–413. MR3184168

[GKM02] Angela Gibney, Sean Keel, and Ian Morrison, *Towards the ample cone of* $\overline{M}_{g,n}$, J. Amer. Math. Soc. **15** (2002), no. 2, 273–294, DOI 10.1090/S0894-0347-01-00384-8. MR1887636

[Gru09] Samuel Grushevsky, *Geometry of* $\mathcal{A}_g$ *and its compactifications*, Algebraic geometry—Seattle 2005. Part 1, Proc. Sympos. Pure Math., vol. 80, Amer. Math. Soc., Providence, RI, 2009, pp. 193–234, DOI 10.1090/pspum/080.1/2483936. MR2483936

[Har84] J. Harris, *On the Kodaira dimension of the moduli space of curves. II. The even-genus case*, Invent. Math. **75** (1984), no. 3, 437–466, DOI 10.1007/BF01388638. MR735335

[Har13] Joe Harris, *Parameter spaces of curves*, Handbook of moduli. Vol. II, Adv. Lect. Math. (ALM), vol. 25, Int. Press, Somerville, MA, 2013, pp. 1–27. MR3184172

[HK00] Yi Hu and Sean Keel, *Mori dream spaces and GIT*, Michigan Math. J. **48** (2000), 331–348, DOI 10.1307/mmj/1030132722. Dedicated to William Fulton on the occasion of his 60th birthday. MR1786494

[HM82] Joe Harris and David Mumford, *On the Kodaira dimension of the moduli space of curves*, Invent. Math. **67** (1982), no. 1, 23–88, DOI 10.1007/BF01393371. With an appendix by William Fulton. MR664324

[HS02] Klaus Hulek and G. K. Sankaran, *The geometry of Siegel modular varieties*, Higher dimensional birational geometry (Kyoto, 1997), Adv. Stud. Pure Math., vol. 35, Math. Soc. Japan, Tokyo, 2002, pp. 89–156. MR1929793

[Igu67] Jun-ichi Igusa, *A desingularization problem in the theory of Siegel modular functions*, Math. Ann. **168** (1967), 228–260, DOI 10.1007/BF01361555. MR0218352

[ILGS09] Elham Izadi, Marco Lo Giudice, and Gregory Kumar Sankaran, *The moduli space of étale double covers of genus 5 curves is unirational*, Pacific J. Math. **239** (2009), no. 1, 39–52, DOI 10.2140/pjm.2009.239.39. MR2449010

[Kan87] V. Kanev, *Principal polarizations of Prym-Tjurin varieties*, Compositio Math. **64** (1987), no. 3, 243–270. MR918413

[Kap93] M. M. Kapranov, *Veronese curves and Grothendieck-Knudsen moduli space* $\overline{M}_{0,n}$, J. Algebraic Geom. **2** (1993), no. 2, 239–262. MR1203685

[Kee92] Sean Keel, *Intersection theory of moduli space of stable n-pointed curves of genus zero*, Trans. Amer. Math. Soc. **330** (1992), no. 2, 545–574, DOI 10.2307/2153922. MR1034665

[Knu83] Finn F. Knudsen, *The projectivity of the moduli space of stable curves. III. The line bundles on* $M_{g,n}$*, and a proof of the projectivity of* $\overline{M}_{g,n}$ *in characteristic* 0, Math. Scand. **52** (1983), no. 2, 200–212, DOI 10.7146/math.scand.a-12002. MR702954

[LM00] A. Losev and Y. Manin, *New moduli spaces of pointed curves and pencils of flat connections*, Michigan Math. J. **48** (2000), 443–472, DOI 10.1307/mmj/1030132728. Dedicated to William Fulton on the occasion of his 60th birthday. MR1786500

[Loo94] Eduard Looijenga, *Smooth Deligne-Mumford compactifications by means of Prym level structures*, J. Algebraic Geom. **3** (1994), no. 2, 283–293. MR1257324

[Loo00] Eduard Looijenga, *A minicourse on moduli of curves*, School on Algebraic Geometry (Trieste, 1999), ICTP Lect. Notes, vol. 1, Abdus Salam Int. Cent. Theoret. Phys., Trieste, 2000, pp. 267–291. MR1795865

[Mum74] David Mumford, *Prym varieties. I*, Contributions to analysis (a collection of papers dedicated to Lipman Bers), Academic Press, New York, 1974, pp. 325–350. MR0379510

[Mum83a] David Mumford, *On the Kodaira dimension of the Siegel modular variety*, Algebraic geometry—open problems (Ravello, 1982), Lecture Notes in Math., vol. 997, Springer, Berlin, 1983, pp. 348–375, DOI 10.1007/BFb0061652. MR714757

[Mum83b] David Mumford, *Towards an enumerative geometry of the moduli space of curves*, Arithmetic and geometry, Vol. II, Progr. Math., vol. 36, Birkhäuser Boston, Boston, MA, 1983, pp. 271–328. MR717614

[MV12] Margarida Melo and Filippo Viviani, *Comparing perfect and 2nd Voronoi decompositions: the matroidal locus*, Math. Ann. **354** (2012), no. 4, 1521–1554, DOI 10.1007/s00208-011-0774-9. MR2993003

[Nam76a] Yukihiko Namikawa, *A new compactification of the Siegel space and degeneration of Abelian varieties. I*, Math. Ann. **221** (1976), no. 2, 97–141, DOI 10.1007/BF01433145. MR0480537

[Nam76b] Yukihiko Namikawa, *A new compactification of the Siegel space and degeneration of Abelian varieties. II*, Math. Ann. **221** (1976), no. 3, 201–241, DOI 10.1007/BF01596390. MR0480538

[Nam80] Yukihiko Namikawa, *Toroidal compactification of Siegel spaces*, Lecture Notes in Mathematics, vol. 812, Springer, Berlin, 1980. MR584625 (82a:32034)

[Ols08] Martin C. Olsson, *Compactifying moduli spaces for abelian varieties*, Lecture Notes in Mathematics, vol. 1958, Springer-Verlag, Berlin, 2008. MR2446415

[Oor81] Frans Oort, *Coarse and fine moduli spaces of algebraic curves and polarized abelian varieties*, Symposia Mathematica, Vol. XXIV (Sympos., INDAM, Rome, 1979), Academic Press, London-New York, 1981, pp. 293–313. MR619252

[Opi16] Morgan Opie, *Extremal divisors on moduli spaces of rational curves with marked points*, Michigan Math. J. **65** (2016), no. 2, 251–285, DOI 10.1307/mmj/1465329013. MR3510907

[Oxl11] James Oxley, *Matroid theory*, 2nd ed., Oxford Graduate Texts in Mathematics, vol. 21, Oxford University Press, Oxford, 2011. MR2849819

[SB06] N. I. Shepherd-Barron, *Perfect forms and the moduli space of abelian varieties*, Invent. Math. **163** (2006), no. 1, 25–45, DOI 10.1007/s00222-005-0453-0. MR2208417

[Ser62] J.-P. Serre, *Rigidité du foncteur de Jacobi d'échelon $n \geq 3$*, Appendix of Séminaire Henri Cartan, 13 no. 2 (1960-1961), Expose No. 17, 1962.

[Tai82] Yung-Sheng Tai, *On the Kodaira dimension of the moduli space of abelian varieties*, Invent. Math. **68** (1982), no. 3, 425–439, DOI 10.1007/BF01389411. MR669424

[Val03] F. Vallentin, *Sphere coverings, lattices, and tilings (in low dimensions)*, Ph.D. thesis, Center for Mathematical Sciences, Munich University of Technology, Munich, Germany, 2003.

[vdG99] Gerard van der Geer, *Cycles on the moduli space of abelian varieties*, Moduli of curves and abelian varieties, Aspects Math., E33, Friedr. Vieweg, Braunschweig, 1999, pp. 65–89, DOI 10.1007/978-3-322-90172-9_4. MR1722539

[vdG13] Gerard van der Geer, *The cohomology of the moduli space of abelian varieties*, Handbook of moduli. Vol. I, Adv. Lect. Math. (ALM), vol. 24, Int. Press, Somerville, MA, 2013, pp. 415–457. MR3184169

[vdGO99] Gerard van der Geer and Frans Oort, *Moduli of abelian varieties: a short introduction and survey*, Moduli of curves and abelian varieties, Aspects Math., E33, Friedr. Vieweg, Braunschweig, 1999, pp. 1–21, DOI 10.1007/978-3-322-90172-9_1. MR1722536

[Ver02] Peter Vermeire, *A counterexample to Fulton's conjecture on $\overline{M}_{0,n}$*, J. Algebra **248** (2002), no. 2, 780–784, DOI 10.1006/jabr.2001.9044. MR1882122

[Wir95] W. Wirtinger, *Untersuchungen über thetafunktionen*, Teubner, Leipzig, 1895.

Department of Mathematical Sciences, Chalmers University of Technology and University of Gothenburg, SE-412 96 Göteborg, Sweden

E-mail address: orsola@chalmers.se

Birational geometry of moduli spaces of sheaves and Bridgeland stability

Jack Huizenga

Abstract. Moduli spaces of sheaves and Hilbert schemes of points have experienced a recent resurgence in interest in the past several years, due largely to new techniques arising from Bridgeland stability conditions and derived category methods. In particular, classical questions about the birational geometry of these spaces can be answered by using new tools such as the positivity lemma of Bayer and Macrì. In this article we first survey classical results on moduli spaces of sheaves and their birational geometry. We then discuss the relationship between these classical results and the new techniques coming from Bridgeland stability, and discuss how cones of ample divisors on these spaces can be computed with these new methods. This survey expands upon the author's talk at the 2015 Bootcamp in Algebraic Geometry preceding the 2015 AMS Summer Research Institute on Algebraic Geometry at the University of Utah.

Contents

1. Introduction

The topic of vector bundles in algebraic geometry is a broad field with a rich history. In the 70's and 80's, one of the main questions of interest was the study of low rank vector bundles on projective spaces $\mathbb{P}^r$. One particularly challenging conjecture in this subject is the following.

2010 *Mathematics Subject Classification.* Primary 14J60; Secondary 14E30, 14J29, 14C05.
Key words and phrases. Moduli spaces of sheaves, Hilbert schemes of points, ample cone, Bridgeland stability.
During the preparation of this article the author was partially supported by a National Science Foundation Mathematical Sciences Postdoctoral Research Fellowship.

CONJECTURE 1.1 (Hartshorne [**Har74**]). *If $r \geq 7$ then any rank 2 bundle on $\mathbb{P}^r_{\mathbb{C}}$ splits as a direct sum of line bundles.*

The Hartshorne conjecture is connected to the study of subvarieties of projective space of small codimension. In particular, the above statement implies that if $X \subset \mathbb{P}^r$ is a codimension 2 smooth subvariety and K_X is a multiple of the hyperplane class then X is a complete intersection. Thus, early interest in the study of vector bundles was born out of classical questions in projective geometry.

Study of these types of questions led naturally to the study of *moduli spaces* of (semistable) vector bundles, parameterizing the isomorphism classes of (semistable) vector bundles with given numerical invariants on a projective variety X (we will define *semistable* later—for now, view it as a necessary condition to get a good moduli space). As often happens in mathematics, these spaces have become interesting in their own right, and their study has become an entire industry. Beginning in the 80's and 90's, and continuing to today, people have studied the basic questions of the geometry of these spaces. Are they smooth? Irreducible? What do their singularities look like? When is the moduli space nonempty? What are divisors on the moduli space? Especially when X is a curve or surface, satisfactory answers to these questions can often be given. We will survey several foundational results of this type in §2-3.

More recently, there has been a great deal of interest in the study of the *birational geometry* of moduli spaces of various geometric objects. Loosely speaking, the goal of such a program is to understand alternate birational models, or *compactifications*, of a moduli space as themselves being moduli spaces for slightly different geometric objects. For instance, the Hassett-Keel program [**HH13**] studies alternate compactifications of the Deligne-Mumford compactification $\overline{M}_g$ of the moduli space of stable curves. Different compactifications can be obtained by studying (potentially unstable) curves with different types of singularities. In addition to being interesting in their own right, moduli spaces provide explicit examples of higher dimensional varieties which can frequently be understood in great detail. We survey the birational geometry of moduli spaces of sheaves from a classical viewpoint in §4.

In the last several years, there has been a great deal of progress in the study of the birational geometry of moduli spaces of sheaves owing to Bridgeland's introduction of the concept of a *stability condition* [**Bri07, Bri08**]. Very roughly, there is a complex manifold $\mathrm{Stab}(X)$, the *stability manifold*, parameterizing stability conditions σ on X. There is a moduli space corresponding to each condition σ, and the stability manifold decomposes into chambers where the corresponding moduli space does not change as σ varies in the chamber. For one of these chambers, the *Gieseker chamber*, the corresponding moduli space is the ordinary moduli space of semistable sheaves. The moduli spaces corresponding to other chambers often happen to be the alternate birational models of the ordinary moduli space. In this way, the birational geometry of a moduli space of sheaves can be viewed in terms of a variation of the moduli problem. In §5 we will introduce Bridgeland stability conditions, and especially study stability conditions on a surface. We study some basic examples on $\mathbb{P}^2$ in §6. Finally, we close the paper in §7 by surveying some recent results on the computation of ample cones of Hilbert schemes of points and moduli spaces of sheaves on surfaces.

Acknowledgements. I would especially like to thank Izzet Coskun and Benjamin Schmidt for many discussions on Bridgeland stability and related topics. In addition, I would like to thank the referee of this article for many valuable comments, as well as Barbara Bolognese, Yinbang Lin, Eric Riedl, Matthew Woolf, and Xialoei Zhao. Finally, I would like to thank the organizers of the 2015 Bootcamp in Algebraic Geometry and the 2015 AMS Summer Research Institute on Algebraic Geometry, as well as the funding organizations for these wonderful events.

2. Moduli spaces of sheaves

The definition of a Bridgeland stability condition is motivated by the classical theory of semistable sheaves. In this section we review the basics of the theory of moduli spaces of sheaves, particularly focusing on the case of a surface. The standard references for this material are Huybrechts-Lehn [**HL10**] and Le Potier [**LeP97**].

2.1. The moduli property.

First we state an idealized version of the moduli problem. Let X be a smooth projective variety with polarization H, and fix a set of discrete numerical invariants of a coherent sheaf E on X. This can be accomplished by fixing the *Hilbert polynomial*

$$P_E(m) = \chi(E \otimes \mathcal{O}_X(mH))$$

of the sheaf.

A *family of sheaves on X over a scheme S* is a (coherent) sheaf $\mathcal{E}$ on $X \times S$ which is S-flat. For a point $s \in S$, we write E_s for the sheaf $\mathcal{E}|_{X\times\{s\}}$. We say $\mathcal{E}$ is a family of *semistable sheaves of Hilbert polynomial P* if E_s is semistable with Hilbert polynomial P for each $s \in S$ (see §2.3 for the definition of semistability). We define a *moduli functor*

$$\mathcal{M}'(P) : \mathrm{Sch}^o \to \mathrm{Set}$$

by defining $\mathcal{M}'(P)(S)$ to be the set of isomorphism classes of families of semistable sheaves on X with Hilbert polynomial P. We will sometimes just write $\mathcal{M}'$ for the moduli functor when the polynomial P is understood.

Let $p : X \times S \to S$ be the projection. If $\mathcal{E}$ is a family of semistable sheaves on X with Hilbert polynomial P and L is a line bundle on S, then $\mathcal{E} \otimes p^*L$ is again such a family. The sheaves E_s and $(\mathcal{E} \otimes p^*L)|_{X\times\{s\}}$ parameterized by any point $s \in S$ are isomorphic, although $\mathcal{E}$ and $\mathcal{E} \otimes p^*L$ need not be isomorphic. We call two families of sheaves on X *equivalent* if they differ by tensoring by a line bundle pulled back from the base, and define a refined moduli functor $\mathcal{M}$ by modding out by this equivalence relation: $\mathcal{M} = \mathcal{M}'/\sim$.

The basic question is whether or not $\mathcal{M}$ can be represented by some nice object, e.g. by a projective variety or a scheme. We recall the following definitions.

DEFINITION 2.1. A functor $\mathcal{F} : \mathrm{Sch}^o \to \mathrm{Set}$ is *represented* by a scheme X if there is an isomorphism of functors $\mathcal{F} \cong \mathrm{Mor}_{\mathrm{Sch}}(-, X)$.

A functor $\mathcal{F} : \mathrm{Sch}^o \to \mathrm{Set}$ is *corepresented* by a scheme X if there is a natural transformation $\alpha : \mathcal{F} \to \mathrm{Mor}_{\mathrm{Sch}}(-, X)$ with the following universal property: if X' is a scheme and $\beta : \mathcal{F} \to \mathrm{Mor}_{\mathrm{Sch}}(-, X')$ a natural transformation, then there is a unique morphism $\pi : X \to X'$ such that β is the composition of α with the transformation $\mathrm{Mor}_{\mathrm{Sch}}(-, X) \to \mathrm{Mor}_{\mathrm{Sch}}(-, X')$ induced by π.

REMARK 2.2. Note that if $\mathcal{F}$ is represented by X then it is also corepresented by X.

If $\mathcal{F}$ is represented by X, then $\mathcal{F}(\operatorname{Spec}\mathbb{C}) \cong \operatorname{Mor}_{\mathrm{Sch}}(\operatorname{Spec}\mathbb{C}, X)$. That is, the points of X are in bijective correspondence with $\mathcal{F}(\operatorname{Spec}\mathbb{C})$. This need not be true if $\mathcal{F}$ is only corepresented by X.

If $\mathcal{F}$ is corepresented by X, then X is unique up to a unique isomorphism.

We now come to the basic definition of moduli space of sheaves.

DEFINITION 2.3. A scheme $M(P)$ is a *moduli space of semistable sheaves with Hilbert polynomial* P if $M(P)$ corepresents $\mathcal{M}(P)$. It is a *fine moduli space* if it represents $\mathcal{M}(P)$.

The most immediate consequence of M being a moduli space is the existence of the *moduli map*. Suppose E is a family of semistable sheaves on X parameterized by S. Then we obtain a morphism $S \to M$ which intuitively sends $s \in S$ to the isomorphism class of the sheaf E_s.

In the special case when the base $\{s\}$ is a point, a family over $\{s\}$ is the isomorphism class of a single sheaf, and the moduli map $\{s\} \to M$ sends that class to a corresponding point. The compatibilities in the definition of a natural transformation ensure that in the case of a family $\mathcal{E}$ parameterized by a base S the image in M of a point $s \in S$ depends only on the isomorphism class of the sheaf E_s parameterized by s.

In the ideal case where the moduli functor $\mathcal{M}$ has a fine moduli space, there is a universal sheaf $\mathcal{U}$ on X parameterized by M. We have an isomorphism

$$\mathcal{M}(M) \cong \operatorname{Mor}_{\mathrm{Sch}}(M, M)$$

and the distinguished identity morphism $M \to M$ corresponds to a family $\mathcal{U}$ of sheaves parameterized by M (strictly speaking, $\mathcal{U}$ is only well-defined up to tensoring by a line bundle pulled back from M). This universal sheaf has the property that if $\mathcal{E}$ is a family of semistable sheaves on X parameterized by S and $f : S \to M$ is the moduli map, then $\mathcal{E}$ and $(\mathrm{id}_X \times f)^*\mathcal{U}$ are equivalent.

2.2. Issues with a naive moduli functor. In this subsection we give some examples to illustrate the importance of the as-yet-undefined *semistability* hypothesis in the definition of the moduli functor. Let $\mathcal{M}^n$ be the *naive* moduli functor of (flat) families of coherent sheaves with Hilbert polynomial P on X, omitting any semistability hypothesis. We might hope that this functor is (co)representable by a scheme M^n with some nice properties, such as the following.

(1) M^n is a scheme of finite type.
(2) The points of M^n are in bijective correspondence with isomorphism classes of coherent sheaves on X with Hilbert polynomial P.
(3) A family of sheaves over a smooth punctured curve $C - \{\mathrm{pt}\}$ can be uniquely completed to a family of sheaves over C.

However, unless some restrictions are imposed on the types of sheaves which are allowed, all three hopes will fail. Properties (2) and (3) will also typically fail for semistable sheaves, but this failure occurs in a well-controlled way.

EXAMPLE 2.4. Consider $X = \mathbb{P}^1$, and let $P = P_{\mathcal{O}_{\mathbb{P}^1}^{\oplus 2}} = 2m + 2$ be the Hilbert polynomial of the rank 2 trivial bundle. Then for any $n \geq 0$, the bundle

$$\mathcal{O}_{\mathbb{P}^1}(n) \oplus \mathcal{O}_{\mathbb{P}^1}(-n)$$

also has Hilbert polynomial $2m+2$, and $h^0(\mathcal{O}_{\mathbb{P}^1}(n) \oplus \mathcal{O}_{\mathbb{P}^1}(-n)) = n+1$. If there is a moduli scheme M^n parameterizing all sheaves on $\mathbb{P}^1$ of Hilbert polynomial P, then M^n cannot be of finite type. Indeed, the loci

$$W_n = \{E : h^0(E) \geq n\} \subset M(P)$$

would then form an infinite decreasing chain of closed subschemes of $M(P)$.

EXAMPLE 2.5. Again consider $X = \mathbb{P}^1$ and $P = 2m+2$. Let $S = \operatorname{Ext}^1(\mathcal{O}_{\mathbb{P}^1}(1), \mathcal{O}_{\mathbb{P}^1}(-1)) = \mathbb{C}$. For $s \in S$, let E_s be the sheaf

$$0 \to \mathcal{O}_{\mathbb{P}^1}(-1) \to E_s \to \mathcal{O}_{\mathbb{P}^1}(1) \to 0$$

defined by the extension class s. One checks that if $s \neq 0$ then $E_s \cong \mathcal{O}_{\mathbb{P}^1}^{\oplus 2}$, but the extension is split for $s = 0$. It follows that the moduli map $S \to M^n$ must be constant, so $\mathcal{O}_{\mathbb{P}^1} \oplus \mathcal{O}_{\mathbb{P}^1}$ and $\mathcal{O}_{\mathbb{P}^1}(1) \oplus \mathcal{O}_{\mathbb{P}^1}(-1)$ are identified in the moduli space M^n.

EXAMPLE 2.6. Suppose X is a smooth variety and F is a coherent sheaf with $\dim \operatorname{Ext}^1(F, F) \geq 1$. Let $S \subset \operatorname{Ext}^1(F, F)$ be a 1-dimensional subspace, and for any $s \in S$ let E_s be the corresponding extension of F by F. Then if $s, s' \in S$ are both not zero, we have

$$E_s \cong E_{s'} \ncong E_0 = F \oplus F.$$

As in the previous example, we see that $F \oplus F$ and a nontrivial extension of F by F must be identified in M^n. Therefore any two extensions of F by F must also be identified in M^n.

If F is semistable, then Example 2.6 is an example of a nontrivial family of S-*equivalent* sheaves. A major theme of this survey is that S-equivalence is the main source of interesting birational maps between moduli spaces of sheaves.

2.3. Semistability. Let E be a coherent sheaf on X. We say that E is *pure* of dimension d if the support of E is d-dimensional and every nonzero subsheaf of E has d-dimensional support.

REMARK 2.7. If $\dim X = n$, then E is pure of dimension n if and only if E is torsion-free.

If E is pure of dimension d then the Hilbert polynomial $P_E(m)$ has degree d. We write it in the form

$$P_E(m) = \alpha_d(E)\frac{m^d}{d!} + \cdots,$$

and define the *reduced Hilbert polynomial* by

$$p_E(m) = \frac{P_E(m)}{\alpha_d(E)}.$$

In the principal case of interest where $d = n = \dim X$, Riemann-Roch gives $\alpha_n(E) = r(E)H^n$ where $r(E)$ is the rank, and

$$p_E(m) = \frac{P_E(m)}{r(E)H^n}.$$

DEFINITION 2.8. A sheaf E is *(semi)stable* if it is pure of dimension d and any proper subsheaf $F \subset E$ has

$$p_F \underset{(-)}{<} p_E,$$

where polynomials are compared at large values. That is, $p_F < p_E$ means that $p_F(m) < p_E(m)$ for all $m \gg 0$.

The above notion of stability is often called Gieseker stability, especially when a distinction from other forms of stability is needed. The foundational result in this theory is that Gieseker semistability is the correct extra condition on sheaves to give well-behaved moduli spaces.

THEOREM 2.9 ([**HL10**, Theorem 4.3.4]). *Let (X, H) be a smooth, polarized projective variety, and fix a Hilbert polynomial P. There is a projective moduli scheme of semistable sheaves on X of Hilbert polynomial P.*

While the definition of Gieseker stability is compact, it is frequently useful to use the Riemann-Roch theorem to make it more explicit. We spell this out in the case of a curve or surface. We define the *slope* of a coherent sheaf E of positive rank on an n-dimensional variety by

$$\mu(E) = \frac{c_1(E).H^{n-1}}{r(E)H^n}.$$

EXAMPLE 2.10 (Stability on a curve). Suppose C is a smooth curve of genus g. The Riemann-Roch theorem asserts that if E is a coherent sheaf on C then

$$\chi(E) = c_1(E) + r(E)(1-g).$$

Polarizing C with $H = p$ a point, we find

$$P_E(m) = \chi(E(m)) = c_1(E(m)) + r(E)(1-g) = r(E)m + (c_1(E) + r(E)(1-g)),$$

and so

$$p_E(m) = m + \frac{c_1(E)}{r(E)} + (1-g).$$

We conclude that if $F \subset E$ then $p_F \underset{(-)}{<} p_E$ if and only if $\mu(F) \underset{(-)}{<} \mu(E)$.

EXAMPLE 2.11 (Stability on a surface). Let X be a smooth surface with polarization H, and let E be a sheaf of positive rank. We define the *total slope* and *discriminant* by

$$\nu(E) = \frac{c_1(E)}{r(E)} \in H^2(X, \mathbb{Q}) \qquad \text{and} \qquad \Delta(E) = \frac{1}{2}\nu(E)^2 - \frac{\mathrm{ch}_2(E)}{r} \in \mathbb{Q}.$$

With this notation, the Riemann-Roch theorem takes the particularly simple form

$$\chi(E) = r(E)(P(\nu(E)) - \Delta(E)),$$

where $P(\nu) = \chi(\mathcal{O}_X) + \frac{1}{2}\nu(\nu - K_X)$ (see [**LeP97**]). The total slope and discriminant behave well with respect to tensor products: if E and F are locally free then

$$\nu(E \otimes F) = \nu(E) + \nu(F)$$
$$\Delta(E \otimes F) = \Delta(E) + \Delta(F).$$

Furthermore, $\Delta(L) = 0$ for a line bundle L; equivalently, in the case of a line bundle the Riemann-Roch formula is $\chi(L) = P(c_1(L))$. Then we compute

$$\begin{aligned}\chi(E(m)) &= r(E)(P(\nu(E) + mH) - \Delta(E)) \\ &= r(E)(\chi(\mathcal{O}_X) + \frac{1}{2}(\nu(E) + mH)(\nu(E) + mH - K_X) - \Delta(E)) \\ &= r(E)(P(\nu(E)) + \frac{1}{2}(mH)^2 + mH.(\nu(E) + \frac{1}{2}K_X) - \Delta(E)) \\ &= \frac{r(E)H^2}{2}m^2 + r(E)H.(\nu(E) + \frac{1}{2}K_X)m + \chi(E),\end{aligned}$$

so

$$p_E(m) = \frac{1}{2}m^2 + \frac{H.(\nu(E) + \frac{1}{2}K_X)}{H^2}m + \frac{\chi(E)}{r(E)H^2}$$

Now if $F \subset E$, we compare the coefficients of p_F and p_E lexicographically to determine when $p_F \underset{(-)}{<} p_E$. We see that $p_F \underset{(-)}{<} p_E$ if and only if either $\mu(F) < \mu(E)$, or $\mu(F) = \mu(E)$ and

$$\frac{\chi(F)}{r(F)H^2} \underset{(-)}{<} \frac{\chi(E)}{r(E)H^2}.$$

EXAMPLE 2.12 (Slope stability). The notion of *slope semistability* has also been studied extensively and frequently arises in the study of Gieseker stability. We say that a torsion-free sheaf E on a variety X with polarization H is *μ-(semi)stable* if every subsheaf $F \subset E$ of strictly smaller rank has $\mu(F) \underset{(-)}{<} \mu(E)$. As we have seen in the curve and surface case, the coefficient of m^{n-1} in the reduced Hilbert polynomial $p_E(m)$ is just $\mu(E)$ up to adding a constant depending only on (X, H). This observation gives the following chain of implications:

$$\mu\text{-stable} \Rightarrow \text{stable} \Rightarrow \text{semistable} \Rightarrow \mu\text{-semistable}.$$

While Gieseker (semi)stability gives the best moduli theory and is therefore the most common to work with, it is often necessary to consider these various other forms of stability to study ordinary stability.

EXAMPLE 2.13 (Elementary modifications). As an example where μ-stability is useful, suppose X is a smooth surface and E is a torsion-free sheaf on X. Let $p \in X$ be a point where X is locally free, and consider sheaves E' defined as kernels of maps $E \to \mathcal{O}_p$, where $\mathcal{O}_p$ is a skyscraper sheaf:

$$0 \to E' \to E \to \mathcal{O}_p \to 0.$$

Intuitively, E' is just E with an additional simple singularity imposed at p. Such a sheaf E' is called an *elementary modification* of E. We have $\mu(E) = \mu(E')$ and $\chi(E') = \chi(E) - 1$, which makes elementary modifications a useful tool for studying sheaves by induction on the Euler characteristic.

Suppose E satisfies one of the four types of stability discussed in Example 2.12. If E is μ-(semi)stable, then it follows that E' is μ-(semi)stable as well. Indeed, if $F \subset E'$ with $r(F) < r(E')$, then also $F \subset E$, so $\mu(F) \underset{(-)}{<} \mu(E)$. But $\mu(E) = \mu(E')$, so $\mu(F) \underset{(-)}{<} \mu(E')$ and E' is μ-(semi)stable.

On the other hand, elementary modifications do not behave as well with respect to Gieseker (semi)stability. For example, take $X = \mathbb{P}^2$. Then $E = \mathcal{O}_{\mathbb{P}^2} \oplus \mathcal{O}_{\mathbb{P}^2}$ is semistable, but any any elementary modification E' of $\mathcal{O}_{\mathbb{P}^2} \oplus \mathcal{O}_{\mathbb{P}^2}$ at a point $p \in \mathbb{P}^2$ is isomorphic to $I_p \oplus \mathcal{O}_{\mathbb{P}^2}$, where I_p is the ideal sheaf of p. Thus E' is not semistable.

It is also possible to give an example of a stable sheaf E such that some elementary modification is not stable. Let $p, q, r \in \mathbb{P}^2$ be distinct points. Then $\text{ext}^1(I_r, I_{\{p,q\}}) = 2$. If E is any non-split extension

$$0 \to I_{\{p,q\}} \to E \to I_r \to 0$$

then E is clearly μ-semistable. In fact, E is stable: the only stable sheaves F of rank 1 and slope 0 with $p_F \leq p_E$ are $\mathcal{O}_{\mathbb{P}^2}$ and I_s for $s \in \mathbb{P}^2$ a point, but $\text{Hom}(I_s, E) = 0$ for any $s \in \mathbb{P}^2$ since the sequence is not split. Now if $s \in \mathbb{P}^2$ is a point distinct from p, q, r and $E \to \mathcal{O}_s$ is a map such that the composition $I_{\{p,q\}} \to E \to \mathcal{O}_s$ is zero, then the corresponding elementary modification

$$0 \to E' \to E \to \mathcal{O}_s \to 0$$

has a subsheaf $I_{\{p,q\}} \subset E'$. We have $p_{I_{\{p,q\}}} = p_{E'}$, so E' is strictly semistable.

Example 2.14 (Chern classes). Let $K_0(X)$ be the Grothendieck group of X, generated by classes $[E]$ of locally free sheaves, modulo relations $[E] = [F] + [G]$ for every exact sequence

$$0 \to F \to E \to G \to 0.$$

There is a symmetric bilinear *Euler pairing* on $K_0(X)$ such that $([E],[F]) = \chi(E \otimes F)$ whenever E, F are locally free sheaves. The *numerical Grothendieck group* $K_{\text{num}}(X)$ is the quotient of $K_0(X)$ by the kernel of the Euler pairing, so that the Euler pairing descends to a nondegenerate pairing on $K_{\text{num}}(X)$.

It is often preferable to fix the Chern classes of a sheaf instead of the Hilbert polynomial. This is accomplished by fixing a class $\mathbf{v} \in K_{\text{num}}(X)$. Any class $\mathbf{v}$ determines a Hilbert polynomial $P_{\mathbf{v}} = (\mathbf{v}, [\mathcal{O}_X(m)])$. In general, a polynomial P can arise as the Hilbert polynomial of several classes $\mathbf{v} \in K_{\text{num}}(X)$. In any family $\mathcal{E}$ of sheaves parameterized by a connected base S the sheaves E_s all have the same class in $K_{\text{num}}(X)$. Therefore, the moduli space $M(P)$ splits into connected components corresponding to the different vectors $\mathbf{v}$ with $P_{\mathbf{v}} = P$. We write $M(\mathbf{v})$ for the connected component of $M(P)$ corresponding to $\mathbf{v}$.

2.4. Filtrations. In addition to controlling subsheaves, stability also restricts the types of maps that can occur between sheaves.

Proposition 2.15.

(1) *(See-saw property) In any exact sequence of pure sheaves*

$$0 \to F \to E \to Q \to 0$$

of the same dimension d, we have $p_F \underset{(-)}{<} p_E$ if and only if $p_E \underset{(-)}{<} p_Q$.

(2) *If F, E are semistable sheaves of the same dimension d and $p_F > p_E$, then* $\text{Hom}(F, E) = 0$.

(3) *If F, E are stable sheaves and $p_F = p_E$, then any nonzero homomorphism $F \to E$ is an isomorphism.*

(4) *Stable sheaves E are* simple: $\text{Hom}(E, E) = \mathbb{C}$.

PROOF.

(1) We have $P_E = P_F + P_Q$, so $\alpha_d(E) = \alpha_d(F) + \alpha_d(Q)$ and
$$p_E = \frac{P_E}{\alpha_d(E)} = \frac{P_F + P_Q}{\alpha_d(E)} = \frac{\alpha_d(F)p_F + \alpha_d(Q)p_Q}{\alpha_d(E)}.$$
Thus p_E is a weighted mean of p_F and p_Q, and the result follows.

(2) Let $f : F \to E$ be a homomorphism, and put $C = \operatorname{Im} f$ and $K = \ker f$. Then C is pure of dimension d since E is, and K is pure of dimension d since F is. By (1) and the semistability of F, we have $p_C \geq p_F > p_E$. This contradicts the semistability of E since $C \subset E$.

(3) Since $p_F = p_E$, F and E have the same dimension. With the same notation as in (2), we instead find $p_C \geq p_F = p_E$, and the stability of E gives $p_C = p_E$ and $C = E$. If f is not an isomorphism then $p_K = p_F$, contradicting stability of F. Therefore f is an isomorphism.

(4) Suppose $f : E \to E$ is any homomorphism. Pick some point $x \in X$. The linear transformation $f_x : E_x \to E_x$ has an eigenvalue $\lambda \in \mathbb{C}$. Then $f - \lambda \operatorname{id}_E$ is not an isomorphism, so it must be zero. Therefore $f = \lambda \operatorname{id}_E$.

□

Harder-Narasimhan filtrations enable us to study arbitrary pure sheaves in terms of semistable sheaves. Proposition 2.15 is one of the important ingredients in the proof of the next theorem.

THEOREM AND DEFINITION 2.16 ([**HL10**]). *Let E be a pure sheaf of dimension d. Then there is a unique filtration*
$$0 = E_0 \subset E_1 \subset \cdots \subset E_\ell = E$$
called the Harder-Narasimhan filtration *such that the quotients* $\operatorname{gr}_i = E_i/E_{i-1}$ *are semistable of dimension d and reduced Hilbert polynomial p_i, where*
$$p_1 > p_2 > \cdots > p_\ell.$$

In order to construct (semi)stable sheaves it is frequently necessary to also work with sheaves that are not semistable. The next example outlines one method for constructing semistable vector bundles. This general method was used by Drézet and Le Potier to classify the possible Hilbert polynomials of semistable sheaves on $\mathbb{P}^2$ [**LeP97, DLP85**].

EXAMPLE 2.17. Let (X, H) be a smooth polarized projective variety. Suppose A and B are vector bundles on X and that the sheaf $\mathcal{H}om(A, B)$ is globally generated. For simplicity assume $r(B) - r(A) \geq \dim X$. Let $S \subset \operatorname{Hom}(A, B)$ be the open subset parameterizing injective sheaf maps; this set is nonempty since $\mathcal{H}om(A, B)$ is globally generated. Consider the family $\mathcal{E}$ of sheaves on X parameterized by S where the sheaf E_s parameterized by $s \in S$ is the cokernel
$$0 \to A \xrightarrow{s} B \to E_s \to 0.$$
Then for general $s \in S$, the sheaf E_s is a vector bundle [**Hui16**, Proposition 2.6] with Hilbert polynomial $P := P_B - P_A$. In other words, restricting to a dense open subset $S' \subset S$, we get a family of locally free sheaves parameterized by S'.

Next, semistability is an open condition in families. Thus there is a (possibly empty) open subset $S'' \subset S'$ parameterizing semistable sheaves. Let $\ell > 0$ be an integer and pick polynomials $P_1, \ldots, P_\ell$ such that $P_1 + \cdots + P_\ell = P$ and the corresponding reduced polynomials $p_1, \ldots, p_\ell$ have $p_1 > \cdots > p_\ell$. Then there is a locally closed subset $S_{P_1,\ldots,P_\ell} \subset S'$ parameterizing sheaves with a Harder-Narasimhan filtration of length ℓ with factors of Hilbert polynomial $P_1, \ldots, P_\ell$. Such loci are called *Shatz strata* in the base S' of the family.

Finally, to show that S'' is nonempty, it suffices to show that the Shatz stratum S_P corresponding to semistable sheaves is dense. One approach to this problem is to show that every Shatz stratum $S_{P_1,\ldots,P_\ell}$ with $\ell \geq 2$ has codimension at least 1. See [**LeP97**, Chapter 16] for an example where this is carried out in the case of $\mathbb{P}^2$.

Just as the Harder-Narasimhan filtration allows us to use semistable sheaves to build up arbitrary pure sheaves, Jordan-Hölder filtrations decompose semistable sheaves in terms of stable sheaves.

THEOREM AND DEFINITION 2.18. [**HL10**] *Let E be a semistable sheaf of dimension d and reduced Hilbert polynomial p. There is a filtration*

$$0 = E_0 \subset E_1 \subset \cdots \subset E_\ell = E$$

called the Jordan-Hölder filtration *such that the quotients* $\operatorname{gr}_i = E_i/E_{i-1}$ *are stable with reduced Hilbert polynomial p. The filtration is not necessarily unique, but the list of stable factors is unique up to reordering.*

We can now precisely state the critical definition of S-equivalence.

DEFINITION 2.19. Semistable sheaves E and F are *S-equivalent* if they have the same list of Jordan-Hölder factors.

We have already seen an example of an S-equivalent family of semistable sheaves in Example 2.6, and we observed that all the parameterized sheaves must be represented by the same point in the moduli space. In fact, the converse is also true, as the next theorem shows.

THEOREM 2.20. *Two semistable sheaves E, F with Hilbert polynomial P are represented by the same point in $M(P)$ if and only if they are S-equivalent. Thus, the points of $M(P)$ are in bijective correspondence with S-equivalence classes of semistable sheaves with Hilbert polynomial P.*

In particular, if there are strictly semistable sheaves of Hilbert polynomial P, then $M(P)$ is not a fine moduli space.

REMARK 2.21. The question of when the open subset $M^s(P)$ parameterizing stable sheaves is a fine moduli space for the moduli functor $\mathcal{M}^s(P)$ of stable families is somewhat delicate; in this case the points of $M^s(P)$ are in bijective correspondence with the isomorphism classes of stable sheaves, but there still need not be a universal family.

One positive result in this direction is the following. Let $\mathbf{v} \in K_{\mathrm{num}}(X)$ be the numerical class of a stable sheaf with Hilbert polynomial P (see Example 2.14). Consider the set of integers of the form $(\mathbf{v}, [F])$, where F is a coherent sheaf and $(-,-)$ is the Euler pairing. If their greatest common divisor is 1, then $M^s(\mathbf{v})$ carries a universal family. (Note that the number-theoretic requirement also guarantees that there are no semistable sheaves of class $\mathbf{v}$.) See [**HL10**, §4.6] for details.

3. Properties of moduli spaces

To study the birational geometry of moduli spaces of sheaves in depth it is typically necessary to have some kind of control over the geometric properties of the space. For example, is the moduli space nonempty? Smooth? Irreducible? What are the divisor classes on the moduli space?

Our original setup of studying a smooth projective polarized variety (X, H) of any dimension is too general to get satisfactory answers to these questions. We first mention some results on smoothness which hold with a good deal of generality, and then turn to more specific cases with far more precise results.

3.1. Tangent spaces, smoothness, and dimension. Let (X, H) be a smooth polarized variety, and let $\mathbf{v} \in K_{\text{num}}(X)$. The tangent space to the moduli space $M = M(\mathbf{v})$ is typically only well-behaved at points $E \in M$ parameterizing stable sheaves E, due to the identification of S-equivalence classes of sheaves in M.

Let $D = \operatorname{Spec} \mathbb{C}[\varepsilon]/(\varepsilon^2)$ be the dual numbers, and let E be a stable sheaf. Then the tangent space to M is the subset of $\operatorname{Mor}(D, M)$ corresponding to maps sending the closed point of D to the point E. By the moduli property, such a map corresponds to a sheaf $\mathcal{E}$ on $X \times D$, flat over D, such that $E_0 = E$.

Deformation theory identifies the set of sheaves $\mathcal{E}$ as above with the vector space $\operatorname{Ext}^1(E, E)$, so there is a natural isomorphism $T_E M \cong \operatorname{Ext}^1(E, E)$. The obstruction to extending a first-order deformation is a class $\operatorname{Ext}^2(E, E)$, and if $\operatorname{Ext}^2(E, E) = 0$ then M is smooth at E.

For some varieties X it is helpful to improve the previous statement slightly, since the vanishing $\operatorname{Ext}^2(E, E) = 0$ can be rare, for example if K_X is trivial. If E is a vector bundle, let

$$\operatorname{tr} : \mathcal{E}nd(E) \to \mathcal{O}_X$$

be the *trace map*, acting fiberwise as the ordinary trace of an endomorphism. Then

$$H^i(\mathcal{E}nd(E)) \cong \operatorname{Ext}^i(E, E),$$

so there are induced maps on cohomology

$$\operatorname{tr}^i : \operatorname{Ext}^i(E, E) \to H^i(\mathcal{O}_X).$$

We write $\operatorname{Ext}^i(E, E)_0 \subset \operatorname{Ext}^i(E, E)$ for $\ker \operatorname{tr}^i$, the subspace of *traceless extensions.* The subspaces $\operatorname{Ext}^i(E, E)_0$ can also be defined if E is just a coherent sheaf, but the construction is more delicate and we omit it.

THEOREM 3.1. *The tangent space to M at a stable sheaf E is canonically isomorphic to* $\operatorname{Ext}^1(E, E)$*, the space of first order deformations of E. If* $\operatorname{Ext}^2(E, E)_0 = 0$*, then M is smooth at E of dimension* $\operatorname{ext}^1(E, E)$.

We now examine several consequences of Theorem 3.1 in the case of curves and surfaces.

EXAMPLE 3.2. Suppose X is a smooth curve of genus g, and let $M(r, d)$ be the moduli space of semistable sheaves of rank r and degree d on X. Then the vanishing $\operatorname{Ext}^2(E, E) = 0$ holds for any sheaf E, so the moduli space $M(P)$ is smooth at every point parameterizing a stable sheaf E. Since stable sheaves are simple, the dimension at such a sheaf is

$$\operatorname{ext}^1(E, E) = 1 - \chi(E, E) = r^2(g - 1) + 1.$$

EXAMPLE 3.3. Let (X, H) be a smooth variety, and let $\mathbf{v} = [\mathcal{O}_X] \in K_{\text{num}}(X)$ be the numerical class of $\mathcal{O}_X$. The moduli space $M(\mathbf{v})$ parameterizes line bundles numerically equivalent to $\mathcal{O}_X$; it is the connected component $\operatorname{Pic}^0 X$ of the Picard scheme $\operatorname{Pic} X$ which contains $\mathcal{O}_X$. For any line bundle $L \in M(\mathbf{v})$, we have $\mathcal{E}nd(L) \cong$

$\mathcal{O}_X$ and the trace map $\mathcal{E}nd(L) \to \mathcal{O}_X$ is an isomorphism. Thus $\operatorname{Ext}^2(L,L)_0 = 0$, and $M(\mathbf{v})$ is smooth of dimension $\operatorname{ext}^1(L,L) = h^1(\mathcal{O}_X) =: q(X)$, the *irregularity* of X.

EXAMPLE 3.4. Suppose (X,H) is a smooth surface and $E \in M^s(P)$ is a stable vector bundle. The sheaf map

$$\operatorname{tr} : \mathcal{E}nd(E) \to \mathcal{O}_X$$

is surjective, so the induced map $\operatorname{tr}^2 : \operatorname{Ext}^2(E,E) \to H^2(\mathcal{O}_X)$ is surjective since X is a surface. Therefore $\operatorname{ext}^2(E,E)_0 = 0$ if and only if $\operatorname{ext}^2(E,E) = h^2(\mathcal{O}_X)$. We conclude that if $\operatorname{ext}^2(E,E)_0 = 0$ then $M(P)$ is smooth at E of local dimension

$$\begin{aligned}\dim_E M(P) = \operatorname{ext}^1(E,E) &= 1 - \chi(E,E) + \operatorname{ext}^2(E,E)\\ &= 1 - \chi(E,E) + h^2(\mathcal{O}_X)\\ &= 2r^2\Delta(E) + \chi(\mathcal{O}_X)(1-r^2) + q(X).\end{aligned}$$

EXAMPLE 3.5. If (X,H) is a smooth surface such that $H.K_X < 0$, then the vanishing $\operatorname{Ext}^2(E,E) = 0$ is automatic. Indeed, by Serre duality,

$$\operatorname{Ext}^2(E,E) \cong \operatorname{Hom}(E, E \otimes K_X)^*.$$

Then

$$\mu(E \otimes K_X) = \mu(E) + \mu(K_X) = \mu(E) + H.K_X < \mu(E),$$

so $\operatorname{Hom}(E, E\otimes K_X) = 0$ by Proposition 2.15.

The assumption $H.K_X < 0$ in particular holds whenever X is a del Pezzo or Hirzebruch surface. Thus the moduli spaces $M(\mathbf{v})$ for these surfaces are smooth at points corresponding to stable sheaves.

EXAMPLE 3.6. If (X,H) is a smooth surface and K_X is trivial (e.g. X is a K3 or abelian surface), then the weaker vanishing $\operatorname{Ext}^2(E,E)_0 = 0$ holds. The trace map $\operatorname{tr}^2 : H^2(\mathcal{E}nd(E)) \to H^2(\mathcal{O}_X)$ is Serre dual to an isomorphism

$$H^0(\mathcal{O}_X) \to H^0(\mathcal{E}nd(E)) = \operatorname{Hom}(E,E),$$

so tr^2 is an isomorphism and $\operatorname{Ext}^2(E,E)_0 = 0$.

3.2. Existence and irreducibility. What are the possible numerical invariants $\mathbf{v} \in K_{\mathrm{num}}(X)$ of a semistable sheaf on X? When the moduli space is nonempty, is it irreducible? As usual, the case of curves is simplest.

3.2.1. *Existence and irreducibility for curves.* Let $M = M(r,d)$ be the moduli space of semistable sheaves of rank r and degree d on a smooth curve X of genus $g \geq 1$. Then M is nonempty and irreducible, and unless X is an elliptic curve and r,d are not coprime then the stable sheaves are dense in M. To show $M(r,d)$ is nonempty one can follow the basic outline of Example 2.17. For more details, see [**LeP97**, Chapter 8].

Irreducibility of $M(r,d)$ can be proved roughly as follows. We may as well assume $r \geq 2$ and $d \geq 2rg$ by tensoring by a sufficiently ample line bundle. Let L denote a line bundle of degree d on X, and consider extensions of the form

$$0 \to \mathcal{O}_X^{r-1} \to E \to L \to 0.$$

As L and the extension class vary, we obtain a family of sheaves $\mathcal{E}$ parameterized by a vector bundle S over the component $\operatorname{Pic}^d(X)$ of the Picard group.

On the other hand, by the choice of d, any semistable $E \in M(r,d)$ is generated by its global sections. A general collection of $r-1$ sections of E will be linearly independent at every $x \in X$, so that the quotient of the corresponding inclusion $\mathcal{O}_X^{r-1} \to E$ is a line bundle. Thus every semistable E fits into an exact sequence as above. The (irreducible) open subset of S parameterizing semistable sheaves therefore maps onto $M(r,d)$, and the moduli space is irreducible.

3.2.2. *Existence for surfaces.* For surfaces the existence question is quite subtle. The first general result in this direction is the Bogomolov inequality.

THEOREM 3.7 (Bogomolov inequality). *If (X,H) is a smooth surface and E is a μ_H-semistable sheaf on X then*

$$\Delta(E) \geq 0.$$

REMARK 3.8. Note that the discriminant $\Delta(E)$ is independent of the particular polarization H, so the inequality holds for any sheaf which is slope-semistable with respect to some choice of polarization.

Recall that line bundles L have $\Delta(L) = 0$, so in a sense the Bogomolov inequality is sharp. However, there are certainly Chern characters $\mathbf{v}$ with $\Delta(\mathbf{v}) \geq 0$ such that there is no semistable sheaf of character $\mathbf{v}$. A refined Bogomolov inequality should bound $\Delta(E)$ from below in terms of the other numerical invariants of E. Solutions to the existence problem for semistable sheaves on a surface can often be viewed as such improvements of the Bogomolov inequality.

3.2.3. *Existence for $\mathbb{P}^2$.* On $\mathbb{P}^2$, the classification of Chern characters $\mathbf{v}$ such that $M(\mathbf{v})$ is nonempty has been carried out by Drézet and Le Potier [**DLP85**, **LeP97**]. A *(semi)exceptional bundle* is a rigid (semi)stable bundle, i.e. a (semi)-stable bundle with $\mathrm{Ext}^1(E,E) = 0$. Examples of exceptional bundles include line bundles, the tangent bundle $T_{\mathbb{P}^2}$, and infinitely more examples obtained by a process of *mutation*. The dimension formula for a moduli space of sheaves on $\mathbb{P}^2$ reads

$$\dim M(\mathbf{v}) = r^2(2\Delta - 1) + 1,$$

so an exceptional bundle has discriminant $\Delta = \frac{1}{2} - \frac{1}{2r^2} < \frac{1}{2}$. The dimension formula suggests an immediate refinement of the Bogomolov inequality: if E is a non-exceptional stable bundle, then $\Delta(E) \geq \frac{1}{2}$.

However, exceptional bundles can provide even stronger Bogomolov inequalities for non-exceptional bundles. For example, suppose E is a semistable sheaf with $0 < \mu(E) < 1$. Then $\mathrm{Hom}(E, \mathcal{O}_X) = 0$ and

$$\mathrm{Ext}^2(E, \mathcal{O}_X) \cong \mathrm{Hom}(\mathcal{O}_X, E \otimes K_X)^* = 0$$

by semistability and Proposition 2.15. Thus $\chi(E, \mathcal{O}_X) \leq 0$. By the Riemann-Roch theorem, this inequality is equivalent to the inequality

$$\Delta(E) \geq P(-\mu(E))$$

where $P(x) = \frac{1}{2}x^2 + \frac{3}{2}x + 1$; this inequality is stronger than the ordinary Bogomolov inequality for any $\mu(E) \in (0,1)$.

Taking all the various exceptional bundles on $\mathbb{P}^2$ into account in a similar manner, one defines a function $\delta : \mathbb{R} \to \mathbb{R}$ with the property that any non-semiexceptional semistable bundle E satisfies $\Delta(E) \geq \delta(\mu(E))$. The graph of δ is Figure 1. Drézet and Le Potier prove the converse theorem: exceptional bundles are the only obstruction to the existence of stable bundles with given numerical invariants.

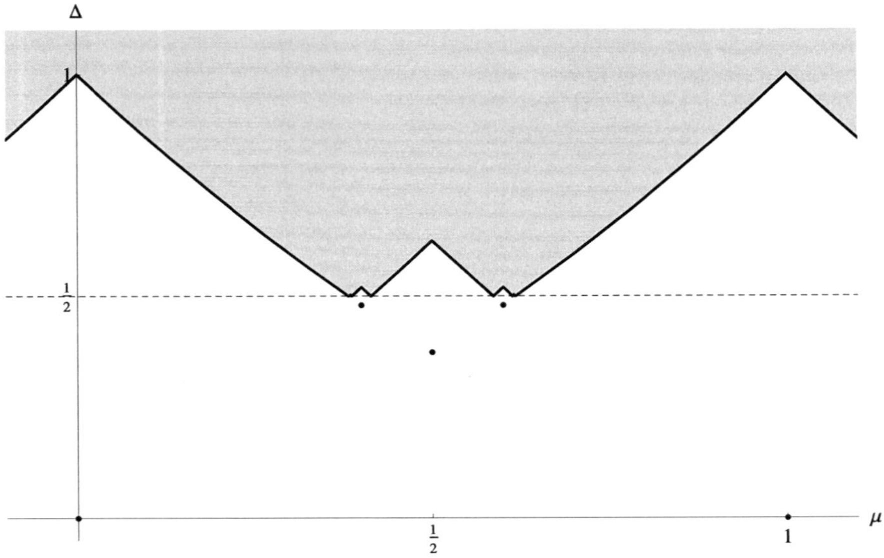

FIGURE 1. The curve $\delta(\mu)$ occurring in the classification of stable bundles on $\mathbb{P}^2$. If (r, μ, Δ) are the invariants of an integral Chern character, then there is a non-exceptional stable bundle E with these invariants if and only if $\Delta \geq \delta(\mu)$. The invariants of the first several exceptional bundles are also displayed.

THEOREM 3.9. *Let* $\mathbf{v}$ *be an integral Chern character on* $\mathbb{P}^2$. *There is a non-exceptional stable vector bundle on* $\mathbb{P}^2$ *with Chern character* $\mathbf{v}$ *if and only if* $\Delta(\mathbf{v}) \geq \delta(\mu(\mathbf{v}))$.

The method of proof follows the outline indicated in Example 2.17.

3.2.4. *Existence for other rational surfaces.* In the case of $X = \mathbb{P}^1 \times \mathbb{P}^1$, Rudakov [**Rud89, Rud94**] gives a solution to the existence problem that is similar to the Drézet-Le Potier result for $\mathbb{P}^2$. However, the geometry of exceptional bundles is more complicated than for $\mathbb{P}^2$, and as a result the classification is somewhat less explicit. To our knowledge a satisfactory answer to the existence problem has not yet been given for a Del Pezzo or Hirzebruch surface.

3.2.5. *Irreducibility for rational surfaces.* For many rational surfaces X it is known that the moduli space $M_H(\mathbf{v})$ is irreducible. One common argument is to introduce a mild relaxation of the notion of semistability and show that the stack parameterizing such objects is irreducible and contains the semistable sheaves as an open dense substack.

For example, Hirschowitz and Laszlo [**HiL93**] introduce the notion of a *prioritary sheaf* on $\mathbb{P}^2$. A torsion-free coherent sheaf E on $\mathbb{P}^2$ is prioritary if

$$\mathrm{Ext}^2(E, E(-1)) = 0.$$

By Serre duality, any torsion-free sheaf whose Harder-Narasimhan factors have slopes that are "not too far apart" will be prioritary, so it is very easy to construct prioritary sheaves. For example, semistable sheaves are prioritary, and sheaves of the form $\mathcal{O}_{\mathbb{P}^2}(a)^{\oplus k} \oplus \mathcal{O}_{\mathbb{P}^2}(a+1)^{\oplus l}$ are prioritary. The class of prioritary sheaves is also closed under elementary modifications, which makes it possible to study them by induction on the Euler characteristic as in Example 2.13.

The Artin stack $\mathcal{P}(\mathbf{v})$ of prioritary sheaves with invariants $\mathbf{v}$ is smooth, essentially because $\mathrm{Ext}^2(E,E)=0$ for any prioritary sheaf. There is a unique prioritary sheaf of a given slope and rank with minimal discriminant, given by a sheaf of the form $\mathcal{O}_{\mathbb{P}^2}(a)^{\oplus k}\oplus\mathcal{O}_{\mathbb{P}^2}(a+1)^{\oplus l}$ with the integers a,k,l chosen appropriately. Hirschowitz and Laszlo show that any connected component of $\mathcal{P}(\mathbf{v})$ contains a sheaf which is an elementary modification of another sheaf. By induction on the Euler characteristic, they conclude that $\mathcal{P}(\mathbf{v})$ is connected, and therefore irreducible. Since semistability is an open property, the stack $\mathcal{M}(\mathbf{v})$ of semistable sheaves is an open substack of $\mathcal{P}(\mathbf{v})$ and therefore dense and irreducible if it is nonempty. Thus the coarse space $M(\mathbf{v})$ is irreducible as well.

Walter [**Wal93**] gives another argument establishing the irreducibility of the moduli spaces $M_H(\mathbf{v})$ on a Hirzebruch surface whenever they are nonempty. The arguments make heavy use of the ruling, and study the stack of sheaves which are prioritary with respect to the fiber class. In more generality, he also studies the question of irreducibility on a geometrically ruled surface, at least under a condition on the polarization which ensures that semistable sheaves are prioritary with respect to the fiber class.

3.2.6. *Existence and irreducibility for K3's.* By work of Yoshioka, Mukai, and others, the existence problem has a particularly simple and beautiful solution when (X,H) is a smooth K3 surface (see [**Yos01**], or [**BM14a, BM14b**] for a simple treatment). Define the *Mukai pairing* $\langle -,-\rangle$ on $K_{\mathrm{num}}(X)$ by $\langle\mathbf{v},\mathbf{w}\rangle=-\chi(\mathbf{v},\mathbf{w})$; we can make sense of this formula by the same method as in Example 2.14. Since X is a K3 surface, K_X is trivial and the Mukai pairing is symmetric by Serre duality. By Example 3.4, if there is a stable sheaf E with invariants $\mathbf{v}$ then the moduli space $M(\mathbf{v})$ has dimension $2+\langle\mathbf{v},\mathbf{v}\rangle$ at E. If E is a stable sheaf of class $\mathbf{v}$ with $\langle\mathbf{v},\mathbf{v}\rangle=-2$, then E is called *spherical* and the moduli space $M_H(\mathbf{v})$ is a single reduced point.

A class $\mathbf{v}\in K_{\mathrm{num}}(X)$ is called *primitive* if it is not a multiple of another class. If the polarization H of X is chosen suitably generically, then $\mathbf{v}$ being primitive ensures that there are no strictly semistable sheaves of class $\mathbf{v}$. Thus, for a generic polarization, a necessary condition for the existence of a stable sheaf is that $\langle\mathbf{v},\mathbf{v}\rangle\geq -2$.

DEFINITION 3.10. A primitive class $\mathbf{v}=(r,c,d)\in K_{\mathrm{num}}(X)$ is called *positive* if $\langle\mathbf{v},\mathbf{v}\rangle\geq-2$ and either

(1) $r>0$, or
(2) $r=0$ and c is effective, or
(3) $r=0$, $c=0$, and $d>0$.

The additional requirements (1)-(3) in the definition are automatically satisfied any time there is a *sheaf* of class $\mathbf{v}$, so they are very mild.

THEOREM 3.11. *Let (X,H) be a smooth K3 surface. Let $\mathbf{v}\in K_{\mathrm{num}}(X)$, and write $\mathbf{v}=m\mathbf{v}_0$, where $\mathbf{v}_0$ is primitive and m is a positive integer.*

If $\mathbf{v}_0$ is positive, then the moduli space $M_H(\mathbf{v})$ is nonempty. If furthermore $m=1$ and the polarization H is sufficiently generic, then $M_H(\mathbf{v})$ is a smooth, irreducible, holomorphic symplectic variety.

If $M_H(\mathbf{v})$ is nonempty and the polarization is sufficiently generic, then $\mathbf{v}_0$ is positive.

The Mukai pairing can be made particularly simple from a computational standpoint by studying it in terms of a different coordinate system. Let

$$H^*_{\mathrm{alg}}(X) = H^0(X,\mathbb{Z}) \oplus \mathrm{NS}(X) \oplus H^4(X,\mathbb{Z}).$$

Then there is an isomorphism $v : K_{\mathrm{num}}(X) \to H^*_{\mathrm{alg}}(X,\mathbb{Z})$ defined by $v(\mathbf{v}) = \mathbf{v}\cdot\sqrt{\mathrm{td}(X)}$. The vector $v(\mathbf{v})$ is called a *Mukai vector*. The Todd class $\mathrm{td}(X) \in H^*_{\mathrm{alg}}(X)$ is $(1,0,2)$, so $\sqrt{\mathrm{td}(X)} = (1,0,1)$ and

$$v(\mathbf{v}) = (\mathrm{ch}_0(\mathbf{v}), \mathrm{ch}_1(\mathbf{v}), \mathrm{ch}_0(\mathbf{v}) + \mathrm{ch}_2(\mathbf{v})) = (r, c_1, r + \frac{c_1^2}{2} - c_2).$$

Suppose $\mathbf{v},\mathbf{w} \in K_{\mathrm{num}}(X)$ have Mukai vectors $v(\mathbf{v}) = (r,c,s)$, $v(\mathbf{w}) = (r',c',s')$. Since $\sqrt{\mathrm{td}(X)}$ is self-dual, the Hirzebruch-Riemann-Roch theorem gives

$$\langle \mathbf{v},\mathbf{w}\rangle = -\chi(\mathbf{v},\mathbf{w}) = -\int_X \mathbf{v}^*\cdot\mathbf{w}\cdot\mathrm{td}(X) = -\int_X (r,-c,s)\cdot(r',c',s') = cc' - rs' - r's.$$

It is worth pointing out that Theorem 3.11 can also be stated as a strong Bogomolov inequality, as in the Drézet-Le Potier result for $\mathbb{P}^2$. Let $\mathbf{v}_0$ be a primitive vector which is the vector of a coherent sheaf. The irregularity of X is $q(X) = 0$ and $\chi(\mathcal{O}_X) = 2$, so as in Example 3.4

$$\langle \mathbf{v}_0, \mathbf{v}_0\rangle = 2r^2\Delta(\mathbf{v}_0) + 2(1-r^2) - 2 = 2r^2(\Delta(\mathbf{v}_0) - 1).$$

Therefore, $\mathbf{v}_0$ is positive and non-spherical if and only if $\Delta(\mathbf{v}_0) \geq 1$.

3.2.7. *General surfaces.* On an arbitrary smooth surface (X,H) the basic geometry of the moduli space is less understood. To obtain good results, it is necessary to impose some kind of additional hypotheses on the Chern character $\mathbf{v}$.

For one possibility, we can take $\mathbf{v}$ to be the character of an ideal sheaf I_Z of a zero-dimensional scheme $Z \subset X$ of length n. Then the moduli space of sheaves of class $\mathbf{v}$ with determinant $\mathcal{O}_X$ is the *Hilbert scheme* of n points on X, written $X^{[n]}$. It parameterizes ideal sheaves of subschemes $Z \subset X$ of length n.

REMARK 3.12. Note that any rank 1 torsion-free sheaf E with determinant $\mathcal{O}_X$ admits an inclusion $E \to E^{**} := \det E = \mathcal{O}_X$, so that E is actually an ideal sheaf. Unless X has irregularity $q(X) = 0$, the Hilbert scheme $X^{[n]}$ and moduli space $M(\mathbf{v})$ will differ, since the latter space also contains sheaves of the form $L \otimes I_Z$, where L is a line bundle numerically equivalent to $\mathcal{O}_X$. In fact, $M(\mathbf{v}) \cong X^{[n]} \times \mathrm{Pic}^0(X)$.

Classical results of Fogarty show that Hilbert schemes of points on a surface are very well-behaved.

THEOREM 3.13 ([**Fog68**]). *The Hilbert scheme of points $X^{[n]}$ on a smooth surface X is smooth and irreducible. It is a fine moduli space, and carries a universal ideal sheaf.*

At the other extreme, if the rank is arbitrary then there are *O'Grady-type* results which show that the moduli space has many good properties if we require the discriminant of our sheaves to be sufficiently large.

THEOREM 3.14 ([**HL10, O'G96**]). *There is a constant C depending on X, H, and r, such that if $\mathbf{v}$ has rank r and $\Delta(\mathbf{v}) \geq C$ then the moduli space $M_H(\mathbf{v})$ is* *nonempty, irreducible, and normal. The μ-stable sheaves E such that* $\mathrm{ext}^2(E,E)_0 = 0$ *are dense in $M_H(\mathbf{v})$, so $M_H(\mathbf{v})$ has the expected dimension*

$$\dim M_H(\mathbf{v}) = 2r^2\Delta(E) + \chi(\mathcal{O}_X)(1-r^2) + q(X).$$

4. Divisors and classical birational geometry

In this section we introduce some of the primary objects of study in the birational geometry of varieties. We then study some simple examples of the birational geometry of moduli spaces from the classical point of view.

4.1. Cones of divisors. Let X be a normal projective variety. Recall that X is *factorial* if every Weil divisor on X is Cartier, and $\mathbb{Q}$-factorial if every Weil divisor has a multiple that is Cartier. To make the discussion in this section easier we will assume that X is $\mathbb{Q}$-factorial. This means that describing a codimension 1 locus on X determines the class of a $\mathbb{Q}$-Cartier divisor.

DEFINITION 4.1. Two Cartier divisors D_1, D_2 (or $\mathbb{Q}$- or $\mathbb{R}$-Cartier divisors) are *numerically equivalent*, written $D_1 \equiv D_2$, if $D_1 \cdot C = D_2 \cdot C$ for every curve $C \subset X$. The *Neron-Severi space* $N^1(X)$ is the real vector space $\mathrm{Pic}(X) \otimes \mathbb{R}/\equiv$.

4.1.1. *Ample and nef cones.* The first object of study in birational geometry is the *ample cone* $\mathrm{Amp}(X)$ of X. Roughly speaking, the ample cone parameterizes the various projective embeddings of X. A Cartier divisor D on X is *ample* if the map to projective space determined by $\mathcal{O}_X(mD)$ is an embedding for sufficiently large m. The Nakai-Moishezon criterion for ampleness says that D is ample if and only if $D^{\dim V}.V > 0$ for every subvariety $V \subset X$. In particular, ampleness only depends on the numerical equivalence class of D. A positive linear combination of ample divisors is also ample, so it is natural to consider the cone spanned by ample classes.

DEFINITION 4.2. The *ample cone* $\mathrm{Amp}(X) \subset N^1(X)$ is the open convex cone spanned by the numerical classes of ample Cartier divisors.

An $\mathbb{R}$-Cartier divisor D is ample if its numerical class is in the ample cone.

From a practical standpoint it is often easier to work with *nef* (i.e. *numerically effective*) divisors instead of ample divisors. We say that a Cartier divisor D is nef if $D.C \geq 0$ for every curve $C \subset X$. This is clearly a numerical condition, so nefness extends easily to $\mathbb{R}$-divisors and they span a cone $\mathrm{Nef}(X)$, the *nef cone* of X. By Kleiman's theorem, the problems of studying ample or nef cones are essentially equivalent.

THEOREM 4.3 ([**Deb01**, Theorem 1.27]). *The nef cone is the closure of the ample cone, and the ample cone is the interior of the nef cone:*

$$\mathrm{Nef}(X) = \overline{\mathrm{Amp}(X)} \qquad \textit{and} \qquad \mathrm{Amp}(X) = \mathrm{Nef}(X)^\circ.$$

Nef divisors are particularly important in birational geometry because they record the behavior of the simplest nontrivial morphisms to other projective varieties, as the next example shows.

EXAMPLE 4.4. Suppose $f : X \to Y$ is any morphism of projective varieties. Let L be a very ample line bundle on Y, and consider the line bundle f^*L. If $C \subset X$ is any irreducible curve, we can find an effective divisor $D \subset Y$ representing L such that the image of C is not contained entirely in D. This implies $C.(f^*L) \geq 0$, so

f^*L is nef. Note that if f contracts some curve $C \subset X$ to a point, then $C.(f^*L) = 0$, so f^*L is on the boundary of the nef cone.

As a partial converse, suppose D is a nef divisor on X such that the linear series $|mD|$ is base point free for some $m > 0$; such a divisor class is called *semiample.* Then for sufficiently large and divisible m, the image of the map $\phi_{|mD|} : X \to |mD|^*$ is a projective variety Y_m carrying an ample line bundle L such that $\phi_{|mD|}^* L = \mathcal{O}_X(mD)$. See [**Laz04**, Theorem 2.1.27] for details and a more precise statement.

EXAMPLE 4.5. Classically, to compute the nef (and hence ample) cone of a variety X one typically first constructs a subcone $\Lambda \subset \mathrm{Nef}(X)$ by finding divisors D on the boundary arising from interesting contractions $X \to Y$ as in Example 4.4. One then dually constructs interesting curves C on X to span a cone $\mathrm{Nef}(X) \subset \Lambda'$ given as the divisors intersecting the curves nonnegatively. If enough divisors and curves are constructed so that $\Lambda = \Lambda'$, then they equal the nef cone.

One of the main features of the positivity lemma of Bayer and Macrì will be that it produces nef divisors on moduli spaces of sheaves M without having to worry about finding a map $M \to Y$ to a projective variety giving rise to the divisor. A priori these nef divisors may not be semiample or have sections at all, so it may or may not be possible to construct these divisors and prove their nefness via more classical constructions. See §7 for more details.

EXAMPLE 4.6. For an easy example of the procedure in Example 4.5, consider the blowup $X = \mathrm{Bl}_p \mathbb{P}^2$ of $\mathbb{P}^2$ at a point p. Then $\mathrm{Pic}\, X \cong \mathbb{Z}H \oplus \mathbb{Z}E$, where H is the pullback of a line under the map $\pi : X \to \mathbb{P}^2$ and E is the exceptional divisor. The Neron-Severi space $N^1(X)$ is the two-dimensional real vector space spanned by H and E. Convex cones in $N^1(X)$ are spanned by two extremal classes.

Since π contracts E, the class H is an extremal nef divisor. We also have a fibration $f : X \to \mathbb{P}^1$, where the fibers are the proper transforms of lines through p. The pullback of a point in $\mathbb{P}^1$ is of class $H - E$, so $H - E$ is an extremal nef divisor. Therefore $\mathrm{Nef}(X)$ is spanned by H and $H - E$.

4.1.2. *(Pseudo)effective and big cones.* The easiest interesting space of divisors to define is perhaps the *effective cone* $\mathrm{Eff}(X) \subset N^1(X)$, defined as the subspace spanned by numerical classes of effective divisors. Unlike nefness and ampleness, however, effectiveness is *not* a numerical property: for instance, on an elliptic curve C, a line bundle of degree 0 has an effective multiple if and only if it is torsion.

The effective cone is in general neither open nor closed. Its closure $\overline{\mathrm{Eff}}(X)$ is less subtle, and called the *pseudo-effective cone.* The interior of the effective cone is the *big cone* $\mathrm{Big}(X)$, spanned by divisors D such that the linear series $|mD|$ defines a map $\phi_{|mD|}$ whose image has the same dimension as X. Thus, big divisors are the natural analog of birational maps. By Kodaira's Lemma [**Laz04**, Proposition 2.2.6], bigness is a numerical property.

EXAMPLE 4.7. The strategy for computing pseudoeffective cones is typically similar to that for computing nef cones. On the one hand, one constructs effective divisors to span a cone $\Lambda \subset \overline{\mathrm{Eff}}(X)$. A *moving curve* is a numerical curve class $[C]$ such that irreducible representatives of the class pass through a general point of X. Thus if D is an effective divisor we must have $D.C \geq 0$; otherwise D would have to contain every irreducible curve of class C. Thus the moving curve classes dually determine a cone $\overline{\mathrm{Eff}}(X) \subset \Lambda'$, and if $\Lambda = \Lambda'$ then they equal the pseudoeffective cone. This approach is justified by the seminal work of Boucksom-Demailly-Păun-Peternell, which establishes a duality between the pseudoeffective cone and the cone of moving curves [**BDPP13**].

EXAMPLE 4.8. On $X = \mathrm{Bl}_p \mathbb{P}^2$, the curve class H is moving and $H.E = 0$. Thus E spans an extremal edge of $\mathrm{Eff}(X)$. The curve class $H - E$ is also moving, and $(H-E)^2 = 0$. Therefore $H - E$ spans the other edge of $\mathrm{Eff}(X)$, and $\mathrm{Eff}(X)$ is spanned by $H - E$ and E.

4.1.3. *Stable base locus decomposition.* The nef cone $\mathrm{Nef}(X)$ is one chamber in a decomposition of the entire pseudoeffective cone $\overline{\mathrm{Eff}}(X)$. By the *base locus* $\mathrm{Bs}(D)$ of a divisor D we mean the base locus of the complete linear series $|D|$, regarded as a subset (i.e. not as a subscheme) of X. By convention, $\mathrm{Bs}(D) = X$ if $|D|$ is empty. The *stable base locus* of D is the subset

$$\mathbb{B}\mathrm{s}(D) = \bigcap_{m>0} \mathrm{Bs}(D)$$

of X. One can show that $\mathbb{B}\mathrm{s}(D)$ coincides with the base locus $\mathrm{Bs}(mD)$ of sufficiently large and divisible multiples mD.

EXAMPLE 4.9. The base locus and stable base locus of D depend on the class of D in $\mathrm{Pic}(X)$, not just on the numerical class of D. For example, if L is a degree 0 line bundle on an elliptic curve X, then $\mathrm{Bs}(L) = X$ unless L is trivial, and $\mathbb{B}\mathrm{s}(L) = X$ unless L is torsion in $\mathrm{Pic}(X)$.

Since (stable) base loci do not behave well with respect to numerical equivalence, for the rest of this subsection we assume $q(X) = 0$ so that linear and numerical equivalence coincide and $N^1(X)_{\mathbb{Q}} = \mathrm{Pic}(X) \otimes \mathbb{Q}$. Then the pseudoeffective cone $\overline{\mathrm{Eff}}(X)$ has a wall-and-chamber decomposition where the stable base locus remains constant on the open chambers. These various chambers control the birational maps from X to other projective varieties. For example, if $f : X \dashrightarrow Y$ is the rational map given by a sufficiently divisible multiple $|mD|$, then the indeterminacy locus of the map is contained in the stable base locus.

EXAMPLE 4.10. Stable base loci decompositions are typically computed as follows. First, one constructs effective divisors in a multiple $|mD|$ and takes their intersection to get a variety Y with $\mathbb{B}\mathrm{s}(D) \subset Y$. In the other direction, one looks for curves C on X such that $C.D < 0$. Then any divisor of class mD must contain C, so $\mathbb{B}\mathrm{s}(D)$ contains every curve numerically equivalent to C.

When the Picard rank of X is two, the chamber decompositions can often be made very explicit. In this case it is notation ally conventient to write, for example, $(D_1, D_2]$ to denote the cone of divisors of the form $a_1 D_1 + a_2 D_2$ with $a_1 \geq 0$ and $a_2 > 0$.

EXAMPLE 4.11. Let $X = \mathrm{Bl}_p \mathbb{P}^2$. The nef cone is $[H, H-E]$, and both $H, H-E$ are basepoint free. Thus the stable base locus is empty in the closed chamber $[H, H-E]$. If $D \in (H, E]$ is an effective divisor, then $D.E < 0$, so D contains E as a component. The stable base locus of divisors in the chamber $(H, E]$ is E.

We now begin to investigate the birational geometry of some of the simplest moduli spaces of sheaves on surfaces from a classical point of view.

4.2. Birational geometry of Hilbert schemes of points. Let X be a smooth surface with irregularity $q(X) = 0$, and let $\mathbf{v}$ be the Chern character of an ideal sheaf I_Z of a collection Z of n points. Then $M(\mathbf{v})$ is the Hilbert scheme $X^{[n]}$ of n points on X, parameterizing zero-dimensional schemes of length n. See §3.2.7 for its basic properties.

4.2.1. *Divisor classes.* Divisor classes on the Hilbert scheme $X^{[n]}$ can be understood entirely in terms of the birational *Hilbert-Chow morphism* $h : X^{[n]} \to X^{(n)}$ to the symmetric product $X^{(n)} = \operatorname{Sym}^n X$. Informally, this map sends the ideal sheaf of Z to the sum of the points in Z, with multiplicities given by the length of the scheme at each point.

REMARK 4.12. The symmetric product $X^{(n)}$ can itself be viewed as the moduli space of 0-dimensional sheaves with Hilbert polynomial $P(m) = n$. Suppose E is a zero-dimensional sheaf with constant Hilbert polynomial ℓ and that E is supported at a single point p. Then E admits a length ℓ filtration where all the quotients are isomorphic to $\mathcal{O}_p$. Thus, E is S-equivalent to $\mathcal{O}_p^{\oplus \ell}$. Since S-equivalent sheaves are identified in the moduli space, the moduli space $M(P)$ is just $X^{(n)}$.

The Hilbert-Chow morphism $h : X^{[n]} \to X^{(n)}$ can now be seen to come from the moduli property for $X^{(n)}$. Let $\mathcal{I}$ be the universal ideal sheaf on $X \times X^{[n]}$. The quotient of the inclusion $\mathcal{I} \to \mathcal{O}_{X \times X^{[n]}}$ is then a family of zero-dimensional sheaves of length n. This family induces a map $X^{[n]} \to X^{(n)}$, which is just the Hilbert-Chow morphism.

The exceptional locus of the Hilbert-Chow morphism is a divisor class B on the Hilbert scheme $X^{[n]}$. Alternately, B is the locus of nonreduced schemes. It is swept out by curves contained in fibers of the Hilbert-Chow morphism. A simple example of such a curve is given by fixing $n-2$ points in X and allowing a length 2 scheme $\operatorname{Spec} \mathbb{C}[\varepsilon]/(\varepsilon^2)$ to "spin" at one additional point.

REMARK 4.13. The divisor class $B/2$ is also Cartier, although it is not effective so it is harder to visualize. Let $\mathcal{Z} \subset X \times X^{[n]}$ denote the universal subscheme of length n, and let $p : \mathcal{Z} \to X$ and $q : \mathcal{Z} \to X^{[n]}$ be the projections. Then the *tautological bundle* $q_* p^* \mathcal{O}_X$ is a rank n vector bundle with determinant of class $-B/2$.

Any line bundle L on X induces a line bundle $L^{(n)}$ on the symmetric product. Pulling back this line bundle by the Hilbert-Chow morphism gives a line bundle $L^{[n]} := h^* L^{(n)}$. This gives an inclusion $\operatorname{Pic}(X) \to \operatorname{Pic}(X^{[n]})$. If L can be represented by a reduced effective divisor D, then $L^{[n]}$ can be represented by the locus

$$D^{[n]} := \{Z \in X^{[n]} : Z \cap D \neq \emptyset\}.$$

Fogarty proves that the divisors mentioned so far generate the Picard group.

THEOREM 4.14 (Fogarty [**Fog73**]). *Let X be a smooth surface with $q(X) = 0$. Then*

$$\operatorname{Pic}(X^{[n]}) \cong \operatorname{Pic}(X) \oplus \mathbb{Z}(B/2).$$

Thus, tensoring by $\mathbb{R}$,

$$N^1(X^{[n]}) \cong N^1(X) \oplus \mathbb{R}B.$$

There is another interesting way to use a line bundle on X to construct effective divisor classes. In examples, many extremal effective divisors can be realized in this way.

EXAMPLE 4.15. Suppose L is a line bundle on X with $m := h^0(L) > n$. If $Z \subset X$ is a general subscheme of length n, then $H^0(L \otimes I_Z) \subset H^0(L)$ is a subspace of codimension n. Thus we get a rational map

$$\phi : X^{[n]} \dashrightarrow G := \mathrm{Gr}(m-n, m)$$

to the Grassmannian G of codimension n subspaces of $H^0(L)$. The line bundle $\widetilde{L}^{[n]} := \phi^*\mathcal{O}_G(1)$ (which is well-defined since the indeterminacy locus of ϕ has codimension at least 2) can be represented by an effective divisor as follows. Let $W \subset H^0(L)$ be a sufficiently general subspace of dimension n; one frequently takes W to be the subspace of sections of L passing through $m-n$ general points. Then the locus

$$\widetilde{D}^{[n]} = \{Z \in X^{[n]} : H^0(L \otimes I_Z) \cap W \neq \{0\}\}$$

is an effective divisor representing $\phi^*\mathcal{O}_G(1)$.

4.2.2. *Curve classes.* Let $C \subset X$ be an irreducible curve. There are two immediate ways that we can induce a curve class on $X^{[n]}$.

EXAMPLE 4.16. Fix $n-1$ points $p_1, \ldots, p_{n-1}$ on X which are not in C. Allowing an nth point p_n to travel along C gives a curve $\widetilde{C}_{[n]} \subset X^{[n]}$.

EXAMPLE 4.17. Suppose C admits a g^1_n. If the g^1_n is base-point free, then we get a degree n map $C \to \mathbb{P}^1$. The fibers of this map induce a rational curve $\mathbb{P}^1 \to X^{[n]}$, and we write $C_{[n]}$ for the class of the image. If the g^1_n is not base-point free, we can first remove the basepoints to get a map $\mathbb{P}^1 \to X^{[m]}$ for some $m < n$, and then glue the basepoints back on to get a map $\mathbb{P}^1 \to X^{[n]}$. The class $C_{[n]}$ doesn't depend on the particular g^1_n used to construct the curve (see for example [**Hui12**, Proposition 3.5] in the case of $\mathbb{P}^2$).

REMARK 4.18. Typically the curve classes $C_{[n]}$ are more interesting than $\widetilde{C}_{[n]}$ and they frequently show up as extremal curves in the cone of curves. However, the class $C_{[n]}$ is only defined if $C_{[n]}$ carries an interesting linear series of degree n, while $\widetilde{C}_{[n]}$ always makes sense; thus curves of class $\widetilde{C}_{[n]}$ are also sometimes used.

Both curve classes $\widetilde{C}_{[n]}$ and $C_{[n]}$ have the useful property that the intersection pairing with divisors is preserved, in the sense that if $D \subset X$ is a divisor then

$$D^{[n]}.\widetilde{C}_{[n]} = D^{[n]}.C_{[n]} = D.C;$$

indeed, it suffices to check the equalities when D and C intersect transversely, and in that case $D^{[n]}$ and $C_{[n]}$ (resp. $\widetilde{C}_{[n]}$) intersect transversely in $D.C$ points.

The intersection with B is more interesting. Clearly

$$\widetilde{C}_{[n]}.B = 0.$$

On the other hand, the nonreduced schemes parameterized by a curve of class $C_{[n]}$ correspond to ramification points of the degree n map $C \to \mathbb{P}^1$. The Riemann-Hurwitz formula then implies

$$C_{[n]}.B = 2g(C) - 2 + 2n.$$

One further curve class is useful; we write C_0 for the class of a curve contracted by the Hilbert-Chow morphism.

4.2.3. *The intersection pairing.* At this point we have collected enough curve and divisor classes to fully determine the intersection pairing between curves and divisors and find relations between the various classes. The classes C_0 and $C_{[n]}$ for C any irreducible curve span $N_1(X)$, so to completely compute the intersection pairing we are only missing the intersection number $C_0.B$. However, since this intersection number is negative, we use the additional curve and divisor classes $\widetilde{C}_{[n]}$ and $\widetilde{D}^{[n]}$ to compute this number. To this end, we compute the intersection numbers of $\widetilde{D}^{[n]}$ with our curve classes.

EXAMPLE 4.19. To compute $\widetilde{D}^{[n]}.C_0$, let $m = h^0(\mathcal{O}_X(D))$, fix $m-n$ general points $p_1, \ldots, p_{m-n}$ in X, and represent $\widetilde{D}^{[n]}$ as the set of schemes Z such that there is a curve on X of class D passing through $p_1, \ldots, p_{m-n}$ and Z. Schemes parameterized by C_0 are supported at $n-1$ general points $q_1, \ldots, q_{n-1}$, with a spinning tangent vector at q_{n-1}. There is a unique curve D' of class D passing through $p_1, \ldots, p_{m-n}, q_1, \ldots, q_{n-1}$, and it is smooth at q_{n-1}, so there is a single point of intersection between C_0 and $\widetilde{D}^{[n]}$, occurring when the tangent vector at q_{n-1} is tangent to D'. Thus $\widetilde{D}^{[n]}.C_0 = 1$.

EXAMPLE 4.20. Next we compute $\widetilde{C}_{[n]}.\widetilde{D}^{[n]}$. Represent $\widetilde{D}^{[n]}$ as in Example 4.19. The curve class $\widetilde{C}_{[n]}$ is represented by fixing $n-1$ points $q_1, \ldots, q_{n-1}$ and letting q_n travel along C. There is a unique curve D' of class D passing through $p_1, \ldots, p_{m-n}, q_1, \ldots, q_{n-1}$, so $\widetilde{C}_{[n]}$ meets $\widetilde{D}^{[n]}$ when $q_n \in C \cap D'$. Thus $\widetilde{C}_{[n]}.\widetilde{D}^{[n]} = C.D$.

EXAMPLE 4.21. For an irreducible curve $C \subset X$, write $\widehat{C}_{[n]}$ for the curve class on $X^{[n]}$ obtained by fixing $n-2$ general points in X, fixing one point on C, and letting one point travel along C and collide with the point fixed on C. It follows immediately that

$$\widehat{C}_{[n]}.D^{[n]} = C.D$$
$$\widehat{C}_{[n]}.\widetilde{D}^{[n]} = C.D - 1.$$

Less immediately, we find $\widehat{C}_{[n]}.B = 2$: while the curve meets B set-theoretically in one point, a tangent space calculation shows this intersection has multiplicity 2.

We now collect our known intersection numbers.

	$D^{[n]}$	$\widetilde{D}^{[n]}$	B
$C_{[n]}$	$C.D$		$2g(C)-2+2n$
$\widetilde{C}_{[n]}$	$C.D$	$C.D$	0
$\widehat{C}_{[n]}$	$C.D$	$C.D-1$	2
C_0	0	1	

As $\widetilde{D}^{[n]}.C_0 \neq 0$, the divisors $\widetilde{D}^{[n]}$ are all not in the codimension one subspace $N^1(X) \subset N^1(X^{[n]})$. Therefore the divisor classes of type $D^{[n]}$ and $\widetilde{D}^{[n]}$ together span $N^1(X^{[n]})$. It now follows that

$$C_0 + \widehat{C}_{[n]} = \widetilde{C}_{[n]}$$

since both sides pair the same with divisors $D^{[n]}$ and $\widetilde{D}^{[n]}$, and thus $C_0.B = -2$. We then also find relations

$$C_{[n]} = \widetilde{C}_{[n]} - (g(C) - 1 + n)C_0$$

and

$$\widetilde{D}^{[n]} = D^{[n]} - \frac{1}{2}B.$$

In particular, the divisors of type $\widetilde{D}^{[n]}$ are all in the half-space of divisors with negative coefficient of B in terms of the Fogarty isomorphism $N^1(X^{[n]}) \cong N^1(X) \oplus \mathbb{R}B$. We can also complete our intersection table.

	$D^{[n]}$	$\widetilde{D}^{[n]}$	B
$C_{[n]}$	$C.D$	$C.D - (g(C) - 1 + n)$	$2g(C) - 2 + 2n$
$\widetilde{C}_{[n]}$	$C.D$	$C.D$	0
$\widehat{C}_{[n]}$	$C.D$	$C.D - 1$	2
C_0	0	1	-2

4.2.4. *Some nef divisors.* Part of the nef cone of $X^{[n]}$ now follows from our knowledge of the intersection pairing. First observe that since $C_0.D^{[n]} = 0$ and $C_0.B < 0$, the nef cone is contained in the half-space of divisors with nonpositive B-coefficient in terms of the Fogarty isomorphism.

If D is an ample divisor on X, then the divisor $D^{(n)}$ on the symmetric product is also ample, so $D^{[n]}$ is nef. Since a limit of nef divisors is nef, it follows that if D is nef on X then $D^{[n]}$ is nef on $X^{[n]}$. Furthermore, if D is on the boundary of the nef cone of X then $D^{[n]}$ is on the boundary of the nef cone of $X^{[n]}$. Indeed, if $C.D = 0$ then $\widetilde{C}_{[n]}.D^{[n]} = 0$ as well. This proves

$$\mathrm{Nef}(X^{[n]}) \cap N^1(X) = \mathrm{Nef}(X),$$

where by abuse of notation we embed $N^1(X)$ in $N^1(X^{[n]})$ by $D \mapsto D^{[n]}$.

Boundary nef divisors which are not contained in the hyperplane $N^1(X)$ are more interesting and more challenging to compute. Bridgeland stability and the positivity lemma will give us a tool for computing and describing these classes.

4.2.5. *Examples.* We close our initial discussion of the birational geometry of Hilbert schemes of points by considering several examples from this classical point of view.

EXAMPLE 4.22 ($\mathbb{P}^{2[n]}$). The Neron-Severi space $N^1(\mathbb{P}^{2[n]})$ of the Hilbert scheme of n points in $\mathbb{P}^2$ is spanned by $H^{[n]}$ and B, where H is the class of a line in $\mathbb{P}^2$. Any divisor in the cone $(H, B]$ is negative on C_0, so the locus B swept out by curves of class C_0 is contained in the stable base locus of any divisor in this chamber. Since $B.\widetilde{H}_{[n]} = 0$ and $\widetilde{H}_{[n]}$ is the class of a moving curve, the divisor B is an extremal effective divisor.

The divisor $H^{[n]}$ is an extremal nef divisor by §4.2.4, so to compute the full nef cone we only need one more extremal nef class. The line bundle $\mathcal{O}_{\mathbb{P}^2}(n-1)$ is *n-very ample*, meaning that if $Z \subset \mathbb{P}^2$ is any zero-dimensional subscheme of length n, then $H^0(I_Z(n-1))$ has codimension n in $H^0(\mathcal{O}_{\mathbb{P}^2}(n-1))$. Consequently, if G is the Grassmannian of codimension-n planes in $H^0(\mathcal{O}_{\mathbb{P}^2}(n-1))$, then the natural map $\phi : \mathbb{P}^{2[n]} \to G$ is a *morphism.* Thus $\phi^*\mathcal{O}_G(1)$ is nef. In our notation for divisors, putting $D = (n-1)H$ we conclude that

$$\widetilde{D}^{[n]} = (n-1)H^{[n]} - \frac{1}{2}B$$

is nef.

Furthermore, $\widetilde{D}^{[n]}$ is not ample. Numerically, simply observe that $\widetilde{D}^{[n]}.H_{[n]} = 0$. More geometrically, if two length n schemes Z, Z' are contained in the same line

L then the subspaces $H^0(I_Z(n-1))$ and $H^0(I_{Z'}(n-1))$ are equal, so ϕ identifies Z and Z'. Note that if Z and Z' are both contained in a single line L then their ideal sheaves can be written as extensions

$$0 \to \mathcal{O}_{\mathbb{P}^2}(-1) \to I_Z \to \mathcal{O}_L(-n) \to 0$$
$$0 \to \mathcal{O}_{\mathbb{P}^2}(-1) \to I_{Z'} \to \mathcal{O}_L(-n) \to 0.$$

This suggests that if we have some new notion of semistability where I_Z is strictly semistable with Jordan-Hölder factors $\mathcal{O}_{\mathbb{P}^2}(-1)$ and $\mathcal{O}_L(-n)$ then the ideal sheaves I_Z and $I_{Z'}$ will be S-equivalent. Thus, in the moduli space of such objects, I_Z and $I_{Z'}$ will be represented by the same point of the moduli space.

EXAMPLE 4.23 ($\mathbb{P}^{2[2]}$). The divisor $\widetilde{H}^{[2]} = H^{[2]} - \frac{1}{2}B$ spanning an edge of the nef cone is also an extremal effective divisor on $\mathbb{P}^{2[2]}$. Indeed, the orthogonal curve class $H_{[2]}$ is a moving curve on $\mathbb{P}^{2[2]}$. Thus there two chambers in the stable base locus decomposition of $\overline{\mathrm{Eff}}(\mathbb{P}^{2[2]})$.

EXAMPLE 4.24 ($\mathbb{P}^{2[3]}$). By Example 4.22, on $\mathbb{P}^{2[3]}$ the divisor $2H^{[3]} - \frac{1}{2}B$ is an extremal nef divisor. The open chambers of the stable base locus decomposition are

$$(H^{[3]}, B), (2H^{[3]} - \frac{1}{2}B, H^{[3]}), \text{ and } (H^{[3]} - \frac{1}{2}B, 2H^{[3]} - \frac{1}{2}B).$$

To establish this, first observe that $H^{[3]} - \frac{1}{2}B$ is the class of the locus D of collinear schemes, since $D.C_0 = 1$ and $D.\widetilde{H}_{[3]} = 1$. The divisor $2H^{[3]} - \frac{1}{2}B$ is orthogonal to curves of class $H_{[3]}$, so the locus of collinear schemes swept out by these curves lies in the stable base locus of any divisor in $(H^{[3]} - \frac{1}{2}B, 2H^{[3]} - \frac{1}{2}B)$. In the other direction, any divisor in $(H^{[3]} - \frac{1}{2}B, 2H^{[3]} - \frac{1}{2}B)$ is the sum of a divisor on the ray spanned by D and an ample divisor. It follows that the stable base locus in this chamber is exactly D.

For many more examples of the stable base locus decomposition of $\mathbb{P}^{2[n]}$, see [**ABCH13**] for explicit examples with $n \leq 9$, [**CH14a**] for a discussion of the chambers where monomial schemes are in the base locus, and [**Hui16, CHW16**] for the effective cone. Alternately, see [**CH14b**] for a deeper survey. Also, see the work of Li and Zhao [**LZ16**] for more recent developments unifying several of these topics.

4.3. Birational geometry of moduli spaces of sheaves. We now discuss some of the basic aspects of the birational geometry of moduli spaces of sheaves. Many of the concepts are mild generalizations of the picture for Hilbert schemes of points.

4.3.1. *Line bundles.* The main method of constructing line bundles on a moduli space of sheaves is by a determinantal construction. First suppose $\mathcal{E}/S$ is a family of sheaves on X parameterized by S. Let $p : S \times X \to S$ and $q : S \times X \to X$ be the projections. The *Donaldson homomorphism* is a map $\lambda_{\mathcal{E}} : K(X) \to \mathrm{Pic}(S)$ defined by the composition

$$\lambda_{\mathcal{E}} : K(X) \xrightarrow{q^*} K^0(S \times X) \xrightarrow{\cdot[\mathcal{E}]} K^0(S \times X) \xrightarrow{p_!} K^0(S) \xrightarrow{\det} \mathrm{Pic}(S)$$

Here $p_! = \sum_i (-1)^i R^i p_*$. Informally, we pull back a sheaf on X to the product, twist by the family $\mathcal{E}$, push forward to S, and take the determinant line bundle. Thus we obtain from any class in $K(X)$ a line bundle on the base S of the family

$\mathcal{E}$. The above discussion is sufficient to define line bundles on a moduli space $M(\mathbf{v})$ of sheaves if there is a universal family $\mathcal{E}$ on $M(\mathbf{v})$: there is then a map $\lambda_{\mathcal{E}} : K(X) \to \operatorname{Pic}(M(\mathbf{v}))$, and the image typically consists of many interesting line bundles on the moduli space.

Things are slightly more delicate in the general case where there is no universal family. As motivation, given a class $\mathbf{w} \in K(X)$, we would like to define a line bundle L on $M(\mathbf{v})$ with the following property. Suppose $\mathcal{E}/S$ is a family of sheaves of character $\mathbf{v}$ and that $\phi : S \to M(\mathbf{v})$ is the moduli map. Then we would like there to be an isomorphism $\phi^* L \cong \lambda_{\mathcal{E}}(\mathbf{w})$, so that the determinantal line bundle $\lambda_{\mathcal{E}}(\mathbf{w})$ on S is the pullback of a line bundle on the moduli space $M(\mathbf{v})$.

In order for this to be possible, observe that the line bundle $\lambda_{\mathcal{E}}(\mathbf{w})$ must be unchanged when it is replaced by $\mathcal{E} \otimes p^* N$ for some line bundle $N \in \operatorname{Pic}(S)$. Indeed, the moduli map $\phi : S \to M(\mathbf{v})$ is not changed when we replace $\mathcal{E}$ by $\mathcal{E} \otimes p^* N$, so $\phi^* L$ is unchanged as well. However, a computation shows that

$$\lambda_{\mathcal{E} \otimes p^* N}(\mathbf{w}) = \lambda_{\mathcal{E}}(\mathbf{w}) \otimes N^{\otimes \chi(\mathbf{v} \otimes \mathbf{w})}.$$

Thus, in order for there to be a chance of defining a line bundle L on $M(\mathbf{v})$ with the desired property we need to assume that $\chi(\mathbf{v} \otimes \mathbf{w}) = 0$.

In fact, if $\chi(\mathbf{v} \otimes \mathbf{w}) = 0$, then there is a line bundle L as above on the stable locus $M^s(\mathbf{v})$, denoted by $\lambda^s(\mathbf{w})$. To handle things rigorously, it is necessary to go back to the construction of the moduli space via GIT. See [**HL10**, §8.1] for full details, as well as a discussion of line bundles on the full moduli space $M(\mathbf{v})$.

THEOREM 4.25 ([**HL10**, Theorem8.1.5]). *Let $\mathbf{v}^\perp \subset K(X)$ denote the orthogonal complement of $\mathbf{v}$ with respect to the Euler pairing $\chi(- \otimes -)$. Then there is a natural homomorphism*

$$\lambda^s : \mathbf{v}^\perp \to \operatorname{Pic}(M^s(\mathbf{v})).$$

In general it is a difficult question to completely determine the Picard group of the moduli space. One of the best results in this direction is the following theorem of Jun Li.

THEOREM 4.26 ([**Li94**]). *Let X be a regular surface, and let $\mathbf{v} \in K(X)$ with $\operatorname{rk} \mathbf{v} = 2$ and $\Delta(\mathbf{v}) \gg 0$. Then the map*

$$\lambda^s : \mathbf{v}^\perp \otimes \mathbb{Q} \to \operatorname{Pic}(M^s(\mathbf{v})) \otimes \mathbb{Q}$$

is a surjection.

More precise results are somewhat rare. We discuss a few of the main such examples here.

EXAMPLE 4.27 (Picard group of moduli spaces of sheaves on $\mathbb{P}^2$). Let $M(\mathbf{v})$ be a moduli space of sheaves on $\mathbb{P}^2$. The Picard group of this space was determined by Drézet [**Dre88**]. The answer depends on the δ-function introduced in the classification of semistable characters in §3.2.3. If $\mathbf{v}$ is the character of an exceptional bundle then $M(\mathbf{v})$ is a point and there is nothing to discuss. If $\delta(\mu(\mathbf{v})) = \Delta(\mathbf{v})$, then $M(\mathbf{v})$ is a moduli space of so-called *height zero* bundles and the Picard group is isomorphic to $\mathbb{Z}$. Finally, if $\delta(\mu(\mathbf{v})) > \Delta(\mathbf{v})$ then the Picard group is isomorphic to $\mathbb{Z} \oplus \mathbb{Z}$. In each case, the Donaldson morphism is surjective.

EXAMPLE 4.28 (Picard group of moduli spaces of sheaves on $\mathbb{P}^1 \times \mathbb{P}^1$). Let $M(\mathbf{v})$ be a moduli space of sheaves on $\mathbb{P}^1 \times \mathbb{P}^1$. Already in this case the Picard group does not appear to be known in every case. See [**Yos96**] for some partial results, as well as results on ruled surfaces in general.

EXAMPLE 4.29 (Picard group of moduli spaces of sheaves on a $K3$ surface). Let X be a $K3$ surface, and let $\mathbf{v} \in K_{\text{num}}(X)$ be a primitive positive vector (see §3.2.6). Let H be a polarization which is generic with respect to $\mathbf{v}$. In this case the story is similar to the computation for $\mathbb{P}^2$, with the Beauville-Bogomolov form playing the role of the δ function. If $\langle \mathbf{v}, \mathbf{v} \rangle = -2$ then $M_H(\mathbf{v})$ is a point. If $\langle \mathbf{v}, \mathbf{v} \rangle = 0$, then the Donaldson morphism $\lambda : \mathbf{v}^\perp \otimes \mathbb{R} \to N^1(M_H(\mathbf{v}))$ is surjective with kernel spanned by $\mathbf{v}$, and $N^1(M_H(\mathbf{v}))$ is isomorphic to $\mathbf{v}^\perp / \mathbf{v}$. Finally, if $\langle \mathbf{v}, \mathbf{v} \rangle > 0$ then the Donaldson morphism is an isomorphism. See [**Yos01**] or [**BM14a**] for details.

EXAMPLE 4.30 (Brill-Noether divisors). For birational geometry it is important to be able to construct sections of line bundles. The determinantal line bundles introduced above frequently have special sections vanishing on *Brill-Noether divisors*. Let (X, H) be a smooth surface, and let $\mathbf{v}$ and $\mathbf{w}$ be an orthogonal pair of Chern characters, i.e. suppose that $\chi(\mathbf{v} \otimes \mathbf{w}) = 0$, and suppose that there is a reasonable, e.g. irreducible, moduli space $M_H(\mathbf{v})$ of semistable sheaves. Suppose F is a vector bundle with $\operatorname{ch} F = \mathbf{w}$, and consider the locus

$$D_F = \{E \in M_H(\mathbf{v}) : H^0(E \otimes F) \neq 0\}.$$

If we assume that $H^2(E \otimes F) = 0$ for every $E \in M_H(\mathbf{v})$ and that $H^0(E \otimes F) = 0$ for a general $E \in M_H(\mathbf{v})$ then the locus D_F will be an effective divisor. Furthermore, its class is $\lambda(\mathbf{w}^*)$.

The assumption that $H^2(E \otimes F) = 0$ often follows easily from stability and Serre duality. For instance, if $\mu_H(\mathbf{v}), \mu_H(\mathbf{w}), \mu_H(K_X) > 0$ and F is a semistable vector bundle then

$$H^2(E \otimes F) = \operatorname{Ext}^2(F^*, E) = \operatorname{Hom}(E, F^*(K_X))^* = 0$$

by stability. On the other hand, it can be quite challenging to verify that $H^0(E \otimes F) = 0$ for a general $E \in M_H(\mathbf{v})$. These types of questions have been studied in [**CHW16**] in the case of $\mathbb{P}^2$ and [**Rya16**] in the case of $\mathbb{P}^1 \times \mathbb{P}^1$. Interesting effective divisors arising in the birational geometry of moduli spaces frequently arise in this way.

4.3.2. *The Donaldson-Uhlenbeck-Yau compactification.* For Hilbert schemes of points $X^{[n]}$, the symmetric product $X^{(n)}$ offered an alternate compactification, with the map $h : X^{[n]} \to X^{(n)}$ being the Hilbert-Chow morphism. Recall that from a moduli perspective the Hilbert-Chow morphism sends the ideal sheaf I_Z to (the S-equivalence class of) the structure sheaf $\mathcal{O}_Z$. Thinking of $\mathcal{O}_X$ as the double-dual of I_Z, the sheaf $\mathcal{O}_Z$ is the cokernel in the sequence

$$0 \to I_Z \to \mathcal{O}_X \to \mathcal{O}_Z \to 0.$$

The Donaldson-Uhlenbeck-Yau compactification can be viewed as analogous to the compactification of the Hilbert scheme by the symmetric product.

Let (X, H) be a smooth surface, and let $\mathbf{v}$ be the Chern character of a semistable sheaf of positive rank. Set-theoretically, the Donaldson-Uhlenbeck-Yau compactification $M_H^{DUY}(\mathbf{v})$ of the moduli space $M_H(\mathbf{v})$ can be defined as follows. Recall that the double dual of any torsion-free sheaf E on X is locally free, and there is a canonical inclusion $E \to E^{**}$. (Note, however, that the double-dual of a Gieseker semistable sheaf is in general only μ_H-semistable). Define T_E as the cokernel

$$0 \to E \to E^{**} \to T_E \to 0,$$

so that T_E is a skyscraper sheaf supported on the singularities of E. In the Donaldson-Uhlenbeck-Yau compactification of $M_H(\mathbf{v})$, a sheaf E is replaced by the pair (E^{**}, T_E) consisting of the μ_H-semistable sheaf E^{**} and the S-equivalence class of T_E, i.e. an element of some symmetric product $X^{(n)}$. In particular, two sheaves which have isomorphic double duals and have singularities supported at the same points (counting multiplicity) are identified in $M_H^{DUY}(\mathbf{v})$, even if the particular singularities are different. The Jun Li morphism $j : M_H(\mathbf{v}) \to M_H^{DUY}(\mathbf{v})$ inducing the Donaldson-Uhlenbeck-Yau compactification arises from the line bundle $\lambda(\mathbf{w})$ associated to the character $\mathbf{w}$ of a 1-dimensional torsion sheaf supported on a curve whose class is a multiple of H. See [**HL10**, §8.2] or [**Li93**] for more details.

4.3.3. *Change of polarization.* Classically, one of the main interesting sources of birational maps between moduli spaces of sheaves is provided by varying the polarization. Suppose that $\{H_t\}$ $(0 \leq t \leq 1)$ is a continuous family of ample divisors on X. Let E be a sheaf which is μ_{H_0}-stable. It may happen for some time $t > 0$ that E is not μ_{H_t}-stable. In this case, there is a smallest time t_0 where E is not $\mu_{H_{t_0}}$-stable, and then E is strictly $\mu_{H_{t_0}}$-semistable. There is then an exact sequence

$$0 \to F \to E \to G \to 0$$

of $\mu_{H_{t_0}}$-semistable sheaves with the same $\mu_{H_{t_0}}$-slope. For $t < t_0$, we have

$$\mu_{H_t}(F) < \mu_{H_t}(E) < \mu_{H_t}(G).$$

On the other hand, in typical examples the inequalities will be reversed for $t > t_0$:

$$\mu_{H_t}(F) > \mu_{H_t}(E) > \mu_{H_t}(G).$$

While E is certainly not μ_{H_t}-semistable for $t > t_0$, if there are sheaves E' fitting as extensions in sequences

$$0 \to G \to E' \to F \to 0$$

then it may happen that E' is μ_{H_t}-stable for $t > t_0$ (although they are certainly not μ_{H_t}-semistable for $t < t_0$).

Thus, the set of H_t-semistable sheaves changes as t crosses t_0, and the moduli space $M_{H_t}(\mathbf{v})$ changes accordingly. It frequently happens that only some very special sheaves become destabilized as t crosses t_0, in which case the expectation would be that the moduli spaces for $t < t_0$ and $t > t_0$ are birational.

To clarify the dependence between the geometry of the moduli space $M_H(\mathbf{v})$ and the choice of polarization H, we partition the cone $\mathrm{Amp}(X)$ of ample divisors on X into chambers where the moduli space remains constant. Let $\mathbf{v}$ be a primitive vector, and suppose E has $\mathrm{ch}(E) = \mathbf{v}$ and is strictly H-semistable for some polarization H. Let $F \subset E$ be an H-semistable subsheaf with $\mu_H(F) = \mu_H(E)$. Then the locus $\Lambda \subset \mathrm{Amp}(X)$ of polarizations H' such that $\mu_{H'}(F) = \mu_{H'}(E)$ is a hyperplane in the ample cone, called a *wall.* The collection of all walls obtained in this way gives the ample cone a locally finite wall-and-chamber decomposition. As H varies within an open chamber, the moduli space $M_H(\mathbf{v})$ remains unchanged. On the other hand, if H crosses a wall then the moduli spaces on either side may be related in interesting ways.

Notice that if say X has Picard rank 1 or we are considering Hilbert schemes of points then no interesting geometry can be obtained by varying the polarization. Recall that in Example 4.24 we saw that even $\mathbb{P}^{2[3]}$ has nontrivial alternate birational models. One of the goals of Bridgeland stability will be to view these alternate models as a variation of the stability condition. Variation of polarization is one of the simplest examples of how a stability condition can be modified in a continuous way, and Bridgeland stability will give us additional "degrees of freedom" with which to vary our stability condition.

5. Bridgeland stability

The definition of a Bridgeland stability condition needs somewhat more machinery than the previous sections. However, we will primarily work with explicit stability conditions where the abstract nature of the definition becomes very concrete. While it would be a good idea to review the basics of derived categories of coherent sheaves, triangulated categories, t-structures, and torsion theories, it is also possible to first develop an appreciation for stability conditions and then go back and fill in the missing details. Good references for background on these topics include [**GM03**] and [**Huy06**].

5.1. Stability conditions in general. Let X be a smooth projective variety. We write $D^b(X)$ for the bounded derived category of coherent sheaves on X. We also write $K_{\mathrm{num}}(X)$ for the Grothendieck group of X modulo numerical equivalence. Following [**Bri07**], we make the following definition.

DEFINITION 5.1. A *Bridgeland stability condition* on X is a pair $\sigma = (Z, \mathcal{A})$ consisting of an $\mathbb{R}$-linear map $Z : K_{\mathrm{num}}(X) \otimes \mathbb{R} \to \mathbb{C}$ (called the *central charge*) and the heart $\mathcal{A} \subset D^b(X)$ of a bounded t-structure (which is an abelian category). Additionally, we require that the following properties be satisfied.

(1) (Positivity) If $0 \neq E \in \mathcal{A}$, then

$$Z(E) \in \mathbb{H} := \{re^{i\theta} : 0 < \theta \leq \pi \text{ and } r > 0\} \subset \mathbb{C}.$$

We define functions $r(E) = \Im Z(E)$ and $d(E) = -\Re Z(E)$, so that $r(E) \geq 0$ and $d(E) > 0$ whenever $r(E) = 0$. Thus r and d are generalizations of the classical rank and degree functions. The *(Bridgeland) σ-slope* is defined by

$$\mu_\sigma(E) = \frac{d(E)}{r(E)} = -\frac{\Re Z(E)}{\Im Z(E)}.$$

(2) (Harder-Narasimhan filtrations) An object $E \in \mathcal{A}$ is called (Bridgeland) σ-(semi)stable if

$$\mu_\sigma(F) \underset{(-)}{<} \mu_\sigma(E)$$

whenever $F \subset E$ is a subobject of E in $\mathcal{A}$. We require that every object of $\mathcal{A}$ has a finite Harder-Narasimhan filtration in $\mathcal{A}$. That is, there is a unique filtration

$$0 = E_0 \subset E_1 \subset \cdots \subset E_\ell = E$$

of objects $E_i \in \mathcal{A}$ such that the quotients $F_i = E_i/E_{i-1}$ are σ-semistable with decreasing slopes $\mu_\sigma(F_1) > \cdots > \mu_\sigma(F_\ell)$.

(3) (Support property) The support property is one final more technical condition which must be satisfied. Fix a norm $\|\cdot\|$ on $K_{\mathrm{num}}(X)\otimes\mathbb{R}$. Then there must exist a constant $C>0$ such that

$$\|E\| \leq C\|Z(E)\|$$

for all semistable objects $E\in\mathcal{A}$.

REMARK 5.2. Let (X,H) be a smooth surface. The subcategory $\operatorname{coh} X\subset D^b(X)$ of sheaves with cohomology supported in degree 0 is the heart of the standard t-structure. We can then try to define a central charge

$$Z(E) = -c_1(E).H + i\,\mathrm{rk}(E)H^2,$$

and the corresponding slope function is the ordinary slope μ_H. However, this does *not* give a Bridgeland stability condition, since $Z(E)=0$ for any finite length torsion sheaf. Thus it is not immediately clear in what way Bridgeland stability generalizes ordinary slope- or Gieseker stability. Nonetheless, for any fixed polarization H and character $\mathbf{v}$ there are Bridgeland stability conditions σ where the σ-(semi)stable objects of character $\mathbf{v}$ are precisely the H-Gieseker (semi)stable sheaves of character $\mathbf{v}$. See §5.4 for more details.

REMARK 5.3. To work with the definition of a stability condition, it is crucial to understand what it means for a map $F\to E$ between objects of the heart $\mathcal{A}$ to be injective. The following exercise is a good test of the definitions involved.

EXERCISE 5.4. Let $\mathcal{A}\subset D^b(X)$ be the heart of a bounded t-structure, and let $\phi: F\to E$ be a map of objects of $\mathcal{A}$. Show that ϕ is injective if and only if the mapping cone $\operatorname{cone}(\phi)$ of ϕ is also in $\mathcal{A}$. In this case, there is an exact sequence

$$0\to F\to E\to \operatorname{cone}(\phi)\to 0$$

in $\mathcal{A}$.

One of the most important features of Bridgeland stability is that the space of all stability conditions on X is a complex manifold in a natural way. In particular, we are able to continuously vary stability conditions and study how the set (or moduli space) of semistable objects varies with the stability condition. Let $\operatorname{Stab}(X)$ denote the space of stability conditions on X. Then Bridgeland proves that there is a natural topology on $\operatorname{Stab}(X)$ such that the forgetful map

$$\begin{aligned}\operatorname{Stab}(X) &\to \operatorname{Hom}_{\mathbb{R}}(K_{\mathrm{num}}(X)\otimes\mathbb{R},\mathbb{C})\\ (Z,\mathcal{A}) &\mapsto Z\end{aligned}$$

is a local homeomorphism. Thus if $\sigma=(Z,\mathcal{A})$ is a stability condition and the linear map Z is deformed by a small amount, there is a unique way to deform the category $\mathcal{A}$ to get a new stability condition.

5.1.1. *Moduli spaces.* Let σ be a stability condition and fix a vector $\mathbf{v}\in K_{\mathrm{num}}(X)$. There is a notion of a flat family $\mathcal{E}/S$ of σ-semistable objects parameterized by an algebraic space S [**BM14a**]. Correspondingly, there is a moduli stack $\mathcal{M}_\sigma(\mathbf{v})$ parameterizing flat families of σ-semistable object of character $\mathbf{v}$. In full generality there are many open questions about the geometry of these moduli spaces. In particular, when is there a projective coarse moduli space $M_\sigma(\mathbf{v})$ parameterizing S-equivalence classes of σ-semistable objects of character $\mathbf{v}$?

Several authors have addressed this question for various surfaces, at least when the stability condition σ does not lie on a *wall* for $\mathbf{v}$ (see §5.3). For instance, there is a projective moduli space $M_\sigma(\mathbf{v})$ when X is $\mathbb{P}^2$ [**ABCH13**], $\mathbb{P}^1 \times \mathbb{P}^1$ or $\mathbb{F}_1$ [**AM16**], an abelian surface [**MYY14**], a $K3$ surface [**BM14a**], or an Enriques surface [**Nue14**]. While projectivity of Gieseker moduli spaces can be shown in great generality, there is no known uniform GIT construction of moduli spaces of Bridgeland semistable objects. Each proof requires deep knowledge of the particular surface.

5.2. Stability conditions on surfaces. Bridgeland [**Bri08**] and Arcara-Bertram [**AB13**] explain how to construct stability conditions on a smooth surface. The construction is very explicit, and these are the only kinds of stability conditions we will consider in this survey. Before beginning we introduce some notation to make the definitions more succinct.

Let X be a smooth surface and let $H, D \in \operatorname{Pic}(X) \otimes \mathbb{R}$ be an ample divisor and an arbitrary *twisting* divisor, respectively. We formally define the *twisted Chern character* $\operatorname{ch}^D = e^{-D} \operatorname{ch}$. Explicitly expanding this definition, this means that

$$\begin{aligned} \operatorname{ch}_0^D &= \operatorname{ch}_0 \\ \operatorname{ch}_1^D &= \operatorname{ch}_1 - D \operatorname{ch}_0 \\ \operatorname{ch}_2^D &= \operatorname{ch}_2 - D \operatorname{ch}_1 + \frac{D^2}{2} \operatorname{ch}_0 . \end{aligned}$$

We can also define twisted slopes and discriminants by the formulas

$$\begin{aligned} \mu_{H,D} &= \frac{H.\operatorname{ch}_1^D}{H^2 \operatorname{ch}_0^D} \\ \Delta_{H,D} &= \frac{1}{2}\mu_{H,D}^2 - \frac{\operatorname{ch}_2^D}{H^2 \operatorname{ch}_0^D}. \end{aligned}$$

For reasons that will become clear in §5.4 it is often useful to add in an additional twist by $K_X/2$. We therefore additionally define

$$\overline{\operatorname{ch}}^D = \operatorname{ch}^{D+\frac{1}{2}K_X} \qquad \overline{\mu}_{H,D} = \mu_{H,D+\frac{1}{2}K_X} \qquad \overline{\Delta}_{H,D} = \Delta_{H,D+\frac{1}{2}K_X}.$$

REMARK 5.5. Note that the twisted slopes $\mu_{H,D}$ and $\overline{\mu}_{H,D}$ are primarily just a notational convenience; they only differ from the ordinary slope by a constant (depending on H and D). On the other hand, twisted discriminants $\Delta_{H,D}$ and $\overline{\Delta}_{H,D}$ do not obey such a simple formula, and are genuinely useful.

REMARK 5.6 (Twisted Gieseker stability). We have already encountered H-Gieseker (semi)stability and the associated moduli spaces $M_H(\mathbf{v})$ of H-Gieseker semistable sheaves. There is a mild generalization of this notion called (H, D)-*twisted Gieseker (semi)stability.* A torsion-free coherent sheaf E is (H, D)-twisted Gieseker (semi)stable if whenever $F \subsetneq E$ we have

(1) $\overline{\mu}_{H,D}(F) \leq \overline{\mu}_{H,D}(E)$ and
(2) whenever $\overline{\mu}_{H,D}(F) = \overline{\mu}_{H,D}(E)$, we have $\overline{\Delta}_{H,D}(F) \underset{(-)}{>} \overline{\Delta}_{H,D}(E)$.

Compare with Example 2.11, which is the case $D = 0$. When H, D are $\mathbb{Q}$-divisors, Matsuki and Wentworth [**MW97**] construct projective moduli spaces $M_{H,D}(\mathbf{v})$ of (H, D)-twisted Gieseker semistable sheaves. Note that any μ_H-stable sheaf is both H-Gieseker stable and (H, D)-twisted Gieseker stable, so that the spaces $M_H(\mathbf{v})$ and $M_{H,D}(\mathbf{v})$ are often either isomorphic or birational.

EXERCISE 5.7. Use the Hodge Index Theorem and the ordinary Bogomolov inequality (Theorem 3.7) to show that if E is μ_H-semistable then

$$\overline{\Delta}_{H,D}(E) \geq 0.$$

We now define a half-plane (or *slice*) of stability conditions on X corresponding to a choice of divisors $H, D \in \operatorname{Pic}(X) \otimes \mathbb{R}$ as above. First fix a number $\beta \in \mathbb{R}$. We define two full subcategories of the category $\operatorname{coh} X$ of coherent sheaves by

$$\mathcal{T}_\beta = \{E \in \operatorname{coh} X : \overline{\mu}_{H,D}(G) > \beta \text{ for every quotient } G \text{ of } E\}$$
$$\mathcal{F}_\beta = \{E \in \operatorname{coh} X : \overline{\mu}_{H,D}(F) \leq \beta \text{ for every subsheaf } F \text{ of } E\}.$$

Note that by convention the (twisted) Mumford slope of a torsion sheaf is ∞, so that $\mathcal{T}_\beta$ contains all the torsion sheaves on X. On the other hand, sheaves in $\mathcal{F}_\beta$ have no torsion subsheaf and so are torsion-free.

For any $\beta \in \mathbb{R}$, the pair of categories $(\mathcal{T}_\beta, \mathcal{F}_\beta)$ form what is called a *torsion pair*. Briefly, this means that $\operatorname{Hom}(T, F) = 0$ for any $T \in \mathcal{T}_\beta$ and $F \in \mathcal{F}_\beta$, and any $E \in \operatorname{coh} X$ can be expressed naturally as an extension

$$0 \to F \to E \to T \to 0$$

of a sheaf $T \in \mathcal{T}_\beta$ by a sheaf $F \in \mathcal{F}_\beta$. Then there is an associated t-structure with heart

$$\mathcal{A}_\beta = \{E^\bullet : \mathrm{H}^{-1}(E^\bullet) \in \mathcal{F}_\beta, \mathrm{H}^0(E^\bullet) \in \mathcal{T}_\beta, \text{ and } \mathrm{H}^i(E^\bullet) = 0 \text{ for } i \neq -1, 0\} \subset D^b(X),$$

where we use a Roman $\mathrm{H}^i(E^\bullet)$ to denote cohomology sheaves.

Some objects of $\mathcal{A}_\beta$ are the sheaves T in $\mathcal{T}_\beta$ (viewed as complexes sitting in degree 0) and shifts $F[1]$ where $F \in \mathcal{F}_\beta$, sitting in degree -1. More generally, every object $E^\bullet \in \mathcal{A}_\beta$ is an extension

$$0 \to \mathrm{H}^{-1}(E^\bullet)[1] \to E^\bullet \to \mathrm{H}^0(E^\bullet) \to 0,$$

where the sequence is exact in the heart $\mathcal{A}_\beta$.

To define stability conditions we now need to define central charges compatible with the hearts $\mathcal{A}_\beta$. Let $\alpha \in \mathbb{R}_{>0}$ be an arbitrary positive real number. We define

$$Z_{\beta,\alpha} = -\overline{\operatorname{ch}}_2^{D+\beta H} + \frac{\alpha^2 H^2}{2} \overline{\operatorname{ch}}_0^{D+\beta H} + iH\overline{\operatorname{ch}}_1^{D+\beta H},$$

and put $\sigma_{\beta,\alpha} = (Z_{\beta,\alpha}, \mathcal{A}_\beta)$. Note that if E is an object of nonzero rank with twisted slope $\overline{\mu}_{H,D}$ and discriminant $\overline{\Delta}_{H,D}$ then the corresponding Bridgeland slope is

$$\mu_{\sigma_{\beta,\alpha}} = -\frac{\Re Z_{\beta,\alpha}}{\Im Z_{\beta,\alpha}} = \frac{(\overline{\mu}_{H,D} - \beta)^2 - \alpha^2 - 2\overline{\Delta}_{H,D}}{\overline{\mu}_{H,D} - \beta}.$$

THEOREM 5.8 ([**AB13**]). *Let X be a smooth surface, and let $H, D \in \operatorname{Pic}(X) \otimes \mathbb{R}$ with H ample. If $\beta, \alpha \in \mathbb{R}$ with $\alpha > 0$, then the pair $\sigma_{\beta,\alpha} = (Z_{\beta,\alpha}, \mathcal{A}_\beta)$ defined above is a Bridgeland stability condition.*

The most interesting part of the theorem is the verification of the Positivity axiom 1 in the Definition 5.1 of a stability condition, which we now sketch. The other parts are quite formal.

SKETCH PROOF OF POSITIVITY. Note that $Z := Z_{\beta,\alpha}$ is an $\mathbb{R}$-linear map. Since the upper half-plane $\mathbb{H} = \{re^{i\theta} : 0 < \theta \leq \pi \text{ and } r > 0\}$ is closed under addition, the exact sequence

$$0 \to \mathrm{H}^{-1}(E^\bullet)[1] \to E^\bullet \to \mathrm{H}^0(E^\bullet) \to 0$$

implies that it is sufficient to check $Z(T) \in \mathbb{H}$ and $Z(F[1]) \in \mathbb{H}$ whenever $T \in \mathcal{T}_\beta$ and $F \in \mathcal{F}_\beta$.

If $T \in \mathcal{T}_\beta$ is not torsion, then $\overline{\mu}_{H,D}(T) > \beta$ is finite. Expanding the definitions immediately gives $H.\overline{\mathrm{ch}}_1^{D+\beta H}(T) > 0$, so $Z(T) \in \mathbb{H}$. If T is torsion with positive-dimensional support, then again $H.\overline{\mathrm{ch}}_1^{D+\beta H}(T) > 0$ and $Z(T) \in \mathbb{H}$. Finally, if $T \neq 0$ has zero-dimensional support then $-\overline{\mathrm{ch}}_2^{D+\beta H}(T) = -\mathrm{ch}_2(T) > 0$ so $Z(T) \in \mathbb{H}$.

Suppose $0 \neq F \in \mathcal{F}_\beta$. If actually $\overline{\mu}_{H,D}(F) < \beta$, then $H.\overline{\mathrm{ch}}_1^{D+\beta H}(F) < 0$ and $Z(F[1]) \in \mathbb{H}$ again follows. So suppose that $\overline{\mu}_{H,D}(F) = \beta$, which gives $\Im Z(F) = 0$. By the definition of $\mathcal{F}_\beta$, the sheaf F is torsion-free and $\overline{\mu}_{H,D+\beta H}$-semistable of $\overline{\mu}_{H,D+\beta H}$ slope 0. By Exercise 5.7 we find that $\overline{\Delta}_{H,D+\beta H}(F) \geq 0$. The formula for the twisted discriminant and the fact that $\alpha > 0$ then gives $\Re Z(F) < 0$, so $\Re Z(F[1]) > 0$. □

To summarize, if we let $\Pi = \{(\beta, \alpha) : \beta, \alpha \in \mathbb{R}, \alpha > 0\}$, the choice of a pair of divisors $H, D \in \mathrm{Pic}(X) \otimes \mathbb{R}$ with H ample defines an embedding

$$\begin{aligned} \Pi &\to \mathrm{Stab}(X) \\ (\beta, \alpha) &\mapsto \sigma_{\beta,\alpha}. \end{aligned}$$

This half-plane of stability conditions is called the (H, D)-*slice* of the stability manifold. We will sometimes abuse notation and write $\sigma \in \Pi$ for a stability condition σ parameterized by the slice. While the stability manifold can be rather large and unwieldy in general (having complex dimension $\dim_{\mathbb{R}} K_{\mathrm{num}}(X) \otimes \mathbb{R}$), much of the interesting geometry can be studied by inspecting the different slices of the manifold.

5.3. Walls. Fix a class $\mathbf{v} \in K_0(X)$. The stability manifold $\mathrm{Stab}(X)$ of X admits a locally finite wall-and-chamber decomposition such that the set of σ-semistable objects of class $\mathbf{v}$ does not vary as σ varies within an open chamber. This is analogous to the wall-and-chamber decomposition of the ample cone $\mathrm{Amp}(X)$ for classical stability, see §4.3.3. If $\mathbf{v}$ is primitive, then a stability condition σ lies on a wall if and only if there is a strictly σ-semistable object of character $\mathbf{v}$.

For computations, the entire stability manifold can be rather unwieldy to work with. One commonly restricts attention to stability conditions in some easily parameterized subset of the stability manifold. Here we focus on the (H, D)-slice $\{\sigma_{\beta,\alpha} : \beta, \alpha \in \mathbb{R}, \alpha > 0\}$ of stability conditions on a smooth surface X determined by a choice of divisors $H, D \in \mathrm{Pic}(X) \otimes \mathbb{R}$ with H ample.

DEFINITION 5.9. Let X be a smooth surface, and fix divisors $H, D \in \mathrm{Pic}(X) \otimes \mathbb{R}$ with H ample. Let $\mathbf{v}, \mathbf{w} \in K_{\mathrm{num}}(X)$ be two classes which have different $\mu_{\sigma_{\beta,\alpha}}$-slopes for some (β, α) with $\alpha > 0$.

(1) The *numerical wall* for $\mathbf{v}$ determined by $\mathbf{w}$ is the subset

$$W(\mathbf{v}, \mathbf{w}) = \{(\beta, \alpha) : \mu_{\sigma_{\beta,\alpha}}(\mathbf{v}) = \mu_{\sigma_{\beta,\alpha}}(\mathbf{w})\} \subset \Pi.$$

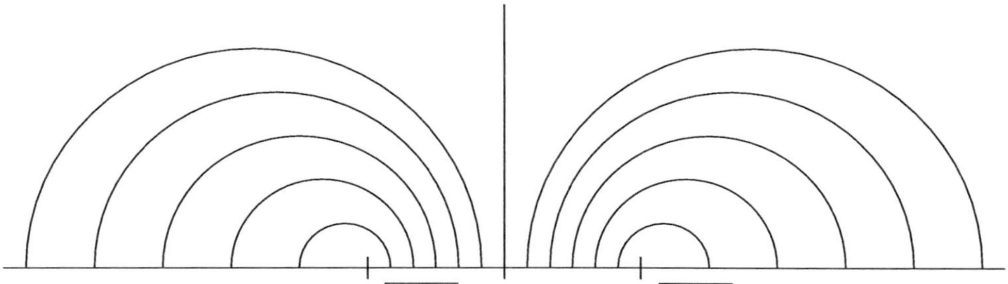

FIGURE 2. Schematic diagram of numerical walls in the (H, D)-slice for a nonzero rank character $\mathbf{v}$ with slope $\overline{\mu}_{H,D}$ and discriminant $\overline{\Delta}_{H,D}$.

(2) The numerical wall for $\mathbf{v}$ determined by $\mathbf{w}$ is a *wall* if there is some $(\beta, \alpha) \in W(\mathbf{v}, \mathbf{w})$ and an exact sequence

$$0 \to F \to E \to G \to 0$$

of $\sigma_{\beta,\alpha}$-semistable objects with $\operatorname{ch} F = \mathbf{w}$ and $\operatorname{ch} E = \mathbf{v}$.

5.3.1. *Geometry of numerical walls.* The geometry of the numerical walls in a slice of the stability manifold is particularly easy to describe. Verifying the following properties is a good exercise in the algebra of Chern classes and the Bridgeland slope function.

(1) First suppose $\mathbf{v}$ has nonzero rank and that the Bogomolov inequality $\Delta_{H,D}(\mathbf{v}) \geq 0$ holds. Then the vertical line $\beta = \overline{\mu}_{H,D}(\mathbf{v})$ is a numerical wall. The other numerical walls form two nested families of semicircles on either side of the vertical wall. These semicircles have centers on the β-axis, and their apexes lie along the hyperbola $\Re Z_{\beta,\alpha}(\mathbf{v}) = 0$ in Π. The two families of semicircles accumulate at the points

$$(\mu_{H,D}(\mathbf{v}) \pm \sqrt{2\Delta_{H,D}(\mathbf{v})}, 0)$$

of intersection of $\Re Z_{\beta,\alpha}(\mathbf{v}) = 0$ with the β-axis. See Figure 2 for an approximate illustration.

(2) If instead $\mathbf{v}$ has rank zero but $c_1(\mathbf{v}) \neq 0$, then the curve $\Re Z_{\beta,\alpha}(\mathbf{v}) = 0$ in Π degenerates to the vertical line

$$\beta = \frac{\overline{\operatorname{ch}}_2^D(\mathbf{v})}{\overline{\operatorname{ch}}_1^D(\mathbf{v}).H}.$$

The numerical walls for $\mathbf{v}$ are all semicircles with center

$$(\overline{\operatorname{ch}}_2^D(\mathbf{v})/(\overline{\operatorname{ch}}_1^D(\mathbf{v}).H), 0)$$

and arbitrary radius.

EXERCISE 5.10. In $\mathbf{v}, \mathbf{w}$ have nonzero rank and different slopes, the numerical semicircular wall $W(\mathbf{v}, \mathbf{w})$ has center $(s_W, 0)$ and radius ρ_W satisfying

$$s_W = \frac{\overline{\mu}_{H,D}(\mathbf{v}) + \overline{\mu}_{H,D}(\mathbf{w})}{2} - \frac{\overline{\Delta}_{H,D}(\mathbf{v}) - \overline{\Delta}_{H,D}(\mathbf{w})}{\overline{\mu}_{H,D}(\mathbf{v}) - \overline{\mu}_{H,D}(\mathbf{w})}$$

$$\rho_W^2 = (s_W - \overline{\mu}_{H,D}(\mathbf{v}))^2 - 2\overline{\Delta}_{H,D}(\mathbf{v}).$$

If $(s_W - \mu_{H,D}(\mathbf{v}))^2 \leq 2\overline{\Delta}_{H,D}(\mathbf{v})$, then the wall is empty.

REMARK 5.11. Let $\mathbf{v}$ be a character of nonzero rank. It follows from the above discussion that if W, W' are numerical walls for $\mathbf{v}$ both lying left of the vertical wall $\beta = \overline{\mu}_{H,D}(\mathbf{v})$ then W is nested inside W' if and only if $s_W > s_{W'}$, where the center of W (resp. W') is $(s_W, 0)$ (resp. $(s_{W'}, 0)$).

5.3.2. *Walls and destabilizing sequences.* In the definition of a wall $W := W(\mathbf{v}, \mathbf{w})$ for $\mathbf{v}$ determined by a character $\mathbf{w}$ we required that there is some point $(\beta, \alpha) \in W$ and a *destabilizing* exact sequence

$$0 \to F \to E \to G \to 0$$

of $\sigma_{\beta,\alpha}$-semistable objects, where $\mathrm{ch}(E) = \mathbf{v}$ and $\mathrm{ch}(F) = \mathbf{w}$. Note that since $(\beta, \alpha) \in W$ we in particular have $\mu_{\sigma_{\beta,\alpha}}(F) = \mu_{\sigma_{\beta,\alpha}}(E) = \mu_{\sigma_{\beta,\alpha}}(G)$. The above sequence is an exact sequence of objects of the categories $\mathcal{A}_\beta$. By the geometry of the numerical walls, the wall W separates the slice Π into two open regions Ω, Ω'. Relabeling the regions if necessary, for $\sigma \in \Omega$ we have $\mu_\sigma(F) > \mu_\sigma(E)$. Therefore E is not σ-semistable for any $\sigma \in \Omega$. On the other hand, E *may* be σ-semistable for $\sigma \in \Omega$; at least the subobject $F \subset E$ does not violate the semistability of E.

Our definition of a wall is perhaps somewhat unsatisfactory due to the dependence on picking some point $(\beta, \alpha) \in W$ where there is a destabilizing exact sequence as above. The next result shows that this definition is equivalent to an a priori stronger definition which appears more natural. Roughly speaking, destabilizing sequences "persist" along the entire wall.

PROPOSITION 5.12 ([**ABCH13**, Lemma 6.3] for $\mathbb{P}^2$, [**Mac14**] in general). *Suppose that*

$$0 \to F \to E \to G \to 0$$

is an exact sequence of $\sigma_{\beta,\alpha}$-semistable objects of the same $\sigma_{\beta,\alpha}$-slope. Put $\mathrm{ch}\, F = \mathbf{w}$ and $\mathrm{ch}\, E = \mathbf{v}$, and suppose $\mathbf{v}$ and $\mathbf{w}$ do not have the same slope everywhere in the (H, D)-slice. Let $W = W(\mathbf{v}, \mathbf{w})$ be the wall defined by these characters. If $(\beta', \alpha') \in W$ is any point on the wall, then the above exact sequence is an exact sequence of $\sigma_{\beta',\alpha'}$-semistable objects of the same $\sigma_{\beta',\alpha'}$-slope.

In particular, each of the objects F, E, G appearing in the above sequence lie in the category $\mathcal{A}_{\beta'}$.

Note that the first part of the proposition is essentially equivalent to the final statement by Exercise 5.4.

5.4. Large volume limit. As mentioned earlier, (twisted) Gieseker moduli spaces of sheaves on surfaces can be recovered as certain moduli spaces of Bridgeland-semistable objects. We say that an object $E^\bullet \in \mathcal{A}_\beta$ is a *sheaf* if it is isomorphic to a sheaf sitting in degree 0. We continue to work in an (H, D)-slice of stability conditions on a smooth surface X.

THEOREM 5.13 ([**ABCH13**, §6] for $\mathbb{P}^2$, [**Mac14**] in general). *Let* $\mathbf{v} \in K_{\mathrm{num}}(X)$ *be a character of positive rank with* $\overline{\Delta}_{H,D}(\mathbf{v}) \geq 0$. *Let* $\beta < \overline{\mu}_{H,D}(\mathbf{v})$, *and suppose* $\alpha \gg 0$ *(depending on* $\mathbf{v}$*). Then an object* $E^\bullet \in \mathcal{A}_\beta$ *is* $\sigma_{\beta,\alpha}$*-semistable if and only if it is an* (H, D)*-semistable sheaf.*

PROOF. Since $\beta < \overline{\mu}_{H,D}(\mathbf{v})$, the stability condition $\sigma_{\beta,\alpha}$ lies left of the vertical wall $\beta = \overline{\mu}_{H,D}(\mathbf{v})$. The walls for $\mathbf{v}$ are locally finite. Considering a neighborhood of a stability condition on the vertical wall shows that there is some largest semicircular wall W left of the vertical wall. The set of σ-semistable objects is constant as σ varies in the chamber between W and the vertical wall.

It is therefore enough to show the following two things. (1) If $E^\bullet \in \mathcal{A}_\beta$ has $\operatorname{ch} E^\bullet = \mathbf{v}$ and is $\sigma_{\beta,\alpha}$-semistable for $\alpha \gg 0$ then $E^\bullet$ is an (H, D)-semistable sheaf. (2) If E is an (H, D)-semistable sheaf of character $\mathbf{v}$, then E is $\sigma_{\beta,\alpha}$-semistable for $\alpha \gg 0$. That is, we may pick α depending on E, and not just depending on $\mathbf{v}$.

(1) First suppose $E^\bullet \in \mathcal{A}_\beta$ is $\sigma_{\beta,\alpha}$-semistable for $\alpha \gg 0$ and $\operatorname{ch} E^\bullet = \mathbf{v}$. If $E^\bullet$ is not a sheaf, then we have an interesting exact sequence

$$0 \to \mathrm{H}^{-1}(E^\bullet)[1] \to E^\bullet \to \mathrm{H}^0(E^\bullet) \to 0$$

in $\mathcal{A}_\beta$. Since $F := \mathrm{H}^{-1}(E^\bullet) \in \mathcal{F}_\beta$, the formula for the Bridgeland slope shows that

$$\mu_{\sigma_{\beta,\alpha}}(F[1]) = \mu_{\sigma_{\beta,\alpha}}(F) \to \infty$$

as $\alpha \to \infty$. On the other hand, since $G := H^0(E^\bullet) \in \mathcal{T}_\beta$ we have $\mu_{\sigma_{\beta,\alpha}}(G) \to -\infty$ as $\alpha \to \infty$, noting that $\operatorname{rk}(G) > 0$ since $\operatorname{rk} \mathbf{v} > 0$. This is absurd since $E^\bullet$ is $\sigma_{\beta,\alpha}$-semistable for $\alpha \gg 0$, and we conclude that $E := E^\bullet \in \mathcal{T}_\beta$ is a sheaf.

Similar arguments show that E is (H, D)-semistable. First suppose E has a $\overline{\mu}_{H,D}$-stable subsheaf F with $\overline{\mu}_{H,D}(F) > \overline{\mu}_{H,D}(E)$. Then the corresponding exact sequence of sheaves

$$0 \to F \to E \to G \to 0$$

is actually a sequence of objects in $\mathcal{T}_\beta$. Indeed, any quotient of an object in $\mathcal{T}_\beta$ is in $\mathcal{T}_\beta$, and $F \in \mathcal{T}_\beta$ by construction. Thus this is actually an exact sequence in $\mathcal{A}_\beta$. The formula for the Bridgeland slope then shows that $\mu_{\sigma_{\beta,\alpha}}(F) > \mu_{\sigma_{\beta,\alpha}}(E)$ for $\alpha \gg 0$, violating the $\sigma_{\beta,\alpha}$-semistability of E. We conclude that E is $\overline{\mu}_{H,D}$-semistable. To see that E is (H, D)-semistable, suppose there is a sequence

$$0 \to F \to E \to G \to 0$$

of sheaves of the same $\overline{\mu}_{H,D}$-slope, but $\overline{\Delta}_{H,D}(F) < \overline{\Delta}_{H,D}(E)$. Then the formula for the Bridgeland slope gives $\mu_{\sigma_{\beta,\alpha}}(F) > \mu_{\sigma_{\beta,\alpha}}(E)$ for *every* α, again contradicting the $\sigma_{\beta,\alpha}$-semistability of E for large α.

(2) Suppose E is (H, D)-semistable of character $\mathbf{v}$, and suppose $F^\bullet$ is a subobject of E in $\mathcal{A}_\beta$. Taking the long exact sequence in cohomology sheaves of the exact sequence

$$0 \to F^\bullet \to E \to G^\bullet \to 0$$

in $\mathcal{A}_\beta$ gives an exact sequence of sheaves

$$0 \to \mathrm{H}^{-1}(F^\bullet) \to 0 \to \mathrm{H}^{-1}(G^\bullet) \to \mathrm{H}^0(F^\bullet) \to E \to H^0(G^\bullet) \to 0.$$

Therefore $\mathrm{H}^{-1}(F^\bullet) = 0$, i.e. $F := F^\bullet$ is a sheaf in $\mathcal{T}_\beta$. The (H, D)-semistability of E then gives $\overline{\mu}_{H,D}(F) \leq \overline{\mu}_{H,D}(E)$, with $\overline{\Delta}_{H,D}(F) \geq \overline{\Delta}_{H,D}(E)$ in case of equality. The formula for $\mu_{\sigma_{\beta,\alpha}}$ then shows that if $\alpha \gg 0$ we have $\mu_{\sigma_{\beta,\alpha}}(F) \leq \mu_{\sigma_{\beta,\alpha}}(E)$. It follows from the finiteness of the walls that E is actually $\sigma_{\beta,\alpha}$-semistable for large α. □

In particular, if $\mathbf{v} \in K_{\text{num}}(X)$ is the character of an (H,D)-semistable sheaf of positive rank, then there is some largest wall W lying to the left of the vertical wall (or, possibly, there are no walls left of the vertical wall). This wall is called the *Gieseker wall*. For stability conditions σ in the open chamber $\mathcal{C}$ bounded by the Gieseker wall and the vertical wall, we have

$$M_\sigma(\mathbf{v}) \cong M_{H,D}(\mathbf{v}).$$

Therefore, any moduli space of twisted semistable sheaves can be recovered as a moduli space of Bridgeland semistable objects.

6. Examples on $\mathbb{P}^2$

In this subsection we investigate a couple of the first interesting examples of Bridgeland stability conditions and their relationship to birational geometry. We focus here on the characters of some small Hilbert schemes of points on $\mathbb{P}^2$. In these cases the definitions simplify considerably, and things can be understood explicitly.

6.1. Notation. Let $X = \mathbb{P}^2$, and fix the standard polarization H. We take $D = 0$; in general, the choice of twisting divisor is only interesting modulo the polarization, as adding a multiple of the polarization to D only translates the (H,D)-slice. The twisting divisor becomes more relevant in examples of higher Picard rank. Additionally, since $K_{\mathbb{P}^2}$ is parallel to H, we may as well work with the ordinary slope and discriminant

$$\mu = \frac{\text{ch}_1}{r} \qquad \Delta = \frac{1}{2}\mu^2 - \frac{\text{ch}_2}{r}$$

instead of the more complicated $\overline{\mu}_{H,0}$ and $\overline{\Delta}_{H,0}$. With these conventions, if $\mathbf{v}$ and $\mathbf{w}$ are characters of positive rank then the wall $W(\mathbf{v},\mathbf{w})$ has center $(s_W, 0)$ and radius ρ_W given by

$$\begin{aligned} s_W &= \frac{\mu(\mathbf{v})+\mu(\mathbf{w})}{2} - \frac{\Delta(\mathbf{v})-\Delta(\mathbf{w})}{\mu(\mathbf{v})-\mu(\mathbf{w})} \\ \rho_W^2 &= (s_W - \mu(\mathbf{v}))^2 - 2\Delta(\mathbf{v}). \end{aligned}$$

If we further let $\mathbf{v} = \text{ch}\, I_Z$ be the character of an ideal of a length n scheme $Z \in \mathbb{P}^{2[n]}$, then the formulas further simplify to

$$\begin{aligned} s_W &= \frac{\mu(\mathbf{w})}{2} + \frac{n-\Delta(\mathbf{w})}{\mu(\mathbf{w})} \\ \rho_W^2 &= s_W^2 - 2n. \end{aligned}$$

The main question to keep in mind is the following.

QUESTION 6.1. *Let I_Z be the ideal sheaf of $Z \in \mathbb{P}^{2[n]}$. For which stability conditions σ in the slice is I_Z a σ-semistable object? What does the destabilizing sequence of I_Z look like along the wall where it is destabilized?*

Note that since I_Z is a Gieseker semistable sheaf, it is $\sigma_{\beta,\alpha}$-semistable if $\alpha \gg 0$ and $\beta < 0 = \mu(I_Z)$. There will be some wall W left of the vertical wall where I_Z is destabilized by some subobject F. For stability conditions σ below this wall, I_Z is never σ-semistable. Thus the region in the slice where I_Z is σ-semistable is bounded by the wall W and the vertical wall. It potentially consists of several of the chambers in the wall-and-chamber decomposition of the slice.

6.2. Types of walls. There are two very different ways in which an ideal sheaf I_Z of length n can be destabilized along a wall. The simplest way I_Z can be destabilized is if it is destabilized by an actual subsheaf, i.e. if there is an exact sequence of sheaves

$$0 \to I_Y(-k) \to I_Z \to T \to 0$$

giving rise to the wall for some zero-dimensional scheme Y of length ℓ. The character $\mathbf{w} = \operatorname{ch} I_W(-k)$ has $(r, \mu, \Delta) = (1, -k, \ell)$, so this wall has center $(s_W, 0)$ with

$$s_W = -\frac{k}{2} - \frac{n-\ell}{k}. \tag{1}$$

A wall obtained in this way is called a *rank one wall.*

On the other hand, subobjects of I_Z in the categories $\mathcal{A}_\beta$ need not be subsheaves of I_Z! In particular, it is entirely possible that I_Z is destabilized by a sequence

$$0 \to F \to I_Z \to G \to 0$$

where $\mathbf{w} = \operatorname{ch} F$ has $\operatorname{rk} \mathbf{w} \geq 2$. Such destabilizing sequences, giving so-called *higher rank walls*, are somewhat more troublesome to deal with. It will be helpful to bound their size, which we now do.

As in the proof of Theorem 5.13, the long exact sequence of cohomology sheaves shows that any subobject $F \subset I_Z$ in a category $\mathcal{A}_\beta$ must actually be a sheaf (but not necessarily a subsheaf). Let K and C be the kernel and cokernel, respectively, of the map of sheaves $F \to I_Z$, so that there is an exact sequence of sheaves

$$0 \to K \to F \to I_Z \to C \to 0.$$

In order for G to be in the categories $\mathcal{A}_\beta$ along the wall $W = W(\mathbf{w}, \mathbf{v})$ (which must be the case by Proposition 5.12), it is necessary and sufficient that we have $K \in \mathcal{F}_\beta$ and $C \in \mathcal{T}_\beta$ for all β along the wall. Indeed, K and C are the cohomology sheaves of the mapping cone of the map $F \to I_Z$, so this follows from Exercise 5.4. The sequence

$$0 \to F \to I_Z \to G \to 0$$

will be exact in the categories along the wall if F is additionally in $\mathcal{T}_\beta$ for β along the wall. These basic considerations lead to the following result.

LEMMA 6.2 ([**ABCH13**], or see [**Bol⁺16**, Lemma 3.1 and Corollary 3.2] for a generalization to arbitrary surfaces). *If an ideal sheaf I_Z of n points in $\mathbb{P}^2$ is destabilized along a wall W given by a subobject F of rank at least 2, then the radius ρ_W of W satisfies*

$$\rho_W^2 \leq \frac{n}{4}.$$

PROOF. We use the notation from above. Since I_Z is rank 1 and torsion-free, a nonzero map $F \to I_Z$ has torsion cokernel. Therefore C is torsion, and it is no condition at all to have $C \in \mathcal{T}_\beta$ along the wall. We further deduce that $c_1(C) \geq 0$, so $c_1(K) \geq c_1(F)$ and $\operatorname{rk}(K) = \operatorname{rk}(F) - 1$. Let $(s_W, 0)$ and ρ_W be the center and radius of W. Since $F \in \mathcal{T}_\beta$ along the wall and $K \in \mathcal{F}_\beta$ along the wall, we have

$$2\rho_W \leq \mu(F) - \mu(K) = \frac{c_1(F)}{\operatorname{rk}(F)} - \frac{c_1(K)}{\operatorname{rk}(K)} \leq -\frac{c_1(F)}{\operatorname{rk}(F)(\operatorname{rk}(F)-1)} \leq -\mu(F) \leq -s_W - \rho_W,$$

so $3\rho_W \le -s_W$. Squaring both sides, $9\rho_W^2 \le s_W^2 = \rho_W^2 + 2n$ by the formula for the radius. The result follows. □

6.3. Small examples. We now consider the stability of ideal sheaves of small numbers of points in $\mathbb{P}^2$ in detail.

EXAMPLE 6.3 (Ideals of 2 points). Let I_Z be the ideal of a length 2 scheme $Z \in \mathbb{P}^{2[2]}$. Such an ideal fits in an exact sequence

$$0 \to \mathcal{O}_{\mathbb{P}^2}(-1) \to I_Z \to \mathcal{O}_L(-2) \to 0$$

where L is the line spanned by Z. If $W = W(\operatorname{ch} \mathcal{O}_{\mathbb{P}^2}(-1), I_Z)$ is the wall defined by this sequence, then Z is certainly not σ-semistable for stability conditions σ inside W. On the other hand, we claim that I_Z is σ-semistable for stability conditions σ on or above W.

To see this, we rule out the possibility that I_Z is destabilized along some wall W' which is larger than W. The wall W has center $(s_W, 0)$ with $s_W = -5/2$ by Equation 1. Its radius is $\rho_W = 3/2$, so the wall W passes through the point $(-1, 0)$. If W' is given by a rank 1 subobject $I_Y(-k)$ then we must have $-k > -1$ in order for $I_Y(-k)$ to be in the categories $\mathcal{A}_\beta$ along the wall W'. This then forces $k = 0$, which means I_Y does not define a semicircular wall. This is absurd.

The other possibility is that W' is a higher-rank wall. But then by Lemma 6.2, W' has radius $\rho_{W'}$ satisfying $\rho_{W'}^2 \le 1/2$. This contradicts that W' is larger than W.

Note that in the above example, if σ is a stability condition on the wall W then I_Z is strictly σ-semistable and S-equivalent to any ideal $I_{Z'}$ where Z' lies on the line spanned by Z. Thus the set of S-equivalence classes of σ-semistable objects is naturally identified with $\mathbb{P}^{2*}$.

EXAMPLE 6.4 (Ideals of 3 collinear points). Let I_Z be the ideal of a length 3 scheme $Z \in \mathbb{P}^{2[3]}$ which is supported on a line. As in Example 6.3, we claim that I_Z is destablized by the sequence

$$0 \to \mathcal{O}_{\mathbb{P}^2}(-1) \to I_Z \to \mathcal{O}_L(-3) \to 0.$$

That is, if W is the wall corresponding to the sequence, then I_Z is σ-semistable for conditions σ on or above the wall. (From the existence of the sequence it is immediately clear that I_Z is not σ-semistable below the wall.)

We compute $s_W = -7/2$ and $\rho_W = \frac{5}{2}$. As in Example 6.3, we conclude that there is no larger rank 1 wall. Any higher rank wall W' would have $\rho_{W'}^2 \le 3/2$, so there can be no larger higher rank wall either. Therefore I_Z is σ-semistable on and above W.

For the next example we will need one additional useful fact.

PROPOSITION 6.5 ([**ABCH13**, Proposition 6.2]). *A line bundle $\mathcal{O}_{\mathbb{P}^2}(-k)$ or a shifted line bundle $\mathcal{O}_{\mathbb{P}^2}(-k)[1]$ is $\sigma_{\beta,\alpha}$-stable whenever it is in the category $\mathcal{A}_\beta$. Thus, $\mathcal{O}_{\mathbb{P}^2}(-k)$ is $\sigma_{\beta,\alpha}$-stable if $\beta < -k$, and $\mathcal{O}_{\mathbb{P}^2}(-k)[1]$ is $\sigma_{\beta,\alpha}$-stable if $\beta \ge -k$.*

In the next example we see our first example of an ideal sheaf destabilized by a higher rank subobject.

EXAMPLE 6.6 (Ideals of 3 general points). Let I_Z be the ideal of a length 3 scheme $Z \in \mathbb{P}^{2[3]}$ which is *not* supported on a line. In this case, the ideal I_Z has a minimal resolution of the form

$$0 \to \mathcal{O}_{\mathbb{P}^2}(-3)^2 \to \mathcal{O}_{\mathbb{P}^2}(-2)^3 \to I_Z \to 0$$

or, equivalently, there is a distinguished triangle

$$\mathcal{O}_{\mathbb{P}^2}(-2)^3 \to I_Z \to \mathcal{O}_{\mathbb{P}^2}(-3)^2[1] \to \cdot.$$

Consider the wall $W = W(\mathcal{O}_{\mathbb{P}^2}(-2), I_Z)$ defined by this sequence. It has center at $(s_W, 0)$ with $s_W = -5/2$, and its radius is $1/2$. By Proposition 6.5 and Exercise 5.4, the above triangle gives an exact sequence

$$0 \to \mathcal{O}_{\mathbb{P}^2}(-2)^3 \to I_Z \to \mathcal{O}_{\mathbb{P}^2}(-3)^2[1] \to 0$$

in the categories $\mathcal{A}_\beta$ along the wall. Then for any σ on the wall, I_Z is an extension of σ-semistable objects of the same slope, and hence is σ-semistable. It follows that I_Z is destabilized precisely along W.

REMARK 6.7 (Correspondence between birational geometry and Bridgeland stability). In [**ABCH13**, §10], this chain of examples is continued in great detail. The regions of σ-semistability of ideal sheaves I_Z of up to 9 points are completely determined by similar methods. A remarkable correspondence between these regions of stability and the stable base locus decomposition was observed and conjectured to hold in general. The following result has since been proved by Li and Zhao.

THEOREM 6.8 ([**LZ16**]). *Let $Z \in \mathbb{P}^{2[n]}$. Let W be the Bridgeland wall where the ideal sheaf I_Z is destabilized. Also, let $yH - \frac{1}{2}B$ be the ray in the Mori cone past which the point $Z \in \mathbb{P}^{2[n]}$ enters the stable base locus. Then*

$$s_W = -y - \frac{3}{2}.$$

Therefore, computations in Bridgeland stability provide a dictionary between semistability and birational geometry. Compare with Examples 4.23, 4.24, 6.3, 6.4, 6.6, which establish the cases $n = 2, 3$ of the result.

More conceptually, Li and Zhao prove that the alternate birational models of any moduli space of sheaves on $\mathbb{P}^2$ can be interpreted as a Bridgeland moduli space, and they match up the walls in the Mori chamber decomposition of the effective cone with the walls in the wall-and-chamber decomposition of the stability manifold. As a consequence, they are able to give new computations of the effective, movable, and ample cones of divisors on these spaces. A crucial ingredient in this program is the smoothness of these Bridgeland moduli spaces, as well as a Drézet-Le Potier type classification of characters $\mathbf{v}$ for which Bridgeland moduli spaces are nonempty [**LZ16**, Theorems 0.1 and 0.2].

The next exercise computes the Gieseker wall for a Hilbert scheme of points on $\mathbb{P}^2$. This is the easiest case of the main problem we will discuss in the next section.

EXERCISE 6.9. Following Examples 6.3 and 6.4, show that the largest wall where some ideal sheaf I_Z of n points is destabilized is the wall $W(\operatorname{ch} \mathcal{O}_{\mathbb{P}^2}(-1), I_Z)$. Furthermore, an ideal I_Z is destabilized along this wall if and only if Z lies on a line.

REMARK 6.10. A similar program to the above has also been undertaken on some other rational surfaces such as Hirzebruch and del Pezzo surfaces. See [**BC13**].

7. The positivity lemma and nef cones

We close the survey by discussing the positivity lemma of Bayer and Macrì and recent applications of this tool to the computation of cones of nef divisors on Hilbert schemes of points and moduli spaces of sheaves. This provides an example where Bridgeland stability provides insight that at present is not understood from a more classical point of view.

7.1. The positivity lemma. The positivity lemma is a tool for constructing nef divisors on moduli spaces of Bridgeland-semistable objects. On a surface, (twisted) Gieseker moduli spaces can themselves be viewed as Bridgeland moduli spaces, so this will also allow us to construct nef divisors on classical moduli spaces. As with the construction of divisors on Gieseker moduli spaces, the starting point is to define a divisor on the base of a family of objects. When the moduli space carries a universal family, the family can be used to define a divisor on the moduli space.

In this direction, let $\sigma = (Z, \mathcal{A})$ be a stability condition on X, and let $\mathcal{E}/S$ be a flat family of σ-semistable objects of character $\mathbf{v}$ parameterized by a proper algebraic space S. We define a numerical divisor class $D_{\sigma,\mathcal{E}} \in N^1(S)$ on S depending on $\mathcal{E}$ and σ by specifying the intersection $D_{\sigma,\mathcal{E}}.C$ with every curve class $C \subset S$. Let $\Phi_{\mathcal{E}} : D^b(S) \to D^b(X)$ be the Fourier-Mukai transform with kernel $\mathcal{E}$, defined by

$$\Phi_{\mathcal{E}}(F) = q_*(p^*F \otimes \mathcal{E}),$$

where $p : S \times X \to S$ and $q : S \times X \to X$ are the projections and all the functors are derived. Then we declare

$$D_{\sigma,\mathcal{E}}.C = \Im\left(-\frac{Z(\Phi_{\mathcal{E}}(\mathcal{O}_C))}{Z(\mathbf{v})}\right).$$

Remark 7.1. Note that if $Z(\mathbf{v}) = -1$ then the formula becomes

$$D_{\sigma,\mathcal{E}}.C = \Im(Z(\Phi_{\mathcal{E}}(\mathcal{O}_C))).$$

If $\Phi_{\mathcal{E}}(\mathcal{O}_C) \in \mathcal{A}$, then $D_{\sigma,\mathcal{E}}.C \geq 0$ would follow from the positivity of the central charge. While it is not necessarily true that $\Phi_{\mathcal{E}}(\mathcal{O}_C) \in \mathcal{A}$, this fact nonetheless plays an important role in the proof of the positivity lemma.

The positivity lemma states that this assignment actually defines a nef divisor on S. Furthermore, there is a simple criterion to detect the curves C meeting the divisor orthogonally.

Theorem 7.2 (Positivity lemma, Theorem 4.1 [**BM14a**]). *The above assignment defines a well-defined numerical divisor class $D_{\sigma,\mathcal{E}}$ on S. This divisor is nef, and a complete, integral curve $C \subset S$ satisfies $D_{\sigma,\mathcal{E}}.C = 0$ if and only if the objects parameterized by two general points of C are S-equivalent with respect to σ.*

If the moduli space $M_\sigma(\mathbf{v})$ carries a universal family $\mathcal{E}$, then Theorem 7.2 constructs a nef divisor $D_{\sigma,\mathcal{E}}$ on the moduli space. In fact, the divisor does not depend on the choice of $\mathcal{E}$; we will see this in the next subsection.

Remark 7.3. If multiplies of $D_{\sigma,\mathcal{E}}$ define a morphism from S to projective space, then the curves C contracted by this morphism are characterized as the curves with $D_{\sigma,\mathcal{E}}.C = 0$. Thus, in a sense, all the interesting birational geometry coming from such a nef divisor $D_{\sigma,\mathcal{E}}$ is due to S-equivalence.

Unfortunately, in general, a nef divisor does not necessarily give rise to a morphism—multiples of the divisor do not necessarily have any sections at all. However, in such cases the positivity lemma is especially interesting. Indeed, one of the easiest ways to construct nef divisors is to pull back ample divisors by a morphism (recall Examples 4.4 and 4.5). The positivity lemma can potentially produce nef divisors not corresponding to any any map at all, in which case nefness is classically more difficult to check.

7.2. Computation of divisors. It is interesting to relate the Bayer-Macrì divisors $D_{\sigma,\mathcal{E}}$ with the determinantal divisors on a base S arising from a family $\mathcal{E}/S$. Now would be a good time to review §4.3.1. Recall that the Donaldson homomorphism is a map

$$\lambda_{\mathcal{E}} : \mathbf{v}^{\perp} \to N^1(S)$$

depending on a choice of family $\mathcal{E}/S$, where $\mathbf{v}^{\perp} \subset K_{\text{num}}(X)_{\mathbb{R}}$. Reviewing the definition of $\lambda_{\mathcal{E}}$, the definition only actually depends on the class of $\mathcal{E} \in K^0(S \times X)$, so it immediately extends to the case where $\mathcal{E}$ is a family of σ-semistable objects.

Since the Euler pairing $(-,-)$ is nondegenerate on $K_{\text{num}}(X)_{\mathbb{R}}$, any linear functional on $K_{\text{num}}(X)_{\mathbb{R}}$ vanishing on $\mathbf{v}$ can be represented by a vector in $\mathbf{v}^{\perp}$. In particular, there is a unique vector $\mathbf{w}_Z \in \mathbf{v}^{\perp}$ such that

$$\Im\left(-\frac{Z(\mathbf{w})}{Z(\mathbf{v})}\right) = (\mathbf{w}_Z, \mathbf{w})$$

holds for all $\mathbf{w} \in K_{\text{num}}(X)_{\mathbb{R}}$. Note that the definition of $\mathbf{w}_Z$ is essentially purely linear-algebraic, and makes no reference to S or $\mathcal{E}$. The next result shows that the Bayer-Macrì divisors are all determinantal.

PROPOSITION 7.4 ([**BM14a**, Proposition 4.4]). *We have*

$$D_{\sigma,\mathcal{E}} = \lambda_{\mathcal{E}}(\mathbf{w}_Z).$$

If N is any line bundle on S, then we have

$$D_{\sigma,\mathcal{E}\otimes p^*N} = \lambda_{\mathcal{E}\otimes p^*N}(\mathbf{w}_Z) = \lambda_{\mathcal{E}}(\mathbf{w}_Z) = D_{\sigma,\mathcal{E}}.$$

In particular, if S is a moduli space $M_\sigma(\mathbf{v})$ with a universal family $\mathcal{E}$, then the divisor $D_\sigma := D_{\sigma,\mathcal{E}}$ does not depend on the choice of universal family.

REMARK 7.5. See [**BM14a**, §4] for less restrictive hypotheses under which a divisor can be defined on the moduli space.

In explicit cases, it can be useful to compute the character $\mathbf{w}_Z$ in more detail. The next result does this in the case of an (H, D)-slice of divisors on a smooth surface X (review §5.2).

LEMMA 7.6 ([**Bol+16**, Proposition 3.8]). *Let X be a smooth surface and let $H, D \in \operatorname{Pic}(X) \otimes \mathbb{R}$, with H ample. If σ is a stability condition in the (H, D)-slice with center $(s_W, 0)$, then the character $\mathbf{w}_Z$ is a multiple of*

$$(-1, -\frac{1}{2}K_X + s_W H + D, m) \in \mathbf{v}^{\perp},$$

where we write Chern characters as $(\mathrm{ch}_0, \mathrm{ch}_1, \mathrm{ch}_2)$. Here the number m is determined by the property that the character is in $\mathbf{v}^{\perp}$.

7.3. Gieseker walls and nef cones. For the rest of the survey we let X be a smooth surface and fix an (H,D)-slice Π of stability conditions. Let $\mathbf{v} \in K_{\mathrm{num}}(\mathbf{v})$ be the Chern character of an (H,D)-semistable sheaf of positive rank. Additionally assume for simplicity that $M_{H,D}(\mathbf{v})$ has a universal family $\mathcal{E}$, so that in particular every (H,D)-semistable sheaf is (H,D)-stable. Recall that the *Gieseker wall* W for $\mathbf{v}$ in the (H,D)-slice is, by definition, the largest wall where an (H,D)-semistable sheaf of character $\mathbf{v}$ is destabilized. For conditions σ on or above W, every (H,D)-semistable sheaf is σ-semistable. Therefore, for any such σ, the universal family $\mathcal{E}$ is a family of σ-semistable objects parameterized by $M_{H,D}(\mathbf{v})$. Each condition σ on or above the wall therefore gives a nef divisor $D_\sigma = D_{\sigma,\mathcal{E}}$ on the moduli space.

COROLLARY 7.7. *With notation as above, if $s \leq s_W$, then the divisor on $M_{H,D}(\mathbf{v})$ corresponding to the class*

$$(-1, -\frac{1}{2}K_X + sH + D, m) \in \mathbf{v}^\perp$$

under the Donaldson homomorphism is nef.

Now let σ be a stability condition on the Gieseker wall. It is natural to wonder whether the "final" nef divisor D_σ produced by this method is a boundary nef divisor. This may or may not be the case. By Theorem 7.2, the divisor D_σ is on the boundary of $\mathrm{Nef}(M_{H,D}(\mathbf{v}))$ if and only if there is a curve in $M_{H,D}(\mathbf{v})$ parameterizing sheaves which are generically S-equivalent with respect to the stability condition σ. This happens if there is some sheaf $E \in M_{H,D}(\mathbf{v})$ destabilized along W by a sequence

$$0 \to F \to E \to G \to 0$$

where it is possible to vary the extension class in $\mathrm{Ext}^1(G,F)$ to obtain non-isomorphic objects E'. This can be subtle, and typically requires further analysis.

7.4. Nef cones of Hilbert schemes of points on surfaces. In this section we survey the recent results of [**Bol+16**] computing nef divisors on the Hilbert scheme $X^{[n]}$ of points on a smooth surface of irregularity $q(X) = 0$. Let $\mathbf{v} = \mathrm{ch}\, I_Z$, where $Z \in X^{[n]}$. For each pair of divisors (H,D) on X, we can interpret $X^{[n]}$ as the moduli space $M_{H,D}(\mathbf{v})$. A stability condition σ in the (H,D)-slice on a wall W with center $(s_W, 0)$ induces a divisor D_σ on $X^{[n]}$ with class a multiple of

$$\frac{1}{2}K_X^{[n]} - s_W H^{[n]} - D^{[n]} - \frac{1}{2}B.$$

The ray spanned by this class tends to the ray spanned by $H^{[n]}$ as $s_W \to -\infty$. As s_W varies in the above expression we obtain a two-dimensional cone of divisors in $N^1(X^{[n]})$ containing the ray spanned by the nef divisor $H^{[n]}$. The positivity lemma allows us to study the nefness of divisors in this cone by studying the Gieseker wall for $\mathbf{v}$ in the (H,D)-slice. Changing the twisting divisor D changes which two-dimensional cone we look at, and the entire nef cone of $X^{[n]}$ can be studied by the systematic variation of the twisting divisor.

The main result we discuss in this section addresses the computation of the Gieseker wall in an (H,D)-slice, at least assuming the number n of points is sufficiently large. We find that the Gieseker wall, or more precisely the subobject computing it, "stabilizes" once n is sufficiently large.

THEOREM 7.8 ([**Bol**$^+$**16**, various results from §3]). *There is a curve $C \subset X$ (depending on H, D) such that if $n \gg 0$ then the Gieseker wall for $\mathbf{v}$ in the (H, D)-slice is computed by the rank 1 subobject $\mathcal{O}_X(-C)$. The intersection number $C.H$ is minimal among all effective curves C on X. The divisor D_σ corresponding to a stability condition σ on the Gieseker wall is an extremal nef divisor. Orthogonal curves to D_σ can be obtained by letting n points move in a g^1_n on C.*

Note that everything here has already been verified for $X = \mathbb{P}^2$, and in fact $n \geq 2$ is sufficient in this case. The destabilizing subobject is always $\mathcal{O}_{\mathbb{P}^2}(-1)$.

SKETCH PROOF. Consider the character $\mathbf{v}$ as varying with n. Then $\overline{\mu}_{H,D}(\mathbf{v})$ is constant, and $\overline{\Delta}_{H,D}(\mathbf{v})$ is of the form $n + \text{const}$. Consider the wall W' given by a rank 1 object $I_Y(-C)$ with C an effective curve, and put $\mathbf{w} = \operatorname{ch} I_Y(-C)$. The wall W' has center at $(s_{W'}, 0)$ with

$$s_{W'} = \frac{\overline{\mu}_{H,D}(\mathbf{v}) + \overline{\mu}_{H,D}(\mathbf{w})}{2} - \frac{\overline{\Delta}_{H,D}(\mathbf{v}) - \overline{\Delta}_{H,D}(\mathbf{w})}{\overline{\mu}_{H,D}(\mathbf{v}) - \overline{\mu}_{H,D}(\mathbf{w})}.$$

As a function of n, this looks like

$$(2) \qquad s_{W'} = -\frac{n}{\mu_H(\mathbf{v}) - \mu_H(\mathbf{w})} + \text{const} = \frac{n}{\mu_H(\mathbf{w})} + \text{const} = -\frac{n}{C.H} + \text{const},$$

where the constant depends on $\mathbf{w}$. Correspondingly, the radius $\rho_{W'}$ grows approximately linearly in n.

Note that the numerical wall given by $\mathcal{O}_X(-C)$ is always at least as large as the numerical wall given by $I_Y(-C)$, by a discriminant calculation. Furthermore, if $I_Y(-C)$ gives an actual wall, i.e. if there is some $I_Z \in X^{[n]}$ fitting in a sequence

$$0 \to I_Y(-C) \to I_Z \to T \to 0,$$

then $\mathcal{O}_X(-C)$ also gives an actual wall. Thus, if the Gieseker wall is computed by a rank 1 sheaf then it is computed by a line bundle $\mathcal{O}_X(-C)$.

In fact, for $n \gg 0$ the Gieseker wall is computed by a line bundle $\mathcal{O}_X(-C)$ and not by some higher rank subobject. This is because an analog of Lemma 6.2 for arbitrary surfaces shows that any higher rank wall for $\mathbf{v}$ in the (H, D)-slice has radius *squared* bounded by n times a constant depending on H, D. On the other hand, as soon as we know there is *some* wall given by a rank 1 subobject it follows that there are walls with radius which is linear in n, implying that the Gieseker wall is not a higher rank wall.

To see that there is some rank 1 wall if $n \gg 0$, let C be any effective curve. For some $Z \in X^{[n]}$, there is an exact sequence of sheaves

$$0 \to \mathcal{O}_X(-C) \to I_Z \to I_{Z \subset C} \to 0.$$

We know the numerical wall W' corresponding to the subobject $\mathcal{O}_X(-C)$ has radius which grows linearly with n. In particular, for $n \gg 0$ the wall is nonempty. Furthermore, since $\overline{\mu}_{H,D}(\mathcal{O}_X(-C))$ and $\overline{\mu}_{H,D}(I_Z)$ are constant with n but $\overline{\Delta}_{H,D}(I_Z)$ is unbounded with n, the sheaf $\mathcal{O}_X(-C)$ is eventually in some of the categories along the wall W'. Thus the above exact sequence of sheaves is an exact sequence along the wall, and this wall is larger than any higher rank wall. We conclude that I_Z is either destabilized along W' or destabilized along some possibly larger rank 1 wall. Either way, there is a rank 1 wall, and the Gieseker wall is a rank 1 wall, computed by some line bundle $\mathcal{O}_X(-C)$ with C effective.

More precisely, the curve C such that the subsheaf $\mathcal{O}_X(-C)$ computes the Gieseker wall for $n \gg 0$ is the effective curve which gives the largest numerical (and hence actual) wall. Considering the Formula (2) for the center of the wall determined by $\mathcal{O}_X(-C)$, we find that C must be an effective curve of minimal H-degree. Furthermore, C must be chosen to minimize the constant which appears in that formula (this depends additionally on D). Any such curve C which asymptotically minimizes Formula (2) in this way computes the Gieseker wall for $n \gg 0$. Curves orthogonal to the divisor D_σ given by a stability condition on the Gieseker wall can now be obtained by varying the extension class in the sequence

$$0 \to \mathcal{O}_X(-C) \to I_Z \to I_{Z\subset C} \to 0;$$

this corresponds to letting Z move in a pencil on C, which can certainly be done for $n \gg 0$. □

More care is taken in [**Bol**$^+$**16**] to determine the precise bounds on n which are necessary for the various steps of the proof. The general method is applied to compute nef cones of Hilbert schemes of sufficiently many points on very general surfaces in $\mathbb{P}^3$, very general double covers of $\mathbb{P}^2$, and del Pezzo surfaces of degree 1. The last example provides an example of a surface of higher Picard rank, where the variation of the twisting divisor is exploited. See [**Bol**$^+$**16**, §4-5] for details. We highlight one of the first interesting cases where the answer appears to be unknown.

PROBLEM 7.9. *Let $X \subset \mathbb{P}^3$ be a very general quintic surface, so that the Picard rank is* 1 *by the Noether-Lefschetz theorem. Compute the nef cone of $X^{[2]}$ and $X^{[3]}$.*

Once $n \geq 4$ in the previous example, the nef cone is known by the general methods above. See [**Bol**$^+$**16**, Proposition 4.5].

7.5. Nef cones of moduli spaces of sheaves on surfaces. We close our discussion with a survey of the main result of [**CH16b**] on the cone of nef divisors on a moduli space of sheaves with large discriminant on an arbitrary smooth surface. In the case of $\mathbb{P}^2$, this result was first discovered in the papers [**CH16a, LZ16**]. The picture for an arbitrary surface is a modest simultaneous generalization of the $\mathbb{P}^2$ case as well as the Hilbert scheme case for an arbitrary surface (see §7.4 or [**Bol**$^+$**16**]).

Again let X be a smooth surface and let H, D be divisors giving a slice of stability conditions. Let $\mathbf{v}$ be the character of an (H, D)-semistable sheaf of positive rank. We assume the discriminant $\overline{\Delta}_{H,D}(\mathbf{v}) \gg 0$ is sufficiently large. Suppose the moduli space $M_{H,D}(\mathbf{v})$ carries a (quasi-)universal family. The goal of [**CH16b**] is to compute the Gieseker wall for $\mathbf{v}$ in the (H, D)-slice and to show that the divisor D_σ corresponding to a stability condition σ on the Gieseker wall is a boundary nef divisor.

The basic picture is similar to the case of a Hilbert scheme of points, and indeed Theorem 7.8 will follow as a special case of this more general result. However, the asymptotics can easily be made much more explicit in the Hilbert scheme case. The common thread between the two results is that as the discriminant $\overline{\Delta}_{H,D}(\mathbf{v})$ is increased, the character $\mathbf{w}$ of a destabilizing subobject giving rise to the Gieseker wall stabilizes. It is furthermore easy to give properties which almost uniquely define the character $\mathbf{w}$.

DEFINITION 7.10. Fix an (H, D)-slice. An *extremal Chern character* $\mathbf{w}$ for $\mathbf{v}$ is any character satisfying the following defining properties.

(E1) We have $0 < r(\mathbf{w}) \leq r(\mathbf{v})$, and if $r(\mathbf{w}) = r(\mathbf{v})$, then $c_1(\mathbf{v}) - c_1(\mathbf{w})$ is effective.
(E2) We have $\mu_H(\mathbf{w}) < \mu_H(\mathbf{v})$, and $\mu_H(\mathbf{w})$ is as close to $\mu_H(\mathbf{v})$ as possible subject to (E1).
(E3) The moduli space $M_{H,D}(\mathbf{w})$ is nonempty.
(E4) The discriminant $\overline{\Delta}_{H,D}(\mathbf{w})$ is as small as possible, subject to (E1)-(E3).
(E5) The rank $r(\mathbf{w})$ is as large as possible, subject to (E1)-(E4).

Note that properties (E1)-(E4) uniquely determine the slope $\mu_H(\mathbf{w})$ and discriminant $\overline{\Delta}_{H,D}(\mathbf{w})$, although $c_1(\mathbf{w})$ is not necessarily uniquely determined. Condition (E5) uniquely specifies the rank of $\mathbf{w}$. We then have the following theorem. Furthermore, notice that the definition does not depend on the discriminant $\overline{\Delta}_{H,D}(\mathbf{v})$, so that $\mathbf{w}$ can be held constant as $\overline{\Delta}_{H,D}(\mathbf{v})$ varies.

THEOREM 7.11 (**[CH16b]**). *Suppose $\overline{\Delta}_{H,D}(\mathbf{v}) \gg 0$. Then the Gieseker wall for $\mathbf{v}$ in the (H, D)-slice is computed by a destabilizing subobject of character $\mathbf{w}$, where $\mathbf{w}$ is an extremal Chern character for $\mathbf{v}$. Furthermore, the divisor D_σ corresponding to a stability condition σ on the Gieseker wall is a boundary nef divisor.*

The argument is largely similar to the proof of Theorem 7.8. First one shows that the destabilizing subobject along the Gieseker wall must actually be a subsheaf, and not some higher rank object. This justifies restriction (E1) in the definition of $\mathbf{w}$ (note that if $r(\mathbf{w}) = r(\mathbf{v})$ then the only way there can be an injection of sheaves $F \to E$ with $\operatorname{ch} F = \mathbf{w}$ and $\operatorname{ch} E = \mathbf{v}$ is if the induced map $\det F \to \det E$ is injective, forcing $c_1(\mathbf{v}) - c_1(\mathbf{w})$ to be effective.

Next, one shows that the subsheaf defining the Gieseker wall must actually be an (H, D)-semistable sheaf. Recalling the formula

$$s_W = \frac{\overline{\mu}_{H,D}(\mathbf{v}) + \overline{\mu}_{H,D}(\mathbf{w})}{2} - \frac{\overline{\Delta}_{H,D}(\mathbf{v}) - \overline{\Delta}_{H,D}(\mathbf{w})}{\overline{\mu}_{H,D}(\mathbf{v}) - \overline{\mu}_{H,D}(\mathbf{w})}$$

for the center of a wall, conditions (E2)-(E4) then ensure that the numerical wall defined by $\mathbf{w}$ is as large as possible when $\overline{\Delta}_{H,D}(\mathbf{v}) \gg 0$. Therefore, the Gieseker wall for $\mathbf{v}$ is no larger than the wall defined by the extremal character $\mathbf{w}$.

REMARK 7.12. Actually computing the extremal character $\mathbf{w}$ can be extremely challenging. Minimizing the discriminant of $\mathbf{w}$ subject to the condition that the moduli space $M_{H,D}(\mathbf{w})$ is nonempty essentially requires knowing the sharpest possible Bogomolov inequalities for semistable sheaves on X. Conversely, if the nef cones of moduli spaces of sheaves on X are known, strong Bogomolov-type inequalities can be deduced. On surfaces such as $\mathbb{P}^2$ and $K3$ surfaces, the extremal character can be computed mechanically using the classification of semistable sheaves; recall for example §3.2.3 and §3.2.6.

The proof of Theorem 7.11 diverges from the Hilbert scheme case when we need to show that the numerical wall for $\mathbf{v}$ defined by an extremal character $\mathbf{w}$ is an actual wall. In the Hilbert scheme case, it is trivial to produce ideal sheaves I_Z which are destabilized by a rank 1 object $\mathcal{O}_X(-C)$: we simply put Z on C, and get an exact sequence

$$0 \to \mathcal{O}_X(-C) \to I_Z \to I_{Z\subset C} \to 0$$

which is an exact sequence in the categories along the wall if the number of points is sufficiently large.

To prove Theorem 7.11, we instead need to produce (H, D)-semistable sheaves E of character $\mathbf{v}$ fitting in sequences of the form

$$0 \to F \to E \to G \to 0$$

where F is (H, D)-semistable of character $\mathbf{w}$. This is somewhat technical. Let $\mathbf{u} = \operatorname{ch} G$; then $\mathbf{u}$ has $r(\mathbf{u}) < r(\mathbf{v})$, and $\overline{\Delta}_{H,D}(\mathbf{u}) \gg 0$. Therefore, by induction on the rank, we may assume the Gieseker wall of $\mathbf{u}$ has been computed. We then show that the Gieseker walls for $\mathbf{w}$ and $\mathbf{u}$ are nested inside $W := W(\mathbf{v}, \mathbf{w})$ if $\overline{\Delta}_{H,D}(\mathbf{v}) \gg 0$. Therefore, any sheaves $F \in M_{H,D}(\mathbf{w})$ and $G \in M_{H,D}(\mathbf{u})$ are actually σ-semistable for any stability condition σ on W. Then any extension E of G by F is σ-semistable, and it can further be shown that a general such extension is actually (H, D)-stable. By varying the extension class, we can produce curves in $M_{H,D}(\mathbf{v})$ parameterizing non-isomorphic (H, D)-stable sheaves; these curves are orthogonal to the nef divisor given by the Gieseker wall. See [**CH16b**, §5-6] for details.

REMARK 7.13. Several applications of Theorem 7.11 to simple surfaces are given in [**CH16b**, §7].

References

[AB13] Daniele Arcara and Aaron Bertram, *Bridgeland-stable moduli spaces for K-trivial surfaces*, J. Eur. Math. Soc. (JEMS) **15** (2013), no. 1, 1–38, DOI 10.4171/JEMS/354. With an appendix by Max Lieblich. MR2998828

[ABCH13] Daniele Arcara, Aaron Bertram, Izzet Coskun, and Jack Huizenga, *The minimal model program for the Hilbert scheme of points on $\mathbb{P}^2$ and Bridgeland stability*, Adv. Math. **235** (2013), 580–626, DOI 10.1016/j.aim.2012.11.018. MR3010070

[AM16] D. Arcara, E. Miles, Projectivity of Bridgeland moduli spaces on del Pezzo surfaces of Picard rank 2, Int. Math. Res. Not., to appear. (2016)

[BM14a] Arend Bayer and Emanuele Macrì, *Projectivity and birational geometry of Bridgeland moduli spaces*, J. Amer. Math. Soc. **27** (2014), no. 3, 707–752, DOI 10.1090/S0894-0347-2014-00790-6. MR3194493

[BM14b] Arend Bayer and Emanuele Macrì, *MMP for moduli of sheaves on K3s via wall-crossing: nef and movable cones, Lagrangian fibrations*, Invent. Math. **198** (2014), no. 3, 505–590, DOI 10.1007/s00222-014-0501-8. MR3279532

[BC13] Aaron Bertram and Izzet Coskun, *The birational geometry of the Hilbert scheme of points on surfaces*, Birational geometry, rational curves, and arithmetic, Springer, New York, 2013, pp. 15–55, DOI 10.1007/978-1-4614-6482-2_2. MR3114922

[Bol+16] Barbara Bolognese, Jack Huizenga, Yinbang Lin, Eric Riedl, Benjamin Schmidt, Matthew Woolf, and Xiaolei Zhao, *Nef cones of Hilbert schemes of points on surfaces*, Algebra Number Theory **10** (2016), no. 4, 907–930, DOI 10.2140/ant.2016.10.907. MR3519100

[BDPP13] Sébastien Boucksom, Jean-Pierre Demailly, Mihai Păun, and Thomas Peternell, *The pseudo-effective cone of a compact Kähler manifold and varieties of negative Kodaira dimension*, J. Algebraic Geom. **22** (2013), no. 2, 201–248, DOI 10.1090/S1056-3911-2012-00574-8. MR3019449

[Bri07] Tom Bridgeland, *Stability conditions on triangulated categories*, Ann. of Math. (2) **166** (2007), no. 2, 317–345, DOI 10.4007/annals.2007.166.317. MR2373143

[Bri08] Tom Bridgeland, *Stability conditions on K3 surfaces*, Duke Math. J. **141** (2008), no. 2, 241–291, DOI 10.1215/S0012-7094-08-14122-5. MR2376815

[CH14a] Izzet Coskun and Jack Huizenga, *Interpolation, Bridgeland stability and monomial schemes in the plane* (English, with English and French summaries), J. Math. Pures Appl. (9) **102** (2014), no. 5, 930–971, DOI 10.1016/j.matpur.2014.02.010. MR3271294

[CH14b] Izzet Coskun and Jack Huizenga, *The birational geometry of the moduli spaces of sheaves on* $\mathbb{P}^2$, Proceedings of the Gökova Geometry-Topology Conference 2014, Gökova Geometry/Topology Conference (GGT), Gökova, 2015, pp. 114–155. MR3381441

[CH16a] Izzet Coskun and Jack Huizenga, *The ample cone of moduli spaces of sheaves on the plane*, Algebr. Geom. **3** (2016), no. 1, 106–136, DOI 10.14231/AG-2016-005. MR3455422

[CH16b] Izzet Coskun and Jack Huizenga, *The ample cone of moduli spaces of sheaves on the plane*, Algebr. Geom. **3** (2016), no. 1, 106–136, DOI 10.14231/AG-2016-005. MR3455422

[CHW16] Izzet Coskun and Jack Huizenga, *The ample cone of moduli spaces of sheaves on the plane*, Algebr. Geom. **3** (2016), no. 1, 106–136, DOI 10.14231/AG-2016-005. MR3455422

[Deb01] Olivier Debarre, *Higher-dimensional algebraic geometry*, Universitext, Springer-Verlag, New York, 2001. MR1841091

[Dre88] Jean-Marc Drezet, *Groupe de Picard des variétés de modules de faisceaux semi-stables sur* $\mathbf{P}_2(\mathbf{C})$ (French), Ann. Inst. Fourier (Grenoble) **38** (1988), no. 3, 105–168. MR976687

[DLP85] J.-M. Drezet and J. Le Potier, *Fibrés stables et fibrés exceptionnels sur* $\mathbf{P}_2$ (French, with English summary), Ann. Sci. École Norm. Sup. (4) **18** (1985), no. 2, 193–243. MR816365

[Fog68] John Fogarty, *Algebraic families on an algebraic surface*, Amer. J. Math **90** (1968), 511–521, DOI 10.2307/2373541. MR0237496

[Fog73] J. Fogarty, *Algebraic families on an algebraic surface. II. The Picard scheme of the punctual Hilbert scheme*, Amer. J. Math. **95** (1973), 660–687, DOI 10.2307/2373734. MR0335512

[GM03] Sergei I. Gelfand and Yuri I. Manin, *Methods of homological algebra*, 2nd ed., Springer Monographs in Mathematics, Springer-Verlag, Berlin, 2003. MR1950475

[Har74] Robin Hartshorne, *Varieties of small codimension in projective space*, Bull. Amer. Math. Soc. **80** (1974), 1017–1032, DOI 10.1090/S0002-9904-1974-13612-8. MR0384816

[HH13] Brendan Hassett and Donghoon Hyeon, *Log minimal model program for the moduli space of stable curves: the first flip*, Ann. of Math. (2) **177** (2013), no. 3, 911–968, DOI 10.4007/annals.2013.177.3.3. MR3034291

[HiL93] André Hirschowitz and Yves Laszlo, *Fibrés génériques sur le plan projectif* (French), Math. Ann. **297** (1993), no. 1, 85–102, DOI 10.1007/BF01459489. MR1238408

[Hui12] Jack W. Huizenga, *Restrictions of Steiner bundles and divisors on the Hilbert scheme of points in the plane*, ProQuest LLC, Ann Arbor, MI, 2012. Thesis (Ph.D.)–Harvard University. MR3054910

[Hui16] Jack Huizenga, *Effective divisors on the Hilbert scheme of points in the plane and interpolation for stable bundles*, J. Algebraic Geom. **25** (2016), no. 1, 19–75, DOI 10.1090/jag/652. MR3419956

[Huy06] D. Huybrechts, *Fourier-Mukai transforms in algebraic geometry*, Oxford Mathematical Monographs, The Clarendon Press, Oxford University Press, Oxford, 2006. MR2244106

[HL10] Daniel Huybrechts and Manfred Lehn, *The geometry of moduli spaces of sheaves*, 2nd ed., Cambridge Mathematical Library, Cambridge University Press, Cambridge, 2010. MR2665168

[Laz04] Robert Lazarsfeld, *Positivity in algebraic geometry. II*, Ergebnisse der Mathematik und ihrer Grenzgebiete. 3. Folge. A Series of Modern Surveys in Mathematics [Results in Mathematics and Related Areas. 3rd Series. A Series of Modern Surveys in Mathematics], vol. 49, Springer-Verlag, Berlin, 2004. Positivity for vector bundles, and multiplier ideals. MR2095472

[LeP97] J. Le Potier, *Lectures on vector bundles*, Cambridge Studies in Advanced Mathematics, vol. 54, Cambridge University Press, Cambridge, 1997. Translated by A. Maciocia. MR1428426

[Li93] Jun Li, *Algebraic geometric interpretation of Donaldson's polynomial invariants*, J. Differential Geom. **37** (1993), no. 2, 417–466. MR1205451

[Li94] Jun Li, *Picard groups of the moduli spaces of vector bundles over algebraic surfaces*, Moduli of vector bundles (Sanda, 1994; Kyoto, 1994), Lecture Notes in Pure and Appl. Math., vol. 179, Dekker, New York, 1996, pp. 129–146. MR1397985

[LZ16] C. Li and X. Zhao, Birational models of moduli spaces of coherent sheaves on the projective plane, preprint. (2016)

[Mac14] Antony Maciocia, *Computing the walls associated to Bridgeland stability conditions on projective surfaces*, Asian J. Math. **18** (2014), no. 2, 263–279, DOI 10.4310/AJM.2014.v18.n2.a5. MR3217637

[MW97] Kenji Matsuki and Richard Wentworth, *Mumford-Thaddeus principle on the moduli space of vector bundles on an algebraic surface*, Internat. J. Math. **8** (1997), no. 1, 97–148, DOI 10.1142/S0129167X97000068. MR1433203

[MYY14] Hiroki Minamide, Shintarou Yanagida, and Kōta Yoshioka, *Some moduli spaces of Bridgeland's stability conditions*, Int. Math. Res. Not. IMRN **19** (2014), 5264–5327. MR3267372

[Nue14] Howard Nuer, *Projectivity and birational geometry of Bridgeland moduli spaces on an Enriques surface*, Proc. Lond. Math. Soc. (3) **113** (2016), no. 3, 345–386, DOI 10.1112/plms/pdw033. MR3551850

[O'G96] Kieran G. O'Grady, *Moduli of vector bundles on projective surfaces: some basic results*, Invent. Math. **123** (1996), no. 1, 141–207, DOI 10.1007/BF01232371. MR1376250

[Rud89] A. N. Rudakov, *Exceptional vector bundles on a quadric* (Russian), Izv. Akad. Nauk SSSR Ser. Mat. **52** (1988), no. 4, 788–812, 896; English transl., Math. USSR-Izv. **33** (1989), no. 1, 115–138. MR966985

[Rud94] Alexei N. Rudakov, *A description of Chern classes of semistable sheaves on a quadric surface*, J. Reine Angew. Math. **453** (1994), 113–135, DOI 10.1515/crll.1994.453.113. MR1285782

[Rya16] T. Ryan, The effective cone of moduli spaces of sheaves on $\mathbb{P}^1 \times \mathbb{P}^1$, University of Illinois-Chicago PhD thesis, 2016.

[Wal93] Charles Walter, *Irreducibility of moduli spaces of vector bundles on birationally ruled surfaces*, Algebraic geometry (Catania, 1993/Barcelona, 1994), Lecture Notes in Pure and Appl. Math., vol. 200, Dekker, New York, 1998, pp. 201–211. MR1651095

[Yos96] Kōta Yoshioka, *The Picard group of the moduli space of stable sheaves on a ruled surface*, J. Math. Kyoto Univ. **36** (1996), no. 2, 279–309. MR1411332

[Yos01] Kōta Yoshioka, *Moduli spaces of stable sheaves on abelian surfaces*, Math. Ann. **321** (2001), no. 4, 817–884, DOI 10.1007/s002080100255. MR1872531

Department of Mathematics, The Pennsylvania State University, University Park, Pennsylvania 16802

E-mail address: huizenga@psu.edu

Gromov–Witten theory: From curve counts to string theory

Emily Clader

ABSTRACT. Enumerative gemoetry is a discipline with roots dating back to antiquity, but it experienced an unexpected revolution in the late twentieth century when theoretical physicists became interested in its connection to string theory. The structure of physical systems led to striking predictions about curve-counting, and the field of Gromov-Witten theory grew up around mathematicians' attempts to explain rigorously why such predictions should hold. This chapter is a broad overview of the subject of Gromov-Witten theory, giving a tour of some of the fundamental definitions, methods, and open questions in the field.

Though its methods are decidedly modern, the problems addressed by Gromov–Witten theory have historical roots dating back hundreds, if not thousands, of years. These are questions of *enumerative geometry*, of which prototypical examples include:

- Given five points in $\mathbb{P}^2$, how many conics pass through all five?
- Given four lines in $\mathbb{P}^3$, how many lines pass through all four?
- How many rational curves of a fixed degree d are there on the quintic threefold,
$$V := \{x_0^5 + x_1^5 + x_2^5 + x_3^5 + x_4^5 = 0\} \subset \mathbb{P}^4?$$

The answer to the first of these questions was known to the ancient Greeks: there is exactly one conic through five (sufficiently general) points in the plane. In keeping with the style of classical Greek mathematics, one can arrive at this solution by simply computing an explicit equation for the conic.

Such explicit computation, while satisfying when successful, tends to be cumbersome as a method of enumeration. A different perspective on enumerative geometry has been in vogue since the eighteenth century: general enumerative questions should be answered in families, often by reducing to a particularly simple degenerate case. This method culminated in the techniques of Schubert calculus. As formulated by Hermann Schubert in his 1879 manuscript [**31**], enumerative geometry should obey the "principle of conservation of number", an insensitivity to continuous variations of the input data. If one wishes to know the number of lines through four fixed lines $\ell_1, \ell_2, \ell_3, \ell_4$ in $\mathbb{P}^3$, for example, then the answer should not depend on the particular choice of the ℓ_i, so long as they are chosen in such a way that the answer is finite. In particular, one may assume that ℓ_1 and ℓ_2 intersect in a point P and that ℓ_3 and ℓ_4 intersect in a point Q. Then there are manifestly two

2010 *Mathematics Subject Classification.* Primary 14H10, 14H81.

lines passing through all four: one joining P to Q, and another where the plane spanned by ℓ_1 and ℓ_2 intersects the plane spanned by ℓ_3 and ℓ_4.

Schubert calculus gave rise to a powerful perspective on enumerative geometry, but its methods were not always mathematically justified. Why should the principle of conservation of number hold, and in what situations might it fail to produce an answer? These issues were important enough to appear on Hilbert's celebrated list of unsolved problems in mathematics, on which the fifteenth problem demands that Schubert calculus be placed on rigorous footing.

The modern era of enumerative geometry— and the solution to Hilbert's fifteenth problem— began with the twin developments of intersection theory and moduli spaces. As we will discuss in the next section, moduli spaces provide a way to geometrically encode families of geometric objects (such as all lines in $\mathbb{P}^3$), so conditions on the objects cut out subspaces of the moduli space. Counting objects satisfying a list of conditions thus amounts to counting intersection points of a collection of subspaces, and the principle of conservation of number is translated into the more familiar fact that intersection numbers, when properly defined, are deformation invariant.

With the advent of intersection and moduli theory, the problems of enumerative geometry could finally be stated in a robust and rigorous way. Still, the actual computation of numerical solutions to those problems remained, in many cases, an unwieldy (if at least well-defined) task. Another breakthrough was needed in order to open the floodgates to such computations, and in this case, the inspiration came from a rather unexpected source: theoretical physics.

Physicists became interested in enumerative geometry because of its connection to string theory, which posits that the fundamental building blocks of the universe are tiny loops. As these loops— or "strings"— travel through spacetime, their motion traces out surfaces, and these real surfaces can be endowed with a complex structure to view them as one-dimensional complex manifolds, or algebraic curves. The probability that a physical system will transform from one state into another is dictated by a count of curves in the spacetime manifold satisfying prescribed conditions.

Physical models are expected to have a rich structure and often a degree of symmetry, if indeed they are accurately describing the world in which we live. This structure, when exploited from a mathematical perspective, predicts surprising relationships between different curve counts, as well as between curve counts and seemingly unrelated mathematical quantities. The field of Gromov–Witten theory grew up around mathematicians' attempts to explain rigorously why such predictions should hold. This task, while formidable and ongoing, has led to the discovery of striking new patterns in enumerative geometric problems, secrets that only emerge when collections of such problems are considered as a whole and packaged together in just the right way.

1. Basic definitions

Throughout what follows, we work over $\mathbb{C}$. By a *smooth curve*, we mean a proper, nonsingular algebraic curve, or in other words, a Riemann surface. Our curves will also be allowed nodal singularities, in which case we define their *genus* as the arithmetic genus $g(C) := h^1(C, \mathcal{O}_C)$. It is useful, though not strictly necessary, if the reader has some familiarity with the moduli space $\overline{\mathcal{M}}_{g,n}$ of curves.

1.1. The moduli space of stable maps. Fix a smooth projective variety X, a curve class $\beta \in H_2(X;\mathbb{Z})$, and non-negative integers g and n. As a set, the moduli space $\overline{\mathcal{M}}_{g,n}(X,\beta)$ consists of isomorphism classes of tuples $(C; x_1, \ldots, x_n; f)$, in which:

(i) C is a (possibly nodal) curve of genus g;
(ii) $x_1, \ldots, x_n \in C$ are distinct nonsingular points;
(iii) $f : C \to X$ is a morphism of "degree" β— that is, $\beta = f_*[C]$;
(iv) the data $(C; x_1, \ldots, x_n; f)$ has finitely many automorphisms.

Here, a *morphism* from $(C; x_1, \ldots, x_n; f)$ to $(C'; x_1', \ldots, x_n'; f')$ is a morphism $s : C \to C'$ such that $s(x_i) = x_i'$ and $f \circ s = f'$.

A tuple $(C; x_1, \ldots, x_n; f)$ satisfying (i) – (iv) is referred to as a *stable map*. The x_i are called *marked points* of C, and points that are either marked points or nodes are called *special points*. The condition of admitting finitely many automorphisms (the "stability" in the definition of a stable map) is equivalent to requiring that, for any irreducible component C_0 of C contracted to a point by f, one has

$$2g(C_0) - 2 + n(C_0) > 0, \tag{1}$$

where $n(C_0)$ is the number of special points on C_0.

As a moduli space, $\overline{\mathcal{M}}_{g,n}(X,\beta)$ has much more structure than that of a set. By definition, a moduli space should be equipped with a geometry that encodes how objects can deform; for example, a path in the moduli space should trace out a one-parameter family of objects.

In order to make the definition of $\overline{\mathcal{M}}_{g,n}(X,\beta)$ precise, then, we need a notion of a family of maps into X. If B is any scheme, a *family of stable maps* parameterized by B is a diagram

$$\begin{array}{ccc} \mathcal{C} & \xrightarrow{f} & X \\ \sigma_1 \uparrow \cdots \uparrow \sigma_n \; \downarrow \pi & & \\ B, & & \end{array}$$

in which π is a flat morphism whose fibers are nodal curves of genus g, the σ_i are disjoint sections of π, and each fiber

$$\left(\pi^{-1}(b); \sigma_1(b), \ldots, \sigma_n(b); f|_{\pi^{-1}(b)}\right)$$

over $b \in B$ is a degree-β stable map. The notion of morphism can be readily generalized to families: it consists of a morphism $s : \mathcal{C} \to \mathcal{C}'$ such that $s \circ \sigma_i = \sigma_i'$ and $f \circ s = f'$. Furthermore, a family over B can be pulled back along a morphism $B' \to B$ to yield a family over B'.

Now, to say that $\overline{\mathcal{M}}_{g,n}(X,\beta)$ is a moduli space for stable maps into X is to say that, for any base scheme B, there is a bijection

$$\begin{array}{c} \{\text{families of stable maps over } B \text{ (up to isomorphism)}\} \\ \updownarrow \\ \{\text{morphisms } B \to \overline{\mathcal{M}}_{g,n}(X,\beta)\}. \end{array} \tag{2}$$

To put it more explicitly, $\overline{\mathcal{M}}_{g,n}(X,\beta)$ admits a *universal family*, a family where the base scheme is the moduli space itself. The bijection (2) associates to a morphism $B \to \overline{\mathcal{M}}_{g,n}(X,\beta)$ the pullback of the universal family to B.

In particular, a stable map is simply a family over $B = \mathrm{Spec}(\mathbb{C})$, so a special case of (2) is the set-theoretic bijection between points of $\overline{\mathcal{M}}_{g,n}(X,\beta)$ and stable maps. But (2) implies much more: it dictates the algebro-geometric structure of the moduli space, assuming that such a space can indeed be constructed.

This brings us to a crucial caveat. A scheme $\overline{\mathcal{M}}_{g,n}(X,\beta)$ for which (2) produces a bijection does not, in fact, exist. The root of the problem lies in the existence of automorphisms of stable maps, which allow one to construct families over which every fiber is isomorphic, but which are nonetheless nontrivial as families. As it turns out, this problem is not entirely devastating, but it requires one to give $\overline{\mathcal{M}}_{g,n}(X,\beta)$ the structure of an *orbifold*, a more general notion than that of a scheme. Orbifold morphisms $B \to \overline{\mathcal{M}}_{g,n}(X,\beta)$ are correspondingly more general (even when B is a scheme), and under this notion of morphism, a bijection (2) into an orbifold $\overline{\mathcal{M}}_{g,n}(X,\beta)$ indeed exists.

1.2. Primary Gromov–Witten invariants. A Gromov–Witten invariant, at its most basic level, should be an artifact of enumerative geometry. More specifically, if $Y_1,\ldots,Y_n$ is a collection of subvarieties of X, then a Gromov–Witten invariant should be a count of the number of curves of genus g and degree β passing through all of the Y_i.

In order to make this precise, one defines *evaluation maps*

$$\mathrm{ev}_i : \overline{\mathcal{M}}_{g,n}(X,\beta) \to X$$

for each $i \in \{1,\ldots,n\}$, sending $(C; x_1,\ldots,x_n; f)$ to $f(x_i)$. A first pass at interpreting the count of genus-g, degree-β curves through all of the Y_i might be as the number of points of intersection

$$\mathrm{ev}_1^{-1}(Y_1) \cap \cdots \cap \mathrm{ev}_n^{-1}(Y_n).$$

A more refined version of this count, capturing its insensitivity to deformations of the subvarieties, is given by evaluating

$$\mathrm{ev}_1^*[Y_1] \cup \cdots \cup \mathrm{ev}_n^*[Y_n], \tag{3}$$

on the fundamental class of the moduli space, where $[Y_i]$ denotes the cohomology class defined by Y_i and we assume that the sum of the codimensions of the Y_i equals the dimension of the moduli space.

There is a problem with this definition, though. The moduli space of stable maps can be singular, and it can have different components of different dimensions. Thus, it is not clear what we mean by the "fundamental class" of the moduli space, nor even by the requirement that the codimensions of the Y_i sum to the dimension of $\overline{\mathcal{M}}_{g,n}(X,\beta)$.

The second of these confusions can be resolved through deformation theory: while $\overline{\mathcal{M}}_{g,n}(X,\beta)$ may not have a well-defined dimension, it does have an "expected" or "virtual" dimension, calculated by studying the space of infinitesimal deformations of a stable map (the tangent space to the moduli space) as well as the obstructions to extending infinitesimal deformations to honest ones. Explicitly, the virtual dimension is

$$\mathrm{vdim} := (\dim X - 3)(1-g) + \int_\beta c_1(T_X) + n. \tag{4}$$

Intuitively, one should understand the virtual dimension by imagining that $\overline{\mathcal{M}}_{g,n}(X,\beta)$ is the zero locus of a section s of a rank-r vector bundle E on some nonsingular ambient space Y. If s does not intersect the zero section of E transversally, then the dimension of $Z(s)$ could be larger than expected, but generically, one expects its dimension to be $\dim(Y) - r$; this is the "virtual" dimension of $Z(s)$.

Equipped with a replacement for the notion of dimension, it is a difficult fact **[2, 22]** that there also exists a replacement for the fundamental class, an element

$$[\overline{\mathcal{M}}_{g,n}(X,\beta)]^{\mathrm{vir}} \in H_{\mathrm{vdim}}(\overline{\mathcal{M}}_{g,n}(X,\beta))$$

known as the *virtual fundamental class*, which agrees with the fundamental class in the case where $\overline{\mathcal{M}}_{g,n}(X,\beta)$ is smooth of the expected dimension. Again, the idea can be explained intuitively by supposing that $\overline{\mathcal{M}}_{g,n}(X,\beta)$ is the zero locus of a section of a vector bundle, which may not meet the zero section transversally. For example, consider the least transverse situation possible, when s is identically zero. Then $[Z(s)] = [Y]$ lies in too-high dimension, but there is a natural way to achieve a homology class in the virtual dimension: take $[Y] \cap e(E)$. This amounts to perturbing $s \equiv 0$ to a transverse section and then taking its zero locus— although, in practice, such a perturbation may not be possible.

We can now give a precise definition of Gromov–Witten invariants.

DEFINITION 1.1. Let $\gamma_1, \ldots, \gamma_n \in H^*(X)$. Then the associated *(primary) Gromov–Witten invariant* is

$$\langle \gamma_1 \cdots \gamma_n \rangle_{g,n,\beta} := \int_{[\overline{\mathcal{M}}_{g,n}(X,\beta)]^{\mathrm{vir}}} \mathrm{ev}_1^*(\gamma_1) \cup \cdots \cup \mathrm{ev}_n^*(\gamma_n). \tag{5}$$

In case the γ_i are Poincaré dual to subvarieties Y_i and the moduli space is smooth of expected dimension, this recovers the intuitive enumeration captured by evaluating (3) on the fundamental class. Unfortunately, the enumerative meaning of the more general quantity (5) is not nearly so clear.

We should warn the reader, further, that (5) is generally an integral over an orbifold, and hence must be suitably interpreted. Orbifolds, roughly speaking, are locally modelled on quotients V/G of a variety V by the action of a finite group G, and integration is defined by pulling back to V and dividing by the order of G. In the case of $\overline{\mathcal{M}}_{g,n}(X,\beta)$, these local groups capture the automorphisms of the stable maps; in particular, this explains the necessity of imposing that such maps have finitely many automorphisms.

1.3. Descendant Gromov–Witten invariants. A generalization of the above-defined Gromov–Witten invariants is useful in order to obtain a more complete picture of the geometry of the moduli space of stable maps.

For each $i \in \{1, \ldots, n\}$, we define a *cotangent line bundle* $\mathbb{L}_i$ on $\overline{\mathcal{M}}_{g,n}(X,\beta)$ to have fiber[1] over a point $(C; x_1, \ldots, x_n; f)$ given by the cotangent line $T^*_{x_i}(C)$. The *psi classes* for $i \in \{1, \ldots, n\}$ are

$$\psi_i := c_1(\mathbb{L}_i) \in H^2(\overline{\mathcal{M}}_{g,n}(X,\beta)).$$

DEFINITION 1.2. Let $\gamma_1, \ldots, \gamma_n \in H^*(X)$, and let $a_1, \ldots, a_n$ be nonnegative integers. Then the associated *descendant Gromov–Witten invariant* is

$$\langle \psi^{a_1}\gamma_1 \cdots \psi^{a_n}\gamma_n \rangle_{g,n,\beta} := \int_{[\overline{\mathcal{M}}_{g,n}(X,\beta)]^{\mathrm{vir}}} \psi_1^{a_1}\mathrm{ev}_1^*(\gamma_1) \cup \cdots \cup \psi_n^{a_n}\mathrm{ev}_n^*(\gamma_n).$$

[1] To be more precise, $\mathbb{L}_i = \sigma_i^*\omega_\pi$, where π is the projection map in the universal family, σ_i is the ith section, and ω_π is the sheaf of relative differentials.

The reason that it is geometrically meaningful to include psi classes in Gromov–Witten invariants is related to a rich recursive structure among the moduli spaces of stable maps.

For any decomposition $g = g_1 + g_2$, $n = n_1 + n_2$, and $\beta = \beta_1 + \beta_2$, there is a divisor $D \subset \overline{\mathcal{M}}_{g,n}(X,\beta)$ whose general element is a curve with two irreducible components, one of genus g_1 containing the first n_1 marked points, on which $\deg(f) = \beta_1$, and the other of genus g_2 containing the last n_2 marked points, on which $\deg(f) = \beta_2$. In fact,

$$D \cong \overline{\mathcal{M}}_{g_1,n_1+1}(X,\beta_1) \times_X \overline{\mathcal{M}}_{g_2,n_2+1}(X,\beta_2), \tag{6}$$

where the fiber product ensures that the last marked point on each component (which is a branch of the node in D) maps to the same point in X. More generally, there are higher-codimension strata in $\overline{\mathcal{M}}_{g,n}(X,\beta)$ parameterizing curves with several components, on which the genus, marked points, and degree are distributed in some specified way. These subvarieties are referred to as *boundary strata.*

Integrals over the boundary strata arise naturally in the computation of Gromov–Witten invariants. For example, the localization formula (which we will discuss in Section 2.5) reduces the Gromov–Witten invariants of toric targets X to integrals over certain very special strata. Due to the existence of expressions like (6) for these strata in terms of simpler moduli spaces of stable maps, this has the effect of producing recursions among Gromov–Witten invariants.

The importance of the psi classes is that they encode the normal bundles to the boundary strata. For example, if D is as in (6), then

$$N_{D/\overline{\mathcal{M}}_{g,n}(X,\beta)} = \mathbb{L}_{n_1+1}^{\vee} \boxtimes \mathbb{L}_{n_2+1}^{\vee}, \tag{7}$$

where $\boxtimes$ indicates that the two bundles are pulled back under the projections to the two factors of (6). Intuitively, the reason for this is that an element of D is a nodal curve, and normal directions to D inside the moduli space (that is, deformations moving away from D) are given by smoothing the node. In local coordinates, the nodal curve can be expressed as $xy = 0$, with x and y giving the two tangent directions at the node. A node-smoothing deformation can be parameterized as $xy = t$, so t gives a local section of the normal bundle $N_{D/\overline{\mathcal{M}}_{g,n}(X,\beta)}$. That is, the normal space is the tensor product of the two tangent spaces at the node, and (7) follows.

1.4. Other curve-counting theories. Before we delve into the methods by which Gromov–Witten invariants are computed, it is worthwhile to revisit the enumerative questions that were our original motivation. After all, we wanted to count things like conics through fixed points in $\mathbb{P}^2$, but what Gromov–Witten theory would have us enumerate is *parameterized* conics— that is, degree-two maps from a curve into $\mathbb{P}^2$. What is more, even if we had hoped to count maps from nonsingular curves, we were forced to allow certain nodal degenerations in order to obtain a compact moduli space, one for which integration and intersection theory are well-behaved. Stable maps are not the only reasonable way to encode curves in a space, nor are they the only solution to the problem of compactification.

Donaldson–Thomas theory, for example, views a curve in a variety X not as a map $C \to X$ but as an ideal of algebraic functions, the defining equations of the curve. While these two perspectives on curves are equivalent when $C \to X$ is an embedding of a nonsingular C, the degenerate objects allowed in the two compactifications are very different. The moduli space of stable maps permits the embedding to degenerate— it can become a multiple cover, or it can contract entire components of the source curve to a point in X— while the moduli space used in Donaldson-Thomas theory, the *Hilbert scheme*, keeps the map an embedding but allows the curve to degenerate, developing nontrivial scheme structure, bad singularities, or isolated points.

There are some drawbacks to Donaldson–Thomas theory; most notably, the Hilbert scheme does not admit a virtual fundamental class unless X is a threefold, so only in this case can invariants be defined. When Donaldson–Thomas (or "DT") invariants are defined, though, they have one intriguing advantage over Gromov–Witten invariants: they are necessarily integers. This is due to the fact that the Hilbert scheme is, in fact, a scheme, as opposed to the moduli space of stable maps, which is only an orbifold. The integrality of DT invariants makes them more suited to honestly enumerative— and, in certain cases, entirely combinatorial[2]— interpretations.

The GW/DT conjecture of Maulik, Nekrasov, Okounkov, and Pandharipande [**23**, **24**] states that the generating functions of Gromov–Witten and Donaldson–Thomas theory should be related, in the cases where both are defined, by an explicit and strikingly simple change of variables, thus lending credence to the claim that both are "counting" the same sorts of objects. The conjecture has been proven for toric threefolds by Maulik–Oblomkov–Okounkov–Pandharipande [**25**], and for a large class of non-toric targets by Pandharipande–Pixton [**27**].

A rather different path to teasing out integers from the geometry of curves is provided by Gopakumar–Vafa invariants, sometimes called BPS states. These were defined physically in terms of a moduli space of "D-branes" (roughly, stable sheaves) of class β in a threefold X. This physical foundation presents the BPS states as integers, but it has not yet been established in precise mathematical terms.

Gopakumar and Vafa also predicted, though, that the BPS states should have a precise relationship to Gromov–Witten invariants. Thus, one can use their conjecture to *define* the Gopakumar–Vafa invariants from a mathematical perspective. From that point-of-view, the Gopakumar–Vafa invariants appear to be a sort of normalization of the Gromov–Witten invariants, accounting for the excess contribution to any count of degree-β maps that arises from k-fold covers of a degree-β' map with $k\beta' = \beta$. We will discuss this further in Section 3.1.

From this definition, the integrality of the Gopakumar–Vafa invariants becomes a mathematical conjecture. It has been proven for toric Calabi–Yau threefolds by Konishi [**17**], and weaker genus-zero results have also been obtained for certain non-toric targets by Kontsevich–Schwarz–Vologodsky [**20**]. By non-algebraic methods, Ionel–Parker proved the integrality of Gopakumar–Vafa invariants for any symplectic Calabi–Yau manifold of real dimension six [**15**], but the general statement in algebraic geometry remains open.

2. Computing Gromov–Witten invariants

In this section, we discuss some of the available methods for computing Gromov–Witten invariants. Our treatment will necessarily be incomplete, but there are many other references from which the interested reader can learn more. These

[2] In particular, when X is toric, the localization procedure discussed in Section 2.5 reduces the computation of DT invariants to counts of subschemes supported at the torus-fixed points of X, which can be described by certain three-dimensional partitions.

include the excellent introduction [16] to the genus-zero Gromov–Witten theory of projective spaces, the detailed yet highly readable account [34] of the all-genus Gromov–Witten theory of a point, and the wide-ranging tour de force [14].

2.1. Basic properties of primary invariants. Several fundamental properties of primary Gromov–Witten invariants follow from a fairly simple observation: there is a *forgetful map*

$$\tau : \overline{\mathcal{M}}_{g,n+1}(X,\beta) \to \overline{\mathcal{M}}_{g,n}(X,\beta)$$

whenever both of these moduli spaces are nonempty. Essentially, one should view τ as the map

$$(C; x_1, \ldots, x_{n+1}; f) \mapsto (C; x_1, \ldots, x_n; f).$$

However, when the last marked point is forgotten, components of C may no longer satisfy the condition (1) of stability. Thus, we require a *stabilization* procedure, which collapses any unstable components to a point. One can then confirm that, under this procedure, τ indeed defines a morphism of moduli spaces.

2.1.1. *Fundamental class property.* Let $\mathbf{1} \in H^0(X)$ denote the unit in cohomology (the Poincaré dual of the fundamental class). Then

$$\langle \gamma_1 \cdots \gamma_n \cdot \mathbf{1} \rangle_{g,n+1,\beta} = 0$$

unless $(g, n, \beta) = (0, 2, 0)$.

This follows from the fact that

$$\int_{[\overline{\mathcal{M}}_{g,n+1}(X,\beta)]^{\text{vir}}} \text{ev}_1^*(\gamma_1) \cdots \text{ev}_n^*(\gamma_n) = \int_{\tau_*[\overline{\mathcal{M}}_{g,n+1}(X,\beta)]^{\text{vir}}} \text{ev}_1^*(\gamma_1) \cdots \text{ev}_n^*(\gamma_n),$$

which is an application of the projection formula. From here, we use that

$$\tau_*[\overline{\mathcal{M}}_{g,n+1}(X,\beta)]^{\text{vir}} = 0. \tag{8}$$

If these spaces were smooth and the virtual fundamental classes were the ordinary ones, then (8) would be immediate from the dimension computation (4), since the pushforward would live in homological degree larger than the dimension of $\overline{\mathcal{M}}_{g,n}(X,\beta)$. More generally, a study of the deformation theory is required.[3]

2.1.2. *Divisor equation.* For a divisor class $[D] \in H^2(X)$, we have

$$\langle \gamma_1 \cdots \gamma_n \cdot [D] \rangle_{g,n+1,\beta} = \left(\int_\beta [D] \right) \cdot \langle \gamma_1 \cdots \gamma_n \rangle_{g,n,\beta}$$

unless $(g, n, \beta) = (0, 2, 0)$. As above, let us see why this is the case if the virtual fundamental classes are the ordinary ones, relying on deformation theory to show that the same is true in the virtual situation. The equation is again an application of the projection formula, together with the fact that

$$\tau_*\left(\text{ev}_{n+1}^*[D] \cap [\overline{\mathcal{M}}_{g,n+1}(X,\beta)]\right) = \tau_*[\text{ev}_{n+1}^{-1}(D)] = \int_\beta [D].$$

The second equality follows from the observation that $\tau|_{\text{ev}_{n+1}^{-1}(D)}$ is generically finite of degree equal to $\int_\beta [D]$; indeed, once an n-pointed stable map $f : C \to X$ has

[3]The essential point is that the deformation theory is "pulled back" under τ. As a simple example to gain some insight, suppose that there is a vector bundle E for which $[\overline{\mathcal{M}}_{g,n}(X,\beta)]^{\text{vir}} = [\overline{\mathcal{M}}_{g,n}(X,\beta)] \cap e(E)$. Then E encodes the deformation theory, and one has $[\overline{\mathcal{M}}_{g,n+1}(X,\beta)]^{\text{vir}} = [\overline{\mathcal{M}}_{g,n+1}(X,\beta)] \cap e(\tau^*E)$. From here, it is straightforward to see that (8) holds.

been chosen, its image generically intersects D in $\int_\beta [D]$ points, and the $(n+1)$st marked point can be placed at any of these.

2.1.3. *Degree-zero invariants.* The above two equations exclude the case $(g, n+1, \beta) = (0, 3, 0)$, since there is no forgetful map in this situation: stable maps with $(g, n, \beta) = (0, 2, 0)$ do not exist. Still, we can compute the Gromov–Witten invariants directly. We have:

$$\langle \gamma_1 \; \gamma_2 \; \gamma_3 \rangle_{0,3,0} = \int_X \gamma_1 \cup \gamma_2 \cup \gamma_3.$$

This follows from the fact that

$$\overline{\mathcal{M}}_{0,3}(X, 0) \cong X,$$

since there is a unique isomorphism $C \cong \mathbb{P}^1$ sending the three marked points to $0, 1$, and ∞, so all that must be chosen to specify a point in $\overline{\mathcal{M}}_{0,3}(X, 0)$ is the image point of the constant map $f : C \to X$. Furthermore, the virtual class actually *is* the ordinary fundamental class in this case, so no deformation-theoretic argument is required.

2.2. WDVV equations and splitting. In genus zero, a different sort of forgetful map also implies useful relations: if $n \geq 4$, we have

$$\phi : \overline{\mathcal{M}}_{0,n}(X, \beta) \to \overline{\mathcal{M}}_{0,4},$$

given by

$$(C; x_1, \ldots, x_n; f) \mapsto (C; x_1, \ldots, x_4)$$

(modulo the same discussion of stabilization mentioned for the morphism τ). Here, $\overline{\mathcal{M}}_{0,4}$ is the moduli space of genus-zero, four-pointed curves without a map to a target; in other words, $\overline{\mathcal{M}}_{0,4} = \overline{\mathcal{M}}_{0,4}(\text{point}, 0)$.

Genus-zero, four-pointed curves can be understood very concretely. First of all, when the curve C is smooth, there is a unique isomorphism $C \cong \mathbb{P}^1$ sending x_1, x_2, and x_3 to $0, 1$, and $\infty \in \mathbb{P}^1$, as remarked above. This isomorphism sends x_4 to some point $q \in \mathbb{P}^1 \setminus \{0, 1, \infty\}$ (the *cross ratio* of x_1, x_2, x_3), and q uniquely specifies the isomorphism class of $(C; x_1, \ldots, x_4)$ in $\overline{\mathcal{M}}_{0,4}$. Thus, the locus of smooth curves in $\overline{\mathcal{M}}_{0,4}$ is isomorphic to $\mathbb{P}^1 \setminus \{0, 1, \infty\}$. From here, it is not a long leap to see that the compactification is $\overline{\mathcal{M}}_{0,4} \cong \mathbb{P}^1$. Indeed, there are three possible reducible 4-pointed curves, depicted in Figure 2.2, and these give the three boundary points.[4]

Each of these three points is a boundary divisor in $\overline{\mathcal{M}}_{0,4}$, which we denote by $D(1,2|3,4)$, $D(1,3|2,4)$, and $D(1,4|2,3)$, respectively. Any point in $\mathbb{P}^1$ gives the same divisor, though, up to linear equivalence, so we have

$$D(1,2|3,4) \equiv D(1,3|2,4) \equiv D(1,4|2,3) \tag{9}$$

in $H^2(\overline{\mathcal{M}}_{0,4})$. Pulling back the linear equivalences (9) under the morphism ϕ, we find a linear equivalence of three boundary divisors in $\overline{\mathcal{M}}_{0,n}(X, \beta)$. These equivalences are referred to— after Witten, Dijkgraaf, Verlinde, and Verlinde— as the *WDVV equations.*

---

[4] Of course, we have really only succeeded in describing $\overline{\mathcal{M}}_{0,4}$ as a set; a true proof that it is isomorphic to $\mathbb{P}^1$ as a scheme would require consideration of the universal curve and the bijection (2) on families.

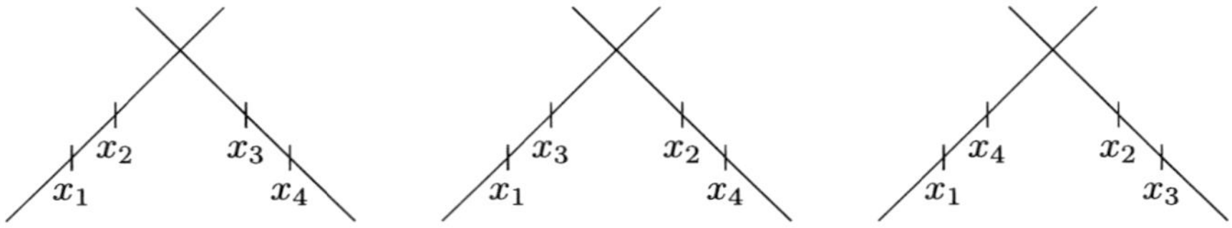

FIGURE 1. Real cartoons of the three singular curves in $\overline{\mathcal{M}}_{0,4}$.

The reason the WDVV equations are useful is that integrals over boundary divisors can be expressed as Gromov–Witten invariants. The proof of this involves lifting the isomorphism (6) to the level of virtual fundamental cycles and interpreting the fiber product explicitly. We find that, if $D \subset \overline{\mathcal{M}}_{0,n}(X,\beta)$ is a boundary divisor corresponding to the decomposition $n = n_1 + n_2$ and $\beta = \beta_1 + \beta_2$ of the marked points and degree, then

$$\int_D \mathrm{ev}_1^*(\gamma_1) \cup \cdots \cup \mathrm{ev}_n^*(\gamma_n) = \tag{10}$$
$$\sum_i \langle \gamma_1 \cdots \gamma_{n_1} \cdot \phi_i \rangle_{0,n_1+1,\beta_1} \cdot \langle \phi^i \cdot \gamma_{n_1+1} \cdots \gamma_n \rangle_{0,n_2+1,\beta_2}.$$

Here, the sum runs over a basis $\{\phi_1, \ldots, \phi_k\}$ for $H^*(X)$, and ϕ^i denotes the dual of ϕ_i under the Poincaré pairing, meaning that $\int_X \phi_i \phi^j = \delta_i^j$.

2.3. A sample computation. Using only the properties outlined in this section, we can already perform many computations. As an example, we can reproduce the answer to the second question posed at the very beginning of the chapter. Namely, when $X = \mathbb{P}^3$, we will compute

$$\langle H^2\ H^2\ H^2\ H^2 \rangle_{0,4,1} = 2,$$

where $H \in H^2(\mathbb{P}^3)$ is the hyperplane class and hence H^2 is the class of a line. Here, we identify $H_2(\mathbb{P}^3) \cong \mathbb{Z}$, so the index $\beta = 1$ denotes the class of a line in homology.

The trick is to consider a moduli space with one more marked point, $\overline{\mathcal{M}}_{0,5}(\mathbb{P}^3, 1)$, and use the WDVV equations to achieve a relation between integrals over boundary divisors there. Specifically, the linear equivalence $D(1,2|3,4) \equiv D(1,3|2,4)$ pulls back under the forgetful map ϕ to the following linear equivalence of boundary divisors in $\overline{\mathcal{M}}_{0,5}(\mathbb{P}^3, 1)$:

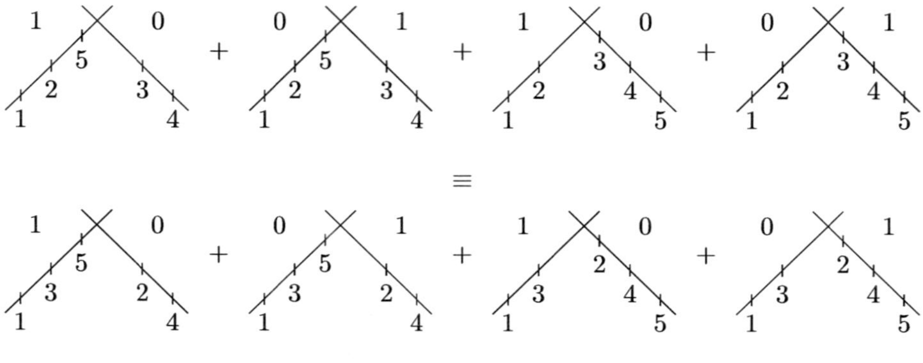

Each marked point is labeled with its number, and each irreducible component with the degree of the restriction of f.

Now, we equate the integrals of

$$\mathrm{ev}_1^*(H) \cup \mathrm{ev}_2^*(H) \cup \mathrm{ev}_3^*(H^2) \cup \mathrm{ev}_4^*(H^2) \cup \mathrm{ev}_5^*(H^2)$$

over these two divisors. We apply the splitting property (10) to each of the four integrals appearing on either side, taking the basis $\phi_i = H^i$ for $i \in \{0, 1, 2, 3\}$. In each of these eight integrals, only one choice of i will yield a possibly nonzero invariant, since the sums of the codimensions of the classes must equal $\mathrm{vdim}(\overline{\mathcal{M}}_{0,n}(\mathbb{P}^3, d)) = 4d+n$. Thus, for example, the integral over the second term on the left-hand side of the above linear equivalence yields

$$\langle H\ H\ H^2\ H^0 \rangle_{0,4,0} \langle H^3\ H^2\ H^2 \rangle_{0,3,1}.$$

In some cases, such as the first term on the left-hand side, the invariants immediately vanish, since no ϕ_i can make the codimensions sum to the virtual dimension. We find:

$$\begin{aligned}&\langle H\ H\ H^2\ H^0 \rangle_{0,4,0} \langle H^3\ H^2\ H^2 \rangle_{0,3,1} + \langle H\ H\ H \rangle_{0,3,0} \langle H^2\ H^2\ H^2\ H^2 \rangle_{0,4,1}\\ =&\langle H\ H^2\ H^2\ H^3 \rangle_{0,4,1} \langle H^0\ H\ H^2 \rangle_{0,3,0} + \langle H\ H^2\ H^0 \rangle_{0,3,0} \langle H^3\ H^2\ H\ H^2 \rangle_{0,4,1}.\end{aligned}$$

Applying the three basic properties from Section 2.1 (together with the invariance of Gromov–Witten invariants under permutation of the inputs), this becomes

$$\langle H^2\ H^2\ H^2\ H^2 \rangle_{0,4,1} = 2 \langle H^2\ H^2\ H^3 \rangle_{0,3,1}.$$

A similar argument, this time applied to integrals of

$$\mathrm{ev}_1^*(H) \cup \mathrm{ev}_2^*(H) \cup \mathrm{ev}_3^*(H^2) \cup \mathrm{ev}_4^*(H^3)$$

over linearly equivalent boundary divisors in $\overline{\mathcal{M}}_{0,4}(\mathbb{P}^3, 1)$, shows that

$$\langle H^2\ H^2\ H^3 \rangle_{0,3,1} = \langle H^3\ H^3 \rangle_{0,2,1}.$$

It should be intuitively clear that $\langle H^3\ H^3 \rangle_{0,2,1} = 1$, since H^3 is the Poincaré dual of a point in $\mathbb{P}^3$, and hence this invariant should be interpreted enumeratively as the number of lines through two fixed points. In fact, with a bit more care— checking that the virtual fundamental class is an ordinary fundamental class, for example— one can verify that this naïve interpretation actually coincides with the Gromov–Witten invariant, thus completing the calculation.

2.4. Basic properties of descendant invariants. We have focused thus far on the computation of primary Gromov–Witten invariants, but adding descendants only enriches the structure.

The basic properties of descendants still follow easily from the existence of the forgetful map $\tau : \overline{\mathcal{M}}_{g,n+1}(X, \beta) \to \overline{\mathcal{M}}_{g,n}(X, \beta)$. However, there is a new complication: the psi classes do not pull back under τ. Rather, one has

$$\psi_i = \tau^* \psi_i + D_i, \tag{11}$$

where D_i is the boundary divisor corresponding to curves with one genus-zero, degree-zero component carrying the marked points i and $n+1$, and another genus-g, degree-β component carrying all of the other marked points. The proof of (11) is not difficult; the essential point is that the curves in D_i become unstable under τ and hence are collapsed, but these are the only curves on which ψ_i and $\tau^*\psi_i$ differ.

Using this comparison result (and very little else about the geometry of the moduli spaces), one can prove the *string equation*,

$$\langle \psi^{a_1}\gamma_1 \cdots \psi^{a_n}\gamma_n \cdot \mathbf{1}\rangle_{g,n+1,\beta} = \sum_{i=1}^{n} \langle \psi^{a_1}\gamma_1 \cdots \psi^{a_i-1}\gamma_i \cdots \psi^{a_n}\gamma_n\rangle_{g,n,\beta},$$

and the *dilaton equation*,

$$\langle \psi^{a_1}\gamma_1 \cdots \psi^{a_n}\gamma_n \cdot \psi\mathbf{1}\rangle_{g,n+1,\beta} = (2g-2+n)\langle \psi^{a_1}\gamma_1 \cdots \psi^{a_n}\gamma_n\rangle_{g,n,\beta},$$

where $\mathbf{1}$ denotes the unit in $H^0(X)$.

Furthermore, in genus zero, there are more complicated equations called the *topological recursion relations* allowing one to reduce powers of the psi classes in general:

(12)

$$\langle \psi^{a_1}\gamma_1 \cdots \psi^{a_n}\gamma_n \cdot \psi^{k+1}\phi_\alpha \cdot \psi^l\phi_\gamma \cdot \psi^m\phi_\delta\rangle_{0,n+3,\beta} =$$

$$\sum_{\substack{[n]=I\sqcup J\\ \beta=\beta_1+\beta_2\\ \mu}} \left\langle \prod_{i\in I}\psi^{a_i}\gamma_i \cdot \psi^k\phi_\alpha \cdot \phi_\mu \right\rangle_{0,|I|+2,\beta_1} \left\langle \phi^\mu \cdot \prod_{j\in J}\psi^{a_j}\gamma_j \cdot \psi^l\phi_\gamma \cdot \psi^m\phi_\delta \right\rangle_{0,|J|+3,\beta_2}.$$

Here, as before, $\{\phi_\mu\}$ denotes a basis for $H^*(X)$, and ϕ^μ is dual to ϕ_μ under the Poincaré pairing. The key point in proving the topological recursion relations is to consider the morphism

$$\varphi : \overline{\mathcal{M}}_{0,n+3}(X,\beta) \to \overline{\mathcal{M}}_{0,3}$$

forgetting the map $f: C \to X$ and all but the last three marked points. Analogously to (11), one has a comparison result on psi classes:

$$\psi_{n+1} = \varphi^*\psi_1 + D(n+1|n+2,n+3), \tag{13}$$

in which $D(n+1|n+2,n+3)$ is the sum of all of the boundary divisors on which the marked point $n+1$ is on one component and the marked points $n+2$ and $n+3$ are on the other. Since $\overline{\mathcal{M}}_{0,3}$ is simply a point, we have $\varphi^*\psi_1 = 0$, so (13) expresses ψ_{n+1} as a sum of boundary divisors. Applying this expression to one of the $k+1$ copies of ψ_{n+1} on the left-hand side of (12) gives the desired expression.

The string equation, dilaton equation, and topological recursion relations together determine many of the descendant invariants from a small subset; we will return to this in Section 2.6.

2.5. Localization. Another method— and a very powerful one— by which certain Gromov–Witten invariants can be reduced to simpler ones is the Atiyah–Bott localization formula. A full discussion of localization would take us too far afield, but we will summarize the idea, referring the reader to [**1**] (or, for the specific case of Gromov–Witten theory, [**12**] or [**14**]) for more detailed information.

Let M be a smooth projective variety equipped with an algebraic action of a torus $\mathbb{T} = (\mathbb{C}^*)^r$. The *equivariant cohomology* $H^*_{\mathbb{T}}(M)$ is an enhanced cohomology theory that takes into account not only the topology of M but also the structure of its $\mathbb{T}$-orbits. When M is a point with the only possible $\mathbb{T}$-action, the equivariant cohomology is a polynomial ring:

$$H^*_{\mathbb{T}}(\text{point}) \cong \mathbb{C}[\lambda_1, \ldots, \lambda_r].$$

More generally, for any M as above, pullback under the map to a point induces a homomorphism

$$H^*_{\mathbb{T}}(\text{point}) \to H^*_{\mathbb{T}}(M),$$

which makes $H^*_{\mathbb{T}}(M)$ into a $\mathbb{C}[\lambda_1, \ldots, \lambda_r]$-module. All of the usual operations on cohomology (pullback, integration, Chern classes, et cetera) have equivariant analogues.

According to the *Atiyah–Bott localization formula*, all of the information about the equivariant cohomology of M is contained in the equivariant cohomology of its $\mathbb{T}$-fixed locus. More precisely, the theorem is the following. Let $\{F_j\}$ be the connected components of the fixed locus, and let $i_j : F_j \hookrightarrow M$ be their inclusions. Then, in the localized ring

$$H^*_{\mathbb{T}}(M) \otimes \mathbb{C}(\lambda_1, \ldots, \lambda_n),$$

the equivariant Euler classes $e_{\mathbb{T}}(N_{F_j/M})$ of the normal bundles are invertible, and one has

$$\int_M \phi = \sum_j \int_{F_j} \frac{i_j^* \phi}{e_{\mathbb{T}}(N_{F_j/M})} \tag{14}$$

for any class $\phi \in H^*_{\mathbb{T}}(M)$.

To apply this to the setting of Gromov–Witten theory, one requires a generalization that allows for integration against virtual fundamental classes. This *virtual localization formula* was proved by Graber–Pandharipande [**12**]. It involves the notion of a "virtual normal bundle", but ultimately, its form is exactly the same as (14).

Now, suppose that X is a smooth projective variety with a $\mathbb{T}$-action. Then the moduli space $\overline{\mathcal{M}}_{g,n}(X, \beta)$ inherits a $\mathbb{T}$-action of its own, given by post-composing a map $f : C \to X$ with the action on X. The fixed loci $F_j \subset \overline{\mathcal{M}}_{g,n}(X, \beta)$ can then be calculated. They need not map entirely into the fixed locus of X (C might map to a $\mathbb{T}$-invariant curve $C' \subset X$ in such a way that the $\mathbb{T}$-action on C' can be "undone" by an automorphism of C), but still, being fixed puts very strong constraints on their topology.

For example, $X = \mathbb{P}^{r-1}$ admits a tautological action of $\mathbb{T} = (\mathbb{C}^*)^r$ through the expression $X = \mathbb{P}(\mathbb{C}^r)$. The torus-fixed stable maps are those where all of the marked points and all of the positive genus occur on irreducible components of C that are collapsed by f to one of the r fixed points, and where these contracted components are connected by rational curves mapping to $\mathbb{T}$-invariant lines in X via a degree-d cover ramified over the two fixed points. The choice of these ramified covers is discrete: it is specified by the degree and the two fixed points over which it ramifies. Thus, integrating over a $\mathbb{T}$-fixed locus in $\overline{\mathcal{M}}_{g,n}(\mathbb{P}^{r-1}, \beta)$ amounts to integrating over a product of moduli spaces $\overline{\mathcal{M}}_{g_i,n_i}$ of curves corresponding to the contracted components of C.

After calculating the Euler classes of the virtual normal bundles, then, one can apply the localization formula to express any Gromov–Witten invariant of $\mathbb{P}^{r-1}$ as a summation, indexed by certain decorated graphs picking out the topology and discrete data, of the much simpler Gromov–Witten invariants of a point.

2.6. Re-packaging the redundancy. As this section has revealed, there is a great deal of redundancy in Gromov–Witten invariants. This leads to some natural questions: what is the minimal amount of information needed to determine all of the Gromov–Witten invariants of a variety X? And, given this base information, how can the rest of the invariants be efficiently read off?

One way to answer these questions is to package Gromov–Witten invariants into a generating function, and to re-phrase the relations among invariants as differential equations that the generating function satisfies. As a first (but still difficult and interesting) example, consider the generating function for descendant invariants of a point:

$$F(t_0, t_1, \ldots) = \sum_{\substack{g\geq 0, n\geq 1\\ 2g-2+n>0}} \sum_{a_1,\ldots,a_n} \frac{t_{a_1}\cdots t_{a_n}}{n!} \int_{\overline{\mathcal{M}}_{g,n}} \psi_1^{a_1}\cdots\psi_n^{a_n},$$

a formal function in infinitely many variables. The string equation is equivalent to the differential equation

$$\frac{\partial F}{\partial t_0} = \frac{1}{2}t_0^2 + \sum_{k=0}^{\infty} t_{k+1}\frac{\partial F}{\partial t_k}, \tag{15}$$

as the reader can easily check. In 1991, Witten conjectured [**33**] that F furthermore satisfies

$$\frac{\partial^2 F}{\partial t_0 \partial t_1} = \frac{1}{2}\left(\frac{\partial^2 F}{\partial t_0^2}\right)^2 + \frac{1}{12}\frac{\partial^4 F}{\partial t_0^4}, \tag{16}$$

the so-called *KdV equation*. Equations (15) and (16), together with the leading term $F = t_0^3/6 + \cdots$, are sufficient to uniquely determine the entire generating function. Witten's conjecture was proved by Kontsevich [**19**] shortly after its announcement, so psi integrals on $\overline{\mathcal{M}}_{g,n}$ are now all effectively known.

Matters get more complicated, of course, when the target becomes more interesting, but many differential equations have been conjectured and some have been proven. This leads to the *Virasoro conjecture* (see [**7, 21, 26**], among many others), and more generally, to the fascinating subject of integrable hierarchies, a topic of much current research.

A very different way to view the redundancy of Gromov–Witten theory was suggested by Givental [**5, 9, 11**]. The idea is to form an infinite-dimensional vector space $\mathcal{H} := H^*(X)((z^{-1}))$ and view the genus-zero invariants of X as functions on the subspace $\mathcal{H}_+ := H^*(X)[z]$. Namely, if $\{\phi_\mu\}$ is a basis for $H^*(X)$, we set

$$\mathbf{t}(z) = \sum_{a,\mu} t_a^\mu z^a \phi_\mu.$$

Then the function

$$F_0^X(\mathbf{t}) = \sum_{n,\beta} \frac{1}{n!}\langle \mathbf{t}(\psi)\cdots\mathbf{t}(\psi)\rangle_{0,n,\beta}$$

is a generating function for all genus-zero descendant invariants of X. Consider the subspace

$$\mathcal{L}_X := \left\{ -z + \mathbf{t}(z) + \sum_{n,\beta,\mu,a} \frac{1}{n!}\langle \mathbf{t}(\psi)\cdots\mathbf{t}(\psi)\cdot\psi^a\phi_\mu\rangle_{0,n+1,\beta}\frac{\phi^\mu}{(-z)^{a+1}} \right\}$$

of $\mathcal{H}$, where $\mathbf{t}(z)$ ranges over all elements of $\mathcal{H}_+$ and ϕ^μ, as always, denotes the Poincaré dual of ϕ_μ. This subspace can be viewed as the graph of the derivative of F_0^X, after making an identification of $\mathcal{H}$ with the cotangent bundle to $\mathcal{H}_+$.

Givental's insight was that the string equation, dilaton equation, and topological recursion relations are equivalent to geometric properties of $\mathcal{L}_X$. Specifically, $\mathcal{L}_X$ is a cone swept out by a finite-dimensional ruling. The upshot of this statement is that $\mathcal{L}_X$, though it lies in an infinite-dimensional vector space, can be uniquely recovered from a finite-dimensional slice. One such slice, known as the *J-function* of X, is given by restricting to points in $\mathcal{L}_X$ for which $\mathbf{t}(z) = t_0^\mu \phi_\mu$ has no psi classes.[5] But there are other slices, some that can be written very explicitly in closed form; finding such a slice is a very succint way to encode knowledge of all the genus-zero descendants of X. We will return to these ideas in Section 3.3.

All of this is unique to genus zero, but there is a deep theory by which higher genus can be brought into the picture. The idea is to look for transformations $\Delta : H^*(X)((z^{-1})) \to H^*(Y)((z^{-1}))$ taking the cone $\mathcal{L}_X$ for one variety to the cone $\mathcal{L}_Y$ for another, and to apply a procedure known as "quantization" whose definition originates in theoretical physics. (See, for example, **[3]** for an exposition of the quantization machinery.) It is expected, following a conjecture of Givental **[10]**, that if Δ takes $\mathcal{L}_X$ to $\mathcal{L}_Y$, then the quantization $\hat{\Delta}$ should take the generating function for all-genus Gromov–Witten invariants of X to all-genus Gromov–Witten invariants of Y.

This conjecture is particularly interesting when applied to targets satisfying a condition known as "semisimplicity".[6] Givental proved **[11]** that, given *any* semisimple X and Y for which $H^*(X)$ and $H^*(Y)$ have the same rank, there exists a transformation Δ taking $\mathcal{L}_X$ to $\mathcal{L}_Y$. In particular, one can take X to be a disjoint union of points. Then the resulting Δ recovers the genus-zero Gromov–Witten theory of Y from the genus-zero Gromov–Witten theory of this simplest of targets— and, if Givental's conjecture holds, the same is true in higher genus.

Givental proved his conjecture for the equivariant Gromov–Witten theory of toric targets (all of which are semisimple) **[9, 10]**. The basic point is that the localization procedure expresses the Gromov–Witten invariants of such a target in terms of the Gromov–Witten theory of a disjoint union of points, namely the fixed points of the torus action. The combinatorics needed to turn this heuristic into a proof of the conjecture, however, are intricate and ingeniously packaged.

Teleman later proved that Givental's conjecture holds for all semisimple targets **[32]**. Thus, in the semisimple case, there is a complicated but powerful sense in which genus-zero Gromov–Witten invariants determine the entire theory.

3. A tour of applications and open questions

We conclude with a brief— and, again, necessarily very incomplete— survey of some of the applications of Gromov–Witten theory to algebro-geometric problems. Of course, in answering one question we often open the door to a host of new ones, so this section will also serve as a tour of a few of the open problems in the field.

3.1. Enumerative geometry. We have seen one example in which Gromov–Witten theory successfully reproduces the intuitive enumerative geometric calculations made by Schubert over a century ago. What is more striking, though, is that

[5] Perhaps, after another glance at the equations of Section 2.4, the reader should not be entirely surprised that this piece of $\mathcal{L}_X$ determines all descendant invariants.

[6] Semisimplicity is defined in terms of the quantum product, a family of product structures on $H^*(X)$ discussed further in Section 3.1. By definition, X is semisimple if the generic member of this family gives $H^*(X)$ the structure of a semisimple algebra.

new insights from physics permit Gromov–Witten theory to answer enumerative questions that were previously outside of mathematicians' reach.

As an example, let N_d denote the number of degree-d rational curves in $\mathbb{P}^2$ passing through $3d-1$ prescribed points. More precisely, these are Gromov–Witten invariants

$$N_d = \langle H^2 \cdots H^2 \rangle_{0,3d-1,d}$$

on $\mathbb{P}^2$, where H^2 is the cohomology class of a point. Prior to the advent of Gromov–Witten theory, only the first few values of N_d were known, and even the computation of $N_4 = 620$ required a nearly Herculean level of computational effort.

The connection to theoretical physics, on the other hand, suggests a different path to these numbers. Based on their role in string theory, Gromov–Witten invariants were expected to fit together into a *quantum product*. This is a family of product structures on $H^*(X)$, parameterized by $\mathbf{t} \in H^*(X)$, where the product $*_{\mathbf{t}}$ is defined by

$$\phi_i *_{\mathbf{t}} \phi_j := \sum_k \sum_{n,\beta} \frac{1}{n!} \langle \mathbf{t} \cdots \mathbf{t} \cdot \phi_i \cdot \phi_j \cdot \phi_k \rangle_{0,n+3,\beta} \cdot \phi^k. \tag{17}$$

To get a sense of the connection to string theory, consider the case where $\mathbf{t} = 0$, so only three-point Gromov–Witten invariants appear. Then $H^*(X)$ should be viewed as the space of possible states of a physical system, and the three-point invariant as a probability that states ϕ_i and ϕ_j will interact to give state ϕ^k.

It is not immediately obvious that the product defined by (17) should be associative; this is a consequence of the WDVV equations. Now take the special case of $X = \mathbb{P}^2$ and work at the basepoint $\mathbf{t} = tH^2$ for a formal parameter t. If one expands the associativity statement

$$H *_{tH^2} (H *_{tH^2} H^2) = (H *_{tH^2} H) *_{tH^2} H^2,$$

then a recursion among the numbers N_d falls out.

This recursion, known as *Kontsevich's formula*, is the following:

$$N_d + \sum_{\substack{d_A+d_B=d \\ d_A,d_B \geq 1}} \binom{3d-4}{3d_A-1} d_A^3 d_B N_{d_A} N_{d_B} = \sum_{\substack{d_A+d_B=d \\ d_A,d_B \geq 1}} \binom{3d-4}{3d_A-2} d_A^2 d_B^2 N_{d_A} N_{d_B}.$$

The remarkable consequence of Kontsevich's formula is that it allows one to compute any of the numbers N_d easily, based only on the initial input data of $N_1 = 1$, the number of lines through two points.[7]

One should be careful, here and in general, about deducing enumerative information from Gromov–Witten invariants. If the moduli space has components of excessive dimension, then integrals against the virtual fundamental class can no longer be interpreted naïvely as counts of intersection points among subvarieties. Moreover, counting stable maps is not *a priori* the same thing as counting curves through subvarieties; a map might intersect some subvariety more than once, leading to over-counting coming from re-labeling the marked points of intersection, or it might have automorphisms, causing it to count as only a fraction of a point in the

[7]More generally, for any r, the genus-zero Gromov–Witten invariants of $\mathbb{P}^r$ can be computed recursively from the single intial value $\langle H^r \; H^r \rangle_{0,2,1} = 1$; this is a special case of the Reconstruction Theorem of Kontsevich–Manin [**18**] and Ruan–Tian [**30**]. We have seen this principle in action in Section 2.3, where the computation of $\langle H^2 \; H^2 \; H^2 \; H^2 \rangle_{0,4,1}$ for $\mathbb{P}^3$ was reduced, by repeated application of the WDVV equations, to the enumeration of lines through two points.

intersection number. One can check that, in the case of the N_d, these issues do not arise: the moduli space has the expected dimension, and generically, stable maps that pass through a particular point do so only once and without automorphisms.

More generally, though, the problem of extracting curve counts (or integers at all) from Gromov–Witten invariants is a serious one. For example, consider the third question with which we started the chapter: how many rational curves of degree d lie on the quintic threefold $V \subset \mathbb{P}^4$? Using the adjunction formula, one can check that

$$\mathrm{vdim}(\overline{\mathcal{M}}_{0,n}(V,d)) = n,$$

and from here, the properties in Section 2.1 easily reduce all genus-zero Gromov–Witten invariants of V to the computation of the 0-point invariants

$$I_d := \langle\ \rangle_{0,0,d}.$$

One might hope that I_d is equal to the number of degree-d rational curves on V, but this is not the case. In particular, composing any degree-d' map $f : \mathbb{P}^1 \to V$ with a k-fold cover $g : \mathbb{P}^1 \to \mathbb{P}^1$, where $kd' = d$, yields a degree-d map $f \circ g : \mathbb{P}^1 \to V$. There is a positive-dimensional family of such covers, which produces components in $\overline{\mathcal{M}}_{0,0}(V,d)$ of excessive dimension and hence spoils the enumerativity of I_d.

One can attempt to fix matters by computing the contribution of such multiple covers to I_d by hand. The answer, under the assumption that the image curve has normal bundle $\mathcal{O}_{\mathbb{P}^1}(-1) \oplus \mathcal{O}_{\mathbb{P}^1}(-1)$ (which, according to the Clemens conjecture, should always be the case), is surprisingly simple: degree-k covers contribute $1/k^3$ to I_d. Thus, it is conjectured that the numbers i_d defined by

$$I_d = \sum_{k|d} \frac{1}{k^3} i_{d/k}$$

are, in fact, integers. (These are the Gopakumar–Vafa invariants discussed in Section 1.4.)

This is still not the end of the story. For $d \leq 9$, the numbers i_d have been shown to agree with the number of rational curves of degree d in V, but for $d \geq 10$, it is not even known whether the number of such curves is finite. Furthermore, even if it is finite, an observation of Pandharipande reveals that it will be smaller than i_d in general, since multiple covers of singular curves of degree five contribute to I_d by more than the generic $1/k^3$ accounted for in i_d. Thus, although the numbers i_d can be computed (using the techniques of mirror symmetry described in Section 3.3 below), the translation into enumerative information still contains a wealth of mysteries.

3.2. The moduli space of curves. The rich structure of Gromov–Witten theory can also be used as a tool for studying the moduli space $\overline{\mathcal{M}}_{g,n}$ of curves, a more classical object of algebro-geometric interest.

For any choice of X and β, there is a map

$$p : \overline{\mathcal{M}}_{g,n}(X,\beta) \to \overline{\mathcal{M}}_{g,n},$$

given by forgetting the data of $f : C \to X$ and stabilizing the curve as necessary. Thus, relations between cohomology classes on $\overline{\mathcal{M}}_{g,n}(X,\beta)$ that arise out of Gromov–Witten-theoretic knowledge— the localization formula, for example, or the quantization formula for higher-genus theory in terms of genus zero— can be pushed forward to yield relations between classes on the moduli space of curves.

The same reasoning applies more generally to other moduli spaces, such as the moduli space of stable quasi-maps or the moduli space of curves with r-spin structure, which parameterize n-pointed curves equipped with some additional datum that can be forgotten to yield a map to $\overline{\mathcal{M}}_{g,n}$.

These methods all produce equations satisfied by a particular family of cohomology classes on $\overline{\mathcal{M}}_{g,n}$, the *tautological classes* $R^*(\overline{\mathcal{M}}_{g,n}) \subset H^*(\overline{\mathcal{M}}_{g,n})$. These are defined simultaneously for all g and n as the minimal family of subrings closed under pushforward by the forgetful morphism τ and the two types of "gluing" morphisms

$$\overline{\mathcal{M}}_{g_1,n_1+1} \times \overline{\mathcal{M}}_{g_2,n_2+1} \to \overline{\mathcal{M}}_{g_1+g_2,n_1+n_2},$$

$$\overline{\mathcal{M}}_{g-1,n+2} \to \overline{\mathcal{M}}_{g,n},$$

which attach together two marked points to form a node. Although non-tautological classes have been shown (with some effort) to exist [**8**, **13**], nearly every geometrically-interesting class is tautological.

Gromov–Witten-theoretic methods have been used to deduce relations in the tautological ring by a number of authors, culminating with the proof by Pandharipande, Pixton, and Zvonkine [**28**] of a set of relations previously conjectured by Pixton [**29**], using the quantization formalism on a moduli space of r-spin structures. It is currently conjectured that Pixton's relations are *all* of the relations in the tautological ring; in particular, they have been proven to imply all of the other relations that had previously been found. This far-reaching conjecture, though, remains a topic of intense study.

3.3. Mirror symmetry. Perhaps the most fruitful— and also the most mysterious— connection between the mathematical and physical sides of Gromov–Witten theory is provided by mirror symmetry. This is a duality, which, while natural to expect from the perspective of physics, is mathematically-speaking both startling and still largely open-ended. We refer the reader to [**6**], [**14**], or [**4**] for a more in-depth discussion; in particular, much of what follows is based on the exposition in [**6**].

The physical motivaion for mirror symmetry comes from objects known as *N=2 superconformal field theories* (SCFTs), of which heterotic string theories are an example. More specifically, a heterotic string theory describes physical processes in terms of a *worldsheet*, the real surface traced out by a string as it propagates through spacetime, which is equipped with a conformal structure. The theory is required to be equivalent under conformal equivalence of the worldsheet, as well as under two *supersymmetries* that transform particles known as bosons into fermions and vice versa. In particular, these properties imply that the infinitesimal symmetries of the theory (the Lie algebra of the symmetry group) form a superconformal algebra.

The solutions to the equations of motion in a heterotic string theory decompose into a "left-moving" and "right-moving" part, and the supersymmetries preserve this decomposition. Thus, the superconformal algebra of infinitesimal symmetries contains two distinguished subalgebras, each isomorphic to $u(1)$, acting by infinitesimal rotation on the left-moving and right-moving supersymmetries, respectively. One can choose a generator for each of these copies of $u(1)$, but the choice is only unique up to sign; it amounts to choosing an ordering of the two supersymmetries.

The connection with mathematics arises out of a particular way to construct a heterotic string theory, called the *nonlinear sigma model*, from the input data of a Calabi–Yau manifold X of complex dimension three with a *complexified Kähler class* ω. That is, X is a compact complex manifold with trivial canonical bundle, and $\omega = B + iJ$ for classes $B, J \in H^2(X;\mathbb{R})$, where J is Kähler.

Crucially, the data of (V, ω) determine not just an $N = 2$ SCFT but a choice of ordering for the supersymmetries. Thus, we have a canonical choice of generator for the $u(1) \times u(1)$ subalgebra of the superconformal algebra described above. This generator can be viewed as an operator on the state space of the physical system, and its eigenspaces can be computed mathematically: for $p, q \geq 0$, the (p, q) eigenspace is $H^q(X, \Lambda^p T_X)$ and the $(-p, q)$ eigenspace is $H^q(X, \Omega^p_X)$.

Now, suppose that one reverses the order of the two supersymmetries. This choice does not change the SCFT, but the result no longer arises out of the data of (X, ω). The heart of the mirror conjecture from a physical perspective is that there should exist a different pair $(X^\vee, \omega^\vee)$, the *mirror* of (X, ω), for which the associated nonlinear sigma model is the same SCFT but with the opposite ordering of supersymmetries.

In particular, this implies that the (p, q) eigenspace for (X, ω) will be exchanged with the $(-p, q)$ eigenspace for $(X^\vee, \omega^\vee)$:

$$H^q(X, \Lambda^p T_X) \cong H^q(X^\vee, \Omega^p_{X^\vee})$$
$$H^q(X, \Omega^p_X) \cong H^q(X^\vee, \Lambda^p T_{X^\vee}).$$

This is especially interesting when $p = q = 1$, in which case $H^1(X, T_X)$ can be viewed as the parameter space for infinitesimal deformations of the complex structure on X and $H^1(X, \Omega_X)$ as the parameter space for infinitesimal deformations of the Kähler class. Thus, a more refined version of mirror symmetry suggests that there should be an isomorphism between the moduli space of complex structures on X and the moduli space of complexified Kähler classes $\omega^\vee$ on $X^\vee$, at least locally around the specific choices (X, ω) and $(X^\vee, \omega^\vee)$.

An even deeper level of symmetry between the SCFTs associated to (X, ω) and $(X^\vee, \omega^\vee)$ comes from consideration of their *correlation functions*, which are certain integrals over the space of all possible worldsheets that describe how particles in the theory interact. Two particular types of correlation functions are the *A-model* and *B-model Yukawa couplings*, where the labels "A" and "B" depend on the particular choice of ordering of the supersymmetries. In mathematical terms, these can be viewed as vector bundles on the complex and Kähler moduli spaces, respectively, each equipped with a connection. The A-model connection is defined in terms of the genus-zero Gromov–Witten invariants of X. In the B-model, it is the *Gauss–Manin connection*, an object that has been well-studied and can be computed very explicitly using ideas from the theory of differential equations.

The upshot of mirror symmetry, then, is an equality between the genus-zero Gromov–Witten invariants of X— packaged into a generating function or *quantum connection*— and certain B-model information about $X^\vee$ that can be exactly calculated. As a result, one obtains striking predictions of Gromov–Witten invariants. We use the word "predictions" here, rather than "calculations", because much of the preceding discussion rests on rather shaky mathematical footing. Given a manifold X, how can $X^\vee$ be constructed? What, precisely, do we mean by the A-model and B-model, and can an equivalence exchanging them be proved mathematically?

Some of these questions have now been rigorously answered. For example, physicists used mirror symmetry to predict the genus-zero invariants I_d of the quintic threefold $V \subset \mathbb{P}^4$, yielding an equation relating a generating function for the I_d

to an explicit hypergeometric series arising out of the B-model. Givental provided an entirely mathematical proof of this statement, by showing that the hypergeometric series gives a slice of the cone $\mathcal{L}_X$ described in Section 2.6.

Other ways to interpret the physical data of the A- and B-model in mathematical language have been proposed, such as Kontsevich's *homological mirror symmetry* relating the derived category of coherent sheaves on X to a certain derived category defined in terms of Lagrangian submanifolds of $X^\vee$, or the *Strominger–Yau–Zaslow (SYZ) conjecture* relating fibrations of X and $X^\vee$ by special Lagrangian tori. Understanding the interplay between all of these ideas, and especially how they might manifest in Gromov–Witten theory beyond genus zero, is a subject of active current research and an ongoing mystery.

References

[1] M. F. Atiyah and R. Bott, *The moment map and equivariant cohomology*, Topology **23** (1984), no. 1, 1–28, DOI 10.1016/0040-9383(84)90021-1. MR721448

[2] K. Behrend and B. Fantechi, *The intrinsic normal cone*, Invent. Math. **128** (1997), no. 1, 45–88, DOI 10.1007/s002220050136. MR1437495

[3] Emily Clader, Nathan Priddis, and Mark Shoemaker, *Geometric quantization with applications to Gromov-Witten theory*, arXiv 1309.1150.

[4] Emily Clader and Yongbin Ruan, *Mirror symmetry constructions*, arXiv 1412.1268.

[5] Tom Coates and Alexander Givental, *Quantum Riemann-Roch, Lefschetz and Serre*, Ann. of Math. (2) **165** (2007), no. 1, 15–53, DOI 10.4007/annals.2007.165.15. MR2276766

[6] David A. Cox and Sheldon Katz, *Mirror symmetry and algebraic geometry*, Mathematical Surveys and Monographs, vol. 68, American Mathematical Society, Providence, RI, 1999. MR1677117

[7] Tohru Eguchi, Kentaro Hori, and Chuan-Sheng Xiong, *Quantum cohomology and Virasoro algebra*, Phys. Lett. B **402** (1997), no. 1-2, 71–80, DOI 10.1016/S0370-2693(97)00401-2. MR1454328

[8] C. Faber and R. Pandharipande, *Tautological and non-tautological cohomology of the moduli space of curves*, Handbook of moduli. Vol. I, Adv. Lect. Math. (ALM), vol. 24, Int. Press, Somerville, MA, 2013, pp. 293–330. MR3184167

[9] Alexander B. Givental, *Gromov-Witten invariants and quantization of quadratic Hamiltonians* (English, with English and Russian summaries), Mosc. Math. J. **1** (2001), no. 4, 551–568, 645. Dedicated to the memory of I. G. Petrovskii on the occasion of his 100th anniversary. MR1901075

[10] Alexander B. Givental, *Semisimple Frobenius structures at higher genus*, Internat. Math. Res. Notices **23** (2001), 1265–1286, DOI 10.1155/S1073792801000605. MR1866444

[11] Alexander B. Givental, *Symplectic geometry of Frobenius structures*, Frobenius manifolds, Aspects Math., E36, Friedr. Vieweg, Wiesbaden, 2004, pp. 91–112. MR2115767

[12] T. Graber and R. Pandharipande, *Localization of virtual classes*, Invent. Math. **135** (1999), no. 2, 487–518, DOI 10.1007/s002220050293. MR1666787

[13] T. Graber and R. Pandharipande, *Constructions of nontautological classes on moduli spaces of curves*, Michigan Math. J. **51** (2003), no. 1, 93–109, DOI 10.1307/mmj/1049832895. MR1960923

[14] Kentaro Hori, Sheldon Katz, Albrecht Klemm, Rahul Pandharipande, Richard Thomas, Cumrun Vafa, Ravi Vakil, and Eric Zaslow, *Mirror symmetry*, Clay Mathematics Monographs, vol. 1, American Mathematical Society, Providence, RI; Clay Mathematics Institute, Cambridge, MA, 2003. With a preface by Vafa. MR2003030

[15] Eleny-Nicoleta Ionel and Thomas Parker, *The Gopakumar-Vafa formula for symplectic manifolds*, arXiv 1306.1516.

[16] Joachim Kock and Israel Vainsencher, *An invitation to quantum cohomology*, Progress in Mathematics, vol. 249, Birkhäuser Boston, Inc., Boston, MA, 2007. Kontsevich's formula for rational plane curves. MR2262630

[17] Yukiko Konishi, *Integrality of Gopakumar-Vafa invariants of toric Calabi-Yau threefolds*, Publ. Res. Inst. Math. Sci. **42** (2006), no. 2, 605–648. MR2250076

[18] M. Kontsevich and Yu. Manin, *Gromov-Witten classes, quantum cohomology, and enumerative geometry [MR1291244 (95i:14049)]*, Mirror symmetry, II, AMS/IP Stud. Adv. Math., vol. 1, Amer. Math. Soc., Providence, RI, 1997, pp. 607–653. MR1416351

[19] Maxim Kontsevich, *Intersection theory on the moduli space of curves and the matrix Airy function*, Comm. Math. Phys. **147** (1992), no. 1, 1–23. MR1171758

[20] Maxim Kontsevich, Albert Schwarz, and Vadim Vologodsky, *Integrality of instanton numbers and p-adic B-model*, Phys. Lett. B **637** (2006), no. 1-2, 97–101, DOI 10.1016/j.physletb.2006.04.012. MR2230876

[21] Y.-P. Lee, *Witten's conjecture and the Virasoro conjecture for genus up to two*, Gromov-Witten theory of spin curves and orbifolds, Contemp. Math., vol. 403, Amer. Math. Soc., Providence, RI, 2006, pp. 31–42, DOI 10.1090/conm/403/07594. MR2234883

[22] Jun Li and Gang Tian, *Virtual moduli cycles and Gromov-Witten invariants of algebraic varieties*, J. Amer. Math. Soc. **11** (1998), no. 1, 119–174, DOI 10.1090/S0894-0347-98-00250-1. MR1467172

[23] D. Maulik, N. Nekrasov, A. Okounkov, and R. Pandharipande, *Gromov-Witten theory and Donaldson-Thomas theory. I*, Compos. Math. **142** (2006), no. 5, 1263–1285, DOI 10.1112/S0010437X06002302. MR2264664

[24] D. Maulik, N. Nekrasov, A. Okounkov, and R. Pandharipande, *Gromov-Witten theory and Donaldson-Thomas theory. II*, Compos. Math. **142** (2006), no. 5, 1286–1304, DOI 10.1112/S0010437X06002314. MR2264665

[25] D. Maulik, A. Oblomkov, A. Okounkov, and R. Pandharipande, *Gromov-Witten/Donaldson-Thomas correspondence for toric 3-folds*, Invent. Math. **186** (2011), no. 2, 435–479, DOI 10.1007/s00222-011-0322-y. MR2845622

[26] A. Okounkov and R. Pandharipande, *Virasoro constraints for target curves*, Invent. Math. **163** (2006), no. 1, 47–108, DOI 10.1007/s00222-005-0455-y. MR2208418

[27] R. Pandharipande and A. Pixton, *Gromov-Witten/Pairs correspondence for the quintic 3-fold*, arXiv 1206.5490.

[28] Rahul Pandharipande, Aaron Pixton, and Dimitri Zvonkine, *Relations on $\overline{\mathcal{M}}_{g,n}$ via 3-spin structures*, J. Amer. Math. Soc. **28** (2015), no. 1, 279–309, DOI 10.1090/S0894-0347-2014-00808-0. MR3264769

[29] A. Pixton, *Conjectural relations in the tautological ring of $\overline{M}_{g,n}$*, arXiv 1207.1918.

[30] Yongbin Ruan and Gang Tian, *A mathematical theory of quantum cohomology*, J. Differential Geom. **42** (1995), no. 2, 259–367. MR1366548

[31] Hermann Schubert, *Kalkül der abzählenden Geometrie* (German), Springer-Verlag, Berlin-New York, 1979. Reprint of the 1879 original; With an introduction by Steven L. Kleiman. MR555576

[32] Constantin Teleman, *The structure of 2D semi-simple field theories*, Invent. Math. **188** (2012), no. 3, 525–588, DOI 10.1007/s00222-011-0352-5. MR2917177

[33] Edward Witten, *Two-dimensional gravity and intersection theory on moduli space*, Surveys in differential geometry (Cambridge, MA, 1990), Lehigh Univ., Bethlehem, PA, 1991, pp. 243–310. MR1144529

[34] Dimitri Zvonkine, *An introduction to moduli spaces of curves and their intersection theory*, Handbook of Teichmüller theory. Volume III, IRMA Lect. Math. Theor. Phys., vol. 17, Eur. Math. Soc., Zürich, 2012, pp. 667–716, DOI 10.4171/103-1/12. MR2952773

Department of mathematics, San Francisco State University, San Francisco, California 94132

Teichmüller dynamics in the eyes of an algebraic geometer

Dawei Chen

ABSTRACT. This is an introduction to the algebraic aspect of Teichmüller dynamics, with a focus on its interplay with the geometry of moduli spaces of curves as well as recent advances in the field.

CONTENTS

1. Introduction

An Abelian differential defines a flat structure such that the underlying Riemann surface can be realized as a plane polygon whose edges are pairwise identified via translation. Varying the shape of the polygon by $\mathrm{GL}_2^+(\mathbb{R})$ induces an action on the moduli space of Abelian differentials, called Teichmüller dynamics, see Figure 1.

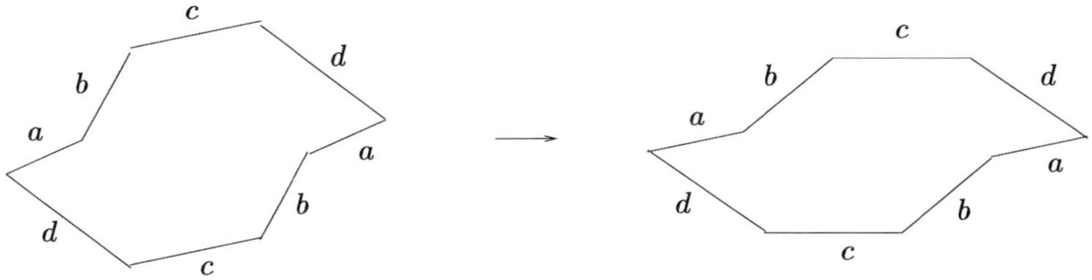

FIGURE 1. $\mathrm{GL}_2^+(\mathbb{R})$-action on a flat surface

2010 *Mathematics Subject Classification.* Primary 14H10, 14H15, 14K20, 30F30, 32G15, 37D40, 37D50.

Key words and phrases. Abelian differential, translation surface, moduli space of curves, affine invariant submanifold, Teichmüller curve, Siegel-Veech constant, Lyapunov exponent, quadratic differential.

During the preparation of this article the author was partially supported by the NSF CAREER Award DMS-1350396 and a Boston College Research Incentive Grant.

The corresponding $\mathrm{GL}_2^+(\mathbb{R})$-orbit closures in the moduli space of Abelian differentials are now known as affine invariant submanifolds. A number of questions about surface geometry boil down to understanding the structures of affine invariant submanifolds. From the viewpoint of algebraic geometry affine invariant submanifolds are of an independent interest, which can provide special subvarieties in the moduli space of curves.

We aim to introduce Teichmüller dynamics from the viewpoint of algebraic geometry. In Section 2 we review background material, including translation surfaces, strata of Abelian differentials, the $\mathrm{GL}_2^+(\mathbb{R})$-action, and affine invariant submanifolds. Section 3 focuses on Teichmüller curves formed by closed $\mathrm{GL}_2^+(\mathbb{R})$-orbits, where we describe their properties, examples, classification, and invariants. In Section 4 we study general affine invariant submanifolds and survey recent breakthroughs about their structures, classification, and boundary behavior. Finally in Section 5 we discuss similar questions for meromorphic and higher order differentials.

This article is written in an expository style. We will often highlight motivations and minimize technical details. For further reading, we refer to [**Z1**, **Mö5**, **Wr2**, **Wr3**] for a number of excellent surveys on related topics.

Acknowledgement. This article is partially based on the lectures given by the author during the Algebraic Geometry Summer Research Institute Bootcamp, July 2015. The author is grateful to the Bootcamp organizers Izzet Coskun, Tommaso de Fernex, Angela Gibney, and Max Lieblich for their invitation and hospitality. The author thanks Alex Wright and the referee for carefully reading the article and many helpful comments.

2. Preliminaries

In this section we introduce basic background material that will be used later.

2.1. Abelian differentials and translation surfaces. A *translation surface* (also called a *flat surface*) is a closed, topological surface X together with a finite set $\Sigma \subset X$ such that:

- There exists an atlas of charts $X \backslash \Sigma \to \mathbb{C}$, where the transition functions are translation.
- For each $p \in \Sigma$, under the Euclidean metric of $\mathbb{C}$ the total angle at p is $(k+1) \cdot (2\pi)$ for some $k \in \mathbb{Z}^+$.

We say that p is a *saddle point of cone angle* $(k+1) \cdot (2\pi)$. Locally one can glue $2k+2$ half-disks consecutively to form a cone of angle $2\pi \cdot (k+1)$, see Figure 2.

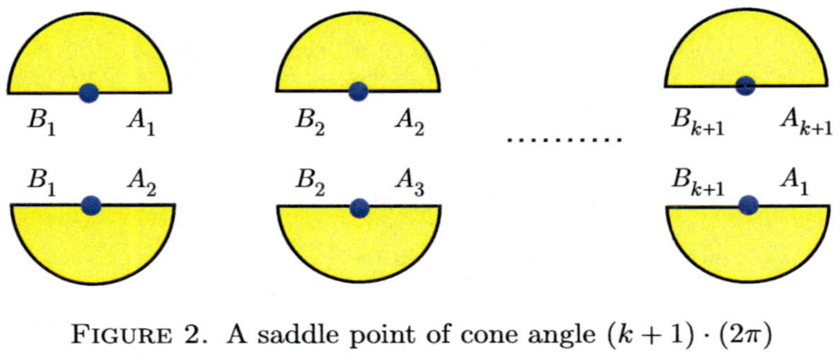

FIGURE 2. A saddle point of cone angle $(k+1) \cdot (2\pi)$

Equivalently, a translation surface is a closed Riemann surface X with an *Abelian differential* ω, not identically zero:

- The set of zeros of ω corresponds to Σ.
- If p is a zero of ω of order k, then the cone angle at p is $(k+1)\cdot(2\pi)$.

For example, take an octagon X with four pairs of parallel edges, see Figure 3. Identifying the edges with the same labels by translation, X becomes a closed surface. All vertices are glued as one point p. By the topological Euler characteristic formula, the genus of X is two. It is moreover a Riemann surface whose complex structure is induced from $\mathbb{C}$. Away from p it admits an atlas of charts with transition functions given by translation: $z' = z + \text{constant}$, see Figure 4.

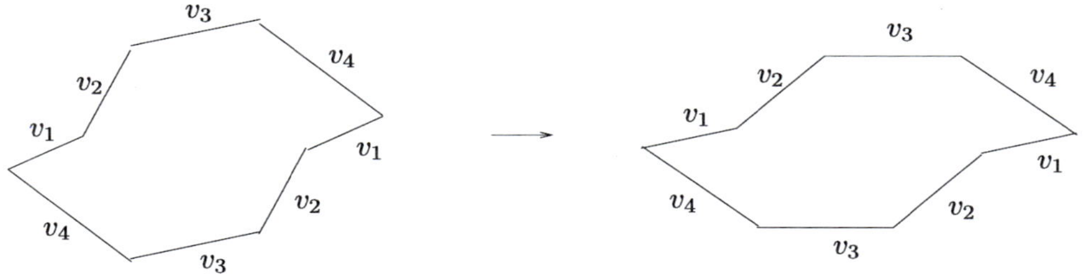

FIGURE 3. An octagon X with four pairs of parallel edges

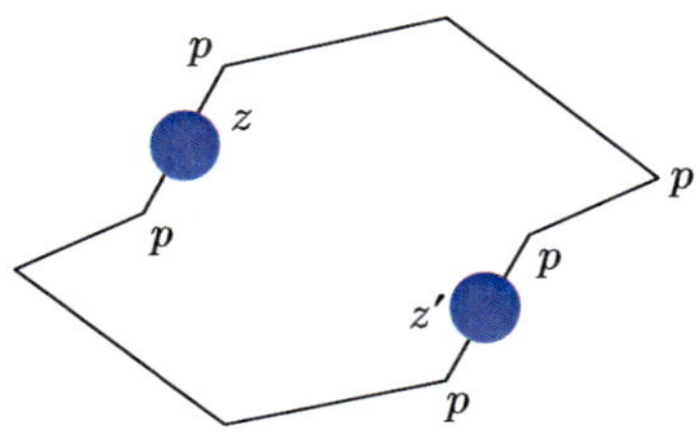

FIGURE 4. Translation structure on $X\backslash p$

The differential $\omega = dz$ is well-defined and nowhere vanishing on $X\backslash p$, which further extends to the entire X. The angle at p is $6\pi = 3\cdot(2\pi)$, hence ω has a local expression $d(z^3) \sim z^2 dz$ at p. In summary, ω is an Abelian differential with a unique zero of order two on a Riemann surface of genus two.

The above example illustrates the equivalence between translation surfaces and Abelian differentials in general. Given a translation surface, away from its saddle points, differentiating local coordinates provides a globally defined Abelian differential. Conversely, integrating an Abelian differential away from its zeros provides an atlas of charts whose transition functions are translation, because antiderivatives differ by constants. In addition, a saddle point p has cone angle $(k+1)\cdot(2\pi)$ if and only if $\omega = d(z^{k+1}) \sim z^k dz$ under a local coordinate z at p, namely, if and only if ω has a zero of order k at p.

Below we provide two more examples. Figure 5 represents a nowhere vanishing differential on a torus. Conversely every Abelian differential on a torus give rises to such a parallelogram presentation. Figure 6 represents an Abelian differential with two simple zeros on a Riemann surface of genus two.

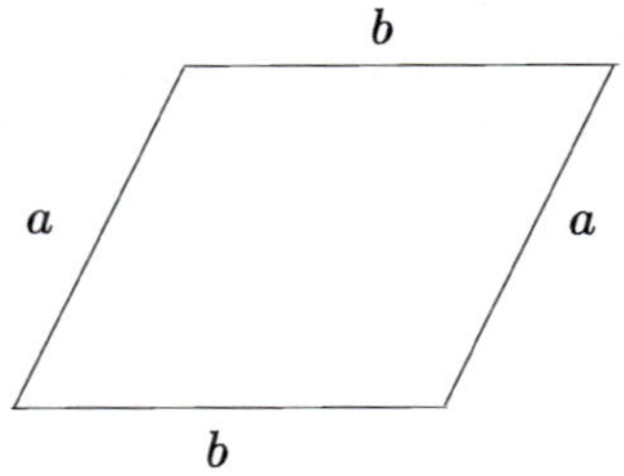

FIGURE 5. A flat torus

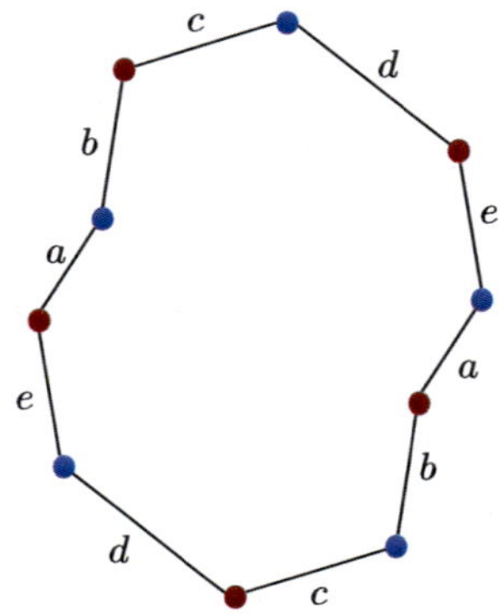

FIGURE 6. A flat surface with two simple zeros

Note that Abelian differentials are sections of the *canonical line bundle*, hence the study of translation surfaces is naturally connected to algebraic geometry.

2.2. Strata of Abelian differentials. We identify Riemann surfaces with smooth complex algebraic curves. Let $\mathcal{M}_g$ be the *moduli space of genus g curves*. Let $\mathcal{H}$ be the *Hodge bundle* over $\mathcal{M}_g$ whose fibers parameterize Abelian differentials on a fixed genus g curve.

Let $\mu = (m_1, \ldots, m_n)$ be a tuple of positive integers such that $\sum_{i=1}^n m_i = 2g-2$. We say that μ is a *partition* of $2g-2$.[1] Define a subset $\mathcal{H}(\mu)$ of $\mathcal{H}$ that parameterizes pairs (X, ω), where X is a Riemann surface of genus g and ω is an Abelian differential on X such that the zero divisor of ω is of type μ:

$$(\omega)_0 = m_1 p_1 + \cdots + m_n p_n.$$

We say that $\mathcal{H}(\mu)$ is the *stratum of Abelian differentials of type* μ. Equivalently, $\mathcal{H}(\mu)$ parameterizes translation surfaces with n saddle points, each having cone angle $(m_i + 1) \cdot (2\pi)$. The union of $\mathcal{H}(\mu)$ over all partitions of $2g-2$ is the Hodge bundle $\mathcal{H}$ (with the zero section removed).

Take a basis $\gamma_1, \ldots, \gamma_{2g+n-1}$ of the relative homology $H_1(X, p_1, \ldots, p_n; \mathbb{Z})$. Integrating ω over each γ_i provides a local coordinate system for $\mathcal{H}(\mu)$, called the *period coordinates*. For instance, the complex vectors v_1, v_2, v_3, and v_4 in Figure 3 above are periods of a translation surface in $\mathcal{H}(2)$. Under the polygon presentation, locally deforming the periods preserves the number of saddle points, their cone

[1] In Section 5 we will consider meromorphic differentials, where the entires m_i are allowed to be negative.

angles, and the way of edge identification. Consequently $\mathcal{H}(\mu)$ is a $(2g+n-1)$-dimensional manifold.[2]

For special partitions μ, $\mathcal{H}(\mu)$ can be *disconnected.* Kontsevich-Zorich ([**KonZor1**]) classified connected components of $\mathcal{H}(\mu)$ for all μ, where extra components arise due to hyperelliptic and spin structures. If a translation surface (X, ω) satisfies that X is hyperelliptic, $(\omega)_0 = (2g-2)z$ or $(\omega)_0 = (g-1)(z_1 + z_2)$, where z is a Weierstrass point of X in the former, or z_1 and z_2 are hyperelliptic conjugate in the latter, we say that (X, ω) is contained in the *hyperelliptic connected component* of the corresponding stratum. For other translation surfaces (X, ω), if $(\omega)_0 = 2k_1 z_1 + \cdots + 2k_n z_n$, then the line bundle $\mathcal{O}_X(\sum_{i=1}^n k_i z_i)$ is a square root of the canonical line bundle, namely, it is a *theta characteristic.* Define its *parity* by $h^0(X, \sum_{i=1}^n k_i z_i) \pmod 2$, which is deformation invariant ([**At, Mum, J**]). A theta characteristic with its parity is called a *spin structure.* In general, $\mathcal{H}(\mu)$ can have *up to three* connected components, distinguished by these hyperelliptic and spin structures.

2.3. $\mathrm{GL}_2^+(\mathbb{R})$-action and affine invariant submanifolds. Given $(X, \omega) \in \mathcal{H}$ and $A \in \mathrm{GL}_2^+(\mathbb{R})$, varying the polygon presentation of (X, ω) by A induces a $\mathrm{GL}_2^+(\mathbb{R})$-action on $\mathcal{H}$, which is called *Teichmüller dynamics.* For example, the $\mathrm{GL}_2^+(\mathbb{R})$-orbit of a flat torus consists of all parallelogram presentations, see Figure 7.

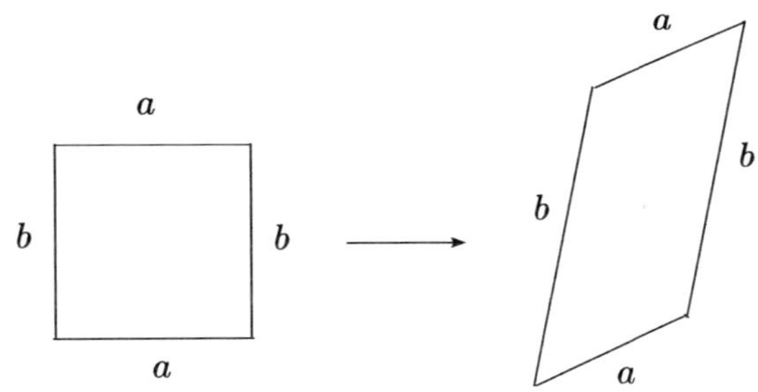

FIGURE 7. $\mathrm{GL}_2^+(\mathbb{R})$-action on a flat torus

The number and cone angles of saddle points are preserved under this action, hence the $\mathrm{GL}_2^+(\mathbb{R})$-action descends to each stratum $\mathcal{H}(\mu)$. Equip $\mathcal{H}(\mu)$ with the standard topology using its period coordinates. For almost all $(X, \omega) \in \mathcal{H}(\mu)$, Masur and Veech ([**Ma, V1**]) showed that its $\mathrm{GL}_2^+(\mathbb{R})$-orbit is *equidistributed* in $\mathcal{H}(\mu)$, hence the orbit closure is the whole stratum (or a connected component if the stratum is disconnected). For special (X, ω), however, its $\mathrm{GL}_2^+(\mathbb{R})$-orbit closure can be a *proper* subset of $\mathcal{H}(\mu)$. Classifying $\mathrm{GL}_2^+(\mathbb{R})$-orbit closures in $\mathcal{H}(\mu)$ is a central question in Teichmüller dynamics.

The recent breakthrough of Eskin-Mirzakhani-Mohammadi ([**EMi, EMM**]) showed that any $\mathrm{GL}_2^+(\mathbb{R})$-orbit closure is an *affine invariant submanifold* in $\mathcal{H}(\mu)$, that is, it is a subspace of $\mathcal{H}(\mu)$ locally defined by real linear homogeneous equations of period coordinates.[3] Filip ([**Fi1**]) further showed that all affine invariant submanifolds are algebraic varieties defined over $\overline{\mathbb{Q}}$. In particular, it means that affine

[2]More precisely it is an orbifold, because special translation surfaces can have extra automorphisms.

[3]Here the term "affine" is different from what it usually means in algebraic geometry. It refers to the linear structure on $\mathbb{C}^n$. Moreover, the closure of an orbit is taken under the standard

invariant submanifolds can be defined and characterized purely in terms of algebraic conditions on the Jacobian. We will elaborate on these results in Section 4.

2.4. Veech group. Let $(X,\omega) \in \mathcal{H}(\mu)$ be a translation surface. Suppose a matrix $A \in \mathrm{SL}_2(\mathbb{R})$ acts on (X,ω). If the resulting translation surface $A \cdot (X,\omega)$ is isomorphic to (X,ω), that is, if the polygon presentation of $A \cdot (X,\omega)$ can be cut into pieces and reassembled via translation to represent (X,ω), we say that A is a *stabilizer* of (X,ω). The subgroup of all stabilizers of (X,ω) in $\mathrm{SL}_2(\mathbb{R})$ is called the *Veech group*, and denoted by $\mathrm{SL}(X,\omega)$. For example, $A = \left(\begin{smallmatrix}1 & 1\\ 0 & 1\end{smallmatrix}\right)$ is in the Veech group of the square torus, see Figure 8.

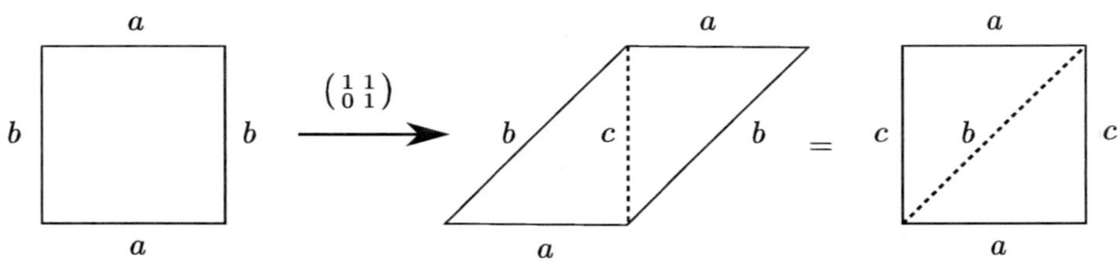

FIGURE 8. An element of the Veech group

A line segment under the flat metric that connects two zeros of ω (not necessarily distinct) is called a *saddle connection*. Since the set of saddle connections of (X,ω) is preserved by a stabilizer, it follows that $\mathrm{SL}(X,\omega)$ is *discrete*. Without loss of generality, suppose (X,ω) has a horizontal saddle connection. Let $g_t = \left(\begin{smallmatrix}e^t & 0\\ 0 & e^{-t}\end{smallmatrix}\right) \in \mathrm{SL}_2(\mathbb{R})$ act on (X,ω). As $t \to \infty$, the horizontal saddle connection becomes arbitrarily long, hence the $\mathrm{SL}_2(\mathbb{R})$-orbit of (X,ω) is *unbounded* in $\mathcal{H}(\mu)$. Consequently $\mathrm{SL}(X,\omega)$ is *not cocompact* in $\mathrm{SL}_2(\mathbb{R})$.

Adding the traces of all $A \in \mathrm{SL}(X,\omega)$ to $\mathbb{Q}$, we obtain a field extension of $\mathbb{Q}$, called the *trace field* of (X,ω). The degree of the trace field over $\mathbb{Q}$ is bounded by the genus of X (see [**Mö5**, Proposition 2.5]).

3. Teichmüller curves

The Hodge bundle $\mathcal{H}$ maps to the moduli space $\mathcal{M}_g$ of smooth genus g curves by forgetting the differentials: $(X,\omega) \mapsto X$. Note that the subgroup $\mathrm{SO}(2)$ acting on an Abelian differential amounts to rotating the corresponding flat surface, hence it does not change the underlying complex structure. Similarly scaling the size of a flat surface preserves the underlying complex structure. It follows that the projection of a $\mathrm{GL}_2^+(\mathbb{R})$-orbit to $\mathcal{M}_g$ factors through the upper half plane $\mathbb{H}$, and the induced map $\mathbb{H} \to \mathcal{M}_g$ (or simply its image) is called a *Teichmüller disk*. On rare occasions a Teichmüller disk forms an algebraic curve in $\mathcal{M}_g$. In that case we call it a *Teichmüller curve*.

---

topology in the Hodge bundle over the interior of the moduli space parameterizing smooth curves. Finally, affine invariant submanifolds can have self crossings.

3.1. Properties of Teichmüller curves. Teichmüller curves are dimensionally minimal affine invariant submanifolds, which possess a number of fascinating properties. To name a few, a Teichmüller curve is a local *isometry* from a curve to $\mathcal{M}_g$ under the Kobayashi/Teichmüller metric ([**SW**, **V3**]). The union of all Teichmüller curves is *dense* in moduli spaces ([**EO1**, **Ch2**]). McMullen ([**Mc7**]) proved that Teichmüller curves are *rigid*, hence they are defined over number fields.[4] Conversely, Ellenberg-McReynolds ([**EMc**]) showed that every curve over a number field is birational to a Teichmüller curve over $\mathbb{C}$. If (X, ω) generates an algebraically primitive Teichmüller curve (see Section 3.3), Möller ([**Mö2**]) showed that the difference of any two zeros of ω is a torsion in the Jacobian of X. Möller ([**Mö1**]) also analyzed the variation of Hodge structures associated to a Teichmüller curve and deduced that it parameterizes curves whose Jacobians have *real multiplication*. These torsion and real multiplication conditions can in fact be used to define (algebraically primitive) Teichmüller curves in purely algebro-geometric terms (see e.g., [**Wr2**, Proposition 2.18]). Teichmüller curves are never complete in $\mathcal{M}_g$, but the closure of a Teichmüller curve only intersects certain boundary divisor of the Deligne-Mumford compactification $\overline{\mathcal{M}}_g$ (see Section 3.4).[5]

3.2. Square-tiled surfaces. We first show that Teichmüller curves exist. One type of Teichmüller curves arises from certain *branched covering construction*. Let $\mu = (m_1, \ldots, m_n)$ be a partition of $2g-2$. Consider a branched cover $\pi : X \to E$ such that

- $\deg \pi = d$,
- $g(X) = g$,
- E is the square torus,
- π has a *unique* branch point $q \in E$,
- π has n ramification points $p_1, \ldots, p_n$ over q, each with ramification order m_i.

Let $\omega = \pi^*(dz)$, where z is the standard coordinate on E. Then by the Riemann-Hurwitz formula

$$(\omega)_0 = m_1 p_1 + \cdots + m_n p_n,$$

hence $(X, \omega) \in \mathcal{H}(\mu)$. Such (X, ω) are called *square-tiled surfaces* (or *origami*). For example, Figure 9 exhibits a degree 5 and genus 2 branched cover of E with a unique ramification point of order 2. The resulting (X, ω) belongs to the stratum $\mathcal{H}(2)$.

The $\mathrm{GL}_2^+(\mathbb{R})$-action on a square-tiled surface amounts to varying the shape of the square, which is exchangeable with varying the shape of the flat torus first, see Figure 10 for an example. Therefore, the Teichmüller disk generated by a square-tiled surface corresponds to the one-dimensional *Hurwitz space* parameterizing degree d, genus g connected covers of all elliptic curves with a unique branch point of ramification type μ. Since Hurwitz spaces are algebraic varieties, it follows that the $\mathrm{GL}_2^+(\mathbb{R})$-orbit of a square-tiled surface gives rise to a Teichmüller

---

[4]Here the rigidity means as a map it does not deform.

[5]Here we restrict to Teichmüller curves generated by Abelian differentials. Special quadratic differentials can also generate Teichmüller curves, see Section 5.1, but they may intersect other boundary divisors of $\overline{\mathcal{M}}_g$.

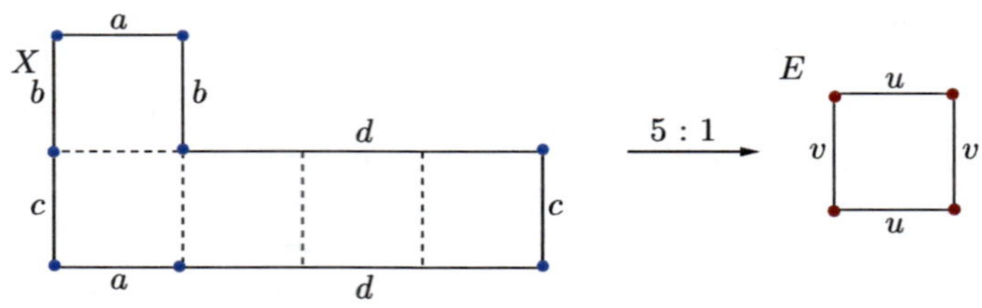

FIGURE 9. A square-tiled surface in $\mathcal{H}(2)$

curve.[6] Since d can be arbitrarily large, this way we indeed obtain *infinitely many* Teichmüller curves in $\mathcal{H}(\mu)$.

FIGURE 10. $\mathrm{GL}_2^+(\mathbb{R})$-action on a square-tiled surface

Square-tiled surfaces correspond to *lattice points* under period coordinates of a stratum ([**EO1**, Lemma 3.1]), see Figure 11. If $\pi : X \to E$ is a square-tiled surface of type $\mu = (m_1, \ldots, m_n)$, for any $\gamma \in H_1(X, p_1, \ldots, p_n; \mathbb{Z})$, $\pi(\gamma)$ represents a closed loop in E, because $p_1, \ldots, p_n$ all map to the unique branch point. Therefore,

$$\int_\gamma \pi^*(dz) = \int_{\pi_*\gamma} dz \in \mathbb{Z} \oplus \mathbb{Z}[i].$$

Conversely if all relative periods of (X, ω) are lattice points, the map $\pi : X \to \mathbb{C}/\mathbb{Z} \oplus \mathbb{Z}[i]$ induced by

$$x \mapsto \int_b^x \omega$$

is well-defined, where b is a fixed base point, hence realizing (X, ω) as a square-tiled surface. This explains density of the union of Teichmüller curves in moduli spaces. It also provides an approach for analyzing *volume growths* of the strata of Abelian differentials by counting the number of such square-tiled surfaces ([**EO1**]).

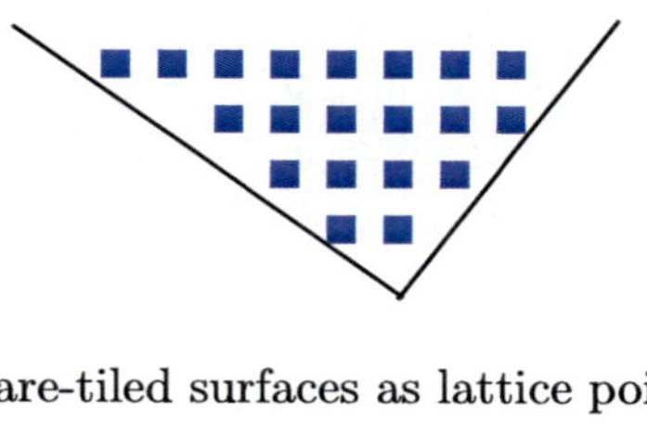

FIGURE 11. Square-tiled surfaces as lattice points in a stratum

[6]When d, g, and μ are fixed, the Hurwitz space can still be disconnected. In that case each of its connected components gives a Teichmüller curve.

Gutkin-Judge ([**GJ**]) showed that the Veech group of a translation surface is *commensurable with* $\mathrm{SL}(2,\mathbb{Z})$ if and only if the surface is tiled by parallelograms (see also [**Mc1**, Theorem 9.8] for a short proof). In this sense Teichmüller curves generated by square-tiled surfaces are called *arithmetic* Teichmüller curves. However, there exist Teichmüller curves of other type that are *not* generated by square-tiled surfaces. In the next section we will survey known results on the classification of Teichmüller curves.

3.3. Classification of Teichmüller curves. By definition, the moduli space $\mathcal{M}_{1,1}$ of elliptic curves is a very first example of Teichmüller curves. Arithmetic Teichmüller curves generated by square-tiled surfaces are coverings of $\mathcal{M}_{1,1}$. On the contrary, a Teichmüller curve is called (geometrically) *primitive* if it does not arise from a curve in a moduli space of lower genus via such a covering construction.

It is a challenging task to find examples of primitive Teichmüller curves. In genus two, Calta and McMullen ([**Ca**, **Mc2**]) independently classified primitive Teichmüller curves in $\mathcal{H}(2)$. They found infinitely many primitive Teichmüller curves whose constructions have several incarnations. In terms of flat geometry, a translation surface (X,ω) generating a primitive Teichmüller curve in $\mathcal{H}(2)$ possesses an L-shaped polygon presentation whose edges satisfy relations in a real quadratic field. In terms of algebraic geometry, the Jacobian of X admits real multiplication, and the corresponding Teichmüller curve lies on a Hilbert modular surface. However for the other stratum $\mathcal{H}(1,1)$ in genus two, McMullen ([**Mc4**]) proved that it contains a unique primitive Teichmüller curve generated by the regular decagon with parallel edges identified. McMullen ([**Mc5**]) further generalized the method of real multiplication by using Prym varieties and discovered infinitely many primitive Teichmüller curves in genus three and four. Using quotients of abelian covers of $\mathbb{P}^1$ by a finite group, in each genus Bouw-Möller ([**BoM**]) constructed (finitely many) primitive Teichmüller curves, generalizing earlier constructions of Veech and Ward ([**V2**, **Wa**]).

Besides the above results, to date only two sporadic examples of primitive Teichmüller curves are known (see e.g., [**Wr1**, p. 782] for an elaboration). It is natural to ask if there are only finitely many primitive Teichmüller curves in a given stratum $\mathcal{H}(\mu)$. There is a stronger notion of primitivity called *algebraically primitive*, if the trace field of a translation surface X in the Teichmüller curve has degree equal to the genus of X (see Section 2.4). For example, the trace field of a square-tiled surface is $\mathbb{Q}$, hence it is far from being algebraically primitive. Indeed, algebraically primitive Teichmüller curves are geometrically primitive, but the converse is not always true (see [**Mö5**, Section 5.1]).

Finiteness results of algebrically primitive Teichmüller curves have been established in various cases. Möller ([**Mö3**]) proved that the hyperelliptic component of $\mathcal{H}(g-1,g-1)$ contains finitely many algebrically primitive Teichmüller curves. The strategy is to track the degeneration of flat surfaces along an algebrically primitive Teichmüller curve in the horizontal and vertical directions and use it to bound the torsion order of the difference of the two zeros. By studying the boundary of the locus of curves with real multiplication, Bainbridge-Möller ([**BaM**]) showed finiteness of algebrically primitive Teichmüller curves in $\mathcal{H}(3,1)$. Bainbridge-Habegger-Möller ([**BHM**]) further established finiteness of algebrically primitive Teichmüller curves for all strata in genus three by a mix of techniques, including the Harder-Narasimhan filtration of the Hodge bundle over Teichmüller curves and height bounds for the boundary points of Teichmüller curves. Matheus-Wright ([**MW**]) proved finiteness of algebrically primitive Teichmüller curves in the minimal stratum $\mathcal{H}(2g-2)$ for each prime genus $g \geq 3$. Their approach is to study orthogonality of Hodge-Teichmüller (real) planes in the Hodge bundle that respect the Hodge decomposition along a Teichmüller curve. Matheus-Nguyen-Wright ([**NW**]) further showed that there are at most finitely many non-arithmetic Teichmüller curves in the hyperelliptic component of $\mathcal{H}(4)$.

One can also study Teichmüller curves contained in an affine invariant submanifold of a stratum. Lanneau-Nguyen ([**LN**]) showed that there are at most finitely many closed $\mathrm{GL}_2^+(\mathbb{R})$-orbits (including primitive Teichmüller curves) in certain Prym loci in genus three. Lanneau-Nguyen-Wright ([**LNW**]) further proved finiteness of closed $\mathrm{GL}_2^+(\mathbb{R})$-orbits in each non-arithmetic rank 1 affine invariant submanifold (see Section 4.2). Apisa ([**Ap**]) showed finiteness of algebrically primitive Teichmüller curves in the hyperelliptic components of each minimal stratum in $g > 2$, as a byproduct of his classification of affine invariant submanifolds in the hyperelliptic components (see Section 4.2).

Although a complete classification of Teichmüller curves is still missing, the seminal work of Eskin-Mirzakhani-Mohammadi ([**EMi, EMM**]) on the structure of $\mathrm{GL}_2^+(\mathbb{R})$-orbit closures provides us a powerful tool. Indeed some of the above results are built on their work. We hope the classification problem of Teichmüller curves (and in general affine invariant submanifolds) can be resolved in the next few decades.

3.4. Slope, Siegel-Veech constant, and Lyapunov exponent. In this section we discuss several invariants of Teichmüller curves and describe a relation between them.

First, the behavior of geodesics on translation surfaces is related to *billiards in polygons*. One can study various counting problems from this viewpoint. Recall that a saddle connection is a line segment connecting two saddle points. Consider counting saddle connections with bounded lengths. For instance, consider the standard torus formed by identifying parallel edges of the unit square and marked at the origin. The number of (non-primitive) saddle connections of length $< L$ (counting with direction) equals the number of lattice points in the disk of radius L, see Figure 12. It has asymptotically *quadratic growth* $\sim \pi L^2$. The leading term π is

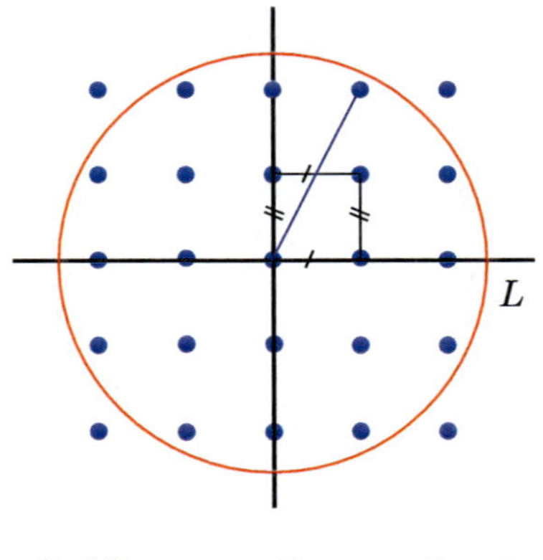

FIGURE 12. Saddle connections on the standard torus

an example of so-called *Siegel-Veech constant.* In general, Siegel-Veech constants relate quadratic growth rates of finite-length trajectories on a translation surface to the volume of the corresponding $\mathrm{SL}_2(\mathbb{R})$-orbit. In what follows we will concentrate on one type of Siegel-Veech constants, called the *area* Siegel-Veech constant, which has a connection to dynamics as well as intersection theory on moduli space.

A real geodesic on a translation surface (X, ω) is called *regular*, if it does not pass through any saddle point, namely, it does not contain any zero of ω. Vary a closed regular geodesic in parallel until it hits a saddle point on both ends. The union of those geodesics fill a *cylinder* cyl, whose boundary circles contain saddle points. For example, the square-tiled surface in Figure 13 has two cylinders in the horizontal direction. Let w and h be the *width* and *height* of a cylinder cyl, respectively. The *area* of the cylinder is $\mathrm{Area(cyl)} = w \cdot h$.

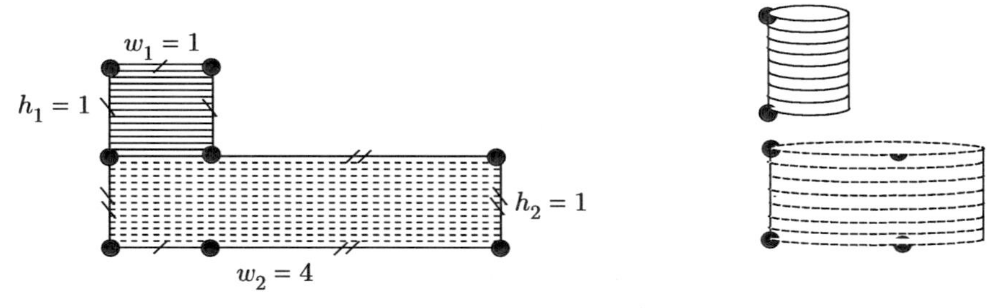

FIGURE 13. Two horizontal cylinders on a square-tiled surface

For $L > 0$, define

$$N(X, L) = \frac{1}{\mathrm{Area}(X)} \cdot \sum_{\substack{\mathrm{cyl} \subset X \\ w(\mathrm{cyl}) < L}} \mathrm{Area(cyl)}.$$

Veech and Eskin-Masur ([**V4**, **EMa**]) showed that for any $\mathrm{GL}_2^+(\mathbb{R})$-orbit closure $\mathcal{N}$, there exists a constant c such that

$$\frac{\pi}{3} \cdot \lim_{L \to \infty} \frac{N(X, L)}{L^2} = c$$

for *generic* $(X, \omega) \in \mathcal{N}$. The constant c is called the *area Siegel-Veech constant* of $\mathcal{N}$.[7] Its value depends on $\mathcal{N}$.

If $\mathcal{N}$ is the orbit of a square-tiled surface, namely, if its projection to $\mathcal{M}_g$ is an arithmetic Teichmüller curve, there is a *combinatorial* way to calculate its area Siegel-Veech constant. Let N be the number of square-tiled surfaces in $\mathcal{N}$. In other words, N is the *Hurwitz number* that counts branched covers of a fixed torus with a unique branch point and prescribed ramification type. Take all square-tiled surfaces in $\mathcal{N}$. For each one, consider all of its horizontal cylinders. Sum up h/w (the *modulus* of a cylinder) over all of them. Denote the total sum by M. For example, the square-tiled surface in Figure 13 above contributes $\frac{h_1}{w_1} + \frac{h_2}{w_2} = \frac{1}{1} + \frac{1}{4}$ to the sum M.

[7]Here our definition of the Siegel-Veech constant differs from its usual definition by a scalar multiple of $\pi^2/3$. We choose to normalize it this way for the convenience of describing its relation with slope and Lyapunov exponent.

For an arithmetic Teichmüller curve generated by a square-tiled surface,

$$c = \frac{M}{N},$$

see [**EKZ1**, Appendix B]. The idea behind the formula is that c measures the average number of cylinders weighted by their moduli h/w in $\mathcal{N}$, where

$$\frac{h}{w} = \frac{hw}{w^2} = \frac{\text{Area(cyl)}}{w^2},$$

hence it is related to the quadratic growth rate of $N(X, L)$ in this case.

For low degree d and low genus g, using monodromy of branched covers one can calculate N and M explicitly. Nevertheless, the enumeration of N as d and g increase is a highly non-trivial problem in symmetric group representations. Eskin-Okounkov ([**EO1**]) analyzed the asymptotic behavior of N and calculated the *volume* growth of strata of Abelian differentials. The enumeration of M is more complicated for large d and g. Joint with Möller and Zagier ([**CMZ**]) we are able to understand the asymptotic growth of M and hence c for arithmetic Teichmüller curves by using techniques of shifted symmetric functions and quasimodular forms.

Next we define an important index associated to a one-dimensional family of stable curves. The *boundary* Δ of the Deligne-Mumford moduli space $\overline{\mathcal{M}}_g$ parameterizes *stable nodal curves*, where $\Delta = \bigcup_{i=0}^{[g/2]} \Delta_i$ is a union of irreducible *boundary divisors* Δ_i. General points of Δ_i parameterize nodal curves of a given topological type, and they can further degenerate, see Figure 14.

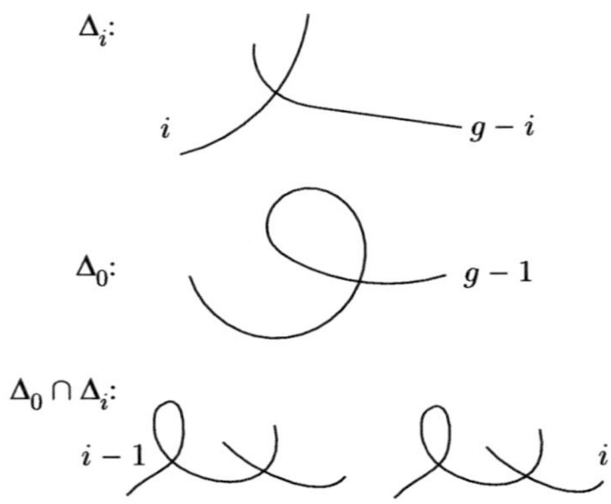

FIGURE 14. Curves in the boundary of $\overline{\mathcal{M}}_g$

The Hodge bundle $\mathcal{H}$ over $\mathcal{M}_g$ extends to a rank g bundle $\overline{\mathcal{H}}$ over $\overline{\mathcal{M}}_g$, whose fiber over a stable curve X parameterizes sections of the dualizing line bundle K_X. Geometrically speaking, the fiber of $\overline{\mathcal{H}}$ over X can be identified with the space of *stable differentials* that have at worst simple pole at each node of X with opposite residues on the two branches of a node ([**HM**, Chapter 3.A]). Denote by λ the first Chern class of $\overline{\mathcal{H}}$ over $\overline{\mathcal{M}}_g$. Given a one-dimensional family B of stable genus g curves, define its *slope* by

$$s = \frac{\deg \Delta|_B}{\deg \lambda|_B}.$$

Morally speaking, $\deg \Delta|_B$ counts the number of nodes (with multiplicity) and $\deg \lambda|_B$ measures the variation of complex structures, see Figure 15.

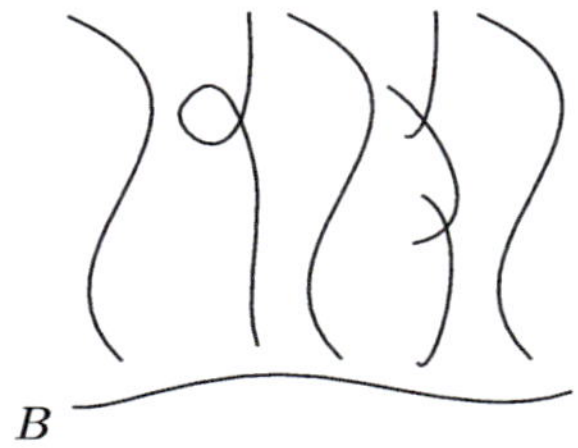

FIGURE 15. A one-dimensional family of stable curves

Let $\mu = (m_1, \ldots, m_n)$ be a partition of $2g-2$. Define

$$\kappa_\mu = \frac{1}{12}\sum_{i=1}^{n}\frac{m_i(m_i+2)}{m_i+1},$$

which depends on μ only. For a Teichmüller curve in $\mathcal{H}(\mu)$, despite that its area Siegel-Veech constant and slope are defined in different contexts, they determine each other ([**Ch1, Ch2, CM1**]):

$$s = \frac{12c}{c+\kappa_\mu}.$$

The upshot to prove the formula consists of the following. First, the constant κ_μ corresponds to the Miller-Morita-Mumford *κ-class*. Next, a cylinder with modulus h/w contributes h/w to the intersection of the Teichmüller curve with the boundary Δ, see Figure 16. Moreover, the numerical relation $12\lambda \equiv \kappa + \Delta$ holds on $\overline{\mathcal{M}}_g$.

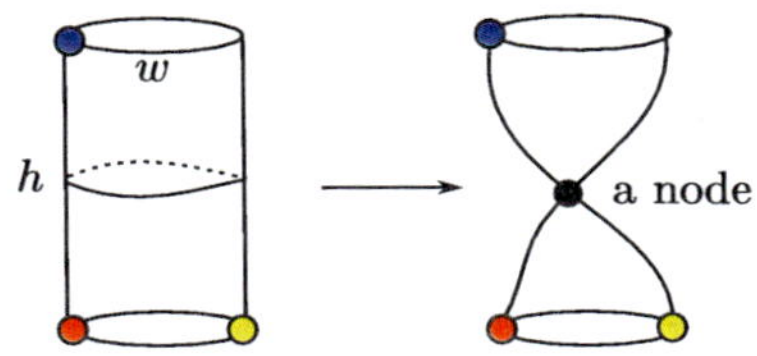

FIGURE 16. Shrinking the core curve of a cylinder

Finally we introduce an important dynamical invariant for an affine invariant submanifold. The diagonal subgroup $\left(\begin{smallmatrix} e^t & 0 \\ 0 & e^{-t}\end{smallmatrix}\right)$ defines the *Teichmüller geodesic flow* on the Hodge bundle $\mathcal{H}$ over an affine invariant submanifold $\mathcal{N}$. Since $\mathcal{H}$ is of rank g, there are g nonnegative *Lyapunov exponents* $\lambda_1 \geq \cdots \geq \lambda_g \geq 0$ as logarithms of mean eigenvalues of monodromy of $\mathcal{H}$ along the flow on $\mathcal{N}$. Morally speaking, Lyapunov exponents measure the separation rates of infinitesimally closed trajectories (see [**Z1**] for more details). Denote the *sum* of the Lyapunov exponents by

$$L = \lambda_1 + \cdots + \lambda_g.$$

In a seminal work, Eskin-Kontsevich-Zorich ([**EKZ1**]) proved that the sum of Lyapunov exponents and the area Siegel-Veech constant determine each other for *any* affine invariant submanifold:

$$L = c + \kappa_\mu.$$

As a corollary, we thus obtain that for a Teichmüller curve, any one of the three numbers s, c, and L determine the other two.

As an application, one can deduce a non-varying phenomenon of Teichmüller curves in low genus. By computer experiments, Kontsevich and Zorich observed that for many strata $\mathcal{H}(\mu)$ in low genus, *all* Teichmüller curves in the same stratum (component) have *non-varying* sums of Lyapunov exponents. They came up with a conjectural list of such strata:

- $g = 2$: $\mathcal{H}(1,1)$, $\mathcal{H}(2)$.
- $g = 3$: $\mathcal{H}(4)$, $\mathcal{H}(3,1)$, $\mathcal{H}(2,2)$, $\mathcal{H}(2,1,1)$.
- $g = 4$: $\mathcal{H}(6)$, $\mathcal{H}(5,1)$, $\mathcal{H}(4,2)$, $\mathcal{H}(3,3)$, $\mathcal{H}(3,2,1)$, $\mathcal{H}^{\mathrm{odd}}(2,2,2)$.
- $g = 5$: $\mathcal{H}(8)$, $\mathcal{H}^{\mathrm{odd}}(6,2)$, $\mathcal{H}(5,3)$, $\mathcal{H}^{\mathrm{hyp}}(4,4)$.

This phenomenon disappears when $g \geq 6$ except for the *hyperelliptic* strata.

Joint with Möller we proved Kontsevich-Zorich's conjecture ([**CM1**]). Let us use $\mathcal{H}(3,1)$ as an example to explain our method. For $(X,\omega) \in \mathcal{H}(3,1)$, it is easy to observe that X is *non-hyperelliptic*. The same holds for any nodal curve X contained in the *boundary* of a Teichmüller curve $\mathcal{T}$ in $\mathcal{H}(3,1)$. Let Hyp be the closure of the locus of hyperelliptic curves in $\overline{\mathcal{M}}_3$, which is a divisor with class

$$\mathrm{Hyp} \equiv 9\lambda - \Delta_0 - 3\Delta_1,$$

see [**HM**, Chapter 3.H]. Since $\mathcal{T}$ is disjoint with Hyp, we conclude that $\mathcal{T} \cdot \mathrm{Hyp} = 0$. Moreover, $\mathcal{T} \cdot \Delta_i = 0$ for $i > 0$, because shrinking the core curve of a cylinder yields a node of type Δ_0 only, see Figure 16 above. By $\mathcal{T} \cdot (9\lambda - \Delta) = 0$, we see that the slope

$$s = \frac{\deg \Delta|_{\mathcal{T}}}{\deg \lambda|_{\mathcal{T}}} = 9.$$

The above calculation is independent of the choice of Teichmüller curves in $\mathcal{H}(3,1)$. In summary, all Teichmüller curves in $\mathcal{H}(3,1)$ have slope 9. By the slope, Siegel-Veech, and Lyapunov exponent formula, sums of Lyapunov exponents are the same for all Teichmüller curves in $\mathcal{H}(3,1)$.

In general, the strategy is to find a geometrically meaningful *divisor* D in $\overline{\mathcal{M}}_g$ (or in $\overline{\mathcal{M}}_{g,n}$ by marking the zeros of ω) such that D is *disjoint* with all Teichmüller curves in $\mathcal{H}(\mu)$. The divisor class of D determines the invariants of those Teichmüller curves.

We remark that Yu-Zuo ([**YZ**]) gave another novel proof of Kontsevich-Zorich conjecture using the *Harder-Narasimhan filtration* of the Hodge bundle over Teichmüller curves. Their idea is the following. Suppose $f : \mathcal{X} \to \mathcal{T}$ is the universal curve over a Teichmüller curve. Let $S_1, \ldots, S_n$ be the disjoint sections of zeros of $(X,\omega) \in \mathcal{T}$. For a tuple of integers $a_1, \ldots, a_n$, if $h^0(X, \sum_{i=1}^n a_i z_i) = k$ for all X, then the direct image sheaf $f_*\mathcal{O}_{\mathcal{X}}(\sum_{i=1}^n a_i S_i)$ is a vector bundle of rank k on $\mathcal{T}$.

Use $\mathcal{H}(3,1)$ as an example again. Since $3z_1 + z_2 \sim K_X$, one checks that $h^0(X, z_1 + z_2) = 1$ and $h^0(X, 2z_1 + z_2) = 2$ for all $X \in \mathcal{T}$ (including the boundary points). Hence we obtain a filtration

$$f_*\mathcal{O}_{\mathcal{X}}(S_1 + S_2) \subset f_*\mathcal{O}_{\mathcal{X}}(2S_1 + S_2) \subset f_*\mathcal{O}_{\mathcal{X}}(3S_1 + S_2),$$

where the last term can be identified with the Hodge bundle twisted by the generating differential of $\mathcal{T}$. Then the sum L of Lyapunov exponents of $\mathcal{T}$ follows from Chern class calculations of the subbundles in the filtration.

We also remark that the slopes of Teichmüller curves can provide information to understand the *cone of effective divisors* of $\overline{\mathcal{M}}_g$. The idea is that any effective divisor on $\overline{\mathcal{M}}_g$ cannot contain all arithmetic Teichmüller curves, because their union is *dense* in $\overline{\mathcal{M}}_g$. Hence the limit of slopes of arithmetic Teichmüller curves as the degree of the coverings approach infinity bounds the numerical class of the divisor. We refer to the survey ([**CFM**]) for more details on the effective cone of moduli spaces.

3.5. Miscellaneous. In this section we collect various results with a flavor in algebraic geometry that are related to the preceding discussions.

Consider the set of all arithmetic Teichmüller curves generated by square-tiled surfaces with a given number of squares and fixed ramification type. McMullen ([**Mc3**]) showed that in $\mathcal{H}(2)$ this set consists of either a single arithmetic Teichmüller curve, or two distinct arithmetic Teichmüller curves. The special case when the number of squares is prime was previously established by Hubert-Lelièver ([**HL**]). For general strata, this is a wide open question. From the viewpoint of algebraic geometry, it amounts to classifying connected components of the Hurwitz space of torus coverings with a unique branch point and prescribed ramification type. Bainbridge ([**Ba**]) calculated Euler characteristics of Teichmüller curves in $\mathcal{H}(2)$. The idea is to calculate the fundamental classes of these Teichmüller curves in certain compactifications of Hilbert modular surfaces. Mukamel ([**Muk**]) further determined the number and type of orbifold points of these Teichmüller curves. Kumar-Mukamel ([**KM**]) described algebraic models of Teichmüller curves in genus two. Weiss ([**We**]) calculated the volume of certain twisted Teichmüller curves on Hilbert modular surfaces and partially classified their connected components.

Siegel-Veech constants and Lyapunov exponents can be defined similarly for any affine invariant submanifold, say the strata themselves. Eskin-Masur-Zorich ([**EMZ**]) analyzed the principal boundary of the moduli space of Abelian differentials and gave a recursive method to compute Siegel-Veech constants of the strata for any Siegel-Veech configuration. For the cylinder configuration, the limit of area Siegel-Veech constants of arithmetic Teichmüller curves in a stratum as the degree of the coverings approach infinity equals the area Siegel-Veech constant of the stratum ([**Ch2**, Appendix A]).

Previously we discussed the sum of Lyapunov exponents. For arithmetic Teichmüller curves generated by cyclic covers of $\mathbb{P}^1$ branched at four points, Eskin-Kontsevich-Zorich ([**EKZ2**]) calculated all individual Lyapunov exponents. Yu ([**Y**]) conjectured that the polygon of Lyapunov spectrum bounds the Harder-Narasimhan polygon on Teichmüller curves, and a proof of this conjecture is recently announced by Eskin-Kontsevich-Möller-Zorich ([**EKMZ**]). The fundamental work of Forni ([**Fo1**]) showed that for a stratum of Abelian differentials no Lyapunov exponent vanishes. The fundamental work of Avila and Viana ([**AV**]) further showed that for a stratum the Lyapunov spectrum is simple, that is, the strict inequality $\lambda_i > \lambda_{i+1}$ holds for all i. For arithmetic Teichmüller curves generated by square-tiled surfaces, Matheus-Möller-Yoccoz ([**MMY**]) gave a Galois-theoretical criterion for the simplicity of their Lyapunov spectra.

The Lyapunov spectrum can degenerate for a special affine invariant submanifold. If $\lambda_2 = \cdots = \lambda_g = 0$, we say that the Lyapunov spectrum is completely degenerate. Forni and Forni-Matheus-Zorich ([**Fo2**, **FMZ**]) found examples of Teichmüller curves with completely degenerate Lyapunov spectrum in genus three and four, respectively. By studying Teichmüller curves that are also Shimura curves, Möller (**[Mö4]**) showed that those examples are the only Teichmüller curves with completely degenerate Lyapunov spectrum with possible exceptions in genus five. Aulicino (**[Au]**) showed that an affine invariant submanifold with completely degenerate Lyapunov spectrum can only be an arithmetic Teichmüller curve in genus at most five, and he further established finiteness of Teichmüller curves with completely degenerate Lyapunov spectrum. Filip (**[Fi2]**) described all situations when an affine invariant submanifold can possess a zero Lyapunov exponent by analyzing monodromy of the corresponding Kontsevich-Zorich cocycle. As a higher dimensional analogue of families of curves, Filip (**[Fi3]**) considered families of K3 surfaces whose second cohomology groups form a local system, and showed that their top Lyapunov exponents are always rational.

To conclude this section we mention two problems that have broad connections to other fields. Kontsevich-Zorich (**[KonZor2]**) conjectured that every connected component of the strata is $K(\pi, 1)$, that is, it has a contractible universal cover and its fundamental group is commensurable with certain mapping class group. For all strata in genus three except $\mathcal{H}(1^4)$, Looijenga-Mondello (**[LM]**) determined their (orbifold) fundamental groups by analyzing geometry of canonical curves. Kontsevich-Soibelman (**[KS]**) speculated that moduli spaces of differentials can be identified with moduli spaces of certain stability conditions, where the $\mathrm{GL}_2^+(\mathbb{R})$-action and saddle connections are analogues of the central charge in the theory of stability conditions. The seminal work of Bridgeland-Smith (**[BS]**) established this identification for moduli spaces of quadratic differentials with simple zeros by relating the finite-length trajectories of such quadratic differentials to the stable objects of the corresponding stability condition.

4. Affine invariant submanifolds

In the preceding section we discussed Teichmüller curves as examples of affine invariant submanifolds. In this section we consider affine invariant submanifolds in general.

4.1. Structure of affine invariant submanifolds. The celebrated Ratner's orbit closure theorem in ergodic theory says that the closures of orbits of unipotent flows on the quotient of a Lie group by a lattice are homogeneous submanifolds. In the context of Teichmüller dynamics, the strata of Abelian differentials do not behave like homogeneous spaces. Hence it is unclear whether $\mathrm{GL}_2^+(\mathbb{R})$-orbit closures can have nice geometric structures. Nevertheless, the recent breakthrough of Eskin-Mirzakhani-Mohammadi (**[EMi, EMM]**) showed that $\mathrm{GL}_2^+(\mathbb{R})$-orbit closures are locally cut out by homogeneous linear equations of period coordinates with real coefficients, thus justifying that $\mathrm{GL}_2^+(\mathbb{R})$-orbit closures are affine invariant submanifolds.[8] Previously this result was only proved in genus two by the fundamental work of McMullen (**[Mc6]**).

Recall that the period coordinates at $(X, \omega) \in \mathcal{H}(\mu)$ are given by integrating ω over a basis of the relative homology $H_1(X, \Sigma; \mathbb{Z})$, where Σ is the set of zeros of ω. Although period coordinates are not canonical, any two choices of period coordinates differ by a matrix in $\mathrm{GL}_n(\mathbb{Z})$, where $n = \dim \mathcal{H}(\mu)$, hence the above

[8] Zorich (**[Z2]**) called it the "Magic Wand Theorem", which is part of Mirzakhani's Fields Medal work.

theorem does not depend on the choice of period coordinates. The proof of the theorem is remarkably long and technical, which involves many ideas from dynamics on homogeneous space, ergodic theory, and measure theory.

Since period coordinates are transcendental, affine invariant submanifolds are apriori only complex-analytic submanifolds. The hidden algebraic nature has been discovered by Filip ([**Fi1**]), who proved that affine invariant submanifolds are algebraic subvarieties of $\mathcal{H}(\mu)$, defined over $\overline{\mathbb{Q}}$. In particular, Filip used tools from variations of Hodge structures and showed that affine invariant submanifolds (except those of full rank, see Section 4.2) parameterize curves with non-trivial endormophisms, such as real multiplication on a factor of the Jacobians, which generalizes Möller's earlier work ([**Mö1, Mö2**]) on torsion and real multiplication for Teichmüller curves. As a corollary, the closure of any Teichmüller disk (in the standard topology) in $\mathcal{M}_g$ is a subvariety of $\mathcal{M}_g$.

4.2. Classification of affine invariant submanifolds. After the structure theorem of Eskin-Mirzakhani-Mohammadi, the classification of affine invariant submanifolds remains to be a central open question in the study of Teichmüller curves.

The tangent space of an affine invariant submanifold $\mathcal{N}$ at (X, ω) can be identified with $H^1(X, \Sigma; \mathbb{C})$ under the period coordinates.[9] Denote by $p : H^1(X, \Sigma; \mathbb{C}) \to H^1(X; \mathbb{C})$ the projection from the relative cohomology to the absolute cohomology. Avila-Eskin-Möller ([**AEM**]) proved that $p(T_{(X,\omega)}\mathcal{N})$ is a complex symplectic vector space, hence it has even dimension. Define the *rank* of $\mathcal{N}$ to be

$$\frac{1}{2}\dim_{\mathbb{C}} p(T_{(X,\omega)}\mathcal{N})$$

, which is at most g by definition. If the rank of $\mathcal{N}$ is bigger than one, we say that $\mathcal{N}$ is of *higher rank*. If it is equal to g, we say that $\mathcal{N}$ has *full rank*. For example, arithmetic Teichmüller curves generated by square-tiled surfaces have rank one.

By analyzing the boundary of affine invariant submanifolds, Mirzakhani-Wright ([**MR**]) proved that if an affine invariant submanifold is of full rank, then it is either a connected component of a stratum or the hyperelliptic locus in a connected component of a stratum. Apisa ([**Ap**]) further showed that all affine invariant submanifolds of higher rank in the hyperelliptic strata arise from covering constructions, which gives a coarse classification of affine invariant submanifolds in the hyperelliptic strata modulo finitely many non-arithmetic Teichmüller curves and their connected components. Based on previously known examples, Mirzakhani conjectured that higher rank affine invariant submanifolds are either connected components of the strata or arise from covering constructions. Recently counterexamples of Mirzakhani's conjecture are announced by McMullen-Mukamel-Wright ([**MMW**]) and Eskin-McMullen-Mukamel-Wright.

4.3. Degeneration of Abelian differentials. Despite the analytic definition of Teichmüller dynamics, a profound algebro-geometric foundation behind the story has already been revealed by many of the preceding results. In order to apply ideas from algebraic geometry, one upshot is to understanding *degeneration* of Abelian differentials, or equivalently, describing a *compactification* of strata of Abelian differentials, analogous to the Deligne-Mumford compactification $\overline{\mathcal{M}}_g$ of the moduli space of curves.

[9] In Teichmüller dynamics $\mathcal{M}$ is commonly used to denote an affine invariant submanifold, but here we reserve $\mathcal{M}$ for the moduli space.

Recall that the Hodge bundle $\mathcal{H}$ extends to a rank g bundle $\overline{\mathcal{H}}$ over $\overline{\mathcal{M}}_g$, parameterizing stable differentials that are sections of the dualizing line bundle. Hence it would be natural to compactify the stratum $\mathcal{H}(\mu)$ by taking its closure in $\overline{\mathcal{H}}$. Nevertheless, a disadvantage of this Hodge bundle compactification is that it can lose information of the limit positions of the zeros of ω, especially if ω vanishes entirely on a component of the underlying reducible curve. Alternatively, up to scaling an Abelian differential is determined by the associated canonical divisor. Hence one can consider the *stratum* $\mathcal{P}(\mu)$ *of canonical divisors of type* μ as the projectivization of $\mathcal{H}(\mu)$. Marking the n zeros of the divisors, $\mathcal{P}(\mu)$ can be regarded as a subvariety of the Deligne-Mumford moduli space $\overline{\mathcal{M}}_{g,n}$ of stable genus g curves with n marked points, hence one can study degeneration of canonical divisors of type μ by taking the closure of $\mathcal{P}(\mu)$ in $\overline{\mathcal{M}}_{g,n}$.

Eisenbud-Harris ([**EH1**]) developed a theory of limit linear series that studies degeneration of line bundles and their sections from smooth curves to nodal curves of compact type. In our context the situation is slightly different, because the zero type μ of the sections is fixed and the underlying curves may fail to be of compact type. Nevertheless, the upshot of *twisting* line bundles by irreducible components of a nodal curve still works. More precisely, define a *twisted canonical divisor of type* μ on a nodal curve X to be a collection of (possibly meromorphic) canonical divisors D_j on each irreducible component X_j of X such that the following conditions hold:

(1) The support of D_j is contained in the set of marked points and the nodes lying in X_j. Moreover, if p_i is a marked point contained in X_j, then $\mathrm{ord}_{p_i}(D_j) = m_i$.
(2) If q is a node of X lying in two irreducible components X_1 and X_2, then $\mathrm{ord}_q(D_1) + \mathrm{ord}_q(D_2) = -2$.
(3) If q is a node of X lying in two irreducible components X_1 and X_2 such that $\mathrm{ord}_q(D_1) = \mathrm{ord}_q(D_2) = -1$, then for any node $q' \in X_1 \cap X_2$, $\mathrm{ord}_{q'}(D_1) = \mathrm{ord}_{q'}(D_2) = -1$. In this case we write $X_1 \sim X_2$.
(4) If q is a node of X lying in two irreducible components X_1 and X_2 such that $\mathrm{ord}_q(D_1) > \mathrm{ord}_q(D_2)$, then for any node $q' \in X_1 \cap X_2$, $\mathrm{ord}_{q'}(D_1) > \mathrm{ord}_{q'}(D_2)$. In this case we write $X_1 \succ X_2$.
(5) There does not exist a directed loop $X_1 \succeq X_2 \succeq \cdots \succeq X_k \succeq X_1$ unless all the relations are $\sim$, where $\succeq$ means $\sim$ or $\succ$.

We briefly explain the motivation behind these conditions. Since the vanishing order along each zero section remains unchanged in a family, it implies condition (1). Since $K_X|_{X_i}$ is locally generated by differentials with a simple pole at a node $q \in X_1 \cap X_2$ for $i = 1, 2$, when twisting by X_i only, the zero or pole order increases by one on one branch of q and decrease by one on the other branch, hence the sum of the vanishing orders does not vary, which implies condition (2). If there is no twist at all, it is the case corresponding to condition (3). Note that twisting by $X_1 + X_2$ does nothing to the nodes lying in their intersection. Hence one can consider twisting, say by a multiple of X_1 only. Then the vanishing orders at all nodes between X_1 and X_2 increase or decrease simultaneously, hence condition (4) follows. By the same token, $X_1 \succeq X_2$ means the twisting coefficient of X_1 is bigger than or equal to that of X_2, which implies the last condition.

By an analytic approach, Gendron ([**Ge**]) implicitly derived the above conditions and used them to study the Kodaira dimension of strata of twisted canonical divisors. Motivated by the theory of limit linear series, the author ([**Ch4**]) considered these conditions for curves of compact type and used them to study Weierstrass point behavior for general elements in the strata. Farkas and Pandharipande ([**FP**]) imposed explicitly these conditions and showed that the corresponding closures in $\overline{\mathcal{M}}_{g,n}$ are reducible in general, containing extra boundary components of dimension one less compared to the main component. It is thus natural to ask what extra conditions can distinguish the main component from the other boundary components in the closure.

Joint with Bainbridge, Gendron, Grushevsky, and Möller ([**BCGGM1**]), we have found the missing condition that arises from a *residue* constraint. Consider the following example. Suppose a family $\mathcal{X}$ of Abelian differentials (X_t, ω_t) degenerate to a nodal curve X at $t = 0$. Suppose X has a separating node q joining two components Y and Z. Without loss of generality, suppose $\lim_{t\to 0} \omega_t|_Y = \eta_Y$ is a holomorphic differential on Y, and $\lim_{t\to 0}(t^{-\ell}\omega_t)|_Z = \eta_Z$ is a meromorphic differential on Z, where $\ell \in \mathbb{Z}^+$. In other words, the twisting at q is given by $\mathcal{O}_{\mathcal{X}}(-\ell Z)$ from the viewpoint of limit linear series. Let v_t be the vanishing cycle on $X_t = Y_t \cup Z_t$ that shrinks to the node q, where $Y_t \to Y$ and $Z_t \to Z$ as $t \to 0$, see Figure 17.

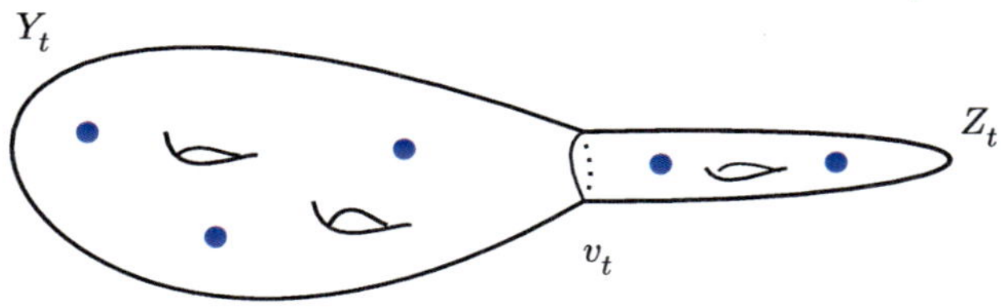

FIGURE 17. A surface with a separating vanishing cycle

Since q is a separating node, $v_t = 0 \in H_1(X_t; \mathbb{Z})$ for t nearby 0. It follows that $\int_{v_t} \omega_t = 0$ for $t \neq 0$ and hence $\int_{v_t} t^{-\ell}\omega_t = 0$. Taking the limit as $t \to 0$ and restricting to the component Z, we conclude that $\mathrm{Res}_q(\eta_Z) = 0$. Conversely, once such a residue condition holds, along with conditions (1)–(4) we are able to prove that the limit differential is smoothable by plumbing techniques in complex-analytic geometry as well as by constructions of flat surfaces.

4.4. Cycle classes of strata of Abelian differentials. As mentioned before, affine invariant submanifolds can provide special subvarieties in the moduli space of curves. It is natural to ask if one can calculate their *cycle classes* in the Chow ring of the moduli space. The first step is to calculate the cycle classes of the strata of Abelian differentials. Tarasca and the author ([**CT**]) calculated the cycle classes of several minimal strata in low genus. Mullane ([**Mul**]) obtained a closed formula for the classes of all strata whose projections are effective divisors in $\overline{\mathcal{M}}_g$. Both results rely on classical intersection theory on the moduli space of curves. Modulo a conjectural relation to Pixton's formula of the double ramification cycle, Janda-Pandharipande-Pixton-Zvonkine ([**FP**, Appendix]) obtained a recursive method to compute the cycle classes of all strata in $\overline{\mathcal{M}}_{g,n}$, which suggests that the strata classes are *tautological*. Their approach relies on analyzing the closure of a stratum of twisted canonical divisors in $\overline{\mathcal{M}}_{g,n}$ by imposing conditions (1)–(4) in the preceding section. Although the closure may have extra components contained in the boundary, those extra components are products of simpler strata of (possibly meromorphic) differentials, hence one can calculate the class of the main component recursively.

One can also consider the cycle class calculation in the Hodge bundle compactification. Korotkin-Zograf ([**KorZog**]) applied Tau functions to compute the divisor class of the closure of the stratum $\mathcal{P}(2, 1^{2g-4})$ in the projectivized Hodge bundle over $\overline{\mathcal{M}}_g$. The author ([**Ch3**]) gave another proof of this divisor class using intersection theory. Recently Sauvaget-Zvonkine have announced that they are able to compute the cycle classes of all strata closures in the Hodge bundle compactification.

5. Meromorphic and higher order differentials

In this section we generalize the discussion of Abelian differentials to higher order differentials possibly with poles.

5.1. Quadratic differentials. First, if q is a quadratic differential, that is, a section of $K^{\otimes 2}$, then it also induces a flat structure, if one allows *rotation by* π in addition to translation as transition functions. The flat structure can be defined locally by taking a square root ω of q, which is up to $\pm$, and that is the reason why rotation by π needs to be part of the transition functions. Moreover if q has at worst simple poles, integrating ω along a path always provides finite length, hence the corresponding flat surface has finite area. In general, quadratic differentials with at worst simple poles are called *half-translation surfaces*. For example, Figure 18 presents a quadratic differential with four simple poles on $\mathbb{P}^1$ as a pillowcase.

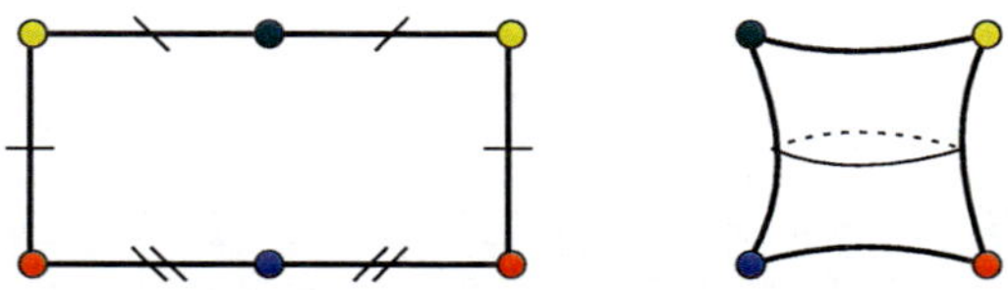

FIGURE 18. A quadratic differential with four simple poles on $\mathbb{P}^1$

Suppose (X, q) is a half-translation surface. One can take the unique *canonical double cover* $\pi : \hat{X} \to X$ branched at the odd singularities of q (zeros of odd order and simple poles). Then there exists a global Abelian differential $\hat{\omega}$ on $\hat{X}$ such that $\pi^* q = \hat{\omega}^2$. From this viewpoint, all questions for Abelian differentials can be similarly asked for quadratic differentials, and indeed many of them can be similarly answered.

In what follows we mention several results analogous to the case of Abelian differentials. Fix a partition ν of $4g-4$ such that all entries of ν are ≥ -1. Let $\mathcal{Q}(\nu)$ be the stratum of quadratic differentials of type ν that are not global squares of Abelian differentials. Lanneau ([**L2**]) classified the connected components of $\mathcal{Q}(\nu)$ for all ν. In $g \geq 5$, it can have at most two connected components, where extra components are caused by hyperelliptic structures. In particular, the spin parity in the Abelian case does not give rise to additional components in the quadratic case ([**L1**]). In genus three and four, there are several *exceptional* disconnected strata. Joint with Möller ([**CM2**]) we found an algebraic parity arising from geometry of canonical curves that distinguishes these exceptional components. We further discovered a number of strata $\mathcal{Q}(\nu)$ in low genus whose Teichmüller curves have non-varying sums of Lyapunov exponents by similar techniques as in the Abelian case. The relation between the area Siegel-Veech constant and the sum of Lyapunov exponents for affine invariant submanifolds in $\mathcal{Q}(\nu)$ also holds by the seminal work of Eskin-Kontsevich-Zorich ([**EKZ1**]). Eskin-Okounkov ([**EO2**]) analyzed the volume growth of $\mathcal{Q}(\nu)$ by enumerating covers of $\mathbb{P}^1$ with certain ramification profile determined by ν. Athreya-Eskin-Zorich ([**AEZ**]) proved an explicit formula for the volume of $\mathcal{Q}(\nu)$ in genus zero. Goujard ([**Go2**]) obtained explicit values for the volume of $\mathcal{Q}(\nu)$ in all low dimensions. Masur-Zorich ([**MZ**]) studied the principal boundary of $\mathcal{Q}(\nu)$, aiming at a recursive way to calculate Siegel-Veech constants of saddle connections. Goujard ([**Go1**]) proved an explicit formula that relates volumes of $\mathcal{Q}(\nu)$ and Siegel-Veech constants. Grivaux-Hubert ([**GH**]) constructed explicit affine invariant submanifolds in $\mathcal{Q}(\nu)$ of arbitrarily large dimension with completely degenerate Lyapunov spectrum.

5.2. Differentials with poles. Previously when we talked about an Abelian differential ω, we assumed that ω is holomorphic, so the corresponding translation surface has finite area. In many cases it would be useful to consider *meromorphic* differentials, in particular when we study the boundary structure of strata of Abelian differentials.

First, we describe the local flat geometry around a pole. Suppose p is a *simple* pole of ω. Then the flat neighborhood of p can be viewed as a *half-infinite cylinder*. The *width* of the cylinder corresponds to the *residue* of ω at p. Figure 19 exhibits a meromorphic differential that has two simple poles with opposite residues, where the poles locate at the positive infinity and negative infinity.

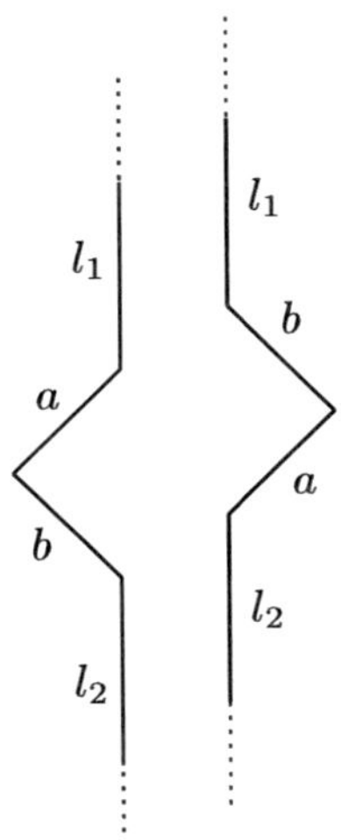

FIGURE 19. A flat surface with two simple poles

If p is a pole of order $m \geq 2$, Boissy ([**Bo**]) showed that one can glue $2m-2$ *broken half-planes* consecutively to form a flat-geometric presentation for a neighborhood of p. The boundary of each broken half-plane consists of a half-line to the left and a half-line to the right, which are connected by finitely many broken line segments (saddle connections). The idea behind Boissy's description is the following. If one glues $2m-2$ half-disks consecutively to form a zero of order $m-2$, the local expression of the differential would be $z^{m-2}dz$. Now change coordinates by $z = 1/w$. Then the differential becomes $\sim w^{-m}dw$, which has a pole of order

m, and the half-disks turn into half-planes. In particular, the residue of ω at p is determined by the complex lengths of the boundary line segments and the gluing pattern of the broken half-planes.

The lower right flat surface of infinite area in Figure 20 below shows the flat-geometric presentation of a meromorphic differential with a double zero and a double pole on a torus, where the pole locates at infinity. It is constructed by removing the interior of a parallelogram Z from the Euclidean plane and then identifying parallel edges by translation. If we slit the plane along a diagonal of Z, we thus recover a pair of broken half-planes as in Boissy's description. In particular, the double pole has no residue.

Building on earlier work of Kontsevich-Zorich and Lanneau ([**KonZor1**, **L2**]), Boissy ([**Bo**]) classified the connected components of strata of meromorphic differentials with prescribed numbers and multiplicities of zeros and poles. Similarly to the holomorphic case, the strata of meromorphic differentials can have at most three connected components, distinguished by hyperelliptic and spin structures.

Let us illustrate an interesting viewpoint using flat geometry of meromorphic differentials to study the boundary of strata of Abelian differentials. The flat surface on the left side of Figure 20 lies in $\mathcal{H}(2)$, which is constructed by removing a parallelogram Z from the interior of a parallelogram Y and identifying parallel edges. If we shrink Z to a point, we obtain a holomorphic differential $(Y, \eta_Y) \in \mathcal{H}(0)$, where the marked point encodes the limit position of the inner square. Alternatively, modulo scaling this procedure amounts to expanding Y to be arbitrarily large, hence the limit object represents a meromorphic differential $(Z, \eta_Z) \in \mathcal{H}(2, -2)$.

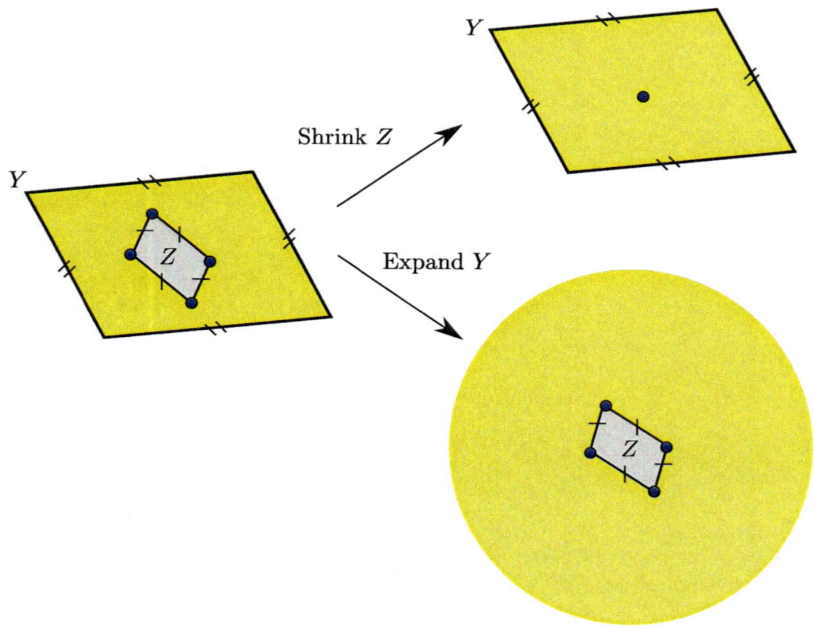

FIGURE 20. Shrinking Z versus expanding Y

Note that these two perspectives correspond to exactly the two *aspects* in the theory of limit linear series, which applied to this case says that a family of curves of genus two with a marked double zero of a canonical divisor (Weierstrass point) degenerates to two elliptic curves joined at one node, where the limit marked point is 2-torsion respect to the node.

From the viewpoint of $\mathrm{GL}_2^+(\mathbb{R})$-action and Teichmüller dynamics, the strata of meromorphic differentials somehow display different properties compared to the case of Abelian differentials (see [**Bo**, Appendix]).

5.3. Higher order differentials. We conclude the survey by describing some future directions. From the viewpoint of algebraic geometry, an Abelian differential is a section of the canonical line bundle K, and a quadratic differential is a section of $K^{\otimes 2}$. Therefore, it is natural to consider *higher order differentials* arising from sections of $K^{\otimes k}$ for a fixed positive integer k. Suppose $\mu = (k_1, \dots, k_n)$ is a partition of $k(2g-2)$, where we allow k_i to be possibly negative. In other words, we want to take meromorphic differentials into account. Let $\mathcal{H}^k(\mu)$ be the *stratum of k-differentials of type μ*, which parameterizes (possibly meromorphic) sections of $K^{\otimes k}$ with zeros and poles of type μ on genus g Riemann surfaces. Since $\mathrm{GL}_2^+(\mathbb{R})$ does not preserve angles in general, there is no natural $\mathrm{GL}_2^+(\mathbb{R})$-action on the strata of k-differentials for $k > 2$. Nevertheless, many other questions regarding Abelian and quadratic differentials can still be asked for k-differentials. In particular, what are the dimension, connected components, compactification, and cycle class of $\mathcal{H}^k(\mu)$? In a forthcoming work ([**BCGGM2**]), we will treat these questions systematically.

References

[Ap] P. Apisa, $\mathrm{GL}(2,\mathbb{R})$ orbit closures in hyperelliptic components of strata, arXiv:1508.05438.

[AEZ] J. Athreya, A. Eskin, and A. Zorich, Right-angled billiards and volumes of moduli spaces of quadratic differentials on $\mathbb{CP}^1$, with an appendix by J. Chaika, *Ann. Sci. École Norm. Sup. (4)* **49** (2016), no. 6.

[At] Michael F. Atiyah, *Riemann surfaces and spin structures*, Ann. Sci. École Norm. Sup. (4) **4** (1971), 47–62. MR0286136

[Au] D. Aulicino, Affine invariant submanifolds with completely degenerate Kontsevich-Zorich spectrum, *Ergodic Theory Dynam. Systems*, to appear.

[AV] Artur Avila and Marcelo Viana, *Simplicity of Lyapunov spectra: proof of the Zorich-Kontsevich conjecture*, Acta Math. **198** (2007), no. 1, 1–56. MR2316268

[AEM] A. Avila, A. Eskin, and M. Möller, Symplectic and isometric $\mathrm{SL}(2,\mathbb{R})$ invariant subbundles of the Hodge bundle, *J. Reine Angew. Math.*, to appear.

[Ba] Matt Bainbridge, *Euler characteristics of Teichmüller curves in genus two*, Geom. Topol. **11** (2007), 1887–2073. MR2350471

[BCGGM1] M. Bainbridge, D. Chen, Q. Gendron, S. Grushevsky, and M. Möller, Compactification of strata of abelian differentials, arXiv:1604.08834.

[BCGGM2] M. Bainbridge, D. Chen, Q. Gendron, S. Grushevsky, and M. Möller, Strata of k-differentials, arXiv:1610.09238.

[BHM] Matt Bainbridge, Philipp Habegger, and Martin Möller, *Teichmüller curves in genus three and just likely intersections in*, Publ. Math. Inst. Hautes Études Sci. **124** (2016), 1–98. MR3578914

[BaM] Matt Bainbridge and Martin Möller, *The Deligne-Mumford compactification of the real multiplication locus and Teichmüller curves in genus 3*, Acta Math. **208** (2012), no. 1, 1–92. MR2910796

[BS] Tom Bridgeland and Ivan Smith, *Quadratic differentials as stability conditions*, Publ. Math. Inst. Hautes Études Sci. **121** (2015), 155–278. MR3349833

[Bo] Corentin Boissy, *Connected components of the strata of the moduli space of meromorphic differentials*, Comment. Math. Helv. **90** (2015), no. 2, 255–286. MR3351745

[BoM] Irene I. Bouw and Martin Möller, *Teichmüller curves, triangle groups, and Lyapunov exponents*, Ann. of Math. (2) **172** (2010), no. 1, 139–185. MR2680418

[Ca] Kariane Calta, *Veech surfaces and complete periodicity in genus two*, J. Amer. Math. Soc. **17** (2004), no. 4, 871–908. MR2083470

[Ch1] Dawei Chen, *Covers of elliptic curves and the moduli space of stable curves*, J. Reine Angew. Math. **649** (2010), 167–205. MR2746470

[Ch2] Dawei Chen, *Square-tiled surfaces and rigid curves on moduli spaces*, Adv. Math. **228** (2011), no. 2, 1135–1162. MR2822219

[Ch3] Dawei Chen, *Strata of abelian differentials and the Teichmüller dynamics*, J. Mod. Dyn. **7** (2013), no. 1, 135–152. MR3071469

[Ch4] D. Chen, Degenerations of Abelian differentials, J. Diff. Geom., to appear.

[CFM] Dawei Chen, Gavril Farkas, and Ian Morrison, *Effective divisors on moduli spaces of curves and abelian varieties*, A celebration of algebraic geometry, Clay Math. Proc., vol. 18, Amer. Math. Soc., Providence, RI, 2013, pp. 131–169. MR3114939

[CM1] Dawei Chen and Martin Möller, *Nonvarying sums of Lyapunov exponents of Abelian differentials in low genus*, Geom. Topol. **16** (2012), no. 4, 2427–2479. MR3033521

[CM2] Dawei Chen and Martin Möller, *Quadratic differentials in low genus: exceptional and non-varying strata* (English, with English and French summaries), Ann. Sci. Éc. Norm. Supér. (4) **47** (2014), no. 2, 309–369. MR3215925

[CMZ] D. Chen, M. Möller, and D. Zagier, Quasimodularity and large genus limits of Siegel-Veech constants, arXiv:1606.04065.

[CT] Dawei Chen and Nicola Tarasca, *Loci of curves with subcanonical points in low genus*, Math. Z. **284** (2016), no. 3-4, 683–714. MR3563252

[EH1] David Eisenbud and Joe Harris, *Limit linear series: basic theory*, Invent. Math. **85** (1986), no. 2, 337–371. MR846932

[EH2] David Eisenbud and Joe Harris, *Existence, decomposition, and limits of certain Weierstrass points*, Invent. Math. **87** (1987), no. 3, 495–515. MR874034

[EMc] Jordan S. Ellenberg and D. B. McReynolds, *Arithmetic Veech sublattices of* $\mathrm{SL}(2,\mathbf{Z})$, Duke Math. J. **161** (2012), no. 3, 415–429. MR2881227

[EKMZ] A. Eskin, M. Kontsevich, M. Möller, and A. Zorich, Lower bounds for Lyapunov exponents of flat bundles on curves, arXiv:1609.01170.

[EKZ1] Alex Eskin, Maxim Kontsevich, and Anton Zorich, *Sum of Lyapunov exponents of the Hodge bundle with respect to the Teichmüller geodesic flow*, Publ. Math. Inst. Hautes Études Sci. **120** (2014), 207–333. MR3270590

[EKZ2] Alex Eskin, Maxim Kontsevich, and Anton Zorich, *Lyapunov spectrum of square-tiled cyclic covers*, J. Mod. Dyn. **5** (2011), no. 2, 319–353. MR2820564

[EMa] Alex Eskin and Howard Masur, *Asymptotic formulas on flat surfaces*, Ergodic Theory Dynam. Systems **21** (2001), no. 2, 443–478. MR1827113

[EMZ] Alex Eskin, Howard Masur, and Anton Zorich, *Moduli spaces of abelian differentials: the principal boundary, counting problems, and the Siegel-Veech constants*, Publ. Math. Inst. Hautes Études Sci. **97** (2003), 61–179. MR2010740

[EMi] A. Eskin and M. Mirzakhani, Invariant and stationary measures for the $\mathrm{SL}(2,\mathbb{R})$ action on Moduli space, arXiv:1302.3320.

[EMM] Alex Eskin, Maryam Mirzakhani, and Amir Mohammadi, *Isolation, equidistribution, and orbit closures for the* $\mathrm{SL}(2,\mathbb{R})$ *action on moduli space*, Ann. of Math. (2) **182** (2015), no. 2, 673–721. MR3418528

[EO1] Alex Eskin and Andrei Okounkov, *Asymptotics of numbers of branched coverings of a torus and volumes of moduli spaces of holomorphic differentials*, Invent. Math. **145** (2001), no. 1, 59–103. MR1839286

[EO2] Alex Eskin and Andrei Okounkov, *Pillowcases and quasimodular forms*, Algebraic geometry and number theory, Progr. Math., vol. 253, Birkhäuser Boston, Boston, MA, 2006, pp. 1–25. MR2263191

[FP] G. Farkas and R. Pandharipande, The moduli space of twisted canonical divisors, with an appendix by F. Janda, R. Pandharipande, A. Pixton, and D.Zvonkine, *J. Inst. Math. Jussieu*, to appear.

[Fi1] Simion Filip, *Splitting mixed Hodge structures over affine invariant manifolds*, Ann. of Math. (2) **183** (2016), no. 2, 681–713. MR3450485

[Fi2] S. Filip, Zero Lyapunov exponents and monodromy of the Kontsevich-Zorich cocycle, arXiv:1410.2129.

[Fi3] S. Filip, Families of K3 surfaces and Lyapunov exponents, arXiv:1412.1779.

[Fo1] Giovanni Forni, *Deviation of ergodic averages for area-preserving flows on surfaces of higher genus*, Ann. of Math. (2) **155** (2002), no. 1, 1–103. MR1888794

[Fo2] Giovanni Forni, *On the Lyapunov exponents of the Kontsevich-Zorich cocycle*, Handbook of dynamical systems. Vol. 1B, Elsevier B. V., Amsterdam, 2006, pp. 549–580. MR2186248

[FMZ] Giovanni Forni, Carlos Matheus, and Anton Zorich, *Zero Lyapunov exponents of the Hodge bundle*, Comment. Math. Helv. **89** (2014), no. 2, 489–535. MR3225454

[Ge] Q. Gendron, The Deligne-Mumford and the incidence variety compactifications of the strata of $\Omega\mathcal{M}_g$, arXiv:1503.03338.

[Go1] Elise Goujard, *Siegel-Veech constants for strata of moduli spaces of quadratic differentials*, Geom. Funct. Anal. **25** (2015), no. 5, 1440–1492. MR3426059

[Go2] E. Goujard, Volumes of strata of moduli spaces of quadratic differentials: getting explicit values, *Ann. Inst. Fourier (Grenoble)*, to appear.

[GH] Julien Grivaux and Pascal Hubert, *Loci in strata of meromorphic quadratic differentials with fully degenerate Lyapunov spectrum*, J. Mod. Dyn. **8** (2014), no. 1, 61–73. MR3296566

[GJ] Eugene Gutkin and Chris Judge, *Affine mappings of translation surfaces: geometry and arithmetic*, Duke Math. J. **103** (2000), no. 2, 191–213. MR1760625

[HM] Joe Harris and Ian Morrison, *Moduli of curves*, Graduate Texts in Mathematics, vol. 187, Springer-Verlag, New York, 1998. MR1631825

[HL] Pascal Hubert and Samuel Lelièvre, *Prime arithmetic Teichmüller discs in $\mathcal{H}(2)$*, Israel J. Math. **151** (2006), 281–321. MR2214127

[J] Dennis Johnson, *Spin structures and quadratic forms on surfaces*, J. London Math. Soc. (2) **22** (1980), no. 2, 365–373. MR588283

[KS] Maxim Kontsevich and Yan Soibelman, *Motivic Donaldson-Thomas invariants: summary of results*, Mirror symmetry and tropical geometry, Contemp. Math., vol. 527, Amer. Math. Soc., Providence, RI, 2010, pp. 55–89. MR2681792

[KorZog] D. Korotkin and P. Zograf, *Tau function and moduli of differentials*, Math. Res. Lett. **18** (2011), no. 3, 447–458. MR2802579

[KonZor1] Maxim Kontsevich and Anton Zorich, *Connected components of the moduli spaces of Abelian differentials with prescribed singularities*, Invent. Math. **153** (2003), no. 3, 631–678. MR2000471

[KonZor2] M. Kontsevich, *Lyapunov exponents and Hodge theory*, The mathematical beauty of physics (Saclay, 1996), Adv. Ser. Math. Phys., vol. 24, World Sci. Publ., River Edge, NJ, 1997, pp. 318–332. MR1490861

[KM] A. Kumar and R. Mukamel, Algebraic models and arithmetic geometry of Teichmüller curves in genus two, *Int. Math. Res. Not.*, to appear.

[L1] Erwan Lanneau, *Parity of the Spin structure defined by a quadratic differential*, Geom. Topol. **8** (2004), 511–538. MR2057772

[L2] Erwan Lanneau, *Connected components of the strata of the moduli spaces of quadratic differentials* (English, with English and French summaries), Ann. Sci. Éc. Norm. Supér. (4) **41** (2008), no. 1, 1–56. MR2423309

[LN] Erwan Lanneau and Duc-Manh Nguyen, *Teichmüller curves generated by Weierstrass Prym eigenforms in genus 3 and genus 4*, J. Topol. **7** (2014), no. 2, 475–522. MR3217628

[LNW] E. Lanneau, D. Nguyen, and A. Wright, Finiteness of Teichmüller curves in non-arithmetic rank 1 orbit closures, *Amer. J. Math.*, to appear.

[LM] Eduard Looijenga and Gabriele Mondello, *The fine structure of the moduli space of abelian differentials in genus 3*, Geom. Dedicata **169** (2014), 109–128. MR3175239

[Ma] Howard Masur, *Interval exchange transformations and measured foliations*, Ann. of Math. (2) **115** (1982), no. 1, 169–200. MR644018

[MZ] Howard Masur and Anton Zorich, *Multiple saddle connections on flat surfaces and the principal boundary of the moduli spaces of quadratic differentials*, Geom. Funct. Anal. **18** (2008), no. 3, 919–987. MR2439000

[MMY] Carlos Matheus, Martin Möller, and Jean-Christophe Yoccoz, *A criterion for the simplicity of the Lyapunov spectrum of square-tiled surfaces*, Invent. Math. **202** (2015), no. 1, 333–425. MR3402801

[MW] Carlos Matheus and Alex Wright, *Hodge-Teichmüller planes and finiteness results for Teichmüller curves*, Duke Math. J. **164** (2015), no. 6, 1041–1077. MR3336840

[Mc1] Curtis T. McMullen, *Teichmüller geodesics of infinite complexity*, Acta Math. **191** (2003), no. 2, 191–223. MR2051398

[Mc2] Curtis T. McMullen, *Billiards and Teichmüller curves on Hilbert modular surfaces*, J. Amer. Math. Soc. **16** (2003), no. 4, 857–885. MR1992827

[Mc3] Curtis T. McMullen, *Teichmüller curves in genus two: discriminant and spin*, Math. Ann. **333** (2005), no. 1, 87–130. MR2169830

[Mc4] Curtis T. McMullen, *Teichmüller curves in genus two: torsion divisors and ratios of sines*, Invent. Math. **165** (2006), no. 3, 651–672. MR2242630

[Mc5] Curtis T. McMullen, *Prym varieties and Teichmüller curves*, Duke Math. J. **133** (2006), no. 3, 569–590. MR2228463

[Mc6] Curtis T. McMullen, *Dynamics of* $\mathrm{SL}_2(\mathbb{R})$ *over moduli space in genus two*, Ann. of Math. (2) **165** (2007), no. 2, 397–456. MR2299738

[Mc7] Curtis T. McMullen, *Rigidity of Teichmüller curves*, Math. Res. Lett. **16** (2009), no. 4, 647–649. MR2525030

[MMW] C. McMullen, R. Mukamel, and A. Wright, Cubic curves and totally geodesic subvarieties of moduli space, preprint, 2016.

[MR] M. Mirzakhani and A. Wright, The boundary of an affine invariant submanifold, arXiv:1508.01446.

[Mö1] Martin Möller, *Variations of Hodge structures of a Teichmüller curve*, J. Amer. Math. Soc. **19** (2006), no. 2, 327–344. MR2188128

[Mö2] Martin Möller, *Periodic points on Veech surfaces and the Mordell-Weil group over a Teichmüller curve*, Invent. Math. **165** (2006), no. 3, 633–649. MR2242629

[Mö3] Martin Möller, *Finiteness results for Teichmüller curves* (English, with English and French summaries), Ann. Inst. Fourier (Grenoble) **58** (2008), no. 1, 63–83. MR2401216

[Mö4] Martin Möller, *Shimura and Teichmüller curves*, J. Mod. Dyn. **5** (2011), no. 1, 1–32. MR2787595

[Mö5] Martin Möller, *Teichmüller curves, mainly from the viewpoint of algebraic geometry*, Moduli spaces of Riemann surfaces, IAS/Park City Math. Ser., vol. 20, Amer. Math. Soc., Providence, RI, 2013, pp. 267–318. MR3114688

[Muk] Ronen E. Mukamel, *Orbifold points on Teichmüller curves and Jacobians with complex multiplication*, Geom. Topol. **18** (2014), no. 2, 779–829. MR3180485

[Mul] S. Mullane, Divisorial strata of abelian differentials, *Int. Math. Res. Not.*, to appear.

[Mum] David Mumford, *Theta characteristics of an algebraic curve*, Ann. Sci. École Norm. Sup. (4) **4** (1971), 181–192. MR0292836

[NW] Duc-Manh Nguyen and Alex Wright, *Non-Veech surfaces in* $\mathcal{H}^{\mathrm{hyp}}(4)$ *are generic*, Geom. Funct. Anal. **24** (2014), no. 4, 1316–1335. MR3248487

[SW] John Smillie and Barak Weiss, *Minimal sets for flows on moduli space*, Israel J. Math. **142** (2004), 249–260. MR2085718

[V1] William A. Veech, *Gauss measures for transformations on the space of interval exchange maps*, Ann. of Math. (2) **115** (1982), no. 1, 201–242. MR644019

[V2] W. A. Veech, *Teichmüller curves in moduli space, Eisenstein series and an application to triangular billiards*, Invent. Math. **97** (1989), no. 3, 553–583. MR1005006

[V3] William A. Veech, *Geometric realizations of hyperelliptic curves*, Algorithms, fractals, and dynamics (Okayama/Kyoto, 1992), Plenum, New York, 1995, pp. 217–226. MR1402493

[V4] William A. Veech, *Siegel measures*, Ann. of Math. (2) **148** (1998), no. 3, 895–944. MR1670061

[Wa] Clayton C. Ward, *Calculation of Fuchsian groups associated to billiards in a rational triangle*, Ergodic Theory Dynam. Systems **18** (1998), no. 4, 1019–1042. MR1645350

[We] Christian Weiß, *Twisted Teichmüller curves*, Lecture Notes in Mathematics, vol. 2104, Springer, Cham, 2014. MR3307268

[Wr1] Alex Wright, *Schwarz triangle mappings and Teichmüller curves: the Veech-Ward-Bouw-Möller curves*, Geom. Funct. Anal. **23** (2013), no. 2, 776–809. MR3053761

[Wr2] Alex Wright, *Translation surfaces and their orbit closures: an introduction for a broad audience*, EMS Surv. Math. Sci. **2** (2015), no. 1, 63–108. MR3354955

[Wr3] Alex Wright, *From rational billiards to dynamics on moduli spaces*, Bull. Amer. Math. Soc. (N.S.) **53** (2016), no. 1, 41–56. MR3403080

[Y] F. Yu, Eigenvalues of curvature, Lyapunov exponents and Harder-Narasimhan filtrations, arXiv:1408.1630.

[YZ] Fei Yu and Kang Zuo, *Weierstrass filtration on Teichmüller curves and Lyapunov exponents*, J. Mod. Dyn. **7** (2013), no. 2, 209–237. MR3106711

[Z1] Anton Zorich, *Flat surfaces*, Frontiers in number theory, physics, and geometry. I, Springer, Berlin, 2006, pp. 437–583. MR2261104

[Z2] A. Zorich, The magic wand theorem of A. Eskin and M. Mirzakhani, *Gaz. Math.* **142** (2014), 39–54.

Department of Mathematics, Boston College, Chestnut Hill, Massachusetts 02467
E-mail address: dawei.chen@bc.edu

Cycles, derived categories, and rationality

Asher Auel and Marcello Bernardara

ABSTRACT. Our main goal is to give a sense of recent developments in the (stable) rationality problem from the point of view of unramified cohomology and 0-cycles as well as derived categories and semiorthogonal decompositions, and how these perspectives intertwine and reflect each other. In particular, in the case of algebraic surfaces, we explain the relationship between Bloch's conjecture, Chow-theoretic decompositions of the diagonal, categorical representability, and the existence of phantom subcategories of the derived category.

CONTENTS

In this text, we explore two potential measures of rationality. The first is the universal triviality of the Chow group of 0-cycles, which is related to the Chow-theoretic decomposition of the diagonal of Bloch and Srinivas. A powerful degeneration method for obstructing the universal triviality of the Chow group of 0-cycles, initiated by Voisin, and developed by Colliot-Thélène and Pirutka, combines techniques from singularity theory and unramified cohomology and has led to a recent breakthrough in the stable rationality problem.

The second is categorical representability, which is defined by the existence of semiorthogonal decompositions of the derived category into components whose dimensions can be bounded. We will give a precise definition of this notion and present many examples, as well as motivate why one should expect categorical representability in codimension 2 for rational varieties.

2010 *Mathematics Subject Classification.* Primary 14C25, 14C30, 14C35, 14E08, 14F05, 14F22, 14J10, 14J28, 14J29, 14J30, 14J35.

Furthermore, we would like to explore how the Chow-theoretic and derived categorical measures of rationality can contrast and reflect each other. One of the motivating topics in this circle of ideas is the relationship, for complex surfaces, between Bloch's conjecture, the universal triviality of the Chow group of 0-cycles, and the existence of phantoms in the derived categories. Another motivating topic is the rationality problem for cubic fourfolds and its connections between derived categories, Hodge theory, as well as Voisin's recent results on the universal triviality of the Chow group of 0-cycles for certain loci of special cubic fourfolds.

Acknowledgements. This text started as notes for lectures given at the Spring 2016 FRG mini-workshop *Unramified cohomology, derived categories, and rationality* at the University of Utah in Salt Lake City, February 26–28, 2016. We are very grateful to the organizers for the invitation to lecture there and for their wonderful hospitality. The original notes were considerably improved on the basis of preliminary talks (which represent the first half of the document) given by M. Mustaţa, I. Shipman, T. de Fernex, and Ch. Hacon, respectively. Discussing with them and other participants during the workshop made for a very fruitful experience. We thank J.-L. Colliot-Thélène and the anonymous referees for numerous helpful comments on the text. The first author was partially funded by NSA Young Investigator Grant H98230-16-1-0321; he would also like to thank the organizers of the conference "New Methods in Birational Geometry" held at the Institut de Mathématiques de Toulouse, June 27–July 1, 2016 for providing excellent working conditions and partial support from ANR-11-LABX-0040-CIMI within the program ANR-11-IDEX-0002-02.

1. Preliminaries on Chow groups

Let k be an arbitrary field. By scheme, we will mean a separated k-scheme of finite type. In this section, we give a quick introduction to the Chow group of algebraic cycles on a scheme up to rational equivalence. See Fulton's book [**84**] for more details.

Denote by $Z_i(X)$ the free $\mathbb{Z}$-module generated by all i-dimensional closed integral subschemes of X. The elements of Z_i are called *algebraic i-cycles*. We will also employ the codimension notation $Z^i(X) = Z_{n-i}(X)$ when X is smooth of pure dimension n. The *support* of an i-cycle $\sum_n a_n[V_n]$ is the union of the closed subschemes V_i in X; it is *effective* if $a_n > 0$ for all n.

Given an $(i+1)$-dimensional closed integral subscheme W of X, and a closed integral subscheme $V \subset W$ of codimension 1, we denote by $\mathscr{O}_{W,V}$ the local ring of W at the generic point of V; it is a local domain of dimension 1 whose field of fractions is the function field $k(W)$. For a nonzero function $f \in \mathscr{O}_{W,V}$, we define the *order* of vanishing $\mathrm{ord}_V(f)$ of f along V to be the length of the $\mathscr{O}_{W,V}$-module $\mathscr{O}_{W,V}/(f)$. The order extends uniquely to a homomorphism $\mathrm{ord}_V : k(W)^\times \to \mathbb{Z}$. If W is normal, when ord_V coincides with the usual discrete valuation on $k(W)$ associated to V. We also define the *divisor* of a rational function $f \in k(W)^\times$ as an i-cycle on X given by

$$[\mathrm{div}(f)] = \sum_{V \subset W} \mathrm{ord}_V(f)[V],$$

where the sum is taken over all closed integral subschemes $V \subset W$ of codimension 1. An i-cycle z on X is *rationally equivalent* to 0 if there exists a finite number of closed integral $(i+1)$-dimensional subschemes $W_j \subset X$ and rational functions

$f_j \in K(W_j)^\times$ such that $z = \sum_j [\mathrm{div}(f_j)]$ in $Z_i(X)$. Note that the set of i-cycles rationally equivalent to 0 forms a subgroup of $Z_i(X)$ since $[\mathrm{div}(f^{-1})] = -[\mathrm{div}(f)]$ for any rational function $f \in K(W)^\times$. Denote the associated equivalence relation on $Z_i(X)$ by $\sim_{\mathrm{rat}}$. Define the *Chow group* of i-cycles on X to be the quotient group $\mathrm{CH}_i(X) = Z_i(X)/\sim_{\mathrm{rat}}$ of algebraic i-cycles modulo rational equivalence.

1.1. Morphisms. Let $f : X \to Y$ be a proper morphism of schemes. Define a *push-forward* map on cycles $f_* : Z_i(X) \to Z_i(Y)$ additively as follows. For a closed integral subscheme $V \subset X$ define

$$f_*([V]) = \begin{cases} 0 & \text{if } \dim(f(V)) < \dim(V) \\ \deg(V/f(V))\,[f(V)] & \text{if } \dim(f(V)) = \dim(V) \end{cases}$$

where $\deg(V/f(V))$ denotes the degree of the finite extension of function fields $k(V)/k(f(V))$ determined by f. This map respects rational equivalence and hence induces a push-forward map on Chow groups $f_* : \mathrm{CH}_i(X) \to \mathrm{CH}_i(Y)$.

Let $f : X \to Y$ be a flat morphism of relative dimension r. Define a *pull-back* map $f^* : Z_i(Y) \to Z_{i+r}(X)$ additively as follows. For a closed integral subscheme $V \subset Y$ define

$$f^*([V]) = [f^{-1}(V)].$$

This map respects rational equivalence and hence induces a pull-back map on Chow groups $f^* : \mathrm{CH}_i(Y) \to \mathrm{CH}_{i+r}(X)$.

A special case of proper push forward is given by considering a closed immersion $\iota : Z \to X$. Letting $j : U \to X$ be the open complement of Z, we note that j is flat. There is an exact excision sequence

$$\mathrm{CH}_i(Z) \xrightarrow{\iota_*} \mathrm{CH}_i(X) \xrightarrow{j^*} \mathrm{CH}_i(U) \longrightarrow 0$$

which comes from an analogous exact sequence on the level of cycles.

Moreover, we have the following compatibility between the proper push-forward and flat pull-back. Given a cartesian diagram

$$\begin{array}{ccc} X' & \xrightarrow{g'} & X \\ {\scriptstyle f'}\downarrow & & \downarrow{\scriptstyle f} \\ Y' & \xrightarrow{g} & Y, \end{array}$$

where g is flat of relative dimension r and f is proper, then $g^* f_* = f'_* g'^*$ as maps $\mathrm{CH}_i(X) \to \mathrm{CH}_{i+r}(Y')$.

A third natural map between Chow groups is the *Gysin map*. Given a regular closed embedding $\iota : X \to Y$ of codimension r, one can define a natural map $\iota^! : \mathrm{CH}_i(Y) \to \mathrm{CH}_{i-r}(X)$. The precise definition of the Gysin map is quite involved, see [**84**, §5.2; 6.2], and in particular, is not induced from a map on cycles. In particular, this map factors through $\mathrm{CH}_i(N)$, where $s : N \to Y$ is the normal cone of X in Y, and can be described by the composition of a Gysin map $s^!$ for vector bundles and the inverse of the pull back f^* by $f : N \to X$.

The Gysin map allows one to write the *excess intersection formula* for a regular closed embedding $\iota : X \to Y$ of codimension r

$$\iota^! \iota_*(\alpha) = c_r(h^* N_{X/Y}) \cap \alpha,$$

for any cycle α in $Z_*(Y)$, where $h : N_{X/Y} \to X$ is the normal bundle of X in Y and c_r denotes the r-th Chern class. In particular, this shows that $\iota^! \iota_* = 0$ whenever $N_{X/Y}$ is trivial.

We now define the Gysin map for any local complete intersection (lci) morphism $f : X \to Y$. Consider the factorization of f as

$$X \xrightarrow{\iota} P \xrightarrow{h} Y,$$

where ι is a regular embedding of codimension r and h is smooth of relative dimension m. Then we can define the Gysin map

$$f^! = \iota^! \circ h^* : \mathrm{CH}_i(Y) \xrightarrow{h^*} \mathrm{CH}_{i-m}(P) \xrightarrow{\iota^!} \mathrm{CH}_{i-m-r}(X).$$

Such a map is independent on the chosen factorization of f and coincides with the flat pull-back f^* whenever f is flat. As a relevant example, any morphism $f : X \to Y$ between smooth k-schemes is lci (indeed, f factors $X \to X \times Y \to Y$ into the regular graph embedding followed by the smooth projection morphism), hence induces a Gysin map $f^! : \mathrm{CH}^i(Y) \to \mathrm{CH}^i(X)$. Finally, even when f is not flat, we often denote $f^* = f^!$, so that f^* is defined for any lci morphism.

1.2. Intersections. Let X be a smooth k-scheme of pure dimension n. Then the Chow group admits an *intersection product* as follows. For closed integral subschemes $V \subset X$ and $W \subset X$ of codimension i and j, respectively, define

$$[V].[W] = \Delta^![V \times W] \in \mathrm{CH}^{i+j}(X)$$

where $\Delta : X \to X \times X$ is the diagonal morphism, a regular embedding of codimension n. This induces a bilinear map $\mathrm{CH}^i(X) \times \mathrm{CH}^j(X) \to \mathrm{CH}^{i+j}(X)$, which makes $\mathrm{CH}(X) = \oplus_{i \geq 0} \mathrm{CH}^i(X)$ into a commutative graded ring with identity the class $[X] \in \mathrm{CH}^0(X)$. Gysin maps between smooth k-schemes are then ring homomorphisms for the intersection product.

One can understand the intersection product in terms of literal intersections of subschemes, via moving lemmas, see [**84**, §11.4] for a discussion of the technicalities involved. Assuming that X is smooth and quasi-projective, given closed integral subschemes $V \subset X$ and $W \subset X$ of codimension i and j, respectively, the moving lemma says that one can replace V by a rationally equivalent cycle $V' = \sum_l a_l [V_l]$ so that V' and W meet properly, i.e., $V_l \cap W$ has codimension $i + j$ for all l. Then one can define $[V].[W] = \sum_l a_l [V_l \cap W]$. A more refined moving lemma is then required to show that the rational equivalence class of this product is independent of the choice of cycle V'.

One easy moving lemma that we will need to use often is the moving lemma for 0-cycles: given a smooth quasi-projective k-scheme X, an open dense subscheme $U \subset X$, and a 0-cycle z on X, there exists a 0-cycle z' on X rationally equivalent to z such that the support of z' is contained in U. See, e.g., [**85**, §2.3], [**62**, p. 599] for a reference to the classical moving lemma, which implies this.

1.3. Correspondences. Let X and Y be smooth k-schemes of pure dimension n and m respectively. We recall some notions from [**84**, §16.1].

DEFINITION 1.3.1. A *correspondence* from X to Y is an element $\alpha \in \mathrm{CH}(X \times Y)$. The same element α, seen in $\mathrm{CH}(Y \times X)$ is called the transpose correspondence α' from Y to X.

Now assume that Y is proper over k and let Z be a smooth equidimensional k-scheme. If $\alpha \in \mathrm{CH}(X \times Y)$ and $\beta \in \mathrm{CH}(Y \times Z)$ are correspondences, we define the composed correspondence

$$\beta \circ \alpha = p_{X\times Z *}(p^*_{X\times Y}(\alpha).p^*_{Y\times Z}(\beta)) \in \mathrm{CH}(X \times Z),$$

where $p_\bullet$ denotes the projection from $X \times Y \times Z$ to $\bullet$, and where we use the intersection product on $\mathrm{CH}(X\times Y\times Z)$. Taking $X = Y = Z$ smooth and proper, the operation of composition of correspondences makes $\mathrm{CH}(X \times X)$ into an associative ring with unit $[\Delta_X]$.

Correspondences between X and Y naturally give rise to morphisms between their Chow groups as follows. If $\alpha \in \mathrm{CH}^{m+i}(X \times Y)$ is a correspondence from X to Y, then we define

$$\begin{aligned} \alpha_* : \mathrm{CH}_j(X) &\longrightarrow \mathrm{CH}_{j-i}(Y) \\ z &\longmapsto q_*(p^*(z).\alpha) \end{aligned} \qquad\qquad \begin{aligned} \alpha^* : \mathrm{CH}^j(Y) &\longrightarrow \mathrm{CH}^{j+i}(X) \\ z &\longmapsto p_*(q^*(z).\alpha) \end{aligned}$$

where p and q denote the projections from $X\times Y$ to X and Y, respectively. If β is a correspondence from Y to Z, then we have $(\beta\circ\alpha)_* = \beta_* \circ \alpha_*$ and $(\beta\circ\alpha)^* = \alpha^* \circ \beta^*$. An important special case are correspondences $\alpha \in \mathrm{CH}^n(X \times X)$, which define maps $\alpha_* : \mathrm{CH}_i(X) \to \mathrm{CH}_i(X)$ and $\alpha^* : \mathrm{CH}^i(X) \to \mathrm{CH}^i(X)$. In particular, the map

$$\mathrm{CH}^n(X \times X) \to \mathrm{End}_{\mathbb{Z}}(\mathrm{CH}_i(X))$$

is a ring homomorphism (see [**84**, Cor. 16.1.2]).

As an example, letting $f : X \to Y$ be a morphism with graph $\Gamma_f \subset X \times Y$, we can consider $\alpha = [\Gamma_f] \in \mathrm{CH}^m(X \times Y)$ as a correspondence from X to Y, and then $\alpha_* = f_*$ and $\alpha^* = f^*$.

1.4. Specialization. Most of the previous intersection theoretic considerations carry over to a more general relative setting, replacing the base field k with a regular base scheme S.

Let X be a scheme that is separated and finite type over S. For a closed integral subscheme $V \subset X$, we define the *relative dimension*

$$\dim_S(V) = \mathrm{tr.deg}(K(V)/K(W)) - \mathrm{codim}_S(W),$$

where W is the closure of the image of V in S. A relative i-cycle on X/S is an integer linear combination of integral subschemes of X of relative dimension i. The notion of rational equivalence of relative i-cycles is as before and we denote by $\mathrm{CH}_i(X/S)$ the group of relative i-cycles on X/S up to rational equivalence. As before, there are push-forwards for proper S-morphisms, pull-backs for flat S-morphisms, and Gysin maps for lci S-morphisms.

Now suppose that $\iota : \overline{S} \to S$ is a regular embedding of codimension r and let $j : S^0 \to S$ be the complement. Consider the following diagram of cartesian squares:

$$\begin{array}{ccccc} \overline{X} & \longrightarrow & X & \longleftarrow & X^0 \\ \downarrow & & \downarrow & & \downarrow \\ \overline{S} & \xrightarrow{\ \iota\ } & S & \xleftarrow{\ j\ } & S^0 \end{array}$$

Noting that $\mathrm{CH}_i(\overline{X}/\overline{S}) = \mathrm{CH}_{i-r}(\overline{X}/S)$ and $\mathrm{CH}_i(X^0/S^0) = \mathrm{CH}_i(X^0/S)$, then the Gysin map $\iota^! : \mathrm{CH}_i(X/S) \to \mathrm{CH}_{i-r}(\overline{X}/S) = \mathrm{CH}_i(\overline{X}/\overline{S})$ gives rise to a diagram

$$\begin{array}{ccccccc} \mathrm{CH}_i(\overline{X}/S) & \xrightarrow{\iota_*} & \mathrm{CH}_i(X/S) & \xrightarrow{j^*} & \mathrm{CH}_i(X^0/S) & \longrightarrow & 0 \\ \| & & \downarrow{\scriptstyle \iota^!} & & \| & & \\ \mathrm{CH}_{i+r}(\overline{X}/\overline{S}) & & \mathrm{CH}_i(\overline{X}/\overline{S}) & \xleftarrow{\sigma} & \mathrm{CH}_i(X^0/S^0) & & \end{array}$$

where the top row is the relative short exact excision sequence. We see that the obstruction to defining a well-defined map *specialization map*

$$\sigma : \mathrm{CH}_i(X^0/S^0) \to \mathrm{CH}_i(\overline{X}/\overline{S})$$

fitting into the diagram is precisely the image of $\iota^!\iota_*$. By the excess intersection formula, if $N_{\overline{S}/S}$ is trivial, then $\iota^!\iota_* = 0$, in which case we arrive at a well-defined specialization map. When it exists, the specialization map is compatible with push-forwards and pull-backs.

An important special case is when $S = \mathrm{Spec}(R)$ for a discrete valuation ring R, so that $S^0 = \mathrm{Spec}(K)$ and $\overline{S} = \mathrm{Spec}(k)$, where k and K denote the residue and the fraction field of R, respectively. Given a separated R-scheme X of finite type, the k-scheme $\overline{X} = X_k$ is the special fiber, the K-scheme $X^0 = X_K$ is the generic fiber, and we arrive at specialization maps $\sigma : \mathrm{CH}_i(X_K) \to \mathrm{CH}_i(X_k)$. For more details, we refer to [**84**, §20.3].

2. Preliminaries on semiorthogonal decompositions

Let k be an arbitrary field. We present here the basic notions of semiorthogonal decompositions and exceptional objects for k-linear triangulated categories, bearing in mind our main application, the derived category of a projective k-variety. We refer to [**104**, Ch. 1, 2, 3] for an introduction to derived categories aimed at algebraic geometers. In particular, we will assume the reader to be familiar with the notions of triangulated and derived categories, and basic homological algebra as well as complexes of coherent sheaves on schemes.

However, a disclaimer here is necessary. The appropriate structure to consider to study derived categories of smooth projective varieties is the structure of k-linear differential graded (dg) category; that is, a category enriched over dg complexes of k-vector spaces (see [**111**] for definitions and main properties). In this perspective, morphisms between two objects in the triangulated structure can be seen as the zeroth cohomology of the complex of morphisms between the same objects in the dg structure. Considering the dg structure is natural under many point of views: above all, all categories we will consider can be endowed with a canonical dg structure (see [**134**]), in such a way that dg functors will correspond to Fourier–Mukai functors (see [**165**]). Moreover, the dg structure allows to define noncommutative motives, which give a motivic framework to semiorthogonal decompositions. Even if related to some of our considerations, we will not treat noncommutative motives in this report. The interested reader can consult [**163**].

2.1. Semiorthogonal decompositions and their mutations.

Let T be a k-linear triangulated category. A full triangulated subcategory A of T is called *admissible* if the embedding functor admits a left and a right adjoint.

DEFINITION 2.1.1 ([**45**]). A *semiorthogonal decomposition* of T is a sequence of admissible subcategories $\mathsf{A}_1, \ldots, \mathsf{A}_n$ of T such that

- $\mathrm{Hom}_{\mathsf{T}}(A_i, A_j) = 0$ for all $i > j$ and any A_i in A_i and A_j in A_j;
- for every object T of T, there is a chain of morphisms
$$0 = T_n \to T_{n-1} \to \ldots \to T_1 \to T_0 = T$$
such that the cone of $T_k \to T_{k-1}$ is an object of A_k for all $k = 1, \ldots, n$.

Such a decomposition will be written
$$\mathsf{T} = \langle \mathsf{A}_1, \ldots, \mathsf{A}_n \rangle.$$

If $\mathsf{A} \subset \mathsf{T}$ is admissible, we have two semiorthogonal decompositions
$$\mathsf{T} = \langle \mathsf{A}^\perp, \mathsf{A} \rangle = \langle \mathsf{A}, {}^\perp \mathsf{A} \rangle,$$
where $\mathsf{A}^\perp$ and ${}^\perp\mathsf{A}$ are, respectively, the left and right orthogonal of A in T (see, for example, [**45**, §3]).

Given a semiorthogonal decomposition $\mathsf{T} = \langle \mathsf{A}, \mathsf{B} \rangle$, Bondal [**44**, §3] defines left and right mutations $L_{\mathsf{A}}(\mathsf{B})$ and $R_{\mathsf{B}}(\mathsf{A})$ of this pair. In particular, there are equivalences $L_{\mathsf{A}}(\mathsf{B}) \simeq \mathsf{B}$ and $R_{\mathsf{B}}(\mathsf{A}) \simeq \mathsf{A}$, and semiorthogonal decompositions
$$\mathsf{T} = \langle L_{\mathsf{A}}(\mathsf{B}), \mathsf{A} \rangle, \qquad \mathsf{T} = \langle \mathsf{B}, R_{\mathsf{B}}(\mathsf{A}) \rangle.$$
We refrain from giving an explicit definition for the mutation functors in general, which can be found in [**44**, §3]. In §2.2 we will give an explicit formula in the case where A and B are generated by exceptional objects.

2.2. Exceptional objects. Very special examples of admissible subcategories, semiorthogonal decompositions, and their mutations are provided by the theory of exceptional objects and collections. The theory of exceptional objects and semiorthogonal decompositions in the case where k is algebraically closed and of characteristic zero was studied in the Rudakov seminar at the end of the 80s, and developed by Rudakov, Gorodentsev, Bondal, Kapranov, Kuleshov, and Orlov among others, see [**91**], [**44**], [**45**], [**48**], and [**154**]. As noted in [**12**], most fundamental properties persist over any base field k.

Let T be a k-linear triangulated category. The triangulated category $\langle \{E_i\}_{i\in I} \rangle$ *generated* by a class of objects $\{E_i\}_{i\in I}$ of T is the smallest thick (that is, closed under direct summands) full triangulated subcategory of T containing the class. We will write $\mathrm{Ext}^r_{\mathsf{T}}(E, F) = \mathrm{Hom}_{\mathsf{T}}(E, F[r])$.

DEFINITION 2.2.1. Let A be a division (not necessarily central) k-algebra (e.g., A could be a field extension of k). An object E of T is called *A-exceptional* if
$$\mathrm{Hom}_{\mathsf{T}}(E, E) = A \quad \text{and} \quad \mathrm{Ext}^r_{\mathsf{T}}(E, E) = 0 \quad \text{for} \quad r \neq 0.$$

An exceptional object in the classical sense [**90**, Def. 3.2] of the term is a k-exceptional object. By *exceptional* object, we mean A-exceptional for some division k-algebra A.

A totally ordered set $\{E_1, \ldots, E_n\}$ of exceptional objects is called an *exceptional collection* if $\mathrm{Ext}^r_{\mathsf{T}}(E_j, E_i) = 0$ for all integers r whenever $j > i$. An exceptional collection is *full* if it generates T, equivalently, if for an object W of T, the vanishing $\mathrm{Ext}^r_{\mathsf{T}}(E_i, W) = 0$ for all $i = 1, \ldots, n$ and all integers r implies $W = 0$. An exceptional collection is *strong* if $\mathrm{Ext}^r_{\mathsf{T}}(E_i, E_j) = 0$ whenever $r \neq 0$.

Exceptional collections provide examples of semiorthogonal decompositions if T is the bounded derived category of a smooth projective scheme.

PROPOSITION 2.2.2 ([**44**, Thm. 3.2]). *Let $\{E_1, \ldots, E_n\}$ be an exceptional collection on the bounded derived category $\mathsf{D}^{\mathrm{b}}(X)$ of a smooth projective k-scheme X. Then there is a semiorthogonal decomposition*

$$\mathsf{D}^{\mathrm{b}}(X) = \langle \mathsf{A}, E_1, \ldots, E_n \rangle,$$

where $\mathsf{A} = \langle E_1, \ldots, E_n \rangle^{\perp}$ is the full subcategory of those objects W such that $\mathrm{Ext}^r_{\mathsf{T}}(E_i, W) = 0$ for all $i = 1, \ldots, n$ and all integers r. In particular, the sequence if full if and only if $\mathsf{A} = 0$.

Given an exceptional pair $\{E_1, E_2\}$ with E_i being A_i-exceptional, consider the admissible subcategories $\langle E_i \rangle$, forming a semiorthogonal pair. We can hence perform right and left mutations, which provide equivalent admissible subcategories.

Recall that mutations provide equivalent admissible subcategories and flip the semiorthogonality condition. It easily follows from the definition that the object $R_{E_2}(E_1)$ is A_1-exceptional, the object $L_{E_1}(E_2)$ is A_2-exceptional, and the pairs $\{L_{E_1}(E_2), E_1\}$ and $\{E_2, R_{E_2}(E_1)\}$ are exceptional. We call $R_{E_2}(E_1)$ the *right mutation* of E_1 through E_2 and $L_{E_1}(E_2)$ the *left mutation* of E_2 through E_1.

In the case of k-exceptional objects, mutations can be explicitly computed.

DEFINITION 2.2.3 ([**90**, §3.4]). Given a k-exceptional pair $\{E_1, E_2\}$ in T, the *left mutation* of E_2 with respect to E_1 is the object $L_{E_1}(E_2)$ defined by the distinguished triangle:

$$\mathrm{Hom}_{\mathsf{T}}(E_1, E_2) \otimes E_1 \xrightarrow{ev} E_2 \longrightarrow L_{E_1}(E_2), \tag{2.1}$$

where ev is the canonical evaluation morphism. The *right mutation* of E_1 with respect to E_2 is the object $R_{E_2}(E_1)$ defined by the distinguished triangle:

$$R_{E_2}(E_1) \longrightarrow E_1 \xrightarrow{coev} \mathrm{Hom}_{\mathsf{T}}(E_1, E_2) \otimes E_2,$$

where $coev$ is the canonical coevaluation morphism.

Given an exceptional collection $\{E_1, \ldots, E_n\}$, one can consider any exceptional pair $\{E_i, E_{i+1}\}$ and perform either right or left mutation to get a new exceptional collection.

Exceptional collections provide an algebraic description of admissible subcategories of T. Indeed, if E is an A-exceptional object in T, the triangulated subcategory $\langle E \rangle \subset \mathsf{T}$ is equivalent to $\mathsf{D}^{\mathrm{b}}(k, A)$. The equivalence $\mathsf{D}^{\mathrm{b}}(k, A) \to \langle E \rangle$ is obtained by sending the complex A concentrated in degree 0 to E. The right adjoint functor is the morphism functor $\mathbf{R}\mathrm{Hom}(-, E)$.

We conclude this section by considering a weaker notion of exceptionality, which depends only on the numerical class and on the bilinear form χ.

DEFINITION 2.2.4. Let X be a smooth projective variety. A *numerically exceptional collection* is a collection $E_1, \ldots, E_n$ of exceptional objects in the derived category $\mathsf{D}^{\mathrm{b}}(X)$ such that $\chi(E_i, E_j) = 0$ for $i > j$ and $\chi(E_i, E_i) = 1$ for all $i = 1, \ldots, n$.

REMARK 2.2.5. It is clear that any exceptional collection is a numerically exceptional collection, while the converse need not to be true.

2.3. How to construct semiorthogonal decompositions? Examples and subtleties. It is quite difficult to describe all semiorthogonal decompositions of a variety X. Moreover, the geometry of X plays a very important rôle in understanding whether the category $\mathsf{D}^{\mathrm{b}}(X)$ has semiorthogonal sequences of admissible subcategories. In general, the most difficult task is to show that they form a generating system for the whole category.

The main motivation for the study of birational geometry via semiorthogonal decompositions is the following famous theorem by Orlov [**146**].

THEOREM 2.3.1 (Orlov). *Let X be a smooth projective variety, $Z \subset X$ a smooth subvariety of codimension $c \geq 2$, and $\sigma : Y \to X$ the blow-up of Z. Then the functor $L\sigma^* : \mathsf{D}^{\mathrm{b}}(X) \to \mathsf{D}^{\mathrm{b}}(Y)$ is fully faithful, and, for $i = 1, \ldots, c-1$ there are fully faithful functors $\Phi_i : \mathsf{D}^{\mathrm{b}}(Z) \to \mathsf{D}^{\mathrm{b}}(Y)$, and a semiorthogonal decomposition*

$$\mathsf{D}^{\mathrm{b}}(Y) = \langle L\sigma^*\mathsf{D}^{\mathrm{b}}(X), \Phi_1\mathsf{D}^{\mathrm{b}}(Z), \ldots, \Phi_{c-1}\mathsf{D}^{\mathrm{b}}(Z)\rangle$$

Notice that Orlov's argument of the fully faithfulness of $L\sigma^*$ extends to the cases of a surjective morphism with rationally connected fibers between smooth and projective varieties, though the description of the orthogonal complement is in general unknown. The fact that $L\sigma^*$ is fully faithful in Theorem 2.3.1 can be seen as a special case of the following Lemma, since a blow up gives a surjective map with the required properties.

LEMMA 2.3.2. *Let X and Y be smooth and projective k-schemes and $\sigma : Y \to X$ a surjective morphism such that $\sigma_*\mathscr{O}_Y = \mathscr{O}_X$ and $R^i\sigma_*\mathscr{O}_Y = 0$ for $i \neq 0$. Then $L\sigma^* : \mathsf{D}^{\mathrm{b}}(X) \to \mathsf{D}^{\mathrm{b}}(Y)$ is fully faithful.*

PROOF. For any A and B objects in $\mathsf{D}^{\mathrm{b}}(X)$, we have

$$\begin{aligned}\mathrm{Hom}_Y(L\sigma^*A, L\sigma^*B) &= \mathrm{Hom}_X(A, R\sigma_*L\sigma^*B)\\ &= \mathrm{Hom}_X(A, B \otimes R\sigma_*\mathscr{O}_Y) = \mathrm{Hom}_X(A, B)\end{aligned}$$

by adjunction, projection formula and by our assumption respectively. □

The canonical bundle and its associated invariants, like the geometric genus and the irregularity, play a central rôle in this theory. First of all it is easy to remark, using Serre duality, that if X has a trivial canonical bundle, then there is no non-trivial semiorthogonal decomposition of $\mathsf{D}^{\mathrm{b}}(X)$. The results obtained by Okawa [**144**] and Kawatani–Okawa [**110**] for low dimensional varieties are also strongly related to the canonical bundle.

THEOREM 2.3.3 (Okawa). *Let C be a smooth projective k-curve of positive genus. Then $\mathsf{D}^{\mathrm{b}}(C)$ has no non-trivial semiorthogonal decompositions.*

THEOREM 2.3.4 (Kawatani–Okawa). *Let k be algebraically closed and S a smooth connected projective minimal surface. Suppose that*

- *either $\kappa(S) = 0$ and S is not a classical Enriques surface, or*
- *$\kappa(S) = 1$ and $p_g(S) > 0$, or*
- *$\kappa(S) = 2$, that $\dim H^1(S, \omega_S) > 1$, and for any one-dimensional connected component of the base locus of ω_S, its intersection matrix is negative definite.*

Then there is no nontrivial semiorthogonal decomposition of $\mathsf{D}^{\mathrm{b}}(S)$.

Roughly speaking, one could say that varieties admitting semiorthogonal decompositions should have cohomological properties which are very close to the ones of a Fano (relatively over some base) variety.

If X is a Fano variety, that is if the canonical bundle ω_X is antiample, any line bundle L on X is a k-exceptional object in $\mathsf{D}^{\mathrm{b}}(X)$, and hence gives a semiorthogonal decomposition $\mathsf{D}^{\mathrm{b}}(X) = \langle \mathsf{A}, L\rangle$, where A consists of objects right orthogonal to L. In the simpler case where X has index Picard rank 1 and index i (that is, $\omega_X = \mathscr{O}(-i)$), and $\mathrm{char}(k) = 0$, one can use Kodaira vanishing theorems to construct a natural k-exceptional sequence, as remarked by Kuznetsov [**122**, Corollary 3.5].

PROPOSITION 2.3.5 (Kuznetsov). *Let X be a smooth Fano variety of Picard rank 1 with ample generator $\mathscr{O}(1)$, and index i. Then there is a semiorthogonal decomposition*

$$\mathsf{D}^{\mathrm{b}}(X) = \langle \mathsf{A}_X, \mathscr{O}_X, \dots \mathscr{O}_X(i-1)\rangle,$$

where $\mathsf{A}_X = \langle \mathscr{O}_X, \dots \mathscr{O}_X(i-1)\rangle^{\perp}$ is the category of objects W satisfying the condition that $\mathrm{Ext}^r(\mathscr{O}(j), W) = 0$ for all $0 \leq j < i$ and for all integers r.

The previous result is easily generalized to the relative case of Mori fiber spaces as in [**12**, Proposition 2.2.2].

PROPOSITION 2.3.6. *Let $\pi : X \to Y$ be a flat surjective fibration between smooth varieties, such that $\mathrm{Pic}(X/Y) \simeq \mathbb{Z}$ with ample generator $\mathscr{O}_{X/Y}(1)$ and such that $\omega_{X/Y} = \mathscr{O}_{X/Y}(-i)$. Set $\mathsf{D}^{\mathrm{b}}(Y)(j) := \pi^*\mathsf{D}^{\mathrm{b}}(Y) \otimes \mathscr{O}_{X/Y}(j)$. For any j, this gives a fully faithful embedding of $\mathsf{D}^{\mathrm{b}}(Y)$ into $\mathsf{D}^{\mathrm{b}}(X)$. Moreover, over a field k of characteristic 0, there is a semiorthogonal decomposition*

$$\mathsf{D}^{\mathrm{b}}(X) = \langle \mathsf{A}_{X/Y}, \mathsf{D}^{\mathrm{b}}(Y)(0), \dots, \mathsf{D}^{\mathrm{b}}(Y)(i-1)\rangle,$$

where

$$\mathsf{A}_{X/Y} = \langle \mathsf{D}^{\mathrm{b}}(Y)(0), \dots, \mathsf{D}^{\mathrm{b}}(Y)(i-1)\rangle^{\perp}$$

is the category of objects W such that $\mathrm{Ext}^r(\pi^*A \otimes \mathscr{O}(j), W) = 0$ for all $0 \leq j < i$, for all integers r, and for all objects A in $\mathsf{D}^{\mathrm{b}}(Y)$.

PROOF. Notice that $\mathsf{D}^{\mathrm{b}}(Y)(j)$ being admissible in $\mathsf{D}^{\mathrm{b}}(X)$ is a consequence of Lemma 2.3.2. The semiorthogonality is given by a relative Kodaira vanishing. Finally, define $\mathsf{A}_{X/Y}$ to be the complement. □

REMARK 2.3.7. The assumption on k having characteristic zero is needed to ensure that the Kodaira vanishing theorem holds on X, but can be weakened. Indeed, Kodaira vanishing theorems hold in characteristic p for varieties that lift to a smooth variety in characteristic 0, see Deligne–Illusie [**78**]. For example, we could consider any complete intersection in projective space of Fano type over a field of characteristic p.

One should consider the decompositions above as the most related to the geometric structure of X, and the category $\mathsf{A}_{X/Y}$ as the best witness of the birational behavior of X.[1] This idea is supported by the following results of Beilinson [**27**] (for the case of $\mathbb{P}^n$) and Orlov [**146**].

PROPOSITION 2.3.8 (Beilinson, Orlov). *Let $\pi : X \to Y$ be a projective bundle of relative dimension r, that is, $X = \mathbb{P}_Y(E)$ for some rank $r+1$ vector bundle E on Y. Then*

$$\mathsf{D}^{\mathrm{b}}(X) = \langle \mathsf{D}^{\mathrm{b}}(Y)(0), \dots, \mathsf{D}^{\mathrm{b}}(Y)(r)\rangle.$$

In other words, $\mathsf{A}_{X/Y} = 0$.

---

[1] Notice, that one can consider the semiorthogonal sequence $\mathsf{D}^{\mathrm{b}}(Y)(j), \dots, \mathsf{D}^{\mathrm{b}}(Y)(j+i-1)$ for any integer j. However, this would give orthogonal complements which are not only equivalent as triangulated categories, but also as dg categories.

The most difficult task in proving the Proposition 2.3.8, already for projective spaces, is to show that a given sequence of categories generates the whole category. This is done in **[27]** using a complex resolving the structure sheaf of the diagonal of $\mathbb{P}^n \times \mathbb{P}^n$. Let us list other known descriptions of $\mathsf{A}_{X/Y}$.

EXAMPLE 2.3.9. Let Y be a smooth projective k-variety and $\pi : X \to Y$ as in Proposition 2.3.6. Then $\mathsf{A}_{X/Y}$ is known in the following cases:

Projective bundles. If $\pi : X \to Y$ is a projective bundle, then $\mathsf{A}_{X/Y} = 0$, **[146]**.

Projective fibrations. Let $\pi : X \to Y$ be a relative Brauer–Severi variety (that is, the geometric fibers of π are projective spaces, but X is not isomorphic to $\mathbb{P}(E)$ for any vector bundle E on Y), and α in $\mathrm{Br}(Y)$ the class of X and r the relative dimension. If $\omega_{X/Y}$ generates $\mathrm{Pic}(X/Y)$, then

$$\mathsf{A}_{X/Y} = \langle \mathsf{D}^{\mathrm{b}}(Y, \alpha), \ldots, \mathsf{D}^{\mathrm{b}}(Y, \alpha^r) \rangle.$$

If $\omega_{X/Y}$ is not primitive, a similar description is possible **[28]**.

Quadric fibrations. Let $\pi : X \to Y$ be a quadric fibration of relative dimension r and let $\mathscr{C}_0$ be the sheaf of even Clifford algebras associated to the quadratic form defining X. Then $\mathsf{A}_{X/Y} = \mathsf{D}^{\mathrm{b}}(Y, \mathscr{C}_0)$, **[116]**.

Fibrations in intersections of quadrics. Let $\{Q_i \to Y\}_{i=0}^s$ be quadric fibrations of relative dimension r and $\pi : X \to Y$ be their intersection (see **[12]** for details), and suppose that $\omega_{X/Y}$ is relatively antiample (that is, $r < 2s$). Then there is a $\mathbb{P}^s$-bundle $Z \to Y$ and a sheaf of Clifford algebras $\mathscr{C}_0$ on Z, and $\mathsf{A}_{X/Y} = \mathsf{D}^{\mathrm{b}}(Z, \mathscr{C}_0)$, **[12]**.

This list is far from being exhaustive, since many specific cases are also known (see, e.g., Table 1 for 3 and 4 dimensional cases).

In the case where k is not algebraically closed, then one can look for semiorthogonal decompositions of $\mathsf{D}^{\mathrm{b}}(X_{\overline{k}})$ and understand whether they can give informations on $\mathsf{A}_{X/Y}$ or, more in general on $\mathsf{D}^{\mathrm{b}}(X)$. This rather challenging problem can be tackled in the simplest case, where $\mathsf{D}^{\mathrm{b}}(X_{\overline{k}})$ is generated by vector bundles, using Galois descent of such vector bundles (see **[11]**). With this in mind one can describe $\mathsf{A}_{X/Y}$ when X is a minimal del Pezzo, Y is a point, and k is any field **[11]**. Other cases of (generalized) Brauer–Severi varieties **[28, 38]** can be treated this way.

On the other hand, even if a geometric description of $\mathsf{A}_{X/Y}$ is not possible, one can calculate its Serre functor.

DEFINITION 2.3.10. Let A be a triangulated k-linear category with finite dimensional morphism spaces. A functor $S : \mathsf{A} \to \mathsf{A}$ is a *Serre functor* if it is a k-linear equivalence inducing a functorial isomorphism

$$\mathrm{Hom}_{\mathsf{A}}(X, Y) \simeq \mathrm{Hom}_{\mathsf{A}}(Y, S(X))^{\vee}$$

of k-vector spaces, for any object X and Y of A.

A category A with a Serre functor S_{A} is a *Calabi–Yau category* (or a *non commutative Calabi–Yau*) of dimension n if $S_{\mathsf{A}} = [n]$. It is a *fractional Calabi–Yau category* of dimension n/c if c is the smallest integer such that the iterate Serre functor is a shift functor and $S_{\mathsf{A}}^c = [n]$. Note that the fractional dimension of A is not a rational number, but a pair of two integer numbers.

The Serre functor generalizes the notion of Serre duality to a more general categorical setting. Indeed, if X is a smooth projective k-variety, then we have that $S_{\mathsf{D}^{\mathrm{b}}(X)}(-) = - \otimes \omega_X[\dim(X)]$ by Serre duality.

As a consequence of the work of Bondal and Kapranov [**45**], if X is a smooth and projective k-scheme and A is an admissible subcategory of $\mathsf{D}^{\mathrm{b}}(X)$, then A has a Serre functor which can be explicitly calculated from the Serre functor of X using adjunctions to the embedding $\mathsf{A} \to \mathsf{D}^{\mathrm{b}}(X)$. Kuznetsov performed explicitly these calculations for Fano hypersurfaces in projective spaces, see [**119**, Cor. 4.3].

PROPOSITION 2.3.11 (Kuznetsov). *Let $X \subset \mathbb{P}^{n+1}$ be a hypersurface of degree $d < n+2$, and set c the greatest common divisor of d and $n+2$. Then A_X is a (fractional) Calabi–Yau category, that is $S_{\mathsf{A}_X}^{d/c} = [\frac{(d-2)(n+2)}{c}]$.*

REMARK 2.3.12. Notice that both d/c and $\frac{(d-2)(n+2)}{c}$ are integers. However, the fractional dimension of A_X is not a simplification of the fraction $\frac{(d-2)(n+2)}{c}$ unless $c = 1$. For example, for a quartic fourfold we obtain $6/2$. However, in the case where d divides $n+2$, A_X is a Calabi–Yau category.

COROLLARY 2.3.13. *If $X \subset \mathbb{P}^5$ is a smooth cubic fourfold, then A_X is a 2-Calabi–Yau category (or a noncommutative K3 surface).*

The categories $\mathsf{A}_{X/Y}$ also admit algebraic descriptions, that is, one can find an equivalence with a triangulated category which arises from purely algebraic constructions. The main examples are Orlov's description via matrix factorizations for Fano complete intersections in projective spaces (see [**148**]) and a rather complicated description based on Homological Projective Duality for fibrations in complete intersections of type $(d, \ldots, d)$ (see [**18**]).

To tackle geometrical problems, we would like a description of $\mathsf{A}_{X/Y}$ by explicit geometric constructions. A first case, which include a lot of Fano varieties, is the case of homogeneous varieties. These are conjectured to always carry a full exceptional sequence, and one can construct a candidate sequence using vanishing theorems and representation of parabolic subgroups, see [**128**]. The hardest part is to prove that such a sequence is full, for which spectral sequences are needed.

The most powerful tool to construct semiorthogonal decompositions is by far Kuznetsov's Homological Projective Duality (HPD). We refrain here to give any definition, for which we refer to the very dense Kuznetsov's original paper [**121**]. In practice, HPD allows to compare semiorthogonal decompositions of dual linear sections of fixed projectively dual varieties. It is in general a hard task to show that two given varieties are HP-dual, and one of the most challenging steps is to deal with singular varieties and their noncommutative resolutions. However, HPD allows one to describe a great amount of semiorthogonal decompositions for Fano varieties or Mori fiber spaces, see [**121**], [**123**], [**120**], [**12**] just to name a few.

On the other hand, (relatively) Fano varieties are not the only class of varieties whose derived category admits a semiorthogonal decompositions. The first natural examples one should consider are surfaces with $p_g = q = 0$, in which case any line bundle is a k-exceptional object. Hence the derived category of such surfaces always admits nontrivial semiorthogonal decompositions. On the other hand, one can argue that, if S is a such a surface, then there is no fully faithful functor $\mathsf{D}^{\mathrm{b}}(C) \to \mathsf{D}^{\mathrm{b}}(S)$ for C a curve of positive genus. Indeed, such a functor would give a nontrivial Albanese variety (or, equivalently, a nontrivial Pic^0), see [**32**]. Another way to present this argument is by noticing that $H^{p,q}(S) = 0$ if $p-q \neq 0$. It follows that the Hochschild homology $HH_i(S) = 0$ for $i \neq 0$. This last fact obstructs the existence of the functor, since $HH_{\pm 1}(C) \neq 0$ for a positive genus curve C.

It is then natural to look for semiorthogonal decompositions of the form:

$$\mathsf{D}^{\mathrm{b}}(S) = \langle \mathsf{A}_S, E_1, \ldots, E_n \rangle,$$

with E_i k-exceptional objects and wonder about the maximal possible value of n and the structure of A_S. Describing A_S is a very challenging question and we will treat examples and their conjectural relation with rationality questions in §6.

Let us conclude by remarking that, studying such surfaces, Böhning, Graf von Bothmer, and Sosna have been able to show that semiorthogonal decompositions do not enjoy, in general, a Jordan–Hölder type property [**41**]. Notice that a further example is explained by Kuznetsov [**124**].

PROPOSITION 2.3.14 (Böhning–Graf von Bothmer–Sosna). *Let X be the classical Godeaux complex surface. The bounded derived category $\mathsf{D}^{\mathrm{b}}(X)$ has two maximal exceptional sequences of different lengths: one of length 11 and one of length 9 which cannot be extended further.*

3. Unramified cohomology and decomposition of the diagonal

Unramified cohomology has emerged in the last four decades as a powerful tool for obstructing (stable) rationality in algebraic geometry. Much of its utility comes from the fact that the theory rests on a combination of tools from scheme theory, birational geometry, and algebraic K-theory. Used notably in the context of Noether's problem in the work of Saltman and Bogomolov, unramified cohomology can be computed purely at the level of the function field, without reference to a specific good model.

3.1. Flavors of rationality. A variety X over a field k is *rational* over k if X is k-birational to the projective space $\mathbb{P}^n$, it is *unirational* over k if there is a dominant rational $\mathbb{P}^N \dashrightarrow X$ for some N, it is *retract rational* over k if there is a dominant rational $\mathbb{P}^N \dashrightarrow X$ with a rational section, it is *stably rational* over k if $X \times \mathbb{P}^N$ is rational for some N. The notion of retract rationality was introduced by Saltman in the context of Noether's problem.

We have the following implications:

rational $\Rightarrow$ stably rational $\Rightarrow$ retract rational $\Rightarrow$ unirational $\Rightarrow$ rationally connected.

Several important motivating problems in the study of rationality in algebraic geometry can be summarized as asking whether these implications are strict.

PROBLEM 3.1.1 (Lüroth problem). *Determine whether a given unirational variety X is rational.*

PROBLEM 3.1.2 (Birational Zariski problem). *Determine whether a given stably rational variety X is rational.*

PROBLEM 3.1.3. *Does there exist a rationally connected variety X, such that $X(k) \neq \varnothing$ and that is not unirational?*

PROBLEM 3.1.4. *Does there exist a retract rational variety X that is not stably rational?*

The Lüroth question has a positive answer in dimension 1 (proved by Lüroth) over an arbitrary field and in dimension 2 over an algebraically closed field of characteristic zero (proved by Castelnuovo [**55**]). There exist counterexamples, i.e., unirational but nonrational surfaces, over the real numbers (as remarked by Segre [**157**]) and over an algebraically closed field of characteristic $p > 0$ (discovered by Zariski [**181**]). The first known counterexamples over $\mathbb{C}$ were in dimension 3, discovered independently by Clemens–Griffiths [**58**], Iskovskih–Manin [**106**], and Artin–Mumford [**9**]. We point out that the example of Artin–Mumford also provided the first example of a unirational variety that is not stably rational over $\mathbb{C}$. The method of the intermediate Jacobian due to Clemens–Griffiths and the method of birational rigidity due to Iskovskih–Manin do not obstruct stable rationality. We will treat the former in more details in §4.

The first known counterexamples to the birational Zariski problem were discovered by Beauville–Colliot-Thélène–Sansuc–Swinnerton-Dyer [**25**] in dimension 2 over non algebraically closed fields and in dimension 3 over $\mathbb{C}$ using the method of the intermediate Jacobian, which we will recall in §4.1.

There exist retract rational tori that are not stably rational over $\mathbb{Q}$ discovered in the context of work by Swan and Voskresenskiĭ on Noether's problem, see [**71**, §8.B, p. 223].

The last two questions are still open over an algebraically closed field!

3.2. Unramified elements. Various notions of the concept of unramified cohomology emerged in the late 1970s and 1980s [**59**], [**60**], [**72**], [**73**], [**68**], mostly motivated by earlier investigations of the Brauer group [**17**], [**94**] and the Gersten conjecture [**36**] in algebraic K-theory. The general notion of "unramified element" of a functor is developed in [**61**, §2]. Rost [**153**, Rem. 5.2] gives a different perspective in terms of cycle modules, also see Morel [**139**, §2]. Let k be a field and denote by Local_k the category of local k-algebras together with local k-algebra homomorphisms. Let Ab be the category of abelian groups and let $M : \mathsf{Local}_k \to \mathsf{Ab}$ be a functor. For any field K/k the group of *unramified elements* of M in K/k is the intersection

$$M_{\mathrm{ur}}(K/k) = \bigcap_{k \subset \mathscr{O} \subset K} \mathrm{im}\big(M(\mathscr{O}) \to M(K)\big)$$

over all rank 1 discrete valuations rings $k \subset \mathscr{O} \subset K$ with $\mathrm{Frac}(\mathscr{O}) = K$.

There is a natural map $M(k) \to M_{\mathrm{ur}}(K/k)$ and we say that the group of unramified elements $M_{\mathrm{ur}}(K/k)$ is *trivial* if this map is surjective.

For X an integral scheme of finite type over a field k, we will often write $M_{\mathrm{ur}}(X/k)$ for $M_{\mathrm{ur}}(k(X)/k)$. By definition, the group $M_{\mathrm{ur}}(X/k)$ is a k-birational invariant of integral schemes of finite type over k.

We will be mostly concerned with the functor $M = H^i_{\text{ét}}(-, \mu)$ with coefficients μ either $\mu_n^{\otimes(i-1)}$ (under the assumption $\mathrm{char}(k) \neq n$) or

$$\mathbb{Q}/\mathbb{Z}(i-1) = \varinjlim \mu_n^{\otimes(i-1)},$$

the direct limit being taken over all integers n coprime to the characteristic of k. In this case, $M_{\mathrm{ur}}(X/k)$ is called the *unramified cohomology* group $H^i_{\mathrm{ur}}(X, \mu)$ of X with coefficients in μ.

The reason why we only consider cohomology of degree i with coefficients that are twisted to degree $(i-1)$ is the following well-known consequence of the norm residue isomorphism theorem proved by Voevodsky, Rost, and Weibel (previously known as the Bloch–Kato conjecture).

THEOREM 3.2.1. *Let K be a field and n a nonnegative integer prime to the characteristic. Then the natural map*

$$H^i(K, \mu_n^{\otimes(i-1)}) \to H^i(K, \mathbb{Q}/\mathbb{Z}(i-1))$$

is injective and the natural map $\varinjlim H^i(K, \mu_n^{\otimes(i-1)}) \to H^i(K, \mathbb{Q}/\mathbb{Z}(i-1))$ *is an isomorphism, where the limit is taken over all n prime to the characteristic.*

REMARK 3.2.2. If k is an algebraically closed field whose characteristic is invertible in μ, then $H^i_{\mathrm{ur}}(X, \mu) = 0$ for all $i > \dim(X)$, since in this case the function field $k(X)$ has cohomological dimension $\dim(X)$.

Another important functor is the Milnor K-theory functor $M = K_i^M(-)$.

Let $\mathcal{H}^i_{\text{ét}}(\mu)$ be the Zariski sheaf on the category of k-schemes Sch_k associated to the functor $H^i_{\text{ét}}(-, \mu)$. The Gersten conjecture, proved by Bloch and Ogus [**36**], allows for the calculation of the cohomology groups of the sheaves $\mathcal{H}^i_{\text{ét}}(\mu)$ on a smooth proper variety X as the cohomology groups of the Gersten complex (also known as the "arithmetic resolution") for étale cohomology:

$$0 \longrightarrow H^i(F(X)) \longrightarrow \bigoplus_{x \in X^{(1)}} H^{i-1}(F(x)) \longrightarrow \bigoplus_{y \in X^{(2)}} H^{i-2}(F(y)) \longrightarrow \cdots$$

where $H^i(-)$ denotes the Galois cohomology group in degree i with coefficients μ either $\mu_n^{\otimes(i-1)}$ or $\mathbb{Q}/\mathbb{Z}(i-1)$, where $X^{(i)}$ is the set of codimension i points x of X with residue field $F(x)$, and where the "residue" morphisms are Gysin boundary maps induced from the spectral sequence associated to the coniveau filtration, see Bloch–Ogus [**36**, Thm. 4.2, Ex. 2.1, Rem. 4.7]. In particular, we have that

$$H^0(X, \mathcal{H}^i_{\text{ét}}(\mu)) = H^i_{\mathrm{ur}}(X, \mu).$$

This circle of ideas is generally called "Bloch–Ogus theory."

Over $\mathbb{C}$, this leads to the following "geometric interpretation" of unramified cohomology, as the direct limit over all Zariski open coverings $\mathscr{U} = \{U_i\}$ of the set

$$\{\{\alpha_i\} \in H^i_B(\mathscr{U}, \mu) \ : \ \alpha_i|_{U_{ij}} = \alpha_j|_{U_{ij}}\}$$

where $H^i_B(\mathscr{U}, \mu) = \prod_i H^i_B(U_i, \mu)$ is Betti cohomology, using the comparison with étale cohomology.

3.3. Purity in low degree. There is a canonical map $H^i_{\text{ét}}(X, \mu) \to H^i_{\mathrm{ur}}(X, \mu)$. If this map is injective, surjective, or bijective we say that the *injectivity*, *weak purity*, or *purity* property hold for étale cohomology in degree i, respectively, see Colliot-Thélène [**61**, §2.2].

For X smooth over a field k, a general cohomological purity theorem for étale cohomology is established by Artin in [**7**, XVI 3.9, XIX 3.2].

THEOREM 3.3.1. *Let X be a smooth variety over a field k and $V \subset X$ a closed subvariety of pure codimension $\geq c$. Then the restriction maps*

$$H^i_{\text{ét}}(X, \mu_n^{\otimes j}) \to H^i_{\text{ét}}(X \smallsetminus V, \mu_n^{\otimes j})$$

are injective for $i < 2c$ and are isomorphisms for $i < 2c - 1$.

An immediate consequence (taking $c = 1$) is that purity holds for étale cohomology in degree ≤ 1, i.e.

$$H^0_{\text{ét}}(X, \mu) = H^0_{\text{ur}}(X, \mu) = \mu, \quad \text{and} \quad H^1_{\text{ét}}(X, \mu) = H^1_{\text{ur}}(X, \mu).$$

See also Colliot-Thélène–Sansuc [**72**, Cor. 3.2, Prop. 4.1] for an extension to any geometrically locally factorial and integral scheme.

Combining this (for $c = 2$) with a cohomological purity result for discrete valuation rings and a Mayer–Vietoris sequence, one can deduce that for X smooth over a field, weak purity holds for étale cohomology in degree 2. Moreover, there's a canonical identification $\text{Br}(X)' = H^2_{\text{ur}}(X, \mathbb{Q}/\mathbb{Z}(1))$ by Bloch–Ogus [**36**] such that the canonical map $H^2_{\text{ét}}(X, \mathbb{Q}/\mathbb{Z}(1)) \to H^2_{\text{ur}}(X, \mathbb{Q}/\mathbb{Z}(1)) = \text{Br}(X)'$ arises from the Kummer exact sequence. Here, $\text{Br}(X)'$ denotes the prime-to-characteristic torsion subgroup of the (cohomological) Brauer group $\text{Br}(X) = H^2_{\text{ét}}(X, \mathbb{G}_m)$. For X a smooth variety over $\mathbb{C}$ (or in fact X any complex analytic space), there is a split exact sequence

$$0 \to \big(H^2(X, \mathbb{Z})/\operatorname{im}(\text{Pic}(X) \to H^2(X, \mathbb{Z})\big) \otimes \mathbb{Q}/\mathbb{Z} \to \text{Br}(X) \to H^3(X, \mathbb{Z})_{\text{tors}} \to 0$$

arising from the exponential sequence. In particular, there is a (noncanonical) isomorphism $\text{Br}(X) \cong (\mathbb{Q}/\mathbb{Z})^{b_2 - \rho} \oplus H^3(X, \mathbb{Z})_{\text{tors}}$, where b_2 is the second Betti number and ρ the Picard rank of X. If X satisfies $H^2(X, \mathcal{O}_X) = 0$ (e.g., X is rationally connected), then $\text{Pic}(X) \to H^2(X, \mathbb{Z})$ is an isomorphism, and hence $\text{Br}(X) = H^3(X, \mathbb{Z})_{\text{tors}}$.

There is a beautiful interpretation of unramified cohomology in degree 3 in terms of cycles of codimension 2, going back to Barbieri-Viale [**19**]. Let X be a smooth projective variety over k. We say that $\text{CH}_0(X)$ is supported in dimension r if there exists a smooth projective variety Y over k of dimension r and a morphism $f : Y \to X$ such that the pushforward $f_* : \text{CH}_0(Y) \to \text{CH}_0(X)$ is surjective. For example, if $\text{CH}_0(X) = \mathbb{Z}$ (e.g., X is rationally connected) then X is supported in dimension 0.

THEOREM 3.3.2 (Colliot-Thélène–Voisin [**74**, Thm. 1.1]). *Let X be a smooth projective variety over $\mathbb{C}$. Assume that $\text{CH}_0(X)$ is supported in dimension 2. Then there is an isomorphism*

$$H^3_{\text{ur}}(X, \mathbb{Q}/\mathbb{Z}(2)) \cong \frac{H^{2,2}(X) \cap H^4(X, \mathbb{Z})}{\operatorname{im}\big(\text{CH}^2(X) \to H^4(X, \mathbb{Z})\big)}.$$

Equivalently, the unramified cohomology of X in degree 3 is the obstruction to the validity of the integral Hodge conjecture for cycles of codimension 2.

More generally, without the assumption that $\text{CH}_0(X)$ is supported in dimension 2, the torsion subgroup of $(H^{2,2}(X) \cap H^4(X, \mathbb{Z}))/\operatorname{im}\big(\text{CH}^2(X) \to H^4(X, \mathbb{Z})\big)$ is a quotient of $H^3_{\text{ur}}(X, \mathbb{Q}/\mathbb{Z}(2))$ by a divisible subgroup.

There is also a version of this result valid over more general fields, in particular over finite fields, due to Colliot-Thélène and Kahn [**67**] and extended to higher codimension cycles by Pirutka [**151**]. Finally, there is a description of $H^4_{\text{ur}}(X, \mathbb{Q}/\mathbb{Z}(3))$ in terms of torsion in $\text{CH}^3(X)$, due to Voisin [**173**].

3.4. Triviality. If F/k is a field extension, we write $X_F = X \times_k F$. If X is geometrically integral over k, we say that $M_{\text{ur}}(X/k)$ is *universally trivial* if

$M_{\mathrm{ur}}(X_F/F)$ is trivial for every field extension F/k. Let N be a positive integer. We say that $M_{\mathrm{ur}}(X/k)$ is *universally N-torsion* if the cokernel of the natural map $M(F) \to M_{\mathrm{ur}}(X_F/F)$ is killed by N for every field extension F/k.

PROPOSITION 3.4.1 ([**61**, §2 and Thm. 4.1.5]). *Let M : $\mathsf{Local}_k \to \mathsf{Ab}$ be a functor satisfying the following conditions:*

- *If $\mathscr{O}$ is a discrete valuation ring containing k, with fraction field K and residue field κ, then $\ker\big(M(\mathscr{O}) \to M(K)\big) \subset \ker\big(M(\mathscr{O}) \to M(\kappa)\big)$.*
- *If A is a regular local ring of dimension 2 containing k, with fraction field K, then $\mathrm{im}\big(M(A) \to M(K)\big) = \bigcap_{\mathsf{ht}(\mathfrak{p})=1} \mathrm{im}\big(M(A_\mathfrak{p}) \to M(K)\big)$.*
- *The group $M_{\mathrm{ur}}(\mathbb{A}^1_k/k)$ is universally trivial.*

Then $M_{\mathrm{ur}}(\mathbb{P}^n_k/k)$ is universally trivial. In particular, if X is a rational variety over k, then $M_{\mathrm{ur}}(X/k)$ is universally trivial.

The functor $H^i_{\mathrm{\acute{e}t}}(-,\mu)$ satisfies the conditions of Proposition 3.4.1 (cf. [**61**, Thm. 4.1.5]), hence if X is a k-rational variety, then $H^i_{\mathrm{ur}}(X,\mu)$ is universally trivial. More generally, $H^i_{\mathrm{ur}}(X,\mu)$ is universally trivial if X is stably rational, see [**68**, Prop. 1.2], or even retract k-rational, which can be proved using the general machinery in [**109**, Cor. RC.12–13], see [**138**, Prop. 2.15].

3.5. Applications: Noether's problem and Artin–Mumford. Here we describe two important examples where unramified cohomology has been used in the rationality problem.

EXAMPLE 3.5.1. Let G be a finite group, V a finite dimensional linear representation over k, and $k(V)$ the field of rational functions on the affine space associated to V. Then Noether's question asks if the field of invariants $k(V)^G$ is purely transcendental over k, equivalently, if the variety V/G is rational. This question was posed by Emmy Noether in 1913, and has endured as one of the most challenging rationality problems in algebraic geometry.

Over the rational numbers, the problem takes on a very arithmetic flavor. Indeed, Noether's original motivation was the inverse Galois problem, see [**162**] for a survey in this direction. So we will focus on the case when k is algebraically closed of characteristic zero.

In this case, the question has a positive answer when G is any abelian group but is still open for $G = A_n$ for $n \geq 6$. Saltman [**155**] gave the first examples of p-groups having a negative answer to Noether's question when k is algebraically closed. While V/G often has terrible singularities and its smooth projective models are not easy to compute, nor feasible to work with, the insight of Saltman was that one could still compute unramified cohomology, in particular, $H^2_{\mathrm{ur}}(k(V)^G/k, \mathbb{Q}/\mathbb{Z}(1))$. By the above purity results, if X were a smooth proper model of $k(V)^G$, then $H^2_{\mathrm{ur}}(k(V)^G/k, \mathbb{Q}/\mathbb{Z}(1)) = \mathrm{Br}(X)$. Bogomolov [**39**] gave a simple group theoretic formula to compute $H^2_{\mathrm{ur}}(k(V)^G/k, \mathbb{Q}/\mathbb{Z}(1))$ purely in terms of G, when k is algebraically closed of characteristic not dividing the order of G. Because of this, this group is often called the "Bogomolov multiplier" in the literature.

We point out that examples of groups G where $H^i_{\mathrm{ur}}(k(V)^G/k, \mathbb{Q}/\mathbb{Z}(1))$ is trivial for $i = 2$ yet nontrivial for $i \geq 3$ were first constructed by Peyre [**150**].

EXAMPLE 3.5.2. Artin and Mumford [**9**] constructed a unirational threefold X over an algebraically closed field of characteristic $\neq 2$ having nontrivial 2-torsion in $\mathrm{Br}(X) = H^3(X,\mathbb{Z})_{\mathrm{tors}}$, which by purity (see §3.3) coincides with $H^2_{\mathrm{ur}}(X, \mathbb{Q}/\mathbb{Z}(1))$, hence such X not retract rational (hence not stably rational). The "Artin–Mumford solid" X is constructed as the desingularization of a double cover of $\mathbb{P}^3$ branched over a certain quartic hypersurface with 10 nodes, and is unirational by construction. The solid X can also be presented as a conic bundle $X \to S$ over a rational surface S. The unramified cohomology perspective on the examples of Artin and Mumford was further investigated by Colliot-Thélène and Ojanguren [**68**].

Denote by $\mathsf{Ab}^\bullet$ the category of graded abelian groups. An important class of functors $M : \mathsf{Local}_k \to \mathsf{Ab}^\bullet$ arise from the general theory of *cycle modules* due to Rost [**153**, Rem. 5.2]. In particular, unramified cohomology arises from the étale cohomology cycle module, and to some extent, the theory of cycle modules is a generalization of the theory of unramified cohomology. Rost's key observation is that classical Chow groups appear as the unramified elements of the Milnor K-theory cycle module. The definition of cohomology groups arising from cycle modules is very parallel to the definition of homology of a CW complex from the singular chain complex. A cycle module M comes equipped with residue maps of graded degree -1

$$M^i(k(X)) \xrightarrow{\partial} \bigoplus_{x \in X^{(1)}} M^{i-1}(k(x))$$

for any integral k-variety X. If X is smooth and proper, then the group of unramified elements $M^i_{\mathrm{ur}}(X/k)$ is defined to be the kernel.

3.6. Decomposition of the diagonal. We say that a smooth proper variety X of dimension n over a field k has an *(integral) decomposition of the diagonal* if we can write

$$\Delta_X = P \times X + Z \tag{3.1}$$

in $\mathrm{CH}^n(X \times X)$, where P is a 0-cycle of degree 1 and Z is a cycle with support in $X \times V$ for some closed subvariety $V \subsetneq X$. We say that X has a *rational decomposition of the diagonal* if there exists $N \geq 1$ such that

$$N\Delta_X = P \times X + Z \tag{3.2}$$

in $\mathrm{CH}^n(X \times X)$, where P is a 0-cycle of degree N and Z is as before. This notion was studied by Bloch and Srinivas [**37**], with the idea going back to Bloch's proof of Mumford's result on 2-forms on surfaces (see [**34**, Lecture 1, Appendix]).

EXAMPLE 3.6.1. The class of the diagonal $\Delta_{\mathbb{P}^n} \in \mathrm{CH}^n(\mathbb{P}^n \times \mathbb{P}^n)$ can be expressed in terms of the pull backs $\alpha, \beta \in \mathrm{CH}^1(\mathbb{P}^n \times \mathbb{P}^n)$ of hyperplane classes from the two projections. The Chow ring can be presented in terms of these classes as

$$\mathrm{CH}(\mathbb{P}^n \times \mathbb{P}^n) = \mathbb{Z}[\alpha, \beta]/(\alpha^{n+1}, \beta^{n+1})$$

and one can compute that

$$\Delta_{\mathbb{P}^n} = \alpha^n + \alpha^{n-1}\beta + \cdots + \alpha\beta^{n-1} + \beta^n$$

in $\mathrm{CH}^n(\mathbb{P}^n \times \mathbb{P}^n)$, see [**84**, Ex. 8.4.2]. The class α^n is the same as the class $P \times \mathbb{P}^n$, for $P \in \mathbb{P}^n$ a rational point, while the classes $\alpha^i\beta^{n-i}$ for $i > 0$ all have support on $\mathbb{P}^n \times H$, where $H \subset \mathbb{P}^n$ is the hyperplane defining β. So any projective space has an integral decomposition of the diagonal.

EXAMPLE 3.6.2. Let $f : Y \to X$ be a proper surjective generically finite morphism of degree N between smooth quasi-projective varieties. Then we have that $(f \times f)_* \Delta_Y = N\Delta_X$. Assume that Y has a decomposition of the diagonal $\Delta_Y = P \times Y + Z$, where P is a 0-cycle of degree 1 on Y and Z is a cycle with support on $Y \times V$. Then

$$N\Delta_X = (f \times f)_* \Delta_Y = (f \times f)_*(P \times Y + Z) = f_*(P) \times X + Z'$$

where $f_*(P)$ is a 0-cycle of degree N and Z' is a cycle on X with support on $X \times f(V)$. Hence X has a rational decomposition of the diagonal.

Let $f : Y \to X$ be a surjective birational morphism between smooth quasi-projective varieties. Given a decomposition of the diagonal $\Delta_X = P \times X + Z$ on X, by the moving lemma for 0-cycles, we can move P, up to rational equivalence, outside of the image of the exceptional locus of f. Then $(f \times f)^* \Delta_X - \Delta_Y$ is a sum of cycles whose projections to Y are contained in the exceptional locus of f. But $(f \times f)^* \Delta_X = (f \times f)^*(P \times X + Z) = f^{-1}(P) \times Y + (f \times f)^*(Z)$, and $(f \times f)^*(Z)$ is a cycle with support on $Y \times f^*(V)$. In total, Y has a decomposition of the diagonal.

We can use this to show that if $\mathbb{P}^n \dashrightarrow X$ is a unirational parameterization of degree N over a field of characteristic zero, then X has a rational decomposition of the diagonal $N\Delta_X = P \times X + Z$. Indeed, by resolution of singularities, we can resolve the rational map to a proper surjective generically finite morphism $Y \to X$ of degree N, where $Y \to \mathbb{P}^n$ is a sequence of blow up maps along smooth centers. By the above considerations, the decomposition of the diagonal on $\mathbb{P}^n$ induces one on Y, which in turn induces the desired rational decomposition of the diagonal on X.

We remark that one can argue without the use of resolution of singularities, using the machinery of [**109**, App. RC], but this is slightly more delicate.

3.7. Decomposition of the diagonal acting on cohomology. A rational decomposition of the diagonal puts strong restrictions on the variety X. For example, the following result is well known.

PROPOSITION 3.7.1. *Let X be a smooth proper geometrically irreducible variety over a field k of characteristic zero. If X has a rational decomposition of the diagonal then $H^0(X, \Omega^i_X) = 0$ and $H^i(X, \mathcal{O}_X) = 0$ for all $i > 0$.*

Over a complex surface, this result goes back to Bloch's proof [**34**, App. Lec. 1] of Mumford's [**141**] result on 2-forms on surfaces, exploiting a decomposition of the diagonal and the action of cycles on various cohomology theories (de Rham and étale). This argument was further developed in [**37**]. A proof over the complex numbers can be found in [**175**, Cor. 10.18, §10.2.2]. A variant of the argument for rigid cohomology in characteristic p is developed by Esnault [**82**, p. 187], in her proof that rationally connected varieties over a finite field have a rational point. A variant of the argument using logarithmic de Rham cohomology over any field is developed by Totaro [**166**, Lem. 2.2] using the cycle class map of Gros.

Let $H^i(-)$ be a cohomology theory with a cycle class map

$$\mathrm{CH}^i(X) \to H^{2i}(X)$$

and a theory of correspondences (basically a Weil cohomology theory), so that for any $\alpha \in H^{2n}(X \times X)$, where $n = \dim(X)$, there is a map

$$\alpha_* = q_*(\alpha . p^*) : H^i(X) \to H^i(X).$$

for any $i \geq 0$. Here p and q are the left and right projections $X \times X \to X$, respectively. When $\alpha = [\Delta_X] \in H^{2n}(X \times X)$, then α_* is the identity map. When $\alpha = [P \times X] \in H^{2n}(X \times X)$, then α_* factors through $H^i(P)$, i.e., there is a commutative diagram

$$\begin{array}{ccc} H^i(X) & \xrightarrow{\alpha_*} & H^i(X) \\ \downarrow & & \| \\ H^i(P) & \xrightarrow{\alpha_*} & H^i(X) \end{array}$$

where the left hand vertical map is the pullback by the inclusion of the zero-dimensional subscheme $P \subset X$ (we have in mind N times a point). Assuming that $H^i(P) = 0$ for $i > 0$, we get that $N[\Delta_X]_* = [Z]_*$ on $H^i(X)$, assuming a rational decomposition of the diagonal as in (3.2).

On the other hand, since Z is a cycle supported on $X \times V$ for $V \subset X$ a proper closed subvariety, the restriction of $[Z]$ to $H^{2n}(X \times X \smallsetminus V)$ is zero. Consider $\alpha = [Z] \in H^{2n}(X \times X)$ and the commutative diagram

$$\begin{array}{ccc} H^i(X) & \xrightarrow{\alpha_*} & H^i(X) \\ \| & & \downarrow \\ H^i(X) & \xrightarrow{0_*} & H^i(X \smallsetminus V) \end{array}$$

where the right hand vertical arrow is the pullback by the inclusion $X \smallsetminus V \subset X$ and the bottom horizontal arrow is the pushforward associated to the restriction of $[Z]$ to $X \times (X \smallsetminus V)$, which is zero. Hence we have that $\alpha_* H^i(X)$ is contained in the kernel of the restriction map $H^i(X) \to H^i(X \smallsetminus V)$. If we additionally assume that the cohomology theory has a localization sequence

$$\cdots \to H^i_V(X) \to H^i(X) \to H^i(X \smallsetminus V) \to \cdots$$

involving cohomology with supports, then we can also conclude that $\alpha_* H^i(X)$ is contained in the image of the map $H^i_V(X) \to H^i(X)$.

Now, applying this to algebraic de Rham cohomology $H^i_{\mathrm{dR}}(-)$, we have that (by the degeneration of the Hodge-to-de Rham spectral sequence) any element $\alpha \in H^{2n}_{\mathrm{dR}}(X \times X)$ in the image of the cycle map lands in

$$H^n(X \times X, \Omega^n_{X \times X}) = H^n(X \times X, \bigoplus_j \Omega^j_X \boxtimes \Omega^{n-j}_X) = \bigoplus_{i,j} H^i(X, \Omega^j_X) \otimes H^{n-i}(X, \Omega^{n-j}_X)$$

so α has a component in $H^0(X, \Omega^i_X) \otimes H^n(X, \Omega^{n-i}_X)$, which is isomorphic (by Serre duality) to $\mathrm{End}(H^0(X, \Omega^i_X))$. Thus this component of the pushforward α_* defines a map $H^0(X, \Omega^i_X) \to H^0(X, \Omega^i_X)$, whose image lands in the kernel of the restriction map $H^0(X, \Omega^i_X) \to H^0(X \smallsetminus V, \Omega^i_{X \smallsetminus V})$. But this kernel is trivial, since restriction to a Zariski open set is injective on global differential forms. We thus conclude that (N times) the identity on $H^0(X, \Omega^i_X)$ coincides with the zero map, hence $H^0(X, \Omega^i_X) = 0$ for all $i > 0$.

Similarly, applying this to the cycle class map of Gros in logarithmic de Rham cohomology, Totaro shows that even in characteristic p, if X has an (integral) decomposition of the diagonal, then $H^0(X, \Omega^i_X) = 0$ for all $i > 0$.

Applying this to the transcendental part of the cohomology

$$H^2(X, \mathbb{Q}_\ell) / \mathrm{im}(\mathrm{NS}(X) \otimes \mathbb{Q}_\ell \to H^2_{\text{ét}}(X, \mathbb{Q}_\ell))$$

of a surface X, Bloch shows that a rational decomposition of the diagonal implies the vanishing of the transcendental part of the cohomology. Via the Hodge-theoretic fact that $p_g(X) = 0$ is equivalent to $b_2(X) - \rho(X) = 0$, this is Bloch's new proof of Mumford's theorem stating that if $p_g(X) > 0$ then the kernel of the degree map $\deg : \mathrm{CH}_0(X) \to \mathbb{Z}$ is not representable.

Applying this to Berthelot's theory of rigid cohomology, Esnault shows that if X is defined over a field of characteristic p and has a rational decomposition of the diagonal, then the Frobenius slope $[0,1)$ part of the rigid cohomology of $H^i(X)$ is trivial for all $i > 0$. If X is defined over a finite field $\mathbb{F}_q$, the Lefschetz trace formula implies that $\#X(\mathbb{F}_q) \equiv 1 \pmod{q}$, in particular, $X(\mathbb{F}_q) \neq \varnothing$.

We remark that over $\mathbb{C}$, an integral decomposition of the diagonal does not imply $H^0(X, \omega_X^{\otimes n}) = 0$ for all $n > 1$. Otherwise, a smooth projective surface X over $\mathbb{C}$ with integral decomposition of the diagonal would, aside from satisfying $p_g(X) = h^0(X, \omega_X) = 0$ and $q = h^1(X, \mathscr{O}_X) = h^0(X, \Omega^1_X) = 0$, would additionally satisfy $P_2(X) = h^0(X, \omega_X^{\otimes 2}) = 0$, hence would be rational by Castelnuovo's criterion. However, there do exist (nonrational) complex surfaces X of general type (e.g., Barlow surfaces) admitting an integral decomposition of the diagonal that emerge in the context of Bloch's conjecture on 0-cycles on surfaces, see §5.3 for a more detailed discussion.

4. Cubic threefolds and special cubic fourfolds

Cubic hypersurfaces of dimension 3 and 4 are some of the most important motivating objects in birational geometry since the last half of the 20th century. An irreducible cubic hypersurface is rational as soon as it has a rational singular point, unless possibly when it is a cone over a cubic hypersurface of lower dimension, see [**75**, Chapter 1, Section 5, Example 1.28]. Working over an algebraically closed field of characteristic not 3, we recall that in dimension 1, a cubic hypersurface is not rational if and only if it is smooth, in which case it is a curve of genus 1. In dimension 2, smooth cubic hypersurfaces are rational, and they are realized geometrically as the blow-up of six points in general position on $\mathbb{P}^2$. In dimension 3, the fact that every smooth cubic hypersurface over $\mathbb{C}$ is not rational is a celebrated theorem of Clemens and Griffith [**58**]. In dimension 4, some families of smooth cubics hypersurfaces are known to be rational, while the very general one is expected to be nonrational, even though not a single one is currently provably nonrational.

Cubic hypersurfaces seem to occupy a space in the birational classification of varieties that is "very close" to rational varieties, in that their familiar cohomological and birational invariants are similar to those of projective space. Proving their nonrationality seems to require the development of much finer techniques. The nonrationality of the cubic threefold was indeed one of the first counterexamples to the Lüroth problem (see Problem 3.1.1) in characteristic zero, and the proof of its nonrationality required a deep study of algebraic cycles and the intermediate Jacobian. The study of the (non)rationality of cubic fourfolds has already attracted Hodge and moduli-theoretic techniques, and is undoubtedly one of the most famous open question in algebraic geometry.

The aim of this section is to introduce "classical" constructions arising in the study of cubic hypersurfaces. Intermediate Jacobians will be presented in the first part. In the second part, we recall the Hodge theoretic approach to moduli spaces of cubic fourfolds, and the known examples. We work here exclusively over $\mathbb{C}$.

4.1. Intermediate Jacobians and cubic threefolds. Recall the definition of the intermediate Jacobians of a smooth complex variety X of dimension n (see [**175**, Ch. 12]). Consider the Betti cohomology group $H^i(X,\mathbb{C})$ together with the Hodge filtration $F^pH^i(X,\mathbb{C})$. If $i = 2j-1$ is odd, the j-th filtered module yields:

$$F^jH^{2j-1}(X,\mathbb{C}) = \bigoplus_{p+q=2j-1,\ p\geq j} H^{p,q}(X).$$

The Hodge structure on Betti cohomology then gives that $H^{2j-1}(X,\mathbb{C})$ is the sum of $F^jH^{2j-1}(X,\mathbb{C})$ and its conjugate, so that

$$H^{2j-1}(X,\mathbb{Z})/\mathrm{Tors} \longrightarrow H^{2j-1}(X,\mathbb{C})/F^jH^{2j-1}(X,\mathbb{C}) = \overline{F^jH^{2j-1}(X,\mathbb{C})}$$

is an injective map (via de Rham cohomology). We define the $(2j-1)$-st intermediate Jacobian $J^{2j-1}(X)$ as the quotient of the $\mathbb{C}$-vector space $\overline{F^jH^{2j-1}(X,\mathbb{C})}$ by this lattice. The Jacobian is in general a complex torus, and not an Abelian variety.

If X is a threefold with $H^1(X,\mathbb{C}) = 0$, then the only nontrivial Jacobian is $J^3(X)$. Indeed, by Poincaré duality $H^1(X,\mathbb{C}) = H^5(X,\mathbb{C}) = 0$, so that $J^1(X) = 0$ and $J^5(X) = 0$. In this case, we denote $J(X) := J^3(X)$. Moreover, assume X is a Fano, or in general, a threefold with $H^1(X,\mathbb{C}) = H^{3,0}(X) = 0$, then the only nontrivial intermediate Jacobian is

$$J(X) = J^3(X) = H^{1,2}(X)/\operatorname{im}(H^3(X,\mathbb{Z})/\mathrm{Tors} \to H^{1,2}(X)).$$

The key idea of Clemens and Griffiths [**58**] is to show that in this case the complex torus $J(X)$ is an abelian variety endowed with a canonical principal polarization. Let us briefly sketch a proof of that fact, loosely following the presentation in [**180**, 3.1]. The cup product gives a unimodular intersection pairing $\langle -,-\rangle$ on $H^3(X,\mathbb{Z})/\mathrm{Tors}$. Moreover, consider any nontrivial $(2,1)$-cohomology classes $\alpha, \beta \in H^{2,1}(X)$. Recall we assume that $H^{1,0}(X) = 0$, so that

$$\langle \alpha, \beta\rangle = 0, \qquad -\sqrt{-1}\langle \alpha, \overline{\alpha}\rangle > 0, \tag{4.1}$$

since the cup product is Hermitian and skew symmetric and respects the Hodge decomposition (see [**175**, 7.2.1] for more details). It follows that $\langle -,-\rangle$ can be identified with the first Chern class of an ample line bundle L on $J(X)$, via the identification $\bigwedge^2 H^1(J(X),\mathbb{Z}) \simeq H^2(J(X),\mathbb{Z})$ (see [**140**, Ch. I, 3] for more details), and the line bundle L is well-defined up to translation. In particular $J(X)$ is an Abelian variety. Moreover, since the cup product is unimodular, $H^0(J(X),L)$ is one dimensional. Hence L is a Theta divisor for $J(X)$, which is then principally polarized. A famous result proved by Clemens and Griffiths [**58**] shows that one can extract a birational invariant from this Abelian variety.

THEOREM 4.1.1 (Clemens–Griffiths [**58**]). *If a complex threefold X is rational, then there exist smooth projective curves $\{C_i\}_{i=1}^r$ and an isomorphism of principally polarized Abelian varieties*

$$J(X) = \bigoplus_{i=1}^r J(C_i).$$

Moreover, if X is a complex threefold with $H^{3,0}(X) = H^{1,0}(X) = 0$, there is a well-defined principally polarized Abelian subvariety $A_X \subset J(X)$ which is a birational invariant: if $X' \dashrightarrow X$ is a birational map, then $A_{X'} \simeq A_X$ as principally polarized Abelian varieties.

SKETCH OF PROOF. It is enough to define A_X and prove the second statement, which is stronger. Indeed, it is easy to see that $J(\mathbb{P}^3) = 0$, so that $A_{\mathbb{P}^3} = 0$. The splitting of the intermediate Jacobian of a rational threefold will then be evident by the definition of A_X.

Clemens and Griffiths show that the category of principally polarized Abelian varieties is semisimple, that is any injective morphism is split (see [**58**, §3]). We work exclusively in this category. Hence we define A_X as follows: any injective map $J(C) \to J(X)$ for C a smooth curve gives a splitting $J(X) = A \oplus J(C)$. There is hence a finite number of curves $\{C_i\}_{i=1}^r$ with $J(C_i) \neq 0$ (i.e., $g(C_i) > 0$) and a splitting $J(X) = A_X \oplus \bigoplus_{i=1}^r J(C_i)$, such that there is no nontrivial morphism $J(C) \to A_X$ for any smooth projective curve C. By semisimplicity of the category of principally polarized abelian varieties, we get that A_X is well defined.

If we consider a birational morphism $\rho : Y \to X$, we can then show that $\rho^* : J(X) \to J(Y)$ is an injective map. Then $J(Y) = J(X) \oplus A$ for some Abelian variety A. If moreover ρ is the blow-up along of a point, then $J(Y) \simeq J(X)$. If ρ is the blow-up along a smooth curve C, then $J(Y) = J(X) \oplus J(C)$.

Consider the birational map $X' \dashrightarrow X$. By Hironaka's resolution of singularities, there is a smooth projective X_1 with birational morphisms $\rho_1 : X_1 \to X'$ and $\pi_1 : X_1 \to X$, such that π_1 is a composition of a finite number of smooth blow-ups. We denote $\{C_i\}_{i=1}^s$ the curves blown-up by π_1. Similarly, there are $\rho_2 : X_2 \to X$ and $\pi_2 : X_2 \to X'$ birational maps with π_2 a composition of a finite number of blow-ups. We denote $\{D_i\}_{i=1}^t$ the curves blown-up by π_2. It follows that looking at the decompositions of $J(X_1)$ and $J(X_2)$ respectively, we have:

$$J(X) \subset J(X') \oplus J(D_1) \oplus \ldots \oplus J(D_t)$$
$$J(X') \subset J(X) \oplus J(C_1) \oplus \ldots \oplus J(C_s)$$

and we conclude, by semisimplicity of the category of principally polarized abelian varieties, that we must have $A_X = A_{X'}$. Indeed the first equation gives $A_X \subset A_{X'}$, and the second one gives $A_{X'} \subset A_X$. □

The first statement of Theorem 4.1.1 provides the *Clemens–Griffiths nonrationality criterion*, namely that if the intermediate Jacobian of a smooth projective threefold X with $h^1 = h^{3,0} = 0$ does not factor (in the category of principally polarized abelian varieties) into a product of Jacobians of curves, then X is not rational. The first application of this criterion is the proof of the nonrationality of a smooth cubic threefold [**58**].

THEOREM 4.1.2 (Clemens–Griffiths). *Let X be a smooth cubic threefold. The principally polarized abelian variety $J(X)$ is not split by Jacobians of curves. In particular, X is not rational.*

We will not give here a proof of Theorem 4.1.2, but just mention that it relies on the careful study of singularities the Theta-divisor of $J(X)$, which is a five-dimensional Abelian variety. Just to mention the huge amount of interesting mathematics appearing in this context, we notice that this question is also related to the *Schottky problem*, that is the study of the moduli of Jacobians inside the moduli space of principally polarized Abelian varieties. Clemens–Griffiths nonrationality criterion applies to any threefold with trivial $H^{1,0}(X)$ and $H^{3,0}(X)$, and has allowed Beauville [**23**], and Shokurov [**160**] to completely classify rational conic bundles over minimal surfaces. We recall that a conic bundle is *standard* if the fiber over any irreducible curve is an irreducible surface (this is equivalent to relative minimality).

THEOREM 4.1.3 (Beauville, Shokurov). *Let $X \to S$ be a relatively minimal conic bundle, with X smooth, over a smooth minimal rational surface S with discriminant divisor $C \subset S$ having at most isolated nodal singularities. Then X is rational if and only if $J(X)$ is split by Jacobians of curves, and this happens only in five cases (besides projective bundles):*

- *S is a plane, and C is a cubic, or a quartic, or a quintic and the discriminant double cover $\widetilde{C} \to C$ is given by an even theta-characteristic in the latter case.*
- *S is a Hirzebruch surface and the fibration $S \to \mathbb{P}^1$ induces either a hyperelliptic or a trigonal structure $C \to \mathbb{P}^1$ on the discriminant divisor.*

The proof of Theorem 4.1.3 relies on the isomorphism $J(X) \simeq \mathrm{Prym}(\widetilde{C}/C)$ as principally polarized Abelian varieties [**23**] and on the study of Prym varieties. Notice that Theorem 4.1.3 recovers Theorem 4.1.2 since the blow-up of a smooth cubic threefold X along any line $l \subset X$ gives a relatively minimal conic bundle $\widetilde{X} \to \mathbb{P}^2$ whose discriminant divisor C is a smooth quintic and $\widetilde{C} \to C$ is given by an odd theta-characteristic. As recalled $J(X) \simeq \mathrm{Prym}(\widetilde{C}/C)$, so one can fairly say then that cubic threefolds are (birationally) the non-rational conic bundles with the smallest intermediate Jacobian.

4.2. Intermediate Jacobians and the Zariski problem. Another important problem where the method of the intermediate Jacobian has been successful is in constructing the first counterexamples to Problem 3.1.2, posed by Zariski in 1949, see [**156**]. Indeed, using Prym variety considerations, Beauville, Colliot-Thélène, Sansuc, and Swinnerton-Dyer [**25**] used the intermediate Jacobian to construct the first example of a nonrational but stably rational variety, a fibration in Châtelet surfaces $V \to \mathbb{P}^1$ with affine model

$$y^2 - \delta(t)z^2 = P(x,t)$$

where $P(x,t) = x^3 + p(t)x + q(t)$ is an irreducible polynomial in $\mathbb{C}[x,t]$ such that its discriminant $\delta(t) = 4p(t)^3 + 27q(t)^2$ has degree ≥ 5. They proved, using the intermediate Jacobian, that V is not rational, yet that $V \times \mathbb{P}^3$ is rational. Shepherd-Barron [**159**] used a slightly different construction to prove that $V \times \mathbb{P}^2$ is rational. It is unknown whether $V \times \mathbb{P}^1$ is rational.

The key point is that the Clemens–Griffiths criterion for irrationality of a threefold using the intermediate Jacobian is not a stable birational invariant. Indeed, it strictly applies to threefolds.

4.3. (Special) cubic fourfolds, Hodge theory and Fano schemes of lines. We turn our attention to smooth cubic fourfolds and the Hodge structure on their middle cohomology. The ideas we present in this section go back to Beauville–Donagi [**26**] and Hassett [**95, 96**]. Let $X \subset \mathbb{P}^5$ be a smooth cubic fourfold. We denote by $h \in H^2(X,\mathbb{Z})$ the Betti cycle class of a hyperplane section of X. In particular, $h^4 = 3$ and

$$H^2(X,\mathbb{Z}) = \mathbb{Z}[h], \qquad H^6(X,\mathbb{Z}) = \mathbb{Z}[h^3/3],$$

by the Lefschetz hyperplane theorem and Poincaré duality. One can, moreover, calculate the Hodge numbers: the (upper half) Hodge diamond of X has the following shape:

$$
\begin{array}{lccccccccc}
H^8(X,\mathbb{C}) & & & & & 1 & & & & \\
H^7(X,\mathbb{C}) & & & & 0 & & 0 & & & \\
H^6(X,\mathbb{C}) & & & 0 & & 1 & & 0 & & \\
H^5(X,\mathbb{C}) & & 0 & & 0 & & 0 & & 0 & \\
H^4(X,\mathbb{C}) & 0 & & 1 & & 21 & & 1 & & 0.
\end{array}
$$

We then focus on the cohomology lattice $H^4(X,\mathbb{Z})$, endowed with the intersection pairing $\langle -,- \rangle$, and we denote by $H^4_0(X,\mathbb{Z})$ the primitive cohomology sublattice. In particular, we have that $h^2 \in H^4(X,\mathbb{Z})$ and that $H^4_0(X,\mathbb{Z}) = \langle h^2 \rangle^{\perp}$.

Rational, algebraic, and homological equivalence all coincide for cycles of codimension 2 on any smooth projective rationally connected variety X over $\mathbb{C}$ satisfying $H^3(X,\mathbb{Z}/l\mathbb{Z}) = 0$ for some prime l, cf. [**63**, Prop. 5.1]. Hence for a smooth cubic fourfold X, the Betti cycle class map $\mathrm{CH}^2(X) \to H^4(X,\mathbb{Z})$ is injective. The image of the cycle class map is contained in the subgroup of Hodge classes $H^{2,2}(X) \cap H^4(X,\mathbb{Z})$. In particular, $\mathrm{CH}^2(X)$, with its intersection product, is a sublattice of $H^{2,2}(X) \cap H^4(X,\mathbb{Z})$, which is positive definite by the Riemann bilinear relations. Furthermore, we have that the cycle class map induces an isomorphism $\mathrm{CH}^2(X) = H^{2,2}(X) \cap H^4(X,\mathbb{Z})$ by the integral Hodge conjecture for cycles of codimension 2 on smooth cubic fourfolds proved by Voisin [**176**, Thm. 18], building on [**142**] and [**182**].

To study the cohomology lattice, we consider the Fano variety of lines $F(X)$, defined to be the subvariety $F(X) \subset \mathrm{Gr}(2,6)$ parameterizing the lines contained in X. Then $F(X)$ is a smooth fourfold. Despite its name, $F(X)$ is an irreducible holomorphic symplectic (IHS) variety, as shown by Beauville and Donagi [**26**, Prop. 2].

The cohomology of $F(X)$ and X are related by an Abel–Jacobi map, as follows. Denote by $Z \subset X \times F(X)$ the universal line over X, and consider the diagram:

$$
\begin{array}{ccc}
 & Z & \\
{}^{p}\swarrow & & \searrow^{q} \\
X & & F(X),
\end{array}
$$

where p and q denote the restrictions to Z of the natural projections from $X \times F(X)$ to X and $F(X)$ respectively. The Abel–Jacobi map $\alpha : H^4(X,\mathbb{Z}) \to H^2(F(X),\mathbb{Z})$ is defined as $\alpha = q_* p^*$. Since $F(X)$ is an IHS variety, $H^2(X,\mathbb{Z})$ is endowed with a bilinear form, which we will denote by $\langle -,- \rangle_{BB}$, the Beauville–Bogomolov form. Moreover, since $F(X) \subset \mathrm{Gr}(2,6)$, we can restrict the class on $\mathrm{Gr}(2,6)$ defining the Plücker embedding to a class $g \in H^2(F(X),\mathbb{Z})$. Now we define the primitive cohomology $H^2_0(F(X),\mathbb{Z}) = \langle g \rangle^{\perp} \subset H^2(F(X),\mathbb{Z})$ to be the orthogonal complement of g with respect to the Beauville–Bogomolov form. One checks that $\alpha(h^2) = g$. Moreover, Beauville and Donagi establish an isomorphism of Hodge structures [**26**].

THEOREM 4.3.1 (Beauville–Donagi [**26**]). *Let X be a smooth cubic fourfold. The Abel–Jacobi map $\alpha : H^4_0(X,\mathbb{Z}) \to H^2_0(F(X),\mathbb{Z})$ satisfies:*

$$\langle \alpha(x), \alpha(y) \rangle_{BB} = -\langle x, y \rangle.$$

In other words, α induces an isomorphism of Hodge structures:

$$H_0^4(X,\mathbb{C}) \simeq H_0^2(F(X),\mathbb{C})(-1).$$

For a smooth projective surface S and a positive integer n, we write $S^{[n]}$ for the Hilbert scheme of length n subscheme on S, which is a smooth projective variety. Beauville and Donagi describe the deformation class of $F(X)$.

THEOREM 4.3.2 (Beauville–Donagi). *The Fano variety of lines $F(X)$ is an irreducible holomorphic symplectic variety deformation equivalent to $S^{[2]}$, where S is a degree 14 K3 surface.*

One possible interpretation of the results of Beauville and Donagi is that the variety $F(X)$ acts as a Hodge-theoretic analogue for the intermediate Jacobian of a cubic threefold.

The proof of Theorem 4.3.2 proceeds via a deformation argument to the case where X is a *Pfaffian cubic fourfold*, as follows. Let V be a 6-dimensional complex vector space and consider $\mathrm{Gr}(2,V) \subset \mathbb{P}(\bigwedge^2 V)$ via the Plücker embedding. The variety $\mathrm{Pf}(4,\bigwedge^2 V^*) \subset \mathbb{P}(\bigwedge^2 V^*)$ is defined as (the projectivization of) the set of degenerate skew-symmetric forms on V, which is isomorphic to the set of skew symmetric 6×6 matrices with rank bounded above by 4. It is a (nonsmooth) cubic hypersurface of $\mathbb{P}(\bigwedge^2 V^*)$ defined by the vanishing of the Pfaffian. Now let $L \subset \mathbb{P}(\bigwedge^2 V)$ be a linear subspace of dimension 8, and denote by $L^* \subset \mathbb{P}(\bigwedge^2 V)$ its orthogonal subspace, which has dimension 5. Then taking L general enough, we have that $X = L^* \cap \mathrm{Pf}(4,\bigwedge^2 V^*)$ is a smooth cubic fourfold in $L^* = \mathbb{P}^5$ and $S = L \cap \mathrm{Gr}(2,V)$ is a smooth K3 surface in $L = \mathbb{P}^8$ with a degree 14 polarization l. Cubic fourfolds arising from this construction are called *Pfaffian* with associated K3 surface S. Then Beauville and Donagi prove Theorem 4.3.2 directly for pfaffian cubic fourfolds.[2]

THEOREM 4.3.3 (Beauville–Donagi). *Let X be a Pfaffian cubic fourfold with associated K3 surface S, not containing a plane. Then X is rational and $F(X)$ is isomorphic to $S^{[2]}$.*

Theorem 4.3.2 is then obtained by a deformation argument from Theorem 4.3.3. The proofs of the two facts stated in Theorem 4.3.3 both rely on the explicit geometric construction of X and S, and do not, on the face of it, seem to be related. However, this result hints at a deep relationship between the Fano variety of lines, K3 surfaces, and the birational geometry of cubic fourfolds.

Hassett's work **[95]** is based on the study of the Hodge structure and the integral cohomology lattice of a smooth cubic fourfold X. A key observation of Beauville and Donagi is that being Pfaffian implies the existence of a rational normal quartic scroll inside X, in fact a two dimensional family of such scrolls parameterized by S. In fact, cubic fourfolds containing rational normal quartic scrolls, and their rationality, were already considered by Fano **[83]**. Hassett's key idea is to consider the class of such a ruled surface in $H^4(X,\mathbb{Z})$ and the lattice-theoretic properties that one can deduce from its existence.

Consider the integral cohomology lattice $H^4(X,\mathbb{Z})$ and its sublattice $H_0^4(X,\mathbb{Z})$. Recall that $F(X)$ is a deformation of $S^{[2]}$ and that $H^2(S^{[2]},\mathbb{Z}) = H^2(S,\mathbb{Z}) \oplus \mathbb{Z}[\delta]$,

[2] The fact that any Pfaffian cubic not containing a plane has the properties required by Beauville and Donagi's proof was proved recently by Bolognesi and Russo **[43]**.

with $\langle \delta, \delta \rangle_{BB} = -2$, is an orthogonal decomposition. In particular,

$$H^2(F(X), \mathbb{Z}) \simeq U^{\oplus 3} \oplus E_8^{\oplus 2} \oplus (-2),$$

where U is the hyperbolic lattice, E_8 is the lattice associated to the Dynkin diagram of type E_8, and (-2) is the rank one primitive sublattice generated by δ. This allows one to calculate the lattice $H_0^4(X, \mathbb{Z})$ via the Abel–Jacobi map.

PROPOSITION 4.3.4 (Hassett [**95**]). *The integral primitive cohomology lattice of a cubic fourfold is*

$$H_0^4(X, \mathbb{Z}) \simeq B \oplus U^{\oplus 2} \oplus E_8^{\oplus 2},$$

where B is a rank 2 lattice with intersection matrix:

$$\begin{pmatrix} 2 & 1 \\ 1 & 2 \end{pmatrix}.$$

In particular, $H_0^4(X, \mathbb{Z})$ has signature $(20, 2)$.

It follows from Proposition 4.3.4 that, though $H_0^4(X, \mathbb{Z})$ has the same rank as a (Tate twist of a) K3 lattice, their signatures differ, since the latter has signature $(19, 3)$. However, one should be tempted to wonder whether, or under which conditions, it is possible to find a K3 surface S and isomorphic sublattices of signature $(19, 2)$ of $H_0^4(X, \mathbb{Z})$ and of $H^2(S, \mathbb{Z})$. On the surface side, there is a very natural (and geometrically relevant) candidate: if l is a polarization on S, then the primitive cohomology $H_0^2(S, \mathbb{Z}) = \langle l \rangle^\perp$ could be a candidate to consider.

For example, let X be a Pfaffian cubic fourfold and and S an associated K3 surface with its polarization l of degree 14. As recalled, X contains a homology class of rational normal quartic scrolls parameterized by S. Let $T \in H^4(X, \mathbb{Z})$ be the cohomology class of this 2-cycle. In particular, T is not homologous to h^2, hence we have a rank 2 primitive sublattice K, generated by T and h^2, of $H^4(X, \mathbb{Z})$. As $T.T = c_2(N_{T/X}) = 10$, we have that the intersection matrix

$$\begin{pmatrix} 3 & 4 \\ 4 & 10 \end{pmatrix}$$

of K has determinant 14, equal to the degree of the polarized K3 surface S associated to the Pfaffian construction of X. The key remark of Hassett is that $K^\perp \subset H^4(X, \mathbb{Z})$ and $l^\perp = H_0^2(X, \mathbb{Z}) \subset H^2(X, \mathbb{Z})$ are isomorphic lattices (up to a Tate twist) in this case. This motivates the following definition.

DEFINITION 4.3.5 (Hassett). A cubic fourfold X is *special* if it contains an algebraic 2-cycle T, not homologous to h^2, i.e., if the rank of $\mathrm{CH}^2(X)$ is at least 2.

Given an abstract rank 2 positive definite lattice K with a distinguished element h^2 of self-intersection 3, a *labeling* of a special cubic fourfold is the choice of a primitive embedding $K \hookrightarrow \mathrm{CH}^2(X)$ identifying the distinguished element with the double hyperplane section h^2. The *discriminant* of a labeled special cubic fourfold (X, K) is defined to be the determinant of the intersection matrix of K; it is a positive integer. Note that a cubic fourfold could have labelings of different discriminants.

Let (X, K) be a labeled special cubic fourfold. A polarized K3 surface (S, l) is *associated* to (X, K) if there is an isomorphism of lattices $K^\perp \simeq H_0^2(S, \mathbb{Z})(-1)$.

EXAMPLE 4.3.6 (Hassett). If X is a Pfaffian cubic fourfold with associated polarized K3 surface (S, l) of degree 14, then X is special, has a labeling K of discriminant 14 defined by the class of the rational normal quartic scrolls parameterized by S, and (S, l) is associated to (X, K).

On the other hand, Bolognesi and Russo [**43**, Thm. 0.2] have shown that any special cubic fourfold of discriminant 14 not containing a plane is Pfaffian. Even more has been proved in [**10**] by completely different techniques: any special cubic fourfold of discriminant 14 that is not Pfaffian must contain two disjoint planes.

EXAMPLE 4.3.7 (Hassett [**95**], [**96**]). Let X be a smooth cubic fourfold containing a plane $P \subset \mathbb{P}^5$. Such X is special, as P is not homologous to h^2. Since $P.P = c_2(N_{P/X}) = 3$, we have that the sublattice K generated by h^2 and P has intersection matrix

$$\begin{pmatrix} 3 & 1 \\ 1 & 3 \end{pmatrix},$$

so defines a labeling of discriminant 8. In general, there is no K3 surface associated to this labeled cubic fourfold (X, K) (see Theorem 4.3.8).

Consider the projection $\mathbb{P}^5 \dashrightarrow \mathbb{P}^2$ from the plane P. Restricting this projection to X gives rise to a rational map $X \dashrightarrow \mathbb{P}^2$ which can be resolved, by blowing up P, into a quadric surface bundle $\pi : \widetilde{X} \to \mathbb{P}^2$, degenerating along a (generically smooth) sextic curve $C \subset \mathbb{P}^2$. The double cover $S \to \mathbb{P}^2$ branched along C is a K3 surface with a polarization of degree 2, which plays a rôle in the Hodge theory of X, but is not associated to (X, K).

Using the period map and the Torelli Theorem (see [**174**]) for cubic fourfolds, one can construct a 20-dimensional (coarse) algebraic moduli space $\mathcal{C}$ of smooth cubic fourfolds, as explained in [**95**, 2.2]. Using this algebraic structure, Hassett shows that the very general cubic fourfold is not special, and that the locus of special cubic fourfolds of fixed discriminant is a divisor of $\mathcal{C}$, which might be empty depending on the value of the discriminant. Hassett also finds further restrictions on the discriminant for special cubic fourfolds having associated K3 surfaces S and for which $F(X)$ is isomorphic to $S^{[2]}$.

THEOREM 4.3.8 (Hassett [**95**]). *Special cubic fourfolds of discriminant d form a nonempty irreducible divisor $\mathcal{C}_d \subset \mathcal{C}$ if and only if $d > 6$ and $d \equiv 0, 2 \bmod 6$.*

Special cubic fourfolds of discriminant $d > 6$ have associated K3 surfaces if and only if d is not divisible by 4, 9, or any odd prime $p \equiv 2$ modulo 3.

Assume that $d = 2(n^2 + n + 1)$ where $n \geq 2$ is an integer, and let X be a generic special cubic fourfold of discriminant d, in which case X has an associated K3 surface S. Then there is an isomorphism $F(X) \simeq S^{[2]}$.

REMARK 4.3.9. The condition on $d > 6$ ensures that X is smooth. For completeness, the low discriminant cases are known: a cubic fourfold of discriminant 2 is determinantal (and hence is singular along a Veronese surface), see [**95**, 4.4]; a cubic fourfold of discriminant 6 has a single ordinary double point, see [**95**, 4.2]. The loci $\mathcal{C}_6$ and $\mathcal{C}_2$ do not lie in the moduli space $\mathcal{C}$, but rather in its boundary (see [**131**] and [**132**]).

The last statement in Theorem 4.3.8 can be made more precise, once one weakens it by asking that $F(X)$ is not isomorphic but just birational to $S^{[2]}$. The numerical necessary and sufficient condition for this was established by Addington [**2**].

THEOREM 4.3.10 (Addington). *Let X be a special cubic fourfold of discriminant d, with associated K3 surface S. Then $F(X)$ is birational to $S^{[2]}$ if and only if d is* *of the form*

$$d = \frac{2n^2 + 2n + 2}{a^2},$$

for some n and a in $\mathbb{Z}$.

As noticed by Addington [**2**], having an associated K3 surface does not necessarily imply that $F(X)$ is birational to $S^{[2]}$. The numerical condition from the second statement of Theorem 4.3.8 is indeed strictly stronger than the numerical condition from Theorem 4.3.10. The smallest value of d for which a special cubic of discriminant d has an associated K3 surface S but $F(X)$ is not birational to $S^{[2]}$ is 74.

Let us recall the known examples of rational cubic fourfolds, in order to consider a Hodge-theoretic expectation about rationality.

EXAMPLE 4.3.11. Let X be a cubic fourfold. If either

2,6) X is singular, e.g. $X \in \mathcal{C}_6$ has a single node or $X \in \mathcal{C}_2$ is determinantal; or

8) X contains a plane P, so that $X \in \mathcal{C}_8$, and the associated quadric surface fibration $\widetilde{X} \to \mathbb{P}^2$ (see Example 4.3.7) admits a multisection of odd degree [**96**]; or

14) X is Pfaffian, so that $X \in \mathcal{C}_{14}$ [**26**];

then X is rational.[3]

In particular, all cubics in $\mathcal{C}_2$ or $\mathcal{C}_6$, and the general cubic in $\mathcal{C}_{14}$ are rational.[4] The cubics in $\mathcal{C}_8$ satisfying condition 8) form a countable union of divisors in $\mathcal{C}_8$ [**96**].

Let X be a cubic containing a plane, and $\widetilde{X} \to \mathbb{P}^2$ the associated quadric fibration. Having an odd section for $\widetilde{X}$ is a sufficient, but not necessary condition for rationality. Indeed, there exist Pfaffian cubics in $\mathcal{C}_8$ such that $\widetilde{X} \to \mathbb{P}^2$ doesn't have any odd section. Such cubics are then rational, they lie in the intersection $\mathcal{C}_8 \cap \mathcal{C}_{14}$ and were constructed in [**13**].

If one imagines that $H^4_0(X, \mathbb{Z})$, with its Hodge structure, plays the rôle that the intermediate Jacobian plays for cubic threefolds, then one would naturally expect that having no associated K3 surface should be an obstruction to rationality. For more on this perspective, see the recent survey [**97**, §3].[5] On the other hand, there is no known example of a nonrational cubic fourfold, and few general families of rational ones. One should be very cautious when wondering whether having an associated K3 surface is a sufficient criterion of rationality. Hassett has asked about the existence of further examples of rational cubic fourfolds [**97**, Question 16].

On the other hand, as we will see in Section 8, Kuznetsov's conjecture [**123**] is equivalent, at least for a generic cubic fourfold, to the statement that the rationality of X is equivalent to the existence of an associated K3, as shown by Addington and Thomas [**3**]. As we will see later, decompositions of the derived category of a cubic

---

[3] A new class of rational cubic fourfolds X has very recently been constructed by Addington, Hassett, Tschinkel, and Várilly-Alvarado [**4**], these are in $\mathcal{C}_{18}$ and are birational to a fibration $\widetilde{X} \to \mathbb{P}^2$ in sextic del Pezzo surfaces admitting a multisection of degree prime to 3.

[4] In fact, every cubic fourfold in $\mathcal{C}_{14}$ is rational, as was proved independently in [**10**] and [**43**].

[5] A sample result showing the interplay between Hodge theory and rationality is provided by Kulikov, who has shown that Hodge-indecomposability of the transcendental cohomology would be a sufficient condition of nonrationality for X, see [**115**]. However, such indecomposability was recently shown not to hold in [**14**].

fourfold increase the amount of evidence motivating the expectation that having an associated K3 surface should be a necessary condition for rationality. Then one should read Kuznetsov conjecture and Hassett's question [**97**, Question 16] as the two most "rational" or "nonrational" expectations for cubic fourfolds.

Let us end this section by recalling Galkin and Shinder's construction [**87**], relying on motivic measures, which aims to describe a criterion of nonrationality. This construction would have given indeed a nonrationality criterion under the hypothesis that the class of the affine line $\mathbb{L}$ in $K_0(\mathrm{Var}(\mathbb{C}))$ is not a zero-divisor (see Chapter 8 for details on this Grothendieck group). Unfortunately, after Galkin and Shinder's paper appeared, Borisov [**50**] proved that $\mathbb{L}$ is indeed a zero-divisor. However, we recall Galkin–Shinder's statement:

Assume that the class of the affine line $\mathbb{L}$ is not a zero divisor in the Grothendieck ring $K_0(\mathrm{Var}(\mathbb{C}))$. If a cubic fourfold X is rational, then $F(X)$ is birational to $S^{[2]}$, where S is a K3 surface.

Though based on a false assumption,[6] the previous statement, together with Theorem 4.3.10, would say that having an associated K3 is not a sufficient condition to rationality, the first examples being cubic with discriminant 74 or 78 (see [**2**]). As a conclusion, we must admit that we are probably facing one of the most intriguing problems of birational geometry: not only proving that the general cubic is not rational, but also classifying the rational ones seems to need much more work and finer invariants.

5. Rationality and 0-cycles

One of the fundamental ingredients in the recent breakthrough in the stable rationality problem was to explicitly tie together the decomposition of the diagonal and the universal triviality of CH_0. Such a link was certainly established in the work of Bloch and Srinivas. In this section, we want to explain this relationship and show how it is useful.

5.1. Diagonals and 0-cycles. We begin with the fact that CH_0 is a birational invariant of smooth proper irreducible varieties, proved by Colliot-Thélène and Coray [**66**, Prop. 6.3] using resolution of singularities and more generally by Fulton [**84**, Ex. 16.1.11] using the theory of correspondences.

LEMMA 5.1.1. *Let X and Y be smooth proper varieties over a field k. If X and Y are k-birationally equivalent then* $\mathrm{CH}_0(X) \cong \mathrm{CH}_0(Y)$.

PROOF. Let $f : Y \dashrightarrow X$ be a birational map and $\alpha \in \mathrm{CH}^n(Y \times X)$ the closure of the graph of f, considered as a correspondence from Y to X. Let $\alpha' \in \mathrm{CH}^n(X \times Y)$ be the transpose correspondence. To verify that α_* and α'_* define inverse bijections, we check that $\alpha' \circ \alpha$ is the sum of the identity (diagonal) correspondence and other correspondences whose projections to Y are contained in proper subvarieties. By the moving lemma for 0-cycles, we can move any element in $\mathrm{CH}_0(Y)$, up to rational equivalence, away from any of these subvarieties, to where $(\alpha' \circ \alpha)_* = \alpha'_* \circ \alpha_*$ is the identity map. □

[6]In fact, less is required, only that $\mathbb{L}$ does not annihilate any sum of varieties of dimension at most 2, a condition which has recently been shown to fail as well, see [**129**] and [**99**].

If X is proper over k, then there is a well-defined degree map $\mathrm{CH}_0(X) \to \mathbb{Z}$. We say that $\mathrm{CH}_0(X)$ is *universally trivial* if $\deg : \mathrm{CH}_0(X_F) \to \mathbb{Z}$ is an isomorphism for every field extension F/k. This notion was first considered by in a paper by Merkurjev [**138**, Thm. 2.11]. Let N be a positive integer. We say that $\mathrm{CH}_0(X)$ is *universally N-torsion* if $\deg : \mathrm{CH}_0(X_F) \to \mathbb{Z}$ is surjective and has kernel killed by N for every field extension F/k.

Note that $\deg : \mathrm{CH}_0(\mathbb{P}^n_k) \simeq \mathbb{Z}$ over any field k, so that $\mathrm{CH}_0(\mathbb{P}^n)$ is universally trivial. By Lemma 5.1.1, if a smooth proper variety X is k-rational then $\mathrm{CH}_0(X)$ is universally trivial. In fact, the same conclusion holds if X is retract k-rational, in particular, stably k-rational, which can be proved using [**109**, Cor. RC.12], see also [**69**].

To check the triviality of $\mathrm{CH}_0(X_F)$ over every field extension F/k seems like quite a burden. However, it suffices to check it over the function field by the following theorem, proved in [**15**, Lemma 1.3].

THEOREM 5.1.2. *Let X be a geometrically irreducible smooth proper variety over a field k. Then the following are equivalent:*

(i) The group $\mathrm{CH}_0(X)$ is universally trivial.
(ii) The variety X has a 0-cycle of degree 1 and $\deg : \mathrm{CH}_0(X_{k(X)}) \to \mathbb{Z}$ is an isomorphism.
(iii) The variety X has an (integral) decomposition of the diagonal.

PROOF. If $\mathrm{CH}_0(X)$ is universally trivial then $\mathrm{CH}_0(X_{k(X)}) = \mathbb{Z}$ and X has a 0-cycle of degree 1, by definition. Let us prove that if X has a 0-cycle P of degree 1 and $\mathrm{CH}_0(X_{k(X)}) = \mathbb{Z}$ then X has a decomposition of the diagonal. Write $n = \dim(X)$. Let $\xi \in X_{k(X)}$ be the $k(X)$-rational point which is the image of the "diagonal morphism" $\mathrm{Spec} k(X) \to X \times_k \mathrm{Spec} k(X)$. By hypothesis, we have $\xi = P_{k(X)}$ in $\mathrm{CH}_0(X_{k(X)})$. The closures of $P_{k(X)}$ and ξ in $X \times X$ are $P \times X$ and the diagonal Δ_X, respectively. By the closure in $X \times X$ of a 0-cycle on $X_{k(X)}$, we mean the sum, taken with multiplicity, of the closures of each closed point in the support of the 0-cycle on $X_{k(X)}$. Hence the class of $\Delta_X - P \times X$ is in the kernel of the map $\mathrm{CH}^n(X \times X) \to \mathrm{CH}^n(X_{k(X)})$. Since $\mathrm{CH}^n(X_{k(X)})$ is the inductive limit of $\mathrm{CH}^n(X \times_k U)$ over all nonempty open subvarieties U of X, we have that $\Delta_X - P \times X$ vanishes in some $\mathrm{CH}^n(X \times U)$. We thus have a decomposition of the diagonal

$$\Delta_X = P \times X + Z$$

in $\mathrm{CH}^n(X \times X)$, where Z is a cycle with support in $X \times V$ for some closed subvariety $X \smallsetminus U = V \subsetneq X$.

Now we prove that if X has a decomposition of the diagonal, then $\mathrm{CH}_0(X)$ is universally trivial. This argument is similar in spirit to the proof of 3.7.1 presented in §3.7. The action of correspondences (from §1.3) on 0-cycles has the following properties: $[\Delta_X]_*$ is the identity map and $[P \times X]_*(z) = \deg(z)P$ for any 0-cycle $z \in \mathrm{CH}_0(X)$. By the easy moving lemma for 0-cycles on a smooth variety recalled at the end of §1.2, for a closed subvariety $V \subsetneq X$, every 0-cycle on X is rationally equivalent to one with support away from V. This implies that $[Z]_* = 0$ for any n-cycle with support on $X \times V$ for a proper closed subvariety $V \subset X$. Thus a decomposition of the diagonal $\Delta_X = P \times X + Z$ as in (3.1) implies that the identity map restricted to the kernel of the degree map $\deg : \mathrm{CH}_0(X) \to \mathbb{Z}$ is zero. For any field extension F/k, we have the base-change $\Delta_{X_F} = P_F \times X_F + Z_F$ of the decomposition of the diagonal (3.1), hence the same argument as above shows that $\mathrm{CH}_0(X_F) = \mathbb{Z}$. We conclude that $\mathrm{CH}_0(X)$ is universally trivial. □

This result is useful because often statements about CH_0 are easier to prove than statements about CH_n. There is also a version with universal N-torsion.

THEOREM 5.1.3. *Let X be a geometrically irreducible smooth proper variety over a field k. Then the following are equivalent:*

(i) The group $\mathrm{CH}_0(X)$ is universally N-torsion.
(ii) The variety X has a 0-cycle of degree 1 and $\deg : \mathrm{CH}_0(X_{k(X)}) \to \mathbb{Z}$ *has kernel killed by N.*
(iii) The variety X has a rational decomposition of the diagonal of the form $N\Delta_X = N(P \times X) + Z$ for a 0-cycle P of degree 1 on X.

Now we mention a result of Merkurjev that helped to inspire the whole theory. Recall, from 3.2, the definition of the group of unramified elements $M_{\mathrm{ur}}(X)$ of a cycle module M and that we say $M_{\mathrm{ur}}(X)$ is trivial when the natural map $M(k) \to M_{\mathrm{ur}}(X)$ is an isomorphism.

THEOREM 5.1.4 (Merkurjev [**138**, Thm. 2.11]). *Let X be a smooth proper variety over a field k. Then the following are equivalent:*

(i) $\mathrm{CH}_0(X)$ is universally trivial.
(ii) $M_{\mathrm{ur}}(X)$ is universally trivial for any cycle module M.

There is also an analogous version of Merkurjev's result for universal N-torsion. The triviality of unramified elements in cycle modules is quite useful.

COROLLARY 5.1.5. *If X is a proper smooth retract rational (e.g., stably rational) variety, then $M_{\mathrm{ur}}(X)$ is universally trivial for all cycle modules M. In particular, all unramified cohomology is universally trivial, e.g., $H^1_{\text{ét}}(X, \mu)$ and $\mathrm{Br}(X)$ are universally trivial and, if $k = \mathbb{C}$, then the integral Hodge conjecture for codimension 2 cycles holds for X.*

EXAMPLE 5.1.6. In the spirit of Mumford's theorem on 2-forms on surfaces, if X is an algebraic surface with $p_g(X) > 0$ (more generally, $\rho(X) < b_2(X)$), then $\mathrm{CH}_0(X)$ is not universally trivial and X does not have a decomposition of the diagonal. Here we use the fact, from §3.3, that $\mathrm{Br}(X) \cong (\mathbb{Q}/\mathbb{Z})^{b_2-\rho} \oplus H$, for H a finite group.

5.2. Rationally connected varieties. A smooth projective variety X over a field k is called *rationally connected* if for every algebraically closed field extension K/k, any two K-points of X can be connected by the image of a K-morphism $\mathbb{P}^1_K \to X_K$.

For example, smooth geometrically unirational varieties are rationally connected. It is a theorem of Campana [**52**] and Kollár–Miyaoka–Mori [**114**] that any smooth projective Fano variety over a field of characteristic zero is rationally connected.

If X is rationally connected, then $\mathrm{CH}_0(X_K) = \mathbb{Z}$ for any algebraically closed field extension K/k. While a standard argument then proves that the kernel of the degree map $\deg : \mathrm{CH}_0(X_F) \to \mathbb{Z}$ is torsion for every field extension F/k, the following more precise result is known.

PROPOSITION 5.2.1. *Let X be a smooth proper connected variety over a field k. Assume that X is rationally connected, or more generally, that $\mathrm{CH}_0(X_K) = \mathbb{Z}$ for all algebraically closed extensions K/k.*

(i) *(Bloch–Srinivas* [**37**, Prop. 1]*) Then X has a rational decomposition of the diagonal.*
(ii) *(Colliot-Thélène* [**62**, Prop. 11]*) Then there exists an integer $N > 0$ such that $\mathrm{CH}_0(X)$ is universally N-torsion.*

Of course, both of these are equivalent by Theorem 5.1.3.

In fact, over $\mathbb{C}$, something more general can be proved.

LEMMA 5.2.2. *Let X be a smooth proper connected variety over an algebraically closed field k of infinite transcendence degree over its prime field (e.g., $k = \mathbb{C}$). If $\mathrm{CH}_0(X) = \mathbb{Z}$ then there exists an integer $N > 0$ such that $\mathrm{CH}_0(X)$ is universally N-torsion.*

PROOF. The variety X is defined over an algebraically closed subfield $L \subset k$, with L algebraic over a field finitely generated over its prime field. That is, there exists a variety X_0 over L with $X \cong X_0 \times_L k$. Let η be the generic point of X_0. Let P be an L-point of X_0. One may embed the function field $F = L(X_0)$ into k, by the transcendence degree hypothesis on k. Let K be the algebraic closure of F inside k. By Lemma 5.2.3 (below) and the hypothesis that $\mathrm{CH}_0(X) = \mathbb{Z}$, we have that $\mathrm{CH}_0(X_0 \times_L F) = \mathbb{Z}$. This implies that there is a finite extension E/F of fields such that $\eta_E - P_E = 0$ in $\mathrm{CH}_0(X_0 \times_L E)$. Taking the corestriction (i.e., pushforward) to F, one finds that $N(\eta_F - P_F) = 0$ in $\mathrm{CH}_0(X_0 \times_L F)$, hence in $\mathrm{CH}_0(X)$ as well. As in the proof of Theorem 5.1.2, we conclude that $\mathrm{CH}_0(X)$ is universally N-torsion. □

LEMMA 5.2.3. *Let X be a smooth projective connected variety over k. If K/k is an extension of fields, then the kernel of the natural map $\mathrm{CH}_0(X) \to \mathrm{CH}_0(X_K)$ is torsion. If k is algebraically closed, then $\mathrm{CH}_0(X) \to \mathrm{CH}_0(X_K)$ is injective.*

PROOF. Let z be a 0-cycle on X that becomes rationally equivalent to zero on X_K. Then there exists a subextension L of K/k that is finitely generated over k, such that z becomes rationally equivalent to zero on X_L. In fact, we can find a finitely generated k-algebra A with fraction field L such that z maps to zero under $\mathrm{CH}_0(X) \to \mathrm{CH}_0(X \times_k U)$ where $U = \operatorname{Spec} A$. When k is algebraically closed, there exists a k-point of U, defining a section of $\mathrm{CH}_0(X) \to \mathrm{CH}_0(X \times_k U)$, showing that z is zero in $\mathrm{CH}_0(X)$. In general, we can find a rational point of U over a finite extension k'/k, so that $z_{k'}$ is zero in $\mathrm{CH}_0(X_{k'})$, from which we conclude that a multiple of z is zero in $\mathrm{CH}_0(X)$ by taking corestriction. □

There exist rationally connected varieties X over an algebraically closed field of characteristic zero with $\mathrm{CH}_0(X)$ not universally trivial. Indeed, let X be a unirational threefold with $H^2_{\mathrm{ur}}(X, \mathbb{Q}/\mathbb{Z}(1)) \cong \mathrm{Br}(X) \neq 0$, see e.g., [**9**]. Then by Theorem 5.1.3, $\mathrm{CH}_0(X)$ is not universally trivial.

However, such examples do not disprove the natural universal generalization of the result of Campana [**52**] and Kollár–Miyaoka–Mori [**114**], and this was posed as a question in [**15**, §1].

QUESTION 5.2.4. *Does there exist a smooth Fano variety X over an algebraically closed field of characteristic* 0 *with $\mathrm{CH}_0(X)$ not universally trivial?*

After this question was posed, Voisin [**177**] constructed the first examples of (smooth) Fano varieties over $\mathbb{C}$ with $\mathrm{CH}_0(X)$ not universally trivial, see §7.3 for more details.

5.3. Surfaces. We recall Bloch's conjecture for a complex surface. Let X be a smooth projective variety. The Albanese map $\mathrm{alb}_X : X \to \mathrm{Alb}(X)$ is universal for morphisms from X to an abelian variety. It extends to the Albanese map

$$\mathrm{alb}_X : A_0(X) \to \mathrm{Alb}(X)$$

where $A_0(X)$ denotes the kernel of the degree map $\mathrm{CH}_0(X) \to \mathbb{Z}$. The Albanese map is surjective on geometric points. In characteristic zero, the dimension of $\mathrm{Alb}(X)$ is $q(X) = h^1(X, \mathscr{O}_X)$. Recall that $p_g(X) = h^0(X, \Omega^n_X)$ where $n = \dim(X)$.

CONJECTURE 5.3.1 (Bloch's conjecture). *Let X be a smooth projective surface over $\mathbb{C}$. If $p_g(X) = 0$ then the Albanese map $\mathrm{alb}_X : A_0(X) \to \mathrm{Alb}(X)$ is injective. In particular, if $p_g(X) = q(X) = 0$, then $A_0(X) = 0$, i.e., $\mathrm{CH}_0(X) = \mathbb{Z}$.*

In fact, Bloch's conjecture is proved for all surfaces that are not of general type by Bloch, Kas, and Lieberman [**35**].

Of course, rational surfaces satisfy $p_g = q = 0$ and have $A_0(X) = 0$. There do exists nonrational surfaces with $p_g = q = 0$ and for which $A_0(X) = 0$. Enriques surfaces were the first examples, extensively studied in [**80**], [**81**, p. 294] with some examples considered earlier in [**152**], see also [**55**]. An Enriques surface has Kodaira dimension 0. We remark that for an Enriques surface X, we have that

$$H^1_{\mathrm{ur}}(X, \mathbb{Z}/2\mathbb{Z}) = H^1_{\text{ét}}(X, \mathbb{Z}/2\mathbb{Z}) = \mathbb{Z}/2\mathbb{Z} \text{ and } H^2_{\mathrm{ur}}(X, \mathbb{Q}/\mathbb{Z}(1)) = \mathrm{Br}(X) = \mathbb{Z}/2\mathbb{Z}.$$

Hence $\mathrm{CH}_0(X)$ is not universally trivial and X does not have a decomposition of the diagonal by Theorem 5.1.4.

The first surfaces of general type with $p_g = q = 0$ were constructed in [**54**] and [**89**]. Simply connected surfaces X of general type for which $p_g = 0$ were constructed by Barlow [**20**], who also proved that $\mathrm{CH}_0(X) = \mathbb{Z}$ for some of them. See also the recent work on Bloch's conjecture by Voisin [**172**].

We want to explore the universal analogue of Bloch's conjecture, i.e., to what extent does $p_g = q = 0$ imply universal triviality of $\mathrm{CH}_0(X)$.

The following result was stated without detailed proof as the last remark of [**37**]. The first proof appeared in [**15**, Prop. 1.19] using results of [**70**] and a different proof appear later in [**178**, Cor. 2.2].

PROPOSITION 5.3.2. *Let X be a smooth proper connected surface over $\mathbb{C}$. Suppose that all groups $H^i_{\mathrm{B}}(X, \mathbb{Z})$ are torsion free and that $\mathrm{CH}_0(X) = \mathbb{Z}$. Then $\mathrm{CH}_0(X)$ is universally trivial and admits a decomposition of the diagonal.*

PROOF. By Lemma 5.2.2, we have that $\mathrm{CH}_0(X)$ is universally N-torsion. Hence by Lemma 3.7.1, we have that $H^i(X, \mathscr{O}_X) = 0$ for all $i \geq 1$. Thus $p_g(X) = q(X) = 0$ and hence $b_3(X) = b_1(X) = 2q(X) = 0$, so that $H^i_{\mathrm{B}}(X, \mathbb{Z})$ consists purely of cycle classes of algebraic cycles.

The torsion-free hypothesis on cohomology allows one to use the work of Colliot-Thélène and Raskind [**70**, Thm. 3.10(d)] on the cohomology of the Milnor K-theory sheaf, to conclude that $\mathrm{CH}_0(X)$ is universally trivial.

The torsion-free hypothesis on cohomology allows Voisin [**178**, Cor. 2.2] to argue using the integral Künneth decomposition of the diagonal (see Remark 5.3.3), that X admits a decomposition of the diagonal. □

We remark that if X is a smooth proper connected surface over $\mathbb{C}$ with torsion free Néron–Severi group $\mathrm{NS}(X)$, then all Betti cohomology groups are torsion free, hence Proposition 5.3.2 applies. Indeed, the torsion in $H^1_{\mathrm{B}}(X,\mathbb{Z})$ is clearly trivial and is dual to the torsion in $H^3_{\mathrm{B}}(X,\mathbb{Z})$, while the torsion in $\mathrm{NS}(X)$ is isomorphic to the torsion in $H^2_{\mathrm{B}}(X,\mathbb{Z})$.

REMARK 5.3.3. If $H^i_{\mathrm{B}}(X,\mathbb{Z})$ is torsion free for all $0 \le i \le n$, then there is an integral Künneth decomposition

$$H^n_{\mathrm{B}}(X \times X,\mathbb{Z}) = \bigoplus_{i+j=n} H^i_{\mathrm{B}}(X,\mathbb{Z}) \otimes H^j_{\mathrm{B}}(X,\mathbb{Z}).$$

This follows from the degeneration of the Künneth spectral sequence with coefficients in $\mathbb{Z}$.

We remark that the simply connected surfaces X of general type with $p_g = q = 0$ and $\mathrm{CH}_0(X) = 0$, e.g., Barlow surfaces, satisfy $\mathrm{Pic}(X) = \mathrm{NS}(X)$ is torsion free, hence Proposition 5.3.2 applies. While the group $\mathrm{CH}_0(X)$ is universally trivial, these surfaces are far from being rational, since they are of general type.

The interested reader can find how to adapt Proposition 5.3.2 over an algebraically closed field of infinite transcendence degree over its prime field.

Finally, we mention that Proposition 5.3.2 has been generalized by Kahn [**107**], and independently by Colliot-Thélène using [**70**], to a determination of the minimal N for which $\mathrm{CH}_0(X)$ is universally N-torsion, which turns out to be the exponent of $\mathrm{NS}(X)$. In general, the minimal $N \geq 1$ for which $\mathrm{CH}_0(X)$ universally N-torsion is a stable birational invariant of smooth proper varieties; its properties are explored in [**56**], where it is called the *torsion order* of X.

6. Categorical representability and rationality, the case of surfaces

This section consists of two main parts. In the first part, we define the notion of categorical representability and begin to classify (or at least, give criteria to discriminate) categories which are representable in low dimension. The second part is devoted to the applications in the case of surfaces.

6.1. Categorical representability. Using semiorthogonal decompositions, one can define a notion of *categorical representability* for a triangulated category. In the case of smooth projective varieties, this is inspired by the classical notions of representability of cycles, see [**30**].

DEFINITION 6.1.1. A k-linear triangulated category T is *representable in dimension m* if it admits a semiorthogonal decomposition

$$\mathsf{T} = \langle \mathsf{A}_1, \ldots, \mathsf{A}_r \rangle,$$

and for each $i = 1, \ldots, r$ there exists a smooth projective connected k-variety Y_i with $\dim Y_i \le m$, such that A_i is equivalent to an admissible subcategory of $\mathsf{D}^{\mathrm{b}}(Y_i)$.

We use the following notation

$$\operatorname{rdim} \mathsf{T} := \min\{m \in \mathbb{N} \mid \mathsf{T} \text{ is representable in dimension } m\},$$

whenever such a finite m exists.

DEFINITION 6.1.2. Let X be a smooth projective k-variety. We say that X is *categorically representable* in dimension (resp. codimension) m if $\mathsf{D}^{\mathrm{b}}(X)$ is representable in dimension m (resp. dimension $\dim(X) - m$).

We will use the following notations:

$$\mathrm{rdim}(X) := \mathrm{rdim}\,\mathsf{D}^{\mathrm{b}}(X) \quad \mathrm{rcodim}(X) := \dim(X) - \mathrm{rdim}\,\mathsf{D}^{\mathrm{b}}(X),$$

and notice that they are both integer numbers.

We notice that, by definition, if $\mathrm{rdim}\,\mathsf{T} = n$, then T is representable in any dimension $m \geq n$.

LEMMA 6.1.3. *Let T be a k-linear triangulated category. If T is representable in dimension n, then T is representable in dimension m for any $m \geq n$.*

REMARK 6.1.4. **Warning!** Suppose that T is representable in dimension n via a semiorthogonal decomposition $\mathsf{T} = \langle \mathsf{A}_1, \ldots, \mathsf{A}_r \rangle$, and let $\mathsf{T} = \langle \mathsf{B}_1 \ldots, \mathsf{B}_s \rangle$ be another semiorthogonal decomposition (that is, B_i is not admissible in A_j and A_j is not admissible in B_i for any i and j). As recalled in Proposition 2.3.14 the Jordan–Hölder property for semiorthogonal decompositions does not hold in general. It follows that one does not know in general whether the B_i are also representable in dimension n, and counterexamples are known: in Bondal–Kuznetsov [**124**] there is a threefold X with a full exceptional sequence $\langle E_1, \ldots, E_6 \rangle$, and another exceptional object F whose complement cannot be generated by exceptional objects.

Let us record a simple corollary of Theorem 2.3.1.

LEMMA 6.1.5. *Let $X \to Y$ be the blow-up of a smooth projective k-variety along a smooth center. Then $\mathrm{rcodim}(X) \geq \max\{\mathrm{rcodim}(Y), 2\}$. In particular, if $\mathrm{rcodim}(Y) \geq 2$, then $\mathrm{rcodim}(X) \geq 2$.*

PROOF. This is a consequence of Theorem 2.3.1. Denoting by $Z \subset Y$ the center of the blow-up, we have $\mathrm{rdim}(X) \leq \max\{\mathrm{rdim}(Y), \mathrm{rdim}(Z)\}$. The statement follows since $\dim(X) = \dim(Y)$ and that Z has codimension at least 2 in Y. □

Inspired by the proof of Theorem 4.1.1, one can consider a birational map $X \dashrightarrow X'$ and its resolutions: by Hironaka's resolution of singularities, there is a smooth projective X_1 with birational morphisms $\rho_1 : X_1 \to X'$ and $\pi_1 : X_1 \to X$, such that π_1 is a composition of a finite number of smooth blow-ups. Similarly, there are $\rho_2 : X_2 \to X$ and $\pi_2 : X_2 \to X'$ birational morphisms with π_2 a composition of a finite number of blow-ups. By Lemma 2.3.2 we have that $\mathsf{D}^{\mathrm{b}}(X')$ is admissible in $\mathsf{D}^{\mathrm{b}}(X_1)$, and $\mathsf{D}^{\mathrm{b}}(X)$ is admissible in $\mathsf{D}^{\mathrm{b}}(X_2)$. Lemma 6.1.5 gives bounds for $\mathrm{rcodim}(X_1)$ and $\mathrm{rcodim}(X_2)$ in terms of $\mathrm{rcodim}(X)$ and $\mathrm{rcodim}(X')$ respectively.

Based on these considerations, Kuznetsov [**126**] argues that if one could properly define an admissible subcategory GK_X of $\mathsf{D}^{\mathrm{b}}(X)$, maximal (with respect to the inclusion ordering) with respect to the property $\mathrm{rdim}\,\mathsf{GK}_X \geq \dim(X) - 1$, then such a category would be a birational invariant, which we would call *the Griffiths–Kuznetsov component* of X. In particular, since $\mathrm{rdim}(\mathbb{P}^n) = 0$, we would have that the Griffiths–Kuznetsov component of a rational variety is trivial.

Even if the Griffiths–Kuznetsov component is not well-defined, we have that if X is rational, then $\mathsf{D}^{\mathrm{b}}(X)$ is admissible in a category with $\mathrm{rdim} \leq \dim(X) - 2$. As we recalled in Remark 6.1.4, there is no known reason to deduce that $\mathrm{rcodim}(X) \geq 2$. However, in the small dimensional cases, we have a stronger understanding of these phenomena. We will come back to this question, giving more detailed arguments for threefolds and examples for fourfolds, in §8.

Representability in dimension 0.

PROPOSITION 6.1.6. *Let* T *be a* k*-linear triangulated category.* $\operatorname{rdim}\mathsf{T} = 0$ *if and only if there exists a semiorthogonal decomposition*

$$\mathsf{T} = \langle \mathsf{A}_1, \dots, \mathsf{A}_r \rangle,$$

such that for each i*, there is a* k*-linear equivalence* $\mathsf{A}_i \simeq \mathsf{D}^{\mathrm{b}}(K_i/k)$ *for an étale* k*-algebra* K_i.

An additive category T is *indecomposable* if for any product decomposition $\mathsf{T} \simeq \mathsf{T}_1 \times \mathsf{T}_2$ into additive categories, we have that $\mathsf{T} \simeq \mathsf{T}_1$ or $\mathsf{T} \simeq \mathsf{T}_2$. Equivalently, T has no nontrivial completely orthogonal decomposition. We remark that for a k-scheme X the category $\mathsf{D}^{\mathrm{b}}(X)$ is indecomposable if and only if X is connected (see [**51**, Ex. 3.2]). More is known if X is the spectrum of a field or a product of fields, see [**11**].

LEMMA 6.1.7. *Let* K *be a* k*-algebra.*

(i) *If* K *is a field and* A *is a nonzero admissible* k*-linear triangulated subcategory of* $\mathsf{D}^{\mathrm{b}}(k, K)$*, then* $\mathsf{A} = \mathsf{D}^{\mathrm{b}}(k, K)$.
(ii) *If* $K \cong K_1 \times \cdots \times K_n$ *is a product of field extensions of* k *and* A *is a nonzero admissible indecomposable* k*-linear triangulated subcategory of* $\mathsf{D}^{\mathrm{b}}(k, K)$*, then* $\mathsf{A} \simeq \mathsf{D}^{\mathrm{b}}(k, K_i)$ *for some* $i = 1, \dots, n$.
(iii) *If* $K \cong K_1 \times \cdots \times K_n$ *is a product of field extensions of* k *and* A *is a nonzero admissible* k*-linear triangulated subcategory of* $\mathsf{D}^{\mathrm{b}}(k, K)$*, then* $\mathsf{A} \simeq \prod_{j \in I} \mathsf{D}^{\mathrm{b}}(k, K_j)$ *for some subset* $I \subset \{1, \dots, n\}$.

PROOF OF PROPOSITION 6.1.6 (SEE ALSO [**11**]). The smooth k-varieties of dimension 0 are precisely the spectra of étale k-algebras. Hence the semiorthogonal decomposition condition is certainly sufficient to get $\operatorname{rdim}\mathsf{T} = 0$. On the other hand, if $\operatorname{rdim}\mathsf{T} = 0$, we have such a semiorthogonal decomposition with each A_i an admissible subcategory of the derived category of an étale k-algebra. By Lemma 6.1.7*(iii)*, we have that A_i is thus itself such a category. □

We have the following corollary of Proposition 6.1.6.

LEMMA 6.1.8. *Let* T *be a* k*-linear triangulated category. If* $\operatorname{rdim}\mathsf{T} = 0$*, then* $K_0(\mathsf{T})$ *is a free* $\mathbb{Z}$*-module of finite rank. In particular, if* X *is smooth and projective and* $\operatorname{rdim}(X) = 0$*, we have that* $\mathrm{CH}^1(X)$ *is torsion-free of finite rank.*

PROOF. The only non-trivial statement is the last one, which can be proved, as in [**88**, Lemma 2.2], using the topological filtration on $K_0(X)$. □

REMARK 6.1.9. Lemma 6.1.8 gives useful criterion: if $K_0(\mathsf{T})$ has torsion elements or if it is not of finite rank, then $\operatorname{rdim}\mathsf{T} > 0$.

In the cases where $\mathsf{T} = \mathsf{D}^{\mathrm{b}}(X)$ for a smooth projective X, if $\operatorname{rdim}(X) = 0$, there are much more consequences that can be obtained using non-commutative motives. For example, when $k \subset \mathbb{C}$ is algebraically closed, the even de Rham cohomology and all the Jacobians are trivial [**32, 135**] and the (rational) Chow motive is of Lefschetz type [**135**].

Representability in dimension 1.

PROPOSITION 6.1.10. *Let* T *be a* k*-linear triangulated category.* $\operatorname{rdim} \mathsf{T} \leq 1$ *if and only if* T *admits a semiorthogonal decomposition whose components belong to the following list:*

(i) categories representable in dimension 0, or
(ii) categories of the form $\mathsf{D}^{\mathrm{b}}(k, \alpha)$*, for* α *in* $\operatorname{Br}(k)$ *the Brauer class of a conic, or*
(iii) categories equivalent to $\mathsf{D}^{\mathrm{b}}(C)$ *for some smooth* k*-curve* C*.*

The main tool in the proof of the previous statement is the indecomposability result for curves due to Okawa that we recalled in Theorem 2.3.3.

PROOF OF PROPOSITION 6.1.10. We already classified triangulated categories representable in dimension 0. We are hence looking for categories A with $\operatorname{rdim} \mathsf{A} = 1$. Of course, if $\mathsf{A} = \mathsf{D}^{\mathrm{b}}(C)$ for some curve, then $\operatorname{rdim} \mathsf{A} \leq 1$ and we are done.

Using Theorem 2.3.3, if A is a nontrivial triangulated category with a full and faithful functor $\phi : \mathsf{A} \to \mathsf{D}^{\mathrm{b}}(C)$ with right and left adjoints, then either ϕ is an equivalence, or $g(C) = 0$. In the latter case, let α be the class of C in $\operatorname{Br}(k)$, which is trivial if and only if $C = \mathbb{P}^1$. By [**28**], there is a semiorthogonal decomposition $\mathsf{D}^{\mathrm{b}}(C) = \langle \mathsf{D}^{\mathrm{b}}(k), \mathsf{D}^{\mathrm{b}}(k, \alpha)\rangle$. It is not difficult to see that this is the only possible semiorthogonal decomposition up to mutations, so we get the proof. □

Let us sketch a criterion of representability in dimension 1, based on the Grothendieck group, in the case where $k = \mathbb{C}$. Given a $\mathbb{Z}$-module M and an integer number $n > 0$, we will denote by $M[n] \subset M$ the kernel of the multiplication by n map $M \xrightarrow{\times n} M$. Such $M[n]$ has a natural structure of $\mathbb{Z}/n\mathbb{Z}$-module. We notice that, if X is a smooth (connected) projective variety of dimension ≤ 1, the modules $K_0(X)[n]$ are well known. Indeed, either X is a point, or X is $\mathbb{P}^1$, or X is a curve of positive genus g. In the first two cases, $K_0(X)$ is free of finite rank, hence $K_0(X)[n] = 0$ for any n. In the latter case, $K_0(X) \simeq \mathbb{Z} \oplus \operatorname{Pic}(X) = \mathbb{Z} \oplus \operatorname{Pic}^0(X) \oplus \mathbb{Z}$ by the Grothendieck–Riemann–Roch Theorem, and the fact that the 1st Chern class is integral. Then $K_0(X)[n] = \operatorname{Pic}^0(X)[n]$. Since $\operatorname{Pic}^0(X)$ is a complex torus of dimension g, we have that $\operatorname{Pic}^0(X)[n] = (\mathbb{Z}/n\mathbb{Z})^{2g}$ (see, e.g., [**140**, § I.1(3)]).

LEMMA 6.1.11. *Suppose that* T *is* $\mathbb{C}$*-linear and* $\operatorname{rdim} \mathsf{T} \leq 1$*. Then, for any integer* n*, we have that* $K_0(\mathsf{T})[n]$ *is a free* $\mathbb{Z}/n\mathbb{Z}$*-module of finite even rank.*

PROOF. Proposition 6.1.10 gives us all the possible components of a semiorthogonal decomposition of T. If A is one of such components, it follows that $K_0(\mathsf{A})[n]$ is either trivial or $(\mathbb{Z}/n\mathbb{Z})^{2g}$ if $\mathsf{A} = \mathsf{D}^{\mathrm{b}}(C)$ and $g = g(C)$. □

Finally, let us just record a very simple remark, as a corollary of Theorem 2.3.3.

COROLLARY 6.1.12. *A smooth projective curve* C *is* k*-rational if and only if* $\operatorname{rdim}(C) = 0$.

A glimpse of representability in dimension 2. It is more difficult to classify categories that are representable in dimension 2. Of course, for any surface S, $\mathsf{T} = \mathsf{D}^{\mathrm{b}}(S)$ satisfies $\operatorname{rdim} \mathsf{T} \leq 2$, but it is a quite challenging question to understand which categories can occur as proper admissible subcategories of surfaces. Using the results of Kawatani–Okawa recalled in Theorem 2.3.4, we need essentially only consider surfaces S that are either ruled or with $p_g = q = 0$, at least in the case where $k = \mathbb{C}$. In the ruled case, say $S \to C$, we have $\mathrm{rdim}(S) = \mathrm{rdim}(C) \leq 1$ with strict inequality holding only for $C = \mathbb{P}^1$. Hence we will focus on surfaces with $p_g = q = 0$.

Notice that, for a smooth projective surface S, any line bundle is k-exceptional if and only if S satisfies $p_g = q = 0$. This is a simple calculation using that line bundles are invertible and the definition of p_g and q. It is then natural to study exceptional collections on such surfaces and describe the corresponding semiorthogonal decompositions. These decompositions are conjecturally related to rationality criteria for surfaces, and we will treat them extensively in §6.2. We also remark that an exceptional collection on a variety X gives a free subgroup of finite rank of $K_0(X)$. Surfaces of general type with $p_g = q = 0$ containing torsion elements in $K_0(S)$ are known, and hence they cannot have a full exceptional sequence. This remark gives rise to the definition of phantom and quasi-phantom categories.

DEFINITION 6.1.13. Let X be a smooth projective variety over k. An admissible subcategory $\mathsf{T} \subset \mathsf{D}^{\mathrm{b}}(X)$ is called a *quasi-phantom* if its Hochschild homology $HH_*(\mathsf{T}) = 0$ vanishes and $K_0(\mathsf{T})$ is a finite abelian group. A quasi-phantom T is a *phantom* if $K_0(\mathsf{T}) = 0$. A phantom is a *geometric phantom* if T_K is still a phantom for any algebraically closed field K containing k. A phantom is a *universal phantom* if, for any smooth and projective variety Y, the admissible subcategory $\mathsf{T} \boxtimes \mathsf{D}^{\mathrm{b}}(Y)$ of $\mathsf{D}^{\mathrm{b}}(X \times Y)$ is a phantom.

Notice that if T is a quasi-phantom, then $\mathrm{rdim}\,\mathsf{T} > 1$.

If k is algebraically closed, we wonder whether every phantom is in fact a geometric phantom, see Remark 6.2.7. On the other hand, if k is general, then more complicated phenomena can arise, already by considering descent of full exceptional sequences from the separable closure of k.

EXAMPLE 6.1.14 (Categories representable in dimension 2). Here is a list of categories T such that $\mathrm{rdim}\,\mathsf{T} = 2$, though T is not equivalent to $\mathsf{D}^{\mathrm{b}}(S)$ for any smooth projective surface S.

Phantoms, $k = \mathbb{C}$. Let S be a determinantal Barlow surface and T the orthogonal complement to an exceptional collection of length 11 [**42**]. Let S be a Dolgachev surface of type $X_9(2,3)$ and T the orthogonal complement to an exceptional collection of length 12 [**57**] (we refer to [**79**] for the notations on Dolgachev surfaces).

Quasi-phantoms, $k = \mathbb{C}$. Since the first example of the classical Godeaux surface [**40**], there are now many examples of quasi-phantoms as orthogonal complements of an exceptional sequence of line bundles of maximal length on surfaces of general type, see [**86**], [**88**], [**112**], [**133**], [**6**], [**76**].

Not quasi-phantoms, $k = \mathbb{C}$. Let S be an Enriques surface [**105**] and T the orthogonal complement to an exceptional collection of length 10. Let S be a classical Godeaux surface [**41**] and T the orthogonal complement to an exceptional collection of length 9. Both categories have $K_0(\mathsf{T}) \otimes \mathbb{Q} \simeq \mathbb{Q}^2$, but $K_0(\mathsf{T})$ not free, and do not admit any exceptional object [**41, 168**].

Not quasi-phantoms, general k. Let $\alpha \in \mathrm{Br}(k)$ be the class of a Severi–Brauer surface or the class of a Severi–Brauer threefold with an involution surface and $\mathsf{T} = \mathsf{D}^{\mathrm{b}}(k, \alpha)$. Let $\mathscr{C}_0$ be the Clifford algebra of an involution surface and $\mathsf{T} = \mathsf{D}^{\mathrm{b}}(k, \mathscr{C}_0)$. Let Q and B are the simple algebras (quadratic over a degree 3 extension of k and cubic over a degree 2 extension of k, respectively) associated to a minimal del Pezzo surface of degree 6 and $\mathsf{T} = \mathsf{D}^{\mathrm{b}}(k, Q)$ or $\mathsf{T} = \mathsf{D}^{\mathrm{b}}(k, B)$. For all these examples, $K_0(\mathsf{T})$ is free of finite rank, see [**11**]. If S is a minimal del Pezzo surface of degree $d < 5$, then T is the semiorthogonal complement of $\langle \mathscr{O}_S \rangle$, and is not of the form $\mathsf{D}^{\mathrm{b}}(K, \alpha)$ for any K/k étale and α in $\mathrm{Br}(K)$, see [**11**].

The result of Kawatani–Okawa, cf. Theorem 2.3.4, suggests that, for $k = \mathbb{C}$, categories representable in dimension exactly 2, and not equivalent to any $\mathsf{D}^{\mathrm{b}}(S)$, should occur only in the case where $p_g = q = 0$. In this case, Bloch's conjecture (see §5.3) would imply that $K_0(S) \otimes \mathbb{Q}$ is a finite vector space, so it is natural to raise the following.

CONJECTURE 6.1.15. *Suppose that* $\mathrm{rdim}\, \mathsf{T} = 2$, *and that* T *is not equivalent to* $\mathsf{D}^{\mathrm{b}}(S)$ *for any surface* S. *Then* $K_0(\mathsf{T})_{\mathbb{Q}}$ *is a finite-dimensional vector space.*

6.2. Rationality questions for surfaces. We turn our attention to the possibility of characterizing rational surfaces via categorical representability. A folklore conjecture by D. Orlov states that a complex surface with a full exceptional collection is rational. We provide here a version for any field k in terms of categorical representability.

CONJECTURE 6.2.1 (Orlov). *A smooth projective surface* S *over a field* k *is* k*-rational if and only if* $\mathrm{rdim}(S) = 0$.

If k is algebraically closed, then being representable in dimension zero is equivalent to having a full exceptional collection. Combining Theorem 2.3.1 and Proposition 2.3.8 it is easy to check that a rational surface has a full exceptional collection, since it is a blow-up along smooth points of either a projective space or a Hirzebruch surface.

If k is not algebraically closed, it is easy to construct rational surfaces without a full exceptional collection, for example a k-rational quadric surface of Picard rank 1, or by blowing up a closed point of degree > 1 on a k-rational surface. In fact, the "only if" part of Conjecture 6.2.1 remains true, thanks to the results of [**11**] for del Pezzo surfaces.

THEOREM 6.2.2. *Let* S *be any smooth* k*-rational surface. Then* $\mathrm{rdim}(S) = 0$.

The converse is more difficult. Let us first recall the following result based on a base change formula by Orlov [**147**].

LEMMA 6.2.3 ([**12**], Lemma 2.9). *Let* X *be a smooth projective variety over* k, *and* K *a finite extension of* k. *Suppose that* $\mathsf{A}_1, \ldots, \mathsf{A}_n$ *are admissible subcategories of* $\mathsf{D}^{\mathrm{b}}(X)$ *such that* $\mathsf{D}^{\mathrm{b}}(X_K) = \langle \mathsf{A}_{1K}, \ldots, \mathsf{A}_{nK} \rangle$. *Then* $\mathsf{D}^{\mathrm{b}}(X) = \langle \mathsf{A}_1, \ldots, \mathsf{A}_n \rangle$.

Using Lemma 6.2.3 and the classification from Proposition 6.1.6, we deduce that, in the case $k \subset \mathbb{C}$, it is enough to check Conjecture 6.2.1 for geometrically rational surfaces and for complex surfaces with $p_g = q = 0$. As remarked earlier, these are the only surfaces where a line bundle is k-exceptional.

Geometrically rational surfaces. The first case is handled in [**11**] for del Pezzo surfaces and in [**168**] for geometrically rational surfaces with a numerically k-exceptional collection of maximal length.

THEOREM 6.2.4. *Let* S *be a geometrically rational surface. If either* S

- *is a blow-up of a del Pezzo and* $\mathrm{rdim}(S) = 0$, *or*

- *has a (numerically) k-exceptional collection of maximal length,*

then S is k-rational.

The results of [**11**] also provide a categorical birational invariant. Recall, from Proposition 2.3.5, the definition of the category A_X for a Fano variety X of Picard rank 1.

THEOREM 6.2.5 ([**11**]). *Let S be a minimal del Pezzo surface of degree d. If $d < 5$, then A_S is a birational invariant. If $d \geq 5$, then the product of components T of A_S with* $\operatorname{rdim} \mathsf{T} > 0$ *is a birational invariant. In particular, there is a well-defined Griffiths–Kuznetsov component.*

Let us conclude by showing that having a full k-exceptional collection is a stronger property than having a decomposition of the diagonal. The following result is a slight generalization of a result of Vial [**168**]. As we will see, nonrational surfaces satisfying the assumptions of Theorem 6.2.6 exist, and are known to have a decomposition of the diagonal, as recalled in Theorem 5.3.2. Here we give a direct proof, adapted from Vial, providing an explicit decomposition of the diagonal from the exceptional objects.

THEOREM 6.2.6. *Let S be a smooth projective surface over k with $p_g = q = 0$ and admitting a semiorthogonal decomposition*

$$\mathsf{D}^{\mathrm{b}}(S) = \langle \mathsf{A}, E_1, \ldots, E_r \rangle, \tag{6.1}$$

where E_i are k-exceptional and A is a geometric phantom. Then S has a decomposition of the diagonal and the integral Chow motive of X is of Lefschetz type.

PROOF. If $\mathsf{A} = 0$, the second statement is a result of Vial [**168**, Thm. 2.7], while the first statement is shown in the course of Vial's proof. We will detail the main steps of Vial's argument to show that having a nontrivial phantom (which Vial does not address) does not affect the proof.

1st Step. First of all, the semiorthogonal decomposition implies that $K_0(S)$ is free of finite rank. Using the topological filtration on K_0, one can show that the integral Chow ring $\mathrm{CH}^*(S)$ is then also free of finite rank (cf. [**168**, Lemma 2.6]). Indeed, we have that $\mathrm{CH}^0(S) = \mathbb{Z}$, $\mathrm{CH}^1(S) = \mathrm{Pic}(S)$ is free of finite rank by Lemma 6.1.8, and $\mathrm{CH}^2(S)$ is free of finite rank since it coincides with the second graded piece of the topological filtration, which is a subgroup of $K_0(S)$, as S has dimension 2. In particular, the group of 0-cycles $A_0(S)$ of degree 0, is free of finite rank. Since A is a geometric phantom, for any algebraically closed field K containing k, the admissible subcategory A_K of $\mathsf{D}^{\mathrm{b}}(S_K)$ is a phantom in the base change of the decomposition (6.1). As before, all remarks above hold for S_K, in particular, $K_0(S_K)$ and $\mathrm{CH}^*(S_K)$ are free of finite rank. However, it is well known that $A_0(S_K)$ is divisible (cf. [**34**, Lec. 1, Lemmas 1.3 and 1.4]), which then implies that $A_0(S_K) = 0$. Since the kernel of the scalar extension map $A_0(S) \to A_0(S_K)$ is torsion (see Lemma 5.2.3), we conclude that $A_0(S) = 0$. We also remark that $r = \mathrm{rk}(K_0(S_K)) = \rho + 2$, where ρ is the geometric Picard rank of S.

2nd Step. Perling [**149**] shows that any numerically k-exceptional collection of maximal length on a surface with $\chi(\mathscr{O}_S) = 1$ can be mutated into a numerically k-exceptional collection of maximal length consisting of objects of rank 1. Given such an exceptional collection, Vial [**168**, Prop. 2.3] provides a $\mathbb{Z}$-basis $D_1, \ldots, D_\rho$ of $\mathrm{CH}^1(S_K) = \mathrm{Pic}(S_K)$ with unimodular intersection matrix M and dual basis

$D_1^\vee, \ldots, D_\rho^\vee$. This is accomplished using Chern classes and the Riemann–Roch formula to compare χ with the intersection pairing.

3rd Step. As shown by Vial [**168**, Cor. 2.5], a surface with $\chi(\mathscr{O}_S) = 1$ and a numerically k-exceptional collection of maximal length has a zero-cycle of degree 1. Hence S_K has always a zero-cycle a of degree 1, then we set $\pi^0 := a \times S_K$ and $\pi^4 := S_K \times a$ as idempotent correspondences in $\mathrm{CH}^2(S_K \times S_K)$. Moreover, Vial defines the correspondences $p_i := D_i \times D_i^\vee$ in $\mathrm{CH}^2(S_K \times S_K)$, which are idempotent since the intersection product is unimodular. It is not difficult to see that all the above correspondences are mutually orthogonal. Set $\Gamma_K := \Delta_{S_K} - \pi^0 - \pi^4 - \sum_{i=1}^{\rho} p_i$.

4th Step. Since π^i and p_i are mutually orthogonal idempotents, Γ_K is idempotent. Moreover, Γ_K acts trivially on $\mathrm{CH}^*(S_K)$, since we have $\mathrm{CH}^2(S_K) = \mathbb{Z}a$ and $\mathrm{CH}^1(S_K)$ generated (over $\mathbb{Z}$) by the D_i's. By [**158**, Prop. 3.7], since K is an algebraically closed extension (universal domain) of k, it follows that Γ is nilpotent. Since Γ is also idempotent, we have $\Gamma = 0$ and the claim holds. □

REMARK 6.2.7. If k is algebraically closed, one can prove that any phantom subcategory of $\mathsf{D}^{\mathrm{b}}(S)$, for S a smooth projective surface over k, is in fact a geometric phantom. Indeed, by rigidity for the Picard group and Lemma 5.2.3, the map $\mathrm{CH}^*(S) \to \mathrm{CH}^*(S_K)$ is an isomorphism for every algebraically closed K/k, and then an argument with the topological filtration implies the same for $K_0(S)$.

When $k = \mathbb{C}$, the proof shows that a semiorthogonal decomposition as in (6.1) implies Bloch's conjecture for S. In this case, Sosna [**161**, Cor. 4.8] (using results of [**92**]) remarks that any phantom category in $\mathsf{D}^{\mathrm{b}}(S)$ is a universal phantom.

Complex surfaces with $p_g = q = 0$. In this paragraph, S has $p_g = q = 0$ and $k = \mathbb{C}$. The study of such surfaces is very rich and already very challenging in the case where $k = \mathbb{C}$. On one hand, it is easy to see that if S has a full exceptional collection, then $K_0(S)$ is a free $\mathbb{Z}$-module of finite rank. This automatically exclude surfaces with torsion line bundles, which would give rise to a torsion class in $K_0(S)$. However, a full understanding of the rôle of (quasi)-phantoms, categories representable in dimension 2, and Conjecture 6.2.1, requires a full understanding of any such surface, independently on the obstruction mentioned above.

First of all, Vial classifies all such S which have a numerically exceptional sequence of maximal length. This is based on the study of the Picard lattice which can be deduced by the numerically exceptional sequence via Riemann–Roch theorem (we refer to [**79**] for the notations on Dolgachev surfaces).

THEOREM 6.2.8 (Vial [**168**]). *Let S be as above. Then S has a numerically exceptional collection of maximal length if and only if it has a numerically exceptional collection of maximal length consisting of line bundles. Moreover, this is the cases if and only if either:*

- *S is not minimal, or*
- *S is rational, or*
- *S is a Dolgachev surface of type $X_9(2,3)$, $X_9(2,4)$, $X_9(3,3)$ or $X_9(2,2,2)$, or*
- $\kappa(S) = 2$.

REMARK 6.2.9. There are cases where such a numerically exceptional collection is actually an exceptional collection, namely rational surfaces [**102**] (in which case it is full), Dolgachev surfaces of type $X_9(2,3)$ [**57**] and many examples of surfaces of general type, see [**86**], [**42**] [**88**], [**112**], [**133**], [**6**], [**76**].

The classification of complex surfaces of general type with these invariants is quite wild, and probably still incomplete, see **[22]** for a recent survey. Proceeding example-wise, one considers many interesting cases to study, but it is not a realistic way to attack Conjecture 6.2.1. On the other hand, such surfaces often come in positive-dimensional moduli, and have ample anticanonical bundle, so that $\mathsf{D}^{\mathrm{b}}(S)$ identifies the isomorphism class of S by the reconstruction theorem of Bondal and Orlov **[47]**. Hence we have a positive dimensional family of equivalence classes of (dg enhanced) triangulated categories, while categories generated by exceptional collections depend on a finite number of countable parameters.

If a dg enhanced triangulated category T is not trivial, then its Hochschild cohomology $HH^*(\mathsf{T})$ is also nontrivial.[7] Moreover, the second Hochschild cohomology encodes deformations of the dg enhanced category T (see **[111**, § 5.4**]**).

Suppose then that we have a family of surfaces S_t of general type depending on a continuous parameter t, and that we can produce, for any of these surfaces, an exceptional collection $\{L_1^t, \ldots, L_n^t\}$ of maximal length consisting of line bundles with A_t its semiorthogonal complement. Since S_t is of general type, the equivalence class of $\mathsf{D}^{\mathrm{b}}(S_t)$ identifies the isomorphism class of S_t by **[47]**. On the other hand, A_t is determined by a dg-quiver with n vertices such that the arrows and relations are determined by the dimensions of the Ext-spaces between the objects L_i^t by **[46]**. Hence, only a countable number of such collections can be full, that is, we could have $\mathsf{A}_t = 0$ only for a discrete set of parameters. It would be natural to expect that the exceptional collection does not vary, while information about the deformation of S_t (hence also about $\mathsf{D}^{\mathrm{b}}(S_t)$) as t varies must be parameterized by A_t. This reasoning is supported, even in cases of Kodaira dimension 1, by two examples, namely a family of Barlow surfaces **[42]** and a family of Dolgachev surfaces of type $X_9(2,3)$ **[57]**, and justifies the following conjectural question.

QUESTION 6.2.10. *Let S_t be a family of minimal non-rational smooth projective surfaces with $p_g = q = 0$ depending on a continuous parameter t. Assume that S_t admits an exceptional sequence $\mathsf{E}_t = \{E_1^t, \ldots, E_n^t\}$ of maximal length for any t.*

1) *Is $\langle \mathsf{E}_t \rangle$ constant? That is, is $\langle \mathsf{E}_t \rangle$ equivalent to $\langle \mathsf{E}_{t'} \rangle$ for any t and t'?*
2) *Is $\mathsf{A}_t = \langle \mathsf{E}_t \rangle^\perp$ nontrivial for all t?*

Notice that, in the case when S_t is of general type, a positive answer to 1) in Question 6.2.10 would imply that A_t is nontrivial for all but possibly one value of t, and that A_t would parameterize the deformations of $\mathsf{D}^{\mathrm{b}}(S_t)$ which reflect the deformations of S_t; a positive answer to 2) would imply Conjecture 6.2.1 for all surfaces of general type with positive dimensional moduli.

7. 0-cycles on cubics

In §5.3, we ended with an essentially complete classification of smooth projective complex universally CH_0-trivial surfaces. In this section we will move from dimension 2 to higher dimension, and discuss the universal CH_0-triviality of complex cubic threefolds and fourfolds. Throughout, we work over $\mathbb{C}$.

---

[7] Denote by $\mathscr{T}$ a dg enhancement of T. The Hochschild cohomology is naturally interpreted as the homology of the complex $\mathscr{H}om(\mathbf{1}_{\mathscr{T}}, \mathbf{1}_{\mathscr{T}})$ computed in the dg category $\mathscr{R}\mathscr{H}om(\mathscr{T}, \mathscr{T})$ where $\mathbf{1}_{\mathscr{T}}$ denotes the identity functor of $\mathscr{T}$, see **[111**, §5.4**]**. It follows that whenever $\mathscr{T}$ is nontrivial, the class of the identity is a nontrivial element of $HH^*(\mathscr{T})$.

7.1. Cubic threefolds. We first provide some background about minimal curve classes on principally polarized abelian varieties. Let C be a smooth projective curve and $(J(C),\Theta)$ its Jacobian with the principal polarization arising from the theta divisor. Choosing a rational point on C, there is an embedding $C \hookrightarrow J(C)$ giving rise to a class $[C] \in \mathrm{CH}_1(J(C))$. This class is related to the theta divisor by means of the *Poincaré formula*

$$[C] = \frac{\Theta^{g-1}}{(g-1)!} \in H^{2g-2}(J(C),\mathbb{Z}),$$

in particular, the class $\Theta^{g-1}/(g-1)!$ is represented by an effective algebraic class in $\mathrm{CH}_1(J(C))$. To some extent, the validity of this formula gives a characterization of Jacobians of curves among principally polarized abelian varieties by the following result of Matsusaka.

Let (A,ϑ) be an irreducible principally polarized abelian variety of dimension g. The class $\vartheta^{g-1}/(g-1)!$ is always an integral Hodge class in $H^{2g-2}(A,\mathbb{Z})$. We will say that this class is algebraic (resp. effective) if it is homologically equivalent to an algebraic (resp. effective) class in $\mathrm{CH}_1(A)$, i.e., is equal to the image of an algebraic (resp. effective) cycle in the image of the class map $\mathrm{CH}_1(A) \to H^{2g-2}(A,\mathbb{Z})$.

THEOREM 7.1.1 (Matsusaka [**137**]). *Let (A,ϑ) be an irreducible principally polarized abelian variety of dimension g. Then there exists a smooth projective curve C such that $(A,\vartheta) \cong (J(C),\Theta)$ as principally polarized abelian varieties if and only if the class $\vartheta^{g-1}/(g-1)!$ is effective.*

If the principally polarized abelian variety (A,ϑ) is not irreducible, then the result of Matsusaka gives a characterization of when (A,ϑ) is a product of Jacobians of curves. In [**58**], this condition is equivalent to (A,ϑ) being of "level one." This characterization gives a nice reformulation of the Clemens–Griffiths criterion for nonrationality (see Theorem 4.1.1) of a smooth projective threefold X satisfying $h^1 = h^{3,0} = 0$ and with intermediate Jacobian $(J(X),\Theta)$:

If $\Theta^{g-1}/(g-1)!$ is not effective, then X is not rational.

One can even interpret the proof of Clemens and Griffiths as showing that $\Theta^4/4!$ is not effective when X is a cubic threefold.

We know that the Clemens–Griffith criterion for nonrationality can fail to detect stable rationality, in particular, can fail to ensure universal CH_0-triviality, see §4.2. In hindsight, a natural question is whether there is a strengthening of the Clemens–Griffiths criterion for obstructing universally CH_0-nontriviality. Voisin [**178**] provides such a strengthened criterion. In the case of cubic threefolds, where universal CH_0-triviality is still an open question, her results are particularly beautiful.

THEOREM 7.1.2 (Voisin [**178**]). *Let X be a smooth cubic threefold with intermediate Jacobian $(J(X),\Theta)$. Then X is universally CH_0-trivial if and only if $\Theta^4/4! \in H^8(J(X),\mathbb{Z})$ is algebraic.*

REMARK 7.1.3. The problem of whether the very general cubic threefold is not stably rational is still open. Voisin [**178**, Thm. 4.5] proves that cubic threefolds are universally CH_0-trivial over a countable union of closed subvarieties of codimension at most 3 in the moduli space. Colliot-Thélène [**65**] has provided a different construction of loci in the moduli space parameterizing universally CH_0-trivial cubic threefolds. Voisin also points out the striking open problem that there is not even a single principally polarized abelian variety (A, ϑ) of dimension $g \geq 4$ known for which $\vartheta^{g-1}/(g-1)!$ is not algebraic!

To give an idea of the ingredients in the proof, we will start with some results of Voisin on the decomposition of the diagonal in various cohomology theories.

PROPOSITION 7.1.4 (Voisin [**178**, Prop. 2.1]). *Let X be a smooth projective variety over a field of characteristic 0. If X admits a decomposition of the diagonal modulo algebraic equivalence, then it admits a decomposition of the diagonal.*

The proof uses a special case of the nilpotence conjecture, proved independently by Voevodsky [**169**] and Voisin [**170**], stating that correspondences in $\mathrm{CH}(X \times X)$ algebraically equivalent to 0 are nilpotent for self-composition.

We point out that the general nilpotence conjecture, for cycles homologically equivalent to 0, is known to imply, in particular, Bloch's conjecture for surfaces with $p_g = 0$, see [**179**, Rem. 3.31].

PROPOSITION 7.1.5. *Let X be a smooth cubic hypersurface such that the group $H^{2i}(X, \mathbb{Z})/\operatorname{im}(\mathrm{CH}^i(X) \to H^{2i}(X, \mathbb{Z}))$ has no 2-torsion for all $i \geq 0$ (e.g., X has odd dimension or dimension 4, or is very general of any dimension). If X admits a decomposition of the diagonal modulo homological equivalence, then it admits a decomposition of the diagonal modulo algebraic equivalence.*

The proof uses the relationship between a decomposition of the diagonal on X and on $X \times X$ and the Hilbert scheme of length 2 subschemes $X^{[2]}$, as well as the fact that $X^{[2]}$ is birational to the total space of a projective bundle over X, cf. [**87**]. There is also a purely topological approach to this result due to Totaro [**167**].

Finally, Voisin provides a general necessary and sufficient condition for the decomposition of the diagonal modulo homological equivalence of a rationally connected threefold.

We recall the Abel–Jacobi map for codimension 2 cycles on a smooth projective threefold X with intermediate Jacobian $J(X)$. The Griffiths Abel–Jacobi map

$$\alpha_X : \mathrm{CH}^2(X)_{\mathrm{hom}} \to J(X)(\mathbb{C})$$

is an isomorphism by the work of Bloch and Srinivas [**37**], since $\mathrm{CH}_0(X) = \mathbb{Z}$.

DEFINITION 7.1.6. The we say that X admits a *universal codimension* 2 *cycle* if there exists $Z \in \mathrm{CH}^2(J(X) \times X)$ such that $Z_a = Z_{a \times X}$ is homologous to 0 for any $a \in J(X)$ and that the morphism $\Phi_Z : J(X) \to J(X)$, induced by $a \mapsto \alpha_X(Z_a)$, is the identity on $J(X)$.

The existence of a universal codimension 2 cycle is equivalent to the tautological class in $\mathrm{CH}^2(X_{\overline{F}})$ being in the image of the map $\mathrm{CH}^2(X_F) \to \mathrm{CH}^2(X_{\overline{F}})$, where $F = \mathbb{C}(J(X))$ is the function field of the intermediate Jacobian, see [**63**, §5.2].

THEOREM 7.1.7 (Voisin [**178**, Thm. 4.1]). *Let X be a rationally connected threefold and $(J(X), \Theta)$ its intermediate Jacobian of dimension g. Then X admits a decomposition of the diagonal modulo homological equivalence if and only if the following properties are satisfied:*

(i) *$H^3(X, \mathbb{Z})$ is torsion free.*
(ii) *X admits a universal codimension 2 cycle.*
(iii) *$\Theta^{g-1}/(g-1)!$ is algebraic.*

REMARK 7.1.8. In fact, Theorem 7.1.7 has the following possible generalizations ([**178**, Thm. 4.2, Rem. 4.3]). If $N\Delta_X$ admits a decomposition modulo homological equivalence then $N^2\Theta^{g-1}/(g-1)!$ is algebraic. Furthermore, if X admits a unirational parameterization of degree N, then $N\Theta^{g-1}/(g-1)!$ is effective.

Finally, we outline the proof of Theorem 7.1.2 due to Voisin. Let X be a cubic threefold with intermediate Jacobian $(J(X), \Theta)$. Propositions 7.1.4 and 7.1.5 imply that a decomposition of the diagonal on X is equivalent to a decomposition of the diagonal modulo homological equivalence. Since $H^3(X, \mathbb{Z})$ is torsion free, by Theorem 7.1.7, it would then suffice to show that X admits a universal codimension 2 cycle. However, this is not known. Instead, Voisin uses results of Markushevich and Tikhomirov [**136**] on parameterizations of $J(X)$ with rationally connected fibers, which implies, using results from [**171**], that if $\Theta^{g-1}/(g-1)!$ is algebraic then X admits a universal codimension 2 cycle.

We point out that Hassett and Tschinkel [**101**] have proved that for every family of smooth Fano threefold not birational to a cubic threefold, either every element in the family is rational or the very general element is not universally CH_0-trivial. They use the degeneration method outlined in §7.3.

7.2. Cubic fourfolds. Cubic fourfolds are rationally connected. Indeed, they are Fano hypersurfaces, hence their rational connectivity is a consequence of the powerful results of [**114**]. A more elementary reason is that they are unirational. This fact that was likely known to M. Noether (cf. [**58**, App. B]).

PROPOSITION 7.2.1. *Let $X \subset \mathbb{P}^{n+1}$ be a cubic hypersurface of dimension $n \geq 2$ containing a line. Then X admits a unirational parameterization of degree 2.*

PROOF. Blowing up a line $\ell \subset X$, we arrive at a conic bundle $\mathrm{Bl}_\ell X \to \mathbb{P}^{n-1}$, for which the exceptional divisor E is a multisection of degree 2. Thus the fiber product $\mathrm{Bl}_\ell X \times_{\mathbb{P}^{n-1}} E \to E$ is a conic bundle with a section, hence a rational variety since E is rational. The map $\mathrm{Bl}_\ell X \times_{\mathbb{P}^{n-1}} E \to X$ is generically finite of degree 2. Thus X admits a unirational parameterization of degree 2. □

Of course every smooth cubic hypersurface of dimension at least 2 over an algebraically closed field contains a line, as one can reduce, by taking hyperplane sections, to the case of a cubic surface. In particular, if X is a cubic fourfold, $\mathrm{CH}_0(X)$ is universally 2-torsion. One way that X could be universally CH_0-trivial is if X admitted a unirational parameterization of odd degree. Indeed, if a variety X admits unirational parameterizations of coprime degrees, then X is universally CH_0-trivial. In fact, the following question is still open.

QUESTION 7.2.2. *Does there exist a nonrational variety with unirational parameterizations of coprime degrees?*

While the existence of cubic fourfolds with unirational parameterizations of odd degree is currently limited, a beautiful result of Voisin [**178**, Thm. 5.6] states that in fact many classes of special cubic fourfolds are universally CH_0-trivial.

We recall, from §4.3, the divisors $\mathcal{C}_d \subset \mathcal{C}$ of special cubic fourfolds of discriminant d in the coarse moduli space $\mathcal{C}$ of cubic fourfolds. These are Noether–Lefschetz type divisors, which are nonempty for $d > 6$ and $d \equiv 0, 2 \bmod 6$, see Theorem 4.3.8. We recall that Voisin [**176**, Thm. 18] has shown the integral Hodge conjecture for cubic fourfolds, i.e., that the cycle class gives rise to an isomorphism

$\mathrm{CH}^2(X) = H^4(X,\mathbb{Z}) \cap H^{2,2}(X)$, and moreover, that every class in $\mathrm{CH}^2(X)$ can be represented by a (possibly singular) rational surface. For X very general in the moduli space, $\mathrm{CH}^2(X)$ is generated by the square of the hyperplane class h^2, and the rank of $\mathrm{CH}^2(X)$ is > 1 if and only if X lies on one of the divisors $\mathcal{C}_d$. For small values of d, the geometry of additional 2-cycles $T \in \mathrm{CH}^2(X)$, for general $X \in \mathcal{C}_d$, is well understood. For $d \leq 20$, this was understood classically. For $d \leq 38$, Nuer [**143**] provides explicit smooth models of the rational surfaces that arise. It is still an open question as to whether $\mathrm{CH}^2(X)$ is always generated by classes of smooth rational surfaces, cf. [**97**, Question 14]. Nuer's approach for $d \leq 38$ provides a unirational parameterization of $\mathcal{C}_d$, while it is known that for $d \gg 0$, the divisors $\mathcal{C}_d$ become of general type, see [**164**].

We are interested in representations of cycle classes for $X \in \mathcal{C}_d$ because of the following result on the existence of unirational parameterizations of odd degree of certain special cubic fourfolds (see [**97**, Cor. 35]), which was initiated in [**100**, §7.5], with corrections by Voisin (see [**97**, Ex. 38]).

PROPOSITION 7.2.3. *Let $X \in \mathcal{C}_d$ be a special cubic fourfold whose additional 2-cycle $T \subset X$ is a rational surface. Assume T has isolated singularities and a smooth normalization. If d is not divisible by 4 then X admits a unirational parameterization of odd degree.*

In general, we do not know if the required rational surface $T \subset X$ can always be chosen with isolated singularities and smooth normalization. The construction of Nuer [**143**] provides smooth rational $T \subset X$ for $d \leq 38$.[8]

However, not assuming the existence of unirational parameterizations of coprime degree, Voisin has the following result.

THEOREM 7.2.4 (Voisin [**178**, Thm. 5.6]). *If $4 \nmid d$ then any $X \in \mathcal{C}_d$ is universally* CH_0*-trivial.*

PROOF. We give a sketch of the proof. The additional class $T \in \mathrm{CH}^2(X)$, such that the discriminant of the sublattice generated by h^2 and T is d, can be represented (after adding multiples of h^2) by a smooth surface (which by abuse of terminology we denote by) $T \subset X$.

First, Voisin proves that (at least under the hypothesis that X is very general in $\mathcal{C}_d$), if there exists any closed subvariety $Y \subsetneq X$ such that $\mathrm{CH}_0(Y) \to \mathrm{CH}_0(X)$ is universally surjective, then X is universally CH_0-trivial.

Second, for a smooth surface $T \subset X$, consider the rational map $T \times T \dashrightarrow X$ defined by sending a pair of points (x, y) to the point residual to the line joining x and y. Voisin proves that if this map is dominant of even degree not divisible by 4, then $\mathrm{CH}_0(T) \to \mathrm{CH}_0(X)$ is universally surjective. Finally, a calculation with Chern classes shows that if $T \subset X$ is a smooth surface in general position then the rational map $T \times T \dashrightarrow X$ is dominant of degree $\equiv d \bmod 4$. Indeed, the degree is equal to twice the number of double points acquired by T after a generic projection from a point, so computations must be made comparing the numerology of the double point formula and the intersection product on X. □

We remark that the universal CH_0-triviality is still open in one of the most interesting classes of special cubic fourfolds, namely that of cubic fourfolds containing a plane, i.e., $X \in \mathcal{C}_8$. One nontrivial consequence of the universal CH_0-triviality

[8]Recently, Lai [**130**] verified that the required rational surface $T \subset X$ is nodal for $d = 42$.

would be the universal triviality of the unramified cohomology in degree 3. For cubic fourfolds containing a plane, this was first proved in [**15**]. For arbitrary cubic fourfolds, this was then proved by Voisin [**177**, Ex. 3.2], with a different proof given by Colliot-Thélène [**63**, Thm. 5.8] (which still relies on Voisin's proof of the integral Hodge conjecture).

7.3. The degeneration method. The degeneration method, initiated by Voisin [**177**, §2] and developed by Colliot-Thélène and Pirutka [**69**], has emerged as a powerful tool for obstructing universal CH_0-triviality for various families of varieties. The idea is that universal CH_0-triviality specializes well in families whose central fiber is mildly singular.

Analogous results for the specialization of rationality in families of threefolds was established by de Fernex and Fusi [**77**]. Already, Beauville [**23**, Lemma 5.6.1] had proved an analogous result for the specialization of the Clemens–Griffiths criterion for nonrationality relying on the Satake compactification of the moduli space of abelian varieties. Also, Kollár [**113**] used a specialization method (to characteristic p) for the existence of differential forms to prove nonrationality of hypersurfaces of large degree, a result that was generalized by Totaro [**166**] using the degeneration method for universally CH_0-triviality. We will outline the degeneration method and some of its applications.

First we define a condition on the resolution of singularities of a singular variety.

DEFINITION 7.3.1. Let X_0 be a proper geometrically integral variety over a field k. We say that a proper birational morphism $f : \widetilde{X}_0 \to X_0$ with $\widetilde{X}_0$ smooth is a *universally* CH_0*-trivial resolution* if $f_* : \mathrm{CH}_0(\widetilde{X}_{0,F}) \to \mathrm{CH}_0(X_F)$ is an isomorphism for all field extensions F/k, and, is a *totally* CH_0*-trivial resolution* if for every scheme-theoretic point x of X_0, the fiber $(\widetilde{X}_0)_x$ is a universally CH_0-trivial variety over the residue field $k(x)$.

The notions of universally and totally CH_0-trivial resolutions are due to Colliot-Thélène and Pirutka [**69**] and define a new class of singularities that should be classified in the spirit of the minimal model program. For example, in characteristic zero, one might ask whether X_0 has rational singularities if it admits a totally CH_0-trivial resolution. It is proved (see [**69**, Prop. 1.8]) that every totally CH_0-trivial resolution is universally CH_0-trivial, but not conversely.

EXAMPLE 7.3.2. Let $X \to \mathbb{P}^2$ be a conic bundle of Artin–Mumford type. Then X has isolated ordinary double points, and the (universally CH_0-trivial) resolution has nontrivial Brauer group.

Let X be a smooth proper geometrically integral variety over k. The *degeneration method* proceeds as follows:

(i) Fit X into a proper flat family $\mathcal{X} \to B$ over a scheme B of finite type, and let X_0 be a possibly singular fiber. Assume for simplicity that the generic fiber is regular.

(ii) Prove that X_0 admits a universally CH_0-trivial resolution $f : \widetilde{X}_0 \to X_0$.

(iii) Prove that $\widetilde{X}_0$ is not universally CH_0-trivial.

The outcome is that the very general fiber of the family $\mathcal{X} \to B$ (though perhaps not X itself) will not be universally CH_0-trivial.

Part *(i)* is, to a large extent, informed by the possibility of achieving part *(iii)*. To this end, one is mostly concerned with finding good singular varieties X_0 whose resolutions have nontrivial unramified cohomological invariants or differential forms. Then one hopes that *(ii)* can be verified for these singular varieties. For example, conic bundles of Artin–Mumford type have been used quite a lot. Kollár [**113**] has constructed hypersurfaces in characteristic p with nontrivial global differential forms.

EXAMPLE 7.3.3.

- In [**177**], a quartic double solid with ≤ 7 nodes is shown to degenerate to an Artin–Mumford example.
- In [**69**], a quartic threefold degenerates to a singular quartic hypersurface model birational to an Artin–Mumford example.
- In [**98**], a conic bundle over a rational surface, whose discriminant curve degenerates to a union of curves of positive genus, is shown to degenerate to an Artin–Mumford example.
- In [**101**], smooth Fano threefolds in a family whose general element is nonrational and not birational to a cubic threefold, are shown to degenerate to an Artin–Mumford example.
- In [**166**], hypersurfaces of large degree were already shown by Kollár to degenerate to singular hypersurfaces in characteristic p with nonzero global differential forms.

We find it striking that all successful instances of the equicharacteristic degeneration method for threefolds over $\mathbb{C}$ use singular central fibers that are birational to threefolds of Artin–Mumford type. However, over arbitrary fields, there are other methods, see [**56**], [**64**].

We shall say a few words about the proof of the degeneration method by Colliot-Thélène and Pirutka [**69**]. First there is a purely local statement about schemes faithfully flat and proper over a discrete valuation ring, to the extent that if the special fiber admits a universally CH_0-trivial resolution, then universal CH_0-triviality of the generic fiber implies the universal CH_0-triviality of the special fiber. This purely local statement uses the specialization homomorphism as developed in §1.4. To get the statement about the very general fiber of a family over a base B, there is a "standard" argument using Chow schemes, see [**69**, App. B].

8. Categorical representability in higher dimension

In this section, we turn to higher dimensional varieties, in particular to varieties of dimension 3 and 4. Even though the constructions work over any field, and considerations related to weak factorization hold over any field of characteristic zero (see [**1**]), we consider here only the case where $k = \mathbb{C}$ (or k algebraically closed of characteristic zero), where the depth of the categorical questions we consider are already quite rich.

The aim of this section is to motivate, by examples and motivic arguments, the following question:

QUESTION 8.0.1. *Is categorical representability in codimension* 2 *a necessary condition for rationality? That is, if* X *is rational, do we have* $\mathrm{rcodim}(X) \geq 2$*?*

Let us first notice that, as soon as we consider varieties of dimension at least 3, we easily find examples of non-rational varieties X with $\mathrm{rcodim}(X) \geq 2$, the easiest example being a projective bundle $X \to C$ of relative dimension at least 2 over a curve C with $g(C) > 0$. We thus restrict our attention to Mori fiber spaces $X \to Y$

over varieties of negative Kodaira dimension. In this case Proposition 2.3.6 gives a natural subcategory $\mathsf{A}_{X/Y}$ as a complement of a finite number of copies of $\mathsf{D}^{\mathrm{b}}(Y)$. We now present some evidence and motivic considerations in order to argue that $\operatorname{rdim} \mathsf{A}_{X/Y}$ should witness obstruction to rationality.

Let $\pi : X \to Y$ be a Mori fiber space of relative dimension m, and let n be the dimension of X. First of all, we remark that if $\operatorname{rdim} \mathsf{A}_{X/Y} \leq d$, then $\operatorname{rdim}(X) \leq \max\{\dim(Y), d\}$. It then follows that to study $\operatorname{rcodim}(X)$ we should focus on $\mathsf{A}_{X/Y}$ and its representability.[9]

8.1. Motivic measures and a rational defect. Let us quickly recall some general motivation for Question 8.0.1. Bondal, Larsen, and Lunts [**49**] defined the Grothendieck ring $PT(k)$ of (dg enhanced) triangulated k-linear categories, by considering the free $\mathbb{Z}$-module generated by equivalence classes of such categories, denoted by $I(-)$, and introducing a scissor-type relation: $I(\mathsf{T}) = I(\mathsf{A}) + I(\mathsf{B})$ if there is a semiorthogonal decomposition $\mathsf{T} = \langle \mathsf{A}, \mathsf{B} \rangle$. The product of this ring is a convolution product, in such a way that the product of $I(\mathsf{D}^{\mathrm{b}}(X))$ and $I(\mathsf{D}^{\mathrm{b}}(Y))$ coincides with $I(\mathsf{D}^{\mathrm{b}}(X \times Y))$. See [**49**] for more details. We notice that the unit $\mathbf{e}$ of $PT(k)$ is the class of $\mathsf{D}^{\mathrm{b}}(k)$ and that if T is generated by r k-exceptional objects, then $I(\mathsf{T}) = r\mathbf{e}$.

One can consider the following subsets of $PT(k)$:

$$PT_d(k) := \langle I(\mathsf{T}) \in PT(k) \mid \operatorname{rdim} \mathsf{T} \leq d \rangle^{+},$$

where $\langle - \rangle^{+}$ is the smallest subset closed under summands. One can show that these subsets are indeed subgroups providing a ring filtration of $PT(k)$.

Notice that, by definition, if $\operatorname{rdim} \mathsf{T} \leq d$, then $I(\mathsf{T})$ is in $PT_d(k)$, but the converse is not true in general, even for $d = 0$, as the following example shows.

EXAMPLE 8.1.1. Recall from Remark 6.1.4 that Kuznetsov has constructed a complex threefold X generated by exceptional objects not satisfying the Jordan–Hölder property [**124**]. In particular, this is based on the description, originally due to Bondal, of a quiver Q with three vertexes and relations, so that there are exceptional objects E_1, E_2, E_3, F in $\mathsf{D}^{\mathrm{b}}(Q)$ and semiorthogonal decompositions:

$$\mathsf{D}^{\mathrm{b}}(Q) = \langle E_1, E_2, E_3 \rangle, \qquad \mathsf{D}^{\mathrm{b}}(Q) = \langle \mathsf{T}, F \rangle,$$

such that $\mathsf{T} = F^{\perp}$ has no exceptional object. It follows that $I(\mathsf{T})$ lies in $PT_0(k)$, since $I(\mathsf{D}^{\mathrm{b}}(Q))$ does, but $\operatorname{rdim} \mathsf{T} > 0$.

It would thus be very interesting to give conditions under which a category T, admissible in some category generated by k-exceptional objects (and hence such that $I(\mathsf{T}) \in PT_0(k)$), admits a full k-exceptional collection.

One can consider the Grothendieck ring $K_0(\mathrm{Var}(k))$ of k-varieties whose unit $1 = [\mathrm{Spec}(k)]$ is the class of the point. If weak factorization holds, then this can be seen as the $\mathbb{Z}$-module generated by isomorphism classes of smooth proper varieties with the relation $[X] - [Z] = [Y] - [E]$ whenever $Y \to X$ is the blow-up

[9] If $m > 1$, this is obvious. If $m = 1$, then the rationality of X implies the rationality of $Y \times \mathbb{P}^1$, and hence the stable rationality of Y. For surfaces over $\mathbb{C}$, stable rationality implies rationality, so if X is a 3-dimensional conic bundle over a rational surface S, then the obstruction is exactly contained in $\mathsf{A}_{X/Y}$, see [**29**] or Theorem 8.2.2. For threefolds, note that if Y is nonrational generic Fano threefold, but not a cubic threefold, then Y is not stably rational, see [**101**]. It follows that if $X \to Y$ is a conic bundle over a stably rational threefold, then the obstruction is again contained in $\mathsf{A}_{X/Y}$, unless perhaps if Y is a cubic threefold.

along the smooth center Z with exceptional divisor E, see [**33**]. Larsen and Lunts have then shown that there is a surjective ring morphism (a *motivic measure*) $\mu : K_0(\mathrm{Var}(k)) \to \mathbb{Z}[SB]$ to the ring generated by stable birational equivalence classes. That is, $\mathbb{Z}[SB]$ is the quotient of the Grothendieck ring $K_0(\mathrm{Var}(k))$ by the stable birational equivalence relation.

Moreover, $\ker\mu = \langle \mathbb{L} \rangle$, the ideal generated by the class $\mathbb{L}$ of the affine line. It follows that (as remarked in [**87**]), if X is rational of dimension n, then:

$$[X] = [\mathbb{P}^n] + \mathbb{L}M_X \tag{8.1}$$

in $K_0(\mathrm{Var}(k))$, where M_X is a $\mathbb{Z}$-linear combination of classes of varieties of dimension bounded above by $n-2$. Galkin and Shinder [**87**] define then

$$([X] - [\mathbb{P}^n])/\mathbb{L} \in K_0(\mathrm{Var}(k))[\mathbb{L}^{-1}]$$

as the *rational defect* of X.

On the other hand, Bondal, Larsen, and Lunts [**49**] show that the assignment

$$\nu : K_0(\mathrm{Var}(k)) \to PT(k), \qquad [X] \mapsto I(\mathsf{D}^{\mathrm{b}}(X))$$

also defines a motivic measure. Moreover, since $I(\mathsf{D}^{\mathrm{b}}(\mathbb{P}^1)) = 2\mathbf{e}$ and $[\mathbb{P}^1] = 1 + \mathbb{L}$ in $K_0(\mathrm{Var}(k))$, we have that $\nu(\mathbb{L}) = \mathbf{e}$. It follows from (8.1) that if X is rational of dimension n, then $I(\mathsf{D}^{\mathrm{b}}(X))$ is in $PT_{n-2}(k)$. We can state the following result, motivating Question 8.0.1.

PROPOSITION 8.1.2. *If X is a smooth projective variety of dimension n such that $I(\mathsf{D}^{\mathrm{b}}(X))$ is not in $PT_{n-2}(k)$, then X is not rational.*

DEFINITION 8.1.3. If X is a smooth and projective variety of dimension n, the class of the element $I(\mathsf{D}^{\mathrm{b}}(X))$ in the group $PT(k)/PT_{n-2}(k)$ is called the *noncommutative motivic rational defect* of X.

We end by remarking that Proposition 8.1.2 is a rather weak result. Indeed, as mentioned above, we have an implication $\mathrm{rdim}(X) \leq i \Longrightarrow I(\mathsf{D}^{\mathrm{b}}(X)) \in PT_i(k)$, but the converse implication is in general not known, even for $i = 0$.

However, Proposition 8.1.2 indicates that in the case of Mori fiber spaces, the category $\mathsf{A}_{X/Y}$ should be the object to consider.

COROLLARY 8.1.4. *Let $X \to Y$ be a Mori fiber space of relative dimension m, and let $n = \dim(X)$. Assume that either Y is rational or $m > 1$. Then $I(\mathsf{D}^{\mathrm{b}}(X))$ is in $PT_{n-2}(k)$ if and only if $I(\mathsf{A}_{X/Y})$ is in $PT_{n-2}(k)$.*

PROOF. The assumptions on Y imply that $I(\mathsf{D}^{\mathrm{b}}(Y))$ is in $PT_{n-2}(k)$. Indeed, first of all note that $\dim(Y) \leq n-1$. Second, if Y is rational, we have $I(\mathsf{D}^{\mathrm{b}}(Y))$ in $PT_{n-3}(k)$. On the other hand, if $\dim(Y) \leq n-2$, we have $I(\mathsf{D}^{\mathrm{b}}(Y))$ in $PT_{n-2}(k)$. We then conclude using the relation $I(\mathsf{D}^{\mathrm{b}}(X)) = mI(\mathsf{D}^{\mathrm{b}}(Y)) + I(\mathsf{A}_{X/Y})$ and the definition of $PT_i(k)$. □

QUESTION 8.1.5. *Let X be a smooth projective variety and $X \to Y$ and $X \to Z$ two Mori fiber space structures. Is $\mathrm{rdim}(\mathsf{A}_{X/Y}) = \mathrm{rdim}(\mathsf{A}_{X/Z})$?*

We can extend our analysis to $X \dashrightarrow Y$, a rational map whose resolution $\widetilde{X} \to Y$ is a Mori fiber space. By abuse of notation, we denote $\mathsf{A}_{X/Y,\rho} := \mathsf{A}_{\widetilde{X}/Y}$ (even though $\mathsf{A}_{X/Y,\rho}$ is not necessarily a subcategory of $\mathsf{D}^{\mathrm{b}}(X)$). For example, X is a cubic threefold and $X \dashrightarrow \mathbb{P}^2$ the projection along any line in X, which is resolved into $\widetilde{X} \to \mathbb{P}^2$, a conic bundle.

COROLLARY 8.1.6. *Suppose that there is a rational map $\rho : X \dashrightarrow Y$ and a commutative diagram:*

$$\begin{array}{ccc} \widetilde{X} & & \\ {\scriptstyle \epsilon}\downarrow & \searrow^{\pi} & \\ X & \overset{\rho}{\dashrightarrow} & Y, \end{array}$$

where $\pi : \widetilde{X} \to Y$ is a Mori fiber space of relative dimension m and $\epsilon : \widetilde{X} \to X$ is a blow up along a smooth center. Assume that either Y is rational or $m > 1$. Then $I(\mathsf{D}^{\mathrm{b}}(X))$ is in $PT_{n-2}(k)$ if and only if $I(\mathsf{A}_{X/Y,\rho})$ is in $PT_{n-2}(k)$.

PROOF. We notice that $\mathsf{D}^{\mathrm{b}}(\widetilde{X})$ has two decompositions, one given by the Mori fiber space map $\widetilde{X} \to Y$ and the other given by the blow-up of X hence containing a copy of X and a finite number of copies of the blown-up loci. Corollary 8.1.4 applies then again, once we write the two decompositions of $I(\mathsf{D}^{\mathrm{b}}(\widetilde{X}))$ □

QUESTION 8.1.7. *Let X be a smooth projective Fano variety of dimension n, and $\rho : X \dashrightarrow Y$ and $\sigma : X \dashrightarrow Z$ be rational Mori fiber spaces as above. Is* $\operatorname{rdim} \mathsf{A}_{X/Y,\rho} > n - 2$ *if and only if* $\operatorname{rdim} \mathsf{A}_{X/Z,\sigma} > n-2$*?*

8.2. Threefolds. In this section, we consider Questions 8.0.1, 8.1.5 and 8.1.7 for threefolds. Let us first notice that we only consider Mori fiber spaces $X \to Y$ with Y of negative Kodaira dimension. Moreover, being interested in rationality, we can exclude the cases where Y is a ruled surface over a curve of positive genus. It follows that we only consider Fano threefolds of Picard rank one, relatively minimal del Pezzo fibrations over $\mathbb{P}^1$ and relatively minimal (or *standard*) conic bundles over rational surfaces.

On one hand, there are now many examples of known semiorthogonal decompositions describing $\mathsf{A}_{X/Y}$ for such varieties, especially for Fano and conic bundles. We recall the most of them in Table 1.

On the other hand, recall from §4.1, that such a threefold X has a unique principally polarized intermediate Jacobian $J(X)$, and that one can define the Griffiths component $A_X \subset J(X)$ to be the maximal component not split by Jacobians of curves. We consider a stronger assumption on $J(X)$, namely that it carries an *incidence polarization* (see [**23**, Déf. 3.2.3]), defined as follows. For any algebraic variety T, and z a cycle in $\mathrm{CH}^2_{\mathbb{Q}}(T \times X)$, the *incidence correspondence $I(z)$ associated to z* is the equivalence class of the cycle $r_*(p^*(z) \cdot q^*(z)) \in \mathrm{CH}^1_{\mathbb{Q}}(T \times T)$, where p, q and r are the projections from $T \times T \times X$ away from the first, the second and the third factor respectively. Recall that $J(X)$ carries a principal polarization, which can be seen as an element θ in $\mathrm{CH}^1(J(X) \times J(X))$. Moreover, we assume that $J(X)$ represents the group $A^2(X) \subset \mathrm{CH}^2(X)$ of (integral) algebraically trivial codimension 2 cycles on X, that is there is a universal regular map $G : A^2(X) \to J(X)$, such that for every regular map $g : A^2(X) \to B$ to an abelian variety B, there is a unique morphism of abelian varieties $u \colon J(X) \to B$ such that $u \circ G = g$. Finally, if all these properties are satisfied, we say that the principal polarization θ of $J(X)$ is an *incidence polarization* if for any algebraic map $f : T \to A^2(X)$ defined by a cycle $z \in \mathrm{CH}^2(T \times X)$ the equality $(G \circ f)^*(\theta) = I(z)$ holds.

Having a principally polarized intermediate Jacobian carrying an incidence polarization may seem a rather restrictive assumption, but this is actually satisfied by most of the (general) Mori fiber spaces over rational bases we are considering in this paragraph. Indeed, in these cases $J(X)$ is known to carry such a polarization, unless X is a Fano of index 2 and degree 1 or a del Pezzo fibration of degree 1 (see [**30**, Rmk. 3.8] for more details).

THEOREM 8.2.1 ([**30**]). *Suppose that X is a threefold with principally polarized intermediate Jacobian $J(X)$ carrying an incidence polarization. Then, assuming* $\mathrm{rcodim}(X) \geq 2$, *the Griffiths component $A_X \subset J(X)$ is trivial.*

SKETCH OF PROOF. The first step in the proof is the classification of categories representable in dimension at most 1, see Proposition 6.1.10. In the complex case, this means that there exist smooth and projective curves $\{C_i\}_{i=1}^s$ and a semiorthogonal decomposition

$$\mathsf{D}^{\mathrm{b}}(X) = \langle \mathsf{D}^{\mathrm{b}}(C_1), \ldots, \mathsf{D}^{\mathrm{b}}(C_s), E_1, \ldots, E_r \rangle,$$

where E_i are exceptional. This implies, via Grothendieck–Riemann–Roch, that

$$\mathrm{CH}^*(X)_{\mathbb{Q}} \simeq \mathbb{Q}^{\oplus r} \oplus \bigoplus_{i=1}^{s} \mathrm{CH}^*(C_i)_{\mathbb{Q}},$$

and that the correspondences giving the maps $\phi_i : \mathrm{CH}^i(C_i)_{\mathbb{Q}} \to \mathrm{CH}^*(X)_{\mathbb{Q}}$ are obtained via the kernel of the Fourier–Mukai functors $\Phi_i : \mathsf{D}^{\mathrm{b}}(C_i) \to \mathsf{D}^{\mathrm{b}}(X)$, which are full and faithful.

One can show that ϕ_i induces an isogeny $J(C_i) \to J(X)$ onto an Abelian subvariety (i.e., its kernel is finite), since it has to send algebraically trivial cycles to algebraically trivial cycles, and since the adjoint to the Fourier–Mukai functor provides a retraction ψ_i of ϕ_i up to torsion. Finally, one can check, using the explicit description of the kernel of the adjoint and the incidence property of the polarization of $J(X)$, that ϕ_i actually preserves the principal polarization. □

Notice that Theorem 8.2.1 has a much stronger generalization relying on the theory of noncommutative motives [**32**] which allows one to define the Jacobian of any admissible subcategory of the derived category $\mathsf{D}^{\mathrm{b}}(X)$ of a smooth projective variety. As recalled, Theorem 8.2.1 applies to almost all general threefolds X under examination, except Fano of index two and degree one or del Pezzo fibrations of degree 1 over $\mathbb{P}^1$, whose polarization of the intermediate Jacobian is knot known to be incidence.

Secondly, by a classification argument, and thanks to the work of many authors (see Table 1), we can state a converse statement for Theorem 8.2.1.

THEOREM 8.2.2. *Let X be a complex threefold. Assume that X is either:*

- *a Fano threefold of Picard rank one, very general in its moduli space, not of index 2 and degree 1; or*
- *a del Pezzo fibration $X \to Y = \mathbb{P}^1$ of degree 4; or*
- *a standard conic bundle $X \to Y$ over a minimal rational surface.*

Then X is rational if and only if $\mathrm{rcodim}(X) \geq 2$. *In particular, this is the case if and only if* $\mathrm{rdim}\,\mathsf{A}_{X/Y} \leq 1$.

PROOF. Suppose that such X is not rational. Then it is known—see, for example, [**24**, Table 1] for the Fano cases (very general quartics and sextic double solids can be treated by the same degeneration arguments as in [**23**, §5]), [**5**] for the del Pezzo fibrations, and [**23, 160**] for the conic bundles—that the rationality defect of X is detected by a nontrivial Griffiths invariant $A_X \subset J(X)$. These Jacobians carry an incidence polarization, so that Theorem 8.2.1 implies that $\mathrm{rcodim}(X) \leq 1$ and also that $\mathrm{rdim}\,\mathsf{A}_X \geq 2$.

Conversely, we know a list of the rational general such varieties: $\mathbb{P}^3$; quadric hypersurfaces; Fano varieties of index 2 and degree 5 or 4 (the latter are intersections of two quadrics); Fano varieties of index 1 and degree 22, 18, 16 or 12; conic bundles over $\mathbb{P}^2$ with discriminant divisor of degree ≤ 4 or of degree 5 and even theta-characteristic; conic bundles over Hirzebruch surfaces with trigonal or hyperelliptic discriminant divisor; del Pezzo fibrations of degree 4 over $\mathbb{P}^1$ that are birational to a conic bundle over a Hirzebruch surface. For those varieties, we recall all the known semiorthogonal decompositions and descriptions of $\mathsf{A}_{X/Y}$, with the corresponding references, in Table 1.

Notice that $\mathbb{P}^1$-bundles over rational surfaces, $\mathbb{P}^2$-bundles and quadric fibrations over $\mathbb{P}^1$ also (trivially) fit the statement. □

Recall that there exist nonrational threefolds with trivial Griffiths invariant, and even with trivial intermediate Jacobian. For example, if X is the Artin–Mumford double solid [**9**] recalled in Example 3.5.2, the obstruction to rationality is not given by a nontrivial Griffiths component, but rather by a nontrivial unramified cohomology class. In this case, X is singular but can be resolved by blowing-up its ten double points $\widetilde{X} \to X$.

PROPOSITION 8.2.3. *Let X be the Artin–Mumford quartic double solid and $\widetilde{X} \to X$ be the blow-up of its ten double points. Then $J(\widetilde{X}) = 0$ and $\widetilde{X}$ is not rational. Moreover, $\mathsf{D}^{\mathrm{b}}(\widetilde{X})$ is a noncommutative resolution of singularities of X and* $\mathrm{rcodim}(\widetilde{X}) = 1$.

PROOF. The fact that $\widetilde{X}$ is nonrational and has trivial Jacobian goes back to the original paper of Artin and Mumford [**9**], where the cohomology groups of X are explicitly calculated. In particular $h^{1,2}(X) = 0$, so that it is easy to get $h^{1,2}(\widetilde{X}) = 0$ which implies $J(\widetilde{X}) = 0$. Finally, just recall that $\mathsf{D}^{\mathrm{b}}(\widetilde{X})$ is a noncommutative resolution of $\mathsf{D}^{\mathrm{b}}(X)$ since $\widetilde{X} \to X$ is a resolution of singularities.

We are going to prove that $\mathrm{rdim}(\widetilde{X}) = 2$ by using an explicit semiorthogonal decomposition: first by showing that $\mathrm{rdim}(\widetilde{X}) \leq 2$, then by showing that the inequality cannot hold strictly.

Hosono and Takagi [**103**] consider the Enriques surface S associated to X (the so-called Reye congruence), and show that there is a semiorthogonal decomposition

$$\mathsf{D}^{\mathrm{b}}(\widetilde{X}) = \langle \mathsf{D}^{\mathrm{b}}(S), E_1, \ldots, E_{12} \rangle,$$

where E_i are exceptional objects. This implies first that $\mathrm{rdim}(\widetilde{X}) \leq 2$.

We want to prove that the inequality cannot hold strictly. First of all, since $J(\widetilde{X}) = 0$, we cannot have admissible subcategories of $\mathsf{D}^{\mathrm{b}}(\widetilde{X})$ equivalent to $\mathsf{D}^{\mathrm{b}}(C)$ for some positive genus curve C. It follows by Proposition 6.1.10 (and $k = \mathbb{C}$) that $\mathrm{rdim}(\widetilde{X}) \leq 2$ implies either $\mathrm{rdim}(\widetilde{X}) = 2$ or $\mathrm{rdim}(\widetilde{X}) = 0$.

Let us exclude the second case. Notice that we have $K_0(\widetilde{X}) = \mathbb{Z}^{12} \oplus K_0(S)$. Moreover, the 2-torsion subgroup $K_0(S)[2]$ of $K_0(S)$ is nontrivial. Indeed, we have $K_0(S)[2] = \mathbb{Z}/2\mathbb{Z}$. Indeed, if S is an Enriques surface, the Chern character is integral and gives an isomorphism between $K_0(S)$ and the singular cohomology of S (similarly, one can argue by using the Bloch conjecture, which is true for S, and the topological filtration of the Grothendieck group of S). In particular,

$K_0(S) = \mathbb{Z} \oplus \mathrm{Pic}(S) \oplus \mathbb{Z}$ and $\mathrm{Pic}(S) \simeq \mathbb{Z}^{10} \oplus \mathbb{Z}/2\mathbb{Z}$ (see, e.g., [**21**, VIII Prop. 15.2]). We conclude that

$$K_0(\widetilde{X}) \simeq \mathbb{Z}^{\oplus 22} \oplus \mathbb{Z}/2\mathbb{Z}.$$

But if $\mathrm{rdim}(\widetilde{X}) = 0$, then $K_0(\widetilde{X})$ is free of finite rank, so we conclude.[10] □

Finally notice that there could be other noncommutative resolution of singularities X, such as a small resolution $X^+ \to X$; note that X^+ is a non projective Moishezon manifold. Similar arguments, based on the semiorthogonal decomposition described by Ingalls and Kuznetsov [**105**], show that $\mathrm{rdim}\, \mathsf{D}^{\mathrm{b}}(X^+) = 2$ as well. In that case, the component $\mathsf{T} \subset \mathsf{D}^{\mathrm{b}}(X^+)$ with $\mathrm{rdim}\, \mathsf{T} = 2$ is a crepant categorical resolution of singularities of A_X. Notice also that, despite the fact that X^+ is not an algebraic variety, it is however an algebraic space, see Artin [**8**], so that the category $\mathsf{D}^{\mathrm{b}}(X^+)$ makes sense.

8.3. Cubic Fourfolds. This section is completely devoted to complex cubic fourfolds. From now on, let X denote a smooth hypersurface of degree 3 in complex projective space $\mathbb{P}^5$. As we have seen in §4.3, one of the most useful tools to investigate the geometry of X is Hodge theory. In particular, we have seen how to construct Noether–Lefschetz type divisors $\mathcal{C}_d$, and to relate the numerical properties of d to the existence of some K3 surface S whose geometry is intimately related to the one of X.

We present results showing how derived categories and semiorthogonal decompositions, in the spirit of Question 8.0.1 and even beyond, provide a new language which superposes, and, hopefully, extends the Hodge theoretic approach. We start with an observation by Kuznetsov that we recalled in Corollary 2.3.13:

The category A_X is a noncommutative K3 surface.

Kuznetsov states the following conjecture [**123**], which we will try to motivate and explore in this section.

CONJECTURE 8.3.1 (Kuznetsov). *A cubic fourfold X is rational if and only if there is a K3 surface S and an equivalence $\mathsf{D}^{\mathrm{b}}(S) \simeq \mathsf{A}_X$.*

We notice that Conjecture 8.3.1 is stronger than Question 8.0.1. Moreover, Addington and Thomas have shown [**3**] that the existence of the equivalence requested in Conjecture 8.3.1 is equivalent to asking X to have an associated K3 surface (in the sense of §4.3), at least if X is general on its Noether–Lefschetz divisor.

THEOREM 8.3.2 (Addington–Thomas [**3**]). *Let X be a smooth cubic fourfold. If there exists a K3 surface S and an equivalence $\mathsf{D}^{\mathrm{b}}(S) \simeq \mathsf{A}_X$, then X is special and has an associated K3 surface. Conversely, if X is special and has an associated K3 surface, and if it is general on some $\mathcal{C}_d$, then there exists a K3 surface S and an equivalence $\mathsf{D}^{\mathrm{b}}(S) \simeq \mathsf{A}_X$.*

The above result can be stated in terms of d, since, as recalled in Theorem 4.3.8, being special with an associated K3 surface is equivalent to lie on $\mathcal{C}_d$ for $d > 6$ not divisible by 4, 9, or any odd prime p which is not 2 modulo 3.

Notice that Theorem 8.3.2 tells that, in Hodge-theoretic terms, Conjecture 8.3.1 could be phrased as "X is rational if and only if it has an associated K3". As we

[10] Notice that we can also use Proposition 6.1.10 since $K_0(X)[2]$ is a one dimensional free $\mathbb{Z}/2\mathbb{Z}$-module.

have seen at the end of §4.3, this seems to be the most "rational" expectation for such varieties. Conjecture 8.3.1 seems then, so far, out of reach. But we want to give, in the last part of this section, a quick idea of the very interesting interplay of information that one can extract this categorical point of view.

First of all, let us briefly sketch how to prove Theorem 8.3.2. The main idea is to have a categorical way to reconstruct the Hodge lattice. This is done by considering the topological K-theory $K_0(X)_{top}$, with the intersection pairing χ, given by the Euler characteristic. Using the semiorthogonal decomposition

$$\mathsf{D}^{\mathrm{b}}(X) = \langle \mathsf{A}_X, \mathcal{O}, \mathcal{O}(1), \mathcal{O}(2)\rangle,$$

it is possible to split off $K_0(\mathsf{A}_X)_{top} \subset K_0(X)_{top}$ as the χ-semiorthogonal complement to the classes $[\mathcal{O}(i)]$, for $i = 0, 1, 2$. If $\mathsf{D}^{\mathrm{b}}(S) \simeq \mathsf{A}_X$ is an equivalence, it is then of Fourier–Mukai type and gives an isomorphism $K_0(S)_{top} \simeq K_0(\mathsf{A}_X)_{top}$ of $\mathbb{Z}$-modules respecting the Euler pairing.

The first observation is obtained using the fact that the Chern character is integral for S, so that $K_0(S)_{top} \otimes \mathbb{C}$ with the pairing χ has a Hodge structure of weight 2 induced by the so-called Mukai lattice structure on cohomology. This is (up to identifying K_0 and the cohomology via the Chern character) the following Hodge structure $\widetilde{H}^{p,q}(S)$:

$$\begin{aligned}
\widetilde{H}^{2,0}(S) &= H^{2,0}(S)\\
\widetilde{H}^{1,1}(S) &= H^{2,2}(S) \oplus H^{1,1}(S) \oplus H^{0,0}(S)\\
\widetilde{H}^{0,2}(S) &= H^{0,2}(S).
\end{aligned}$$

For the cubic fourfold X, Addington and Thomas define the Mukai lattice as follows:

$$\begin{aligned}
\widetilde{H}^{2,0}(X) &= H^{3,1}(X)\\
\widetilde{H}^{1,1}(X) &= H^{4,4}(X) \oplus H^{3,3}(X) \oplus H^{2,2}(X) \oplus H^{1,1}(X) \oplus H^{0,0}(X)\\
\widetilde{H}^{0,2}(X) &= H^{1,3}(X),
\end{aligned}$$

and obtain the corresponding weight 2 Hodge structure on $K_0(\mathsf{A}_X)_{top}$ (up to identifying K_0 and the cohomology via the Chern character). It follows then that if $\mathsf{D}^{\mathrm{b}}(S) \simeq \mathsf{A}_X$, there is a Hodge isometry $K_0(S)_{top} \simeq K_0(\mathsf{A}_X)_{top}$.

For the general X, the numerical properties of $K_0(\mathsf{A}_X)_{top}$ are explored, and the lattice $(K_0(\mathsf{A}_X)_{top}, \chi)$ is related to the Hodge lattice $H^4(X, \mathbb{Z})$. Of particular interest is the result that X has an associated K3 surface if and only if the Mukai lattice on the numerical $K_0(\mathsf{A}_X)_{num}$[11] contains a hyperbolic plane [**3**, Thm. 3.1]. Moreover, one can characterize classes of skyscraper sheaves of points $[\mathcal{O}_x]$ and $[\mathcal{O}_y]$ inside $K_0(S)_{top}$, purely using their behavior under the Euler pairing: the sublattice they generate is a hyperbolic lattice. Hence, it follows that if X has no associated K3, such classes do not exist and hence there cannot be any equivalence $\mathsf{D}^{\mathrm{b}}(S) \simeq \mathsf{A}_X$. A similar result for cubic fourfolds containing a plane with two dimensional group of algebraic 2-cycles was obtained by Kuznetsov [**123**].

On the other hand, consider a divisor $\mathcal{C}_d$ where d is such that X has an associated K3 surface. Then consider the intersection $\mathcal{C}_d \cap \mathcal{C}_8$ with the locus of cubics containing a plane. As we will see in Example 8.3.3 (see also Table 1), in this case there is a degree 2 K3 surface S with a Brauer class α and an equivalence $\mathsf{A}_X \simeq \mathsf{D}^{\mathrm{b}}(S, \alpha)$. Moreover, one can find an X on $\mathcal{C}_d \cap \mathcal{C}_8$ such that α vanishes. This

[11] The numerical K_0 is obtained by taking the quotient of the algebraic K_0 by the kernel of the Euler form.

gives a K3 surface S of degree 2 and an equivalence $\mathsf{A}_X \simeq \mathsf{D}^{\mathrm{b}}(S)$ in this particular case. Now, though one would like to deform this K3 surface, it is not the right thing to do. Instead, one has to construct an appropriate K3 surface S' of degree d as a moduli space of vector bundles on S, so that $\mathsf{D}^{\mathrm{b}}(S') \simeq \mathsf{D}^{\mathrm{b}}(S) \simeq \mathsf{A}_X$. Then the existence of a K3 surface S' and of the equivalence $\mathsf{A}_X \simeq \mathsf{D}^{\mathrm{b}}(S')$ for the general X in $\mathcal{C}_d$ is obtained by a degeneration method.

We turn now to explicit cases supporting Conjecture 8.3.1. Indeed, in all the cases where a cubic fourfold is known to be rational, there is an explicit realization of the K3 surface S and of the equivalence $\mathsf{D}^{\mathrm{b}}(S) \simeq \mathsf{A}_X$.

EXAMPLE 8.3.3 (Kuznetsov [**123**]). Let X be a cubic fourfold containing a plane P. As explained in Example 4.3.7, the blow-up $\widetilde{X} \to X$ of the plane P has a structure of quadric surface bundle $\widetilde{X} \to \mathbb{P}^2$.

Recall, from Example 2.3.9, that there is an identification $\mathsf{A}_{\widetilde{X}/\mathbb{P}^2} \simeq \mathsf{D}^{\mathrm{b}}(\mathbb{P}^2, \mathscr{C}_0)$ (see [**116**]). On the other hand, there is a semiorthogonal decomposition induced by the blow-up $\widetilde{X} \to X$. Comparing the two decompositions via explicit mutations, gives an equivalence $\mathsf{A}_X \simeq \mathsf{D}^{\mathrm{b}}(\mathbb{P}^2, \mathscr{C}_0)$.

Suppose that the degeneration divisor of the quadric fibration $\widetilde{X} \to \mathbb{P}^2$ is smooth, and consider the double cover $S \to \mathbb{P}^2$ ramified along it. Kuznetsov shows [**116**, Prop. 3.13] that $\mathscr{C}_0$ lifts to a sheaf of Clifford algebras with Brauer class α in $\mathrm{Br}(S)$, see also [**16**, Prop. 1.11]. It follows that $\mathsf{A}_X \simeq \mathsf{D}^{\mathrm{b}}(S, \alpha)$.

On the other hand, the classical theory of quadratic forms says that the class α is trivial if and only if $\widetilde{X} \to \mathbb{P}^2$ has an odd section. As described in Example 4.3.7 this is a necessary condition for rationality.

Secondly, Kuznetsov shows that if S has Picard rank one, then α is nontrivial and $\mathsf{D}^{\mathrm{b}}(S, \alpha)$ cannot be equivalent to any $\mathsf{D}^{\mathrm{b}}(S')$, for S' a K3 surface. In particular, one should expect that cubics with a plane with such an S are not rational. On the other hand, there exist rational cubics with a plane, such that α is not trivial but $\mathsf{D}^{\mathrm{b}}(S, \alpha) \simeq \mathsf{D}^{\mathrm{b}}(S')$ for some other K3 surface S', see [**13**] for an explicit examples.

EXAMPLE 8.3.4 (Kuznetsov [**123**]). Let X be a Pfaffian cubic fourfold. As explained in Example 4.3.6, there is a classical duality construction providing a degree 14 K3 surface S associated to X. In this case, the powerful theory of Homological Projective Duality allows to show that $\mathsf{A}_X \simeq \mathsf{D}^{\mathrm{b}}(S)$.

EXAMPLE 8.3.5 (Kuznetsov [**123**]). Let X be a cubic fourfold with a single node x. The projection $\mathbb{P}^5 \dashrightarrow \mathbb{P}^4$ from the point x gives a rational parametrization $X \dashrightarrow \mathbb{P}^4$. The resolution of the latter map is obtained by blowing-up x and is realized as a blow-up $\widetilde{X} \to \mathbb{P}^4$ of a degree 6 K3 surface S, obtained as a complete intersection of a cubic and a quadric. Then one can show that there is a crepant resolution of singularities $\widetilde{\mathsf{A}}_X$ and an equivalence $\widetilde{\mathsf{A}}_X \simeq \mathsf{D}^{\mathrm{b}}(S)$.

EXAMPLE 8.3.6 ([**31**]). Let X be a determinantal cubic fourfold. In this case, Homological Projective Duality can be used to show that there is a crepant resolution of singularities $\widetilde{\mathsf{A}}_X$ which is generated by six exceptional objects. Roughly speaking, one should think of the latter as a crepant resolution of a degeneration of the K3 surface from the previous case.

We can then rephrase Example 4.3.11 listing all known rational cubic fourfolds in terms of Conjecture 8.3.1.

EXAMPLE 8.3.7. Let X be a cubic fourfold. If either

2,6) X is singular, e.g. $X \in \mathcal{C}_6$ has a single node or $X \in \mathcal{C}_2$ is determinantal;

8) X contains a plane P, so that $X \in \mathcal{C}_8$, and the associated quadric surface fibration $\widetilde{X} \to \mathbb{P}^2$ (see Example 4.3.7) admits an odd section [**96**]; or

14) X is Pfaffian, so that $X \in \mathcal{C}_{14}$ [**26**];

then X is rational and (a crepant categorical resolution of) A_X is equivalent to (a crepant categorical resolution of) $\mathsf{D}^{\mathrm{b}}(S)$, for some K3 surface S.

8.4. Other fourfolds. Let us quickly conclude with another example, described in [**12**], of fourfolds whose rationality or nonrationality is conjecturally related to categorical representability via an explicit semiorthogonal decomposition.

Let X be a fourfold with a Mori Fiber Space structure $X \to \mathbb{P}^1$ such that the fibers are complete intersections of two quadrics. This means that there is a projective bundle $\mathbb{P}(E) \to \mathbb{P}^1$ with $\mathrm{rk}(E) = 5$, and two line bundle valued nondegenerate quadratic forms $q_i : L_i \to S_2(E)$, such that $X \subset \mathbb{P}(E)$ is the complete intersection of the two quadric fibrations $Q_i \to \mathbb{P}^1$ given by the forms q_i. Moreover, working over $\mathbb{C}$, we know that $X \to \mathbb{P}^1$ has a smooth section (see, e.g. [**12**, Lemma 1.9.3] for a direct argument, or use [**53**] or [**93**]).

Setting $F := L_1 \oplus L_2$, one has that the linear span q of q_i and q_2 gives a quadric fibration $Q \to \mathbb{P}(F)$ of relative dimension 4 over a Hirzebruch surface $\mathbb{P}(F)$. The smooth section of $X \to \mathbb{P}^1$ gives a smooth section of $Q \to \mathbb{P}(F)$, along which we can perform reduction by hyperbolic splitting. This means that we can split off the form q a hyperbolic lattice, whose complement gives a quadric fibration $Q' \to \mathbb{P}(F)$ of dimension two less than Q. That is, $Q' \to \mathbb{P}(F)$ is a quadric surface fibration.

Homological Projective Duality and Morita equivalence of Clifford algebras under hyperbolic splitting (see [**12**] for details) show that $\mathsf{A}_{X/\mathbb{P}^1} \simeq \mathsf{A}_{Q'/\mathbb{P}(F)}$. The latter is known to be equivalent to $\mathsf{D}^{\mathrm{b}}(\mathbb{P}(F), \mathscr{C}_0)$, where $\mathscr{C}_0$ is the sheaf of even Clifford algebras of the quadric surface fibration $Q' \to \mathbb{P}(F)$. Finally, assuming the degeneration divisor of $Q' \to \mathbb{P}(F)$ to be smooth, we have a smooth double cover $S \to \mathbb{P}(F)$ and a Brauer class α in $\mathrm{Br}(S)$, such that $\mathsf{A}_{X/\mathbb{P}^1} \simeq \mathsf{D}^{\mathrm{b}}(S, \alpha)$. Notice that the composition $S \to \mathbb{P}(F) \to \mathbb{P}^1$ endows S with a fibration into genus 2 curves, since there are 6 degenerate quadrics for each fiber of $\mathbb{P}(F) \to \mathbb{P}^1$.

The following conjecture is inspired by Question 8.0.1 and Kuznetsov's conjecture 8.3.1 for cubic fourfolds.

CONJECTURE 8.4.1 ([**12**], Conj. 5.1.2). *Let $X \to \mathbb{P}^1$ be a fibration in complete intersections of two four-dimensional quadrics.*

- **Weak version.** *The fourfold X is rational if and only if* $\mathrm{rcodim}(X) \geq 2$.
- **Strong version.** *The fourfold X is rational if and only if* $\mathrm{rdim}\, \mathsf{A}_{X/\mathbb{P}^1} \leq 2$.

Y	X	Explicit description of $\mathsf{A}_{X/Y}$	Ref.
		Threefolds	
pt	$\mathbb{P}^3$	0	[27]
	quadric 3fold	$\langle S\rangle$, a spinor bundle	[108]
	index 2, degree 5	$\langle U, V\rangle$, two vector bundles	[145]
	int. of 2 quadrics	$\mathsf{D}^{\mathrm{b}}(C)$ with $g(C) = 2$ and $J(X) = J(C)$	[48]
	cubic 3fold	fractional Calabi-Yau	[119]
	nodal cubic 3fold*	$\langle \mathsf{D}^{\mathrm{b}}(C), L\rangle$, L a l.bd., $g(C) = 2$ and $J(X) = J(C)$	[30]
	det'l. cubic 3fold*	four exceptional objects	[31]
	quartic double solid	noncommutative Enriques surface	[127]
	Artin-Mumford*	complement of exc. collection on Enriques surface	[105]
	index 1, degree 22	$\langle U, V, W\rangle$, three vector bundles	[117]
	index 1, degree 18	$\langle \mathsf{D}^{\mathrm{b}}(C), U\rangle$, U a v.b., $g(C) = 2$ and $J(C) = J(X)$	[120]
	index 1, degree 16	$\langle \mathsf{D}^{\mathrm{b}}(C), U\rangle$, U a v.b., $g(C) = 3$ and $J(C) = J(X)$	[120]
	index 1, degree 14	$\langle \mathsf{B}, U\rangle$, U a v.b., B fractional Calabi-Yau	[119]
	index 1, degree 12	$\langle \mathsf{D}^{\mathrm{b}}(C), U\rangle$, U a v.b., $g(C) = 7$ and $J(C) = J(X)$	[118]
	Hyperell. Gushel–Mukai	noncommutative Enriques surface	[127]
	int. of 3 quadrics	$\mathsf{D}^{\mathrm{b}}(\mathbb{P}^2, \mathscr{C}_0)$, $\mathscr{C}_0$ a Clifford algebra	[116]
	quartic 3fold	fractional Calabi–Yau	[119]
$\mathbb{P}^1$	$\mathbb{P}^2$-bundle	0	[146]
	quadric bundle	$\mathsf{D}^{\mathrm{b}}(C)$, C hyperelliptic and $J(X) = J(C)$	[116]
	dP4 fibration	$\mathsf{D}^{\mathrm{b}}(S, \mathscr{C}_0)$, S a Hirz. surf, $\mathscr{C}_0$ a Cliff. algebra	[12]
	rational dP4 fibration	$\mathsf{D}^{\mathrm{b}}(C)$, with $J(C) = J(X)$	[12]
rat'l. surf.	$\mathbb{P}^1$-bundle	0	[146]
	conic bundle	$\mathsf{D}^{\mathrm{b}}(Y, \mathscr{C}_0)$, $\mathscr{C}_0$ a Clifford algebra	[116]
$\mathbb{P}^2$	rational conic bundle with deg. curve Γ	$\langle \mathsf{D}^{\mathrm{b}}(C), V\rangle$, V v.b and $J(C) = J(X)$ if $g(\Gamma) > 1$	[29]
		$\langle L_1, L_2, L_3\rangle$, L_i line bd. if $g(\Gamma) = 1$	[29]
Hirz.	rational conic bundle with deg. curve Γ	$\langle \mathsf{D}^{\mathrm{b}}(C), V_1, V_2\rangle$, V_i v.b and $J(C) = J(X)$ if $g(\Gamma) = 3$	[29]
		$\langle \mathsf{D}^{\mathrm{b}}(C_1), \mathsf{D}^{\mathrm{b}}(C_2)\rangle$, with $J(X) = J(C_1) \oplus J(C_2)$ if $g(\Gamma) = 2$	[29]
		Fourfolds	
any	$\mathbb{P}$-bundle	0	[146]
	quadric bundle	$\mathsf{D}^{\mathrm{b}}(Y, \mathscr{C}_0)$, $\mathscr{C}_0$ Cliff. algebra	[116]
pt	quadric 4fold	$\langle S_1, S_2\rangle$, spinor bundles	[108]
	cubic fourfold	a noncommutative K3	[123]
	Pfaffian cubic	$\mathsf{D}^{\mathrm{b}}(S)$, with S a degree 14 K3	[123]
	cubic with a plane	$\mathsf{D}^{\mathrm{b}}(S, \alpha)$, with S a degree 2 K3 and $\alpha \in \mathrm{Br}(S)$ $\alpha = 0$ iff the associated quadric fibr. has a section	[123]
	nodal cubic*	$\mathsf{D}^{\mathrm{b}}(S)$, with S a degree 6 K3	[123]
	determinantal cubic*	six exceptional objects	[31]
	general cubic in $\mathcal{C}_d$ with associated K3	$\mathsf{D}^{\mathrm{b}}(S)$, with S a K3 surface of degree d	[3]
	deg 10 ind 2 in $G(2,5)$	a noncommutative K3	[125]
$\mathbb{P}^1$	fibration in intersections of 2 quadrics	$\mathsf{D}^{\mathrm{b}}(T, \alpha)$, $T \to \mathbb{P}^1$ hyperell. fib. $\alpha \in \mathrm{Br}(T)$, $\alpha = 0$ if S has a line/$\mathbb{P}^1$	[12]

TABLE 1. Known descriptions of $\mathsf{A}_{X/Y}$, for $X \to Y$ a Mori fiber space, and $\dim(X) = 3$ or $\dim(X) = 4$ and $\kappa(Y) = -\infty$. In the nonsmooth cases, indicated by *, the description refers to a categorical resolution of $\mathsf{A}_{X/Y}$

References

[1] Dan Abramovich, Kalle Karu, Kenji Matsuki, and Jarosław Włodarczyk, *Torification and factorization of birational maps*, J. Amer. Math. Soc. **15** (2002), no. 3, 531–572, DOI 10.1090/S0894-0347-02-00396-X. MR1896232

[2] Nicolas Addington, *On two rationality conjectures for cubic fourfolds*, Math. Res. Lett. **23** (2016), no. 1, 1–13, DOI 10.4310/MRL.2016.v23.n1.a1. MR3512874

[3] Nicolas Addington and Richard Thomas, *Hodge theory and derived categories of cubic fourfolds*, Duke Math. J. **163** (2014), no. 10, 1885–1927, DOI 10.1215/00127094-2738639. MR3229044

[4] Nicolas Addington, B. Hassett, Y. Tschinkel, and A. Várilly-Alvarado, *Cubic fourfolds fibered in sextic del Pezzo surfaces*, preprint arXiv:1606.05321, 2016.

[5] V. A. Alekseev, *On conditions for the rationality of three-folds with a pencil of del Pezzo surfaces of degree* 4 (Russian), Mat. Zametki **41** (1987), no. 5, 724–730, 766. MR898133

[6] Valery Alexeev and Dmitri Orlov, *Derived categories of Burniat surfaces and exceptional collections*, Math. Ann. **357** (2013), no. 2, 743–759, DOI 10.1007/s00208-013-0917-2. MR3096524

[7] *Théorie des topos et cohomologie étale des schémas. Tome 1: Théorie des topos* (French), Lecture Notes in Mathematics, Vol. 269, Springer-Verlag, Berlin-New York, 1972. Séminaire de Géométrie Algébrique du Bois-Marie 1963–1964 (SGA 4); Dirigé par M. Artin, A. Grothendieck, et J. L. Verdier. Avec la collaboration de N. Bourbaki, P. Deligne et B. Saint-Donat. MR0354652

[8] M. Artin, *Algebraization of formal moduli. II. Existence of modifications*, Ann. of Math. (2) **91** (1970), 88–135, DOI 10.2307/1970602. MR0260747

[9] M. Artin and D. Mumford, *Some elementary examples of unirational varieties which are not rational*, Proc. London Math. Soc. (3) **25** (1972), 75–95, DOI 10.1112/plms/s3-25.1.75. MR0321934

[10] Asher Auel, *The complement of the locus of pfaffian cubic fourfolds*, preprint, 2015.

[11] Asher Auel and Marcello Bernardara, *Semiorthogonal decompositions and birational geometry of del Pezzo surfaces over arbitrary fields*, preprint arXiv:1511.07576, 2015.

[12] Asher Auel, Marcello Bernardara, and Michele Bolognesi, *Fibrations in complete intersections of quadrics, Clifford algebras, derived categories, and rationality problems*, J. Math. Pures Appl. (9) **102** (2014), no. 1, 249–291, DOI 10.1016/j.matpur.2013.11.009. MR3212256

[13] Asher Auel, Marcello Bernardara, Michele Bolognesi, and Anthony Várilly-Alvarado, *Cubic fourfolds containing a plane and a quintic del Pezzo surface*, Algebr. Geom. **1** (2014), no. 2, 181–193, DOI 10.14231/AG-2014-010. MR3238111

[14] Asher Auel, Christian Böhning, and Hans-Christian Graf von Bothmer, *The transcendental lattice of the sextic Fermat surface*, Math. Res. Lett. **20** (2013), no. 6, 1017–1031, DOI 10.4310/MRL.2013.v20.n6.a2. MR3228617

[15] Asher Auel, Jean-Louis Colliot-Thélène, and Raman Parimala, *Universal unramified cohomology of cubic fourfolds containing a plane*, Birkhäuser series Progress in Mathematics vol. 320, 2017.

[16] Asher Auel, Raman Parimala, and Venapally Suresh, *Quadric surface bundles over surfaces*, Doc. Math. (2015), 31–70. MR3404375

[17] Maurice Auslander and Oscar Goldman, *The Brauer group of a commutative ring*, Trans. Amer. Math. Soc. **97** (1960), 367–409, DOI 10.2307/1993378. MR0121392

[18] M. Ballard, D. Deliu, D. Favero, M.U. Isik, and L. Katzarkov, *On the Derived Categories of Degree d Hypersurface Fibrations*, preprint arXiv:1409.5568, 2014.

[19] L. Barbieri-Viale, *Cicli di codimensione 2 su varietà unirazionali complesse*, K-Theory (Strasbourg, 1992), Astérisque **226** (1994), 13–41.

[20] Rebecca Barlow, *Rational equivalence of zero cycles for some more surfaces with* $p_g = 0$, Invent. Math. **79** (1985), no. 2, 303–308, DOI 10.1007/BF01388975. MR778129

[21] Wolf P. Barth, Klaus Hulek, Chris A. M. Peters, and Antonius Van de Ven, *Compact complex surfaces*, 2nd ed., Ergebnisse der Mathematik und ihrer Grenzgebiete. 3. Folge. A Series of Modern Surveys in Mathematics, vol. 4, Springer-Verlag, Berlin, 2004. MR2030225

[22] Ingrid Bauer, Fabrizio Catanese, and Roberto Pignatelli, *Surfaces of general type with geometric genus zero: a survey*, Complex and differential geometry, Springer Proc. Math., vol. 8, Springer, Heidelberg, 2011, pp. 1–48, DOI 10.1007/978-3-642-20300-8_1. MR2964466

[23] Arnaud Beauville, *Variétés de Prym et jacobiennes intermédiaires* (French), Ann. Sci. École Norm. Sup. (4) **10** (1977), no. 3, 309–391. MR0472843

[24] Arnaud Beauville, *The Lüroth problem*, Rationality Problems in Algebraic Geometry, pp. 1-27. Springer Lecture Notes 2172 (2016).

[25] Arnaud Beauville, Jean-Louis Colliot-Thélène, Jean-Jacques Sansuc, and Peter Swinnerton-Dyer, *Variétés stablement rationnelles non rationnelles* (French), Ann. of Math. (2) **121** (1985), no. 2, 283–318, DOI 10.2307/1971174. MR786350

[26] Arnaud Beauville and Ron Donagi, *La variété des droites d'une hypersurface cubique de dimension* 4 (French, with English summary), C. R. Acad. Sci. Paris Sér. I Math. **301** (1985), no. 14, 703–706. MR818549

[27] A. A. Beĭlinson, *Coherent sheaves on* $\mathbf{P}^n$ *and problems in linear algebra* (Russian), Funktsional. Anal. i Prilozhen. **12** (1978), no. 3, 68–69. MR509388

[28] Marcello Bernardara, *A semiorthogonal decomposition for Brauer-Severi schemes*, Math. Nachr. **282** (2009), no. 10, 1406–1413, DOI 10.1002/mana.200610826. MR2571702

[29] Marcello Bernardara and Michele Bolognesi, *Derived categories and rationality of conic bundles*, Compos. Math. **149** (2013), no. 11, 1789–1817, DOI 10.1112/S0010437X13007392. MR3133293

[30] Marcello Bernardara and Michele Bolognesi, *Categorical representability and intermediate Jacobians of Fano threefolds*, EMS Ser. Congr. Rep., Eur. Math. Soc., Zürich, 2012, pp. 1–25. MR3050698

[31] Marcello Bernardara, Michele Bolognesi, and Daniele Faenzi, *Homological projective duality for determinantal varieties*, Adv. Math. **296** (2016), 181–209, DOI 10.1016/j.aim.2016.04.003. MR3490767

[32] Marcello Bernardara and Gonçalo Tabuada, *From semi-orthogonal decompositions to polarized intermediate Jacobians via Jacobians of noncommutative motives*, Mosc. Math. J. **16** (2016), no. 2, 205–235. MR3480702

[33] Franziska Bittner, *The universal Euler characteristic for varieties of characteristic zero*, Compos. Math. **140** (2004), no. 4, 1011–1032, DOI 10.1112/S0010437X03000617. MR2059227

[34] Spencer Bloch, *Lectures on algebraic cycles*, Duke University Mathematics Series, IV, Duke University, Mathematics Department, Durham, N.C., 1980. MR558224

[35] Spencer Bloch, A. Kas, and D. Lieberman, *Zero cycles on surfaces with* $p_g = 0$, Compos. Math. **33** (1976), no. 2, 135–145. MR0435073

[36] Spencer Bloch and Arthur Ogus, *Gersten's conjecture and the homology of schemes*, Ann. Sci. École Norm. Sup. (4) **7** (1974), 181–201 (1975). MR0412191

[37] Spencer Bloch and V. Srinivas, *Remarks on correspondences and algebraic cycles*, Amer. J. Math. **105** (1983), no. 5, 1235–1253, DOI 10.2307/2374341. MR714776

[38] M. Blunk, *A Derived Equivalence for some Twisted Projective Homogeneous Varieties*, preprint arXiv:1204.0537, 2012.

[39] F. A. Bogomolov, *The Brauer group of quotient spaces of linear representations* (Russian), Izv. Akad. Nauk SSSR Ser. Mat. **51** (1987), no. 3, 485–516, 688; English transl., Math. USSR-Izv. **30** (1988), no. 3, 455–485. MR903621

[40] Christian Böhning, Hans-Christian Graf von Bothmer, and Pawel Sosna, *On the derived category of the classical Godeaux surface*, Adv. Math. **243** (2013), 203–231, DOI 10.1016/j.aim.2013.04.017. MR3062745

[41] Christian Böhning, Hans-Christian Graf von Bothmer, and Pawel Sosna, *On the Jordan-Hölder property for geometric derived categories*, Adv. Math. **256** (2014), 479–492, DOI 10.1016/j.aim.2014.02.016. MR3177299

[42] Christian Böhning, Hans-Christian Graf von Bothmer, Ludmil Katzarkov, and Pawel Sosna, *Determinantal Barlow surfaces and phantom categories*, J. Eur. Math. Soc. (JEMS) **17** (2015), no. 7, 1569–1592, DOI 10.4171/JEMS/539. MR3361723

[43] M. Bolognesi and F. Russo, *Some loci of rational cubic fourfolds*, with an appendix by G. Staglianò, preprint arXiv:1504.05863, 2016.

[44] A. I. Bondal, *Representations of associative algebras and coherent sheaves* (Russian), Izv. Akad. Nauk SSSR Ser. Mat. **53** (1989), no. 1, 25–44; English transl., Math. USSR-Izv. **34** (1990), no. 1, 23–42. MR992977

[45] A. I. Bondal and M. M. Kapranov, *Representable functors, Serre functors, and reconstructions* (Russian), Izv. Akad. Nauk SSSR Ser. Mat. **53** (1989), no. 6, 1183–1205, 1337; English transl., Math. USSR-Izv. **35** (1990), no. 3, 519–541. MR1039961

[46] A. I. Bondal and M. M. Kapranov, *Framed triangulated categories* (Russian), Mat. Sb. **181** (1990), no. 5, 669–683; English transl., Math. USSR-Sb. **70** (1991), no. 1, 93–107. MR1055981

[47] Alexei Bondal and Dmitri Orlov, *Reconstruction of a variety from the derived category and groups of autoequivalences*, Compos. Math. **125** (2001), no. 3, 327–344, DOI 10.1023/A:1002470302976. MR1818984

[48] A.I. Bondal and Dmitri Orlov, *Semiorthogonal decomposition for algebraic varieties*, MPIM preprint 1995-15, 1995.

[49] Alexey I. Bondal, Michael Larsen, and Valery A. Lunts, *Grothendieck ring of pretriangulated categories*, Int. Math. Res. Not. **29** (2004), 1461–1495, DOI 10.1155/S1073792804140385. MR2051435

[50] L. Borisov, *Class of the affine line is a zero divisor in the Grothendieck ring*, preprint arXiv:1412.6194, 2015.

[51] Tom Bridgeland, *Equivalences of triangulated categories and Fourier-Mukai transforms*, Bull. London Math. Soc. **31** (1999), no. 1, 25–34, DOI 10.1112/S0024609398004998. MR1651025

[52] F. Campana, *Connexité rationnelle des variétés de Fano* (French), Ann. Sci. École Norm. Sup. (4) **25** (1992), no. 5, 539–545. MR1191735

[53] F. Kampana, T. Peternell, and A. V. Pukhlikov, *The generalized Tsen theorem and rationally connected Fano fibrations* (Russian, with Russian summary), Mat. Sb. **193** (2002), no. 10, 49–74, DOI 10.1070/SM2002v193n10ABEH000685; English transl., Sb. Math. **193** (2002), no. 9-10, 1443–1468. MR1937041

[54] L. Campedelli, *Sopra alcuni piani doppi notevoli con curve di diramazione d el decimo ordine*, Atti Acad. Naz. Lincei **15** (1932), 536–542.

[55] G. Castelnuovo, *Sulle superficie di genere zero*, Memorie della Soc.It. delle Scienze (detta dei XL), ser. III, t. 10, (1896).

[56] A. Chatzistamatiou and M. Levine, *Torsion orders of complete intersections*, preprint arXiv:1605.01913, 2016.

[57] Y. Cho and Y. Lee, *Exceptional collections on Dolgachev surfaces associated with degenerations*, preprint arXiv:1506.05213, 2015.

[58] C. Herbert Clemens and Phillip A. Griffiths, *The intermediate Jacobian of the cubic threefold*, Ann. of Math. (2) **95** (1972), 281–356, DOI 10.2307/1970801. MR0302652

[59] Jean-Louis Colliot-Thélène, *Formes quadratiques multiplicatives et variétés algébriques* (French), Bull. Soc. Math. France **106** (1978), no. 2, 113–151. MR507745

[60] Jean-Louis Colliot-Thélène, *Formes quadratiques multiplicatives et variétés algébriques: deux compléments* (French, with German summary), Bull. Soc. Math. France **108** (1980), no. 2, 213–227. MR606090

[61] Jean-Louis Colliot-Thélène, *Birational invariants, purity and the Gersten conjecture*, Proc. Sympos. Pure Math., vol. 58, Amer. Math. Soc., Providence, RI, 1995, pp. 1–64. MR1327280

[62] Jean-Louis Colliot-Thélène, *Un théorème de finitude pour le groupe de Chow des zéro-cycles d'un groupe algébrique linéaire sur un corps p-adique* (French, with German summary), Invent. Math. **159** (2005), no. 3, 589–606, DOI 10.1007/s00222-004-0393-0. MR2125734

[63] Jean-Louis Colliot-Thélène, *Descente galoisienne sur le second groupe de Chow: mise au point et applications* (French, with French summary), Doc. Math. **Extra vol.: Alexander S. Merkurjev's sixtieth birthday** (2015), 195–220. MR3404380

[64] Jean-Louis Colliot-Thélène, *Non rationalité stable d'hypersurfaces cubiques sur des corps non algébriquement clos*, preprint arXiv:1606.06982, 2016.

[65] Jean-Louis Colliot-Thélène, CH_0*-trivialité universelle d'hypersurfaces cubiques presque diagonales*, arXiv:1607.05673, to appear in Algebraic Geometry.

[66] Jean-Louis Colliot-Thélène and Daniel Coray, *L'équivalence rationnelle sur les points fermés des surfaces rationnelles fibrées en coniques* (French), Compositio Math. **39** (1979), no. 3, 301–332. MR550646

[67] Jean-Louis Colliot-Thélène and Bruno Kahn, *Cycles de codimension 2 et H^3 non ramifié pour les variétés sur les corps finis* (French, with English and French summaries), J. K-Theory **11** (2013), no. 1, 1–53, DOI 10.1017/is012009001jkt194. MR3034282

[68] Jean-Louis Colliot-Thélène and Manuel Ojanguren, *Variétés unirationnelles non rationnelles: au-delà de l'exemple d'Artin et Mumford* (French), Invent. Math. **97** (1989), no. 1, 141–158, DOI 10.1007/BF01850658. MR999316

[69] Jean-Louis Colliot-Thélène and Alena Pirutka, *Hypersurfaces quartiques de dimension 3: non-rationalité stable* (French, with English and French summaries), Ann. Sci. Éc. Norm. Supér. (4) **49** (2016), no. 2, 371–397. MR3481353

[70] Jean-Louis Colliot-Thélène and Wayne Raskind, *$\mathcal{K}_2$-cohomology and the second Chow group*, Math. Ann. **270** (1985), no. 2, 165–199, DOI 10.1007/BF01456181. MR771978

[71] Jean-Louis Colliot-Thélène and Jean-Jacques Sansuc, *La R-équivalence sur les tores* (French), Ann. Sci. École Norm. Sup. (4) **10** (1977), no. 2, 175–229. MR0450280

[72] Jean-Louis Colliot-Thélène and Jean-Jacques Sansuc, *Cohomologie des groupes de type multiplicatif sur les schémas réguliers* (French, with English summary), C. R. Acad. Sci. Paris Sér. A-B **287** (1978), no. 6, A449–A452. MR510771

[73] Jean-Louis Colliot-Thélène and Jean-Jacques Sansuc, *Fibrés quadratiques et composantes connexes réelles* (French), Math. Ann. **244** (1979), no. 2, 105–134, DOI 10.1007/BF01420486. MR550842

[74] Jean-Louis Colliot-Thélène and Claire Voisin, *Cohomologie non ramifiée et conjecture de Hodge entière* (French, with English and French summaries), Duke Math. J. **161** (2012), no. 5, 735–801, DOI 10.1215/00127094-1548389. MR2904092

[75] János Kollár, Karen E. Smith, and Alessio Corti, *Rational and nearly rational varieties*, Cambridge Studies in Advanced Mathematics, vol. 92, Cambridge University Press, Cambridge, 2004. MR2062787

[76] Stephen Coughlan, *Enumerating exceptional collections of line bundles on some surfaces of general type*, Doc. Math. **20** (2015), 1255–1291. MR3424480

[77] Tommaso de Fernex and Davide Fusi, *Rationality in families of threefolds*, Rend. Circ. Mat. Palermo (2) **62** (2013), no. 1, 127–135, DOI 10.1007/s12215-013-0110-1. MR3031573

[78] Pierre Deligne and Luc Illusie, *Relèvements modulo p^2 et décomposition du complexe de de Rham* (French), Invent. Math. **89** (1987), no. 2, 247–270, DOI 10.1007/BF01389078. MR894379

[79] Igor Dolgachev, *Algebraic surfaces with $q = p_g = 0$*, C.I.M.E. Summer Sch., vol. 76, Springer, Heidelberg, 2010, pp. 97–215, DOI 10.1007/978-3-642-11087-0_3. MR2757651

[80] Federigo Enriques, *Sulle irrazionalità da cui può farsi dipendere la risoluzione d'un' equazione algebrica $f(xyz) = 0$ con funzioni razionali di due parametri* (Italian), Math. Ann. **49** (1897), no. 1, 1–23, DOI 10.1007/BF01445357. MR1510950

[81] Federigo Enriques, *Memorie scelte di geometria. Vol. I: 1893–1898* (Italian), Pubblicate a cura dell'Accademia Nazionale dei Lincei, Nicola Zanichelli Editore, Bologna, 1956. MR0212647

[82] Hélène Esnault, *Varieties over a finite field with trivial Chow group of 0-cycles have a rational point*, Invent. Math. **151** (2003), no. 1, 187–191, DOI 10.1007/s00222-002-0261-8. MR1943746

[83] Gino Fano, *Sulle forme cubiche dello spazio a cinque dimensioni contenenti rigate razionali del 4° ordine* (Italian), Comment. Math. Helv. **15** (1943), 71–80, DOI 10.1007/BF02565634. MR0010433

[84] William Fulton, *Intersection theory*, Ergebnisse der Mathematik und ihrer Grenzgebiete (3) [Results in Mathematics and Related Areas (3)], vol. 2, Springer-Verlag, Berlin, 1984. MR732620

[85] William Fulton, *Rational equivalence on singular varieties*, Inst. Hautes Études Sci. Publ. Math. **45** (1975), 147–167. MR0404257

[86] Sergey Galkin and Evgeny Shinder, *Exceptional collections of line bundles on the Beauville surface*, Adv. Math. **244** (2013), 1033–1050, DOI 10.1016/j.aim.2013.06.007. MR3077896

[87] Sergey Galkin and Evgeny Shinder, *The Fano variety of lines and rationality problem for a cubic hypersurface*, preprint arXiv:1405.5154, 2014.

[88] Sergey Galkin, Ludmil Katzarkov, Anton Mellit, and Evgeny Shinder, *Derived categories of Keum's fake projective planes*, Adv. Math. **278** (2015), 238–253, DOI 10.1016/j.aim.2015.03.001. MR3341791

[89] L. Godeaux, *Les involutions cycliques appartenant à une surface algébrique*, Actual. Sci. Ind. **270**, Hermann, Paris, 1935.

[90] A. L. Gorodentsev, *Exceptional bundles on surfaces with a moving anticanonical class* (Russian), Izv. Akad. Nauk SSSR Ser. Mat. **52** (1988), no. 4, 740–757, 895; English transl., Math. USSR-Izv. **33** (1989), no. 1, 67–83. MR966982

[91] A. L. Gorodentsev and A. N. Rudakov, *Exceptional vector bundles on projective spaces*, Duke Math. J. **54** (1987), no. 1, 115–130, DOI 10.1215/S0012-7094-87-05409-3. MR885779

[92] Sergey Gorchinskiy and Dmitri Orlov, *Geometric phantom categories*, Publ. Math. Inst. Hautes Études Sci. **117** (2013), 329–349, DOI 10.1007/s10240-013-0050-5. MR3090263

[93] Tom Graber, Joe Harris, and Jason Starr, *Families of rationally connected varieties*, J. Amer. Math. Soc. **16** (2003), no. 1, 57–67, DOI 10.1090/S0894-0347-02-00402-2. MR1937199

[94] A. Grothendieck, *Le groupe de Brauer. I, II, III*, Dix Exposés sur la Cohomologie des Schémas (J. Giraud et al), North-Holland, 1968.

[95] Brendan Hassett, *Special cubic fourfolds*, Compositio Math. **120** (2000), no. 1, 1–23, DOI 10.1023/A:1001706324425. MR1738215

[96] Brendan Hassett, *Some rational cubic fourfolds*, J. Algebraic Geom. **8** (1999), no. 1, 103–114. MR1658216

[97] Brendan Hassett, *Cubic fourfolds, K3 surfaces, and rationality questions*, lecture notes for a 2015 CIME-CIRM summer school available on `http://www.math.brown.edu/~bhassett/papers.html`.

[98] Brendan Hassett, Andrew Kresch, and Yuri Tschinkel, *Stable rationality and conic bundles*, Math. Ann. **365** (2016), no. 3-4, 1201–1217, DOI 10.1007/s00208-015-1292-y. MR3521088

[99] Brendan Hassett and K.-W. Lai, *Cremona transformations and derived equivalences of K3 surfaces*, preprint arXiv:1612.07751, 2017.

[100] Brendan Hassett and Yuri Tschinkel, *Rational curves on holomorphic symplectic fourfolds*, Geom. Funct. Anal. **11** (2001), no. 6, 1201–1228, DOI 10.1007/s00039-001-8229-1. MR1878319

[101] Brendan Hassett and Yuri Tschinkel, *On stable rationality of Fano threefolds and del Pezzo fibrations*, arXiv:1601.07074, to appear in Journal für die reine und angewandte Mathematik.

[102] Lutz Hille and Markus Perling, *Exceptional sequences of invertible sheaves on rational surfaces*, Compos. Math. **147** (2011), no. 4, 1230–1280, DOI 10.1112/S0010437X10005208. MR2822868

[103] S. Hosono and H. Takagi, *Derived categories of Artin-Mumford double solids*, preprint ArXiv 1506.02744 (2015).

[104] Daniel Huybrechts, *Fourier-Mukai transforms in algebraic geometry*, Oxford Mathematical Monographs, The Clarendon Press, Oxford University Press, Oxford, 2006. MR2244106

[105] Colin Ingalls and Alexander Kuznetsov, *On nodal Enriques surfaces and quartic double solids*, Math. Ann. **361** (2015), no. 1-2, 107–133, DOI 10.1007/s00208-014-1066-y. MR3302614

[106] V. A. Iskovskih and Ju. I. Manin, *Three-dimensional quartics and counterexamples to the Lüroth problem* (Russian), Mat. Sb. (N.S.) **86(128)** (1971), 140–166. MR0291172

[107] B. Kahn, *Torsion order of smooth projective surfaces*, with an appendix by J.-L. Colliot-Thélène, preprint arXiv:1605.01762, 2016.

[108] M. M. Kapranov, *Derived category of coherent bundles on a quadric* (Russian), Funktsional. Anal. i Prilozhen. **20** (1986), no. 2, 67. MR847146

[109] Nikita A. Karpenko and Alexander S. Merkurjev, *On standard norm varieties* (English, with English and French summaries), Ann. Sci. Éc. Norm. Supér. (4) **46** (2013), no. 1, 175–214 (2013). MR3087392

[110] K. Kawatani and S. Okawa *Nonexistence of semiorthogonal decompositions and sections of the canonical bundle*, preprint ArXiv:1508.00682

[111] Bernhard Keller, *On differential graded categories*, International Congress of Mathematicians. Vol. II, Eur. Math. Soc., Zürich, 2006, pp. 151–190. MR2275593

[112] J. Keum, *A vanishing theorem on fake projective planes with enough automorphisms*, preprint: arXiv:1407.7632v3.

[113] János Kollár, *Nonrational hypersurfaces*, J. Amer. Math. Soc. **8** (1995), no. 1, 241–249, DOI 10.2307/2152888. MR1273416

[114] János Kollár, Yoichi Miyaoka, and Shigefumi Mori, *Rational connectedness and boundedness of Fano manifolds*, J. Differential Geom. **36** (1992), no. 3, 765–779. MR1189503

[115] Vik. S. Kulikov, *A remark on the nonrationality of a generic cubic fourfold* (Russian, with Russian summary), Mat. Zametki **83** (2008), no. 1, 61–68, DOI 10.1134/S0001434608010070; English transl., Math. Notes **83** (2008), no. 1-2, 57–64. MR2399998

[116] Alexander Kuznetsov, *Derived categories of quadric fibrations and intersections of quadrics*, Adv. Math. **218** (2008), no. 5, 1340–1369, DOI 10.1016/j.aim.2008.03.007. MR2419925

[117] A. Kuznetsov, *Exceptional Collection of Vector Bundles on V22 Fano Threefolds*, Vestn. Mosk. Univ., Ser. 1: Mat., Mekh., No. 3, 41–44 (1996), english translation in Moscow Univ. Math. Bull. 51 (3), 35–37 (1996).

[118] Alexander Kuznetsov, *Derived categories of the Fano threefolds V_{12}* (Russian, with Russian summary), Mat. Zametki **78** (2005), no. 4, 579–594, DOI 10.1007/s11006-005-0152-6; English transl., Math. Notes **78** (2005), no. 3-4, 537–550. MR2226730

[119] Alexander Kuznetsov, *Derived category of a cubic threefold and the variety V_{14}* (Russian, with Russian summary), Tr. Mat. Inst. Steklova **246** (2004), no. Algebr. Geom. Metody, Svyazi i Prilozh., 183–207; English transl., Proc. Steklov Inst. Math. **3(246)** (2004), 171–194. MR2101293

[120] Alexander Kuznetsov, *Hyperplane sections and derived categories* (Russian, with Russian summary), Izv. Ross. Akad. Nauk Ser. Mat. **70** (2006), no. 3, 23–128, DOI 10.1070/IM2006v070n03ABEH002318; English transl., Izv. Math. **70** (2006), no. 3, 447–547. MR2238172

[121] Alexander Kuznetsov, *Homological projective duality*, Publ. Math. Inst. Hautes Études Sci. **105** (2007), 157–220, DOI 10.1007/s10240-007-0006-8. MR2354207

[122] Alexander Kuznetsov, *Derived categories of Fano threefolds* (Russian, with Russian summary), Tr. Mat. Inst. Steklova **264** (2009), no. Mnogomernaya Algebraicheskaya Geometriya, 116–128, DOI 10.1134/S0081543809010143; English transl., Proc. Steklov Inst. Math. **264** (2009), no. 1, 110–122. MR2590842

[123] Alexander Kuznetsov, *Derived categories of cubic fourfolds*, Progr. Math., vol. 282, Birkhäuser Boston, Inc., Boston, MA, 2010, pp. 219–243, DOI 10.1007/978-0-8176-4934-0_9. MR2605171

[124] Alexander Kuznetsov, *A simple counterexample to the Jordan-Hölder property for derived categories* preprint math.AG/1304.0903 (2013).

[125] Alexander Kuznetsov, *Semiorthogonal decompositions in algebraic geometry*, in Proceedings of the International Congress of Mathematicians (Seoul, 2014), vol. II (2014), pp. 635–660. arXiv:1404.3143.

[126] Alexander Kuznetsov, *Derived categories view on rationality problems*, Lecture notes for the CIME-CIRM summer school (Levico Terme, 2015). arXiv:1509.09115.

[127] Alexander Kuznetsov and Alexander Perry, *Derived categories of cyclic covers and their branch divisors*, Selecta Math. (N.S.) **23** (2017), no. 1, 389–423, DOI 10.1007/s00029-016-0243-0. MR3595897

[128] Alexander Kuznetsov and Alexander Polishchuk, *Exceptional collections on isotropic Grassmannians*, J. Eur. Math. Soc. (JEMS) **18** (2016), no. 3, 507–574, DOI 10.4171/JEMS/596. MR3463417

[129] Alexander Kuznetsov and E. Shinder, *Grothendieck ring of varieties, D- and L-equivalence, and families of quadrics*, preprint arXiv:1612.07193, 2016.

[130] K.-W. Lai, *New cubic fourfolds with odd degree unirational parametrizations*, preprint arXiv:1606.03853, 2016.

[131] Radu Laza, *The moduli space of cubic fourfolds via the period map*, Ann. of Math. (2) **172** (2010), no. 1, 673–711, DOI 10.4007/annals.2010.172.673. MR2680429

[132] Eduard Looijenga, *The period map for cubic fourfolds*, Invent. Math. **177** (2009), no. 1, 213–233, DOI 10.1007/s00222-009-0178-6. MR2507640

[133] Kyoung-Seog Lee, *Derived categories of surfaces isogenous to a higher product*, J. Algebra **441** (2015), 180–195, DOI 10.1016/j.jalgebra.2015.06.022. MR3391925

[134] Valery A. Lunts and Dmitri Orlov, *Uniqueness of enhancement for triangulated categories*, J. Amer. Math. Soc. **23** (2010), no. 3, 853–908, DOI 10.1090/S0894-0347-10-00664-8. MR2629991

[135] Matilde Marcolli and Gonçalo Tabuada, *From exceptional collections to motivic decompositions via noncommutative motives*, J. Reine Angew. Math. **701** (2015), 153–167, DOI 10.1515/crelle-2013-0027. MR3331729

[136] D. Markushevich and A. S. Tikhomirov, *The Abel-Jacobi map of a moduli component of vector bundles on the cubic threefold*, J. Algebraic Geom. **10** (2001), no. 1, 37–62. MR1795549

[137] Teruhisa Matsusaka, *On a characterization of a Jacobian variety*, Memo. Coll. Sci. Univ. Kyoto. Ser. A. Math. **32** (1959), 1–19. MR0108497

[138] Alexander Merkurjev, *Unramified elements in cycle modules*, J. Lond. Math. Soc. (2) **78** (2008), no. 1, 51–64, DOI 10.1112/jlms/jdn011. MR2427051

[139] Fabien Morel, *Milnor's conjecture on quadratic forms and mod 2 motivic complexes*, Rend. Sem. Mat. Univ. Padova **114** (2005), 63–101 (2006). MR2207862

[140] David Mumford, *Abelian varieties*, Tata Institute of Fundamental Research Studies in Mathematics, No. 5, Published for the Tata Institute of Fundamental Research, Bombay; Oxford University Press, London, 1970. MR0282985

[141] David Mumford, *Rational equivalence of* 0*-cycles on surfaces*, J. Math. Kyoto Univ. **9** (1968), 195–204. MR0249428

[142] J. P. Murre, *On the Hodge conjecture for unirational fourfolds*, Nederl. Akad. Wetensch. Proc. Ser. A **80**=Indag. Math. **39** (1977), no. 3, 230–232. MR0457434

[143] H. Nuer, *Unirationality of moduli spaces of special cubic fourfolds and K3 surfaces*, preprint arXiv:1503.05256, 2015.

[144] Shinnosuke Okawa, *Semi-orthogonal decomposability of the derived category of a curve*, Adv. Math. **228** (2011), no. 5, 2869–2873, DOI 10.1016/j.aim.2011.06.043. MR2838062

[145] Dmitri Orlov, *An Exceptional Collection of Vector Bundles on the variety V5*, Vestn. Mosk. Univ., Ser. 1: Mat., Mekh., No. 5, 69–71 (1991), english translation in Moscow Univ. Math. Bull. 46 (5), 48–50 (1991).

[146] Dmitri Orlov, *Projective bundles, monoidal transformations, and derived categories of coherent sheaves* (Russian, with Russian summary), Izv. Ross. Akad. Nauk Ser. Mat. **56** (1992), no. 4, 852–862, DOI 10.1070/IM1993v041n01ABEH002182; English transl., Russian Acad. Sci. Izv. Math. **41** (1993), no. 1, 133–141. MR1208153

[147] Dmitri Orlov, *Derived categories of coherent sheaves on abelian varieties and equivalences between them* (Russian, with Russian summary), Izv. Ross. Akad. Nauk Ser. Mat. **66** (2002), no. 3, 131–158, DOI 10.1070/IM2002v066n03ABEH000389; English transl., Izv. Math. **66** (2002), no. 3, 569–594. MR1921811

[148] Dmitri Orlov, *Derived categories of coherent sheaves and triangulated categories of singularities*, Progr. Math., vol. 270, Birkhäuser Boston, Inc., Boston, MA, 2009, pp. 503–531, DOI 10.1007/978-0-8176-4747-6_16. MR2641200

[149] M. Perling, *Combinatorial aspects of exceptional sequences on (rational) surfaces*, preprint ArXiv 1311.7349.

[150] Emmanuel Peyre, *Unramified cohomology and rationality problems*, Math. Ann. **296** (1993), no. 2, 247–268, DOI 10.1007/BF01445105. MR1219902

[151] A. Pirutka, *Invariants birationnels dans la suite spectrale de Bloch–Ogus*, J. K-theory **10** (2012), 565–582.

[152] T. Reye, *Die Geometrie der Lage*, Vorträge, Baumgärtner, Leipzig, 1882.

[153] Markus Rost, *Chow groups with coefficients*, Doc. Math. **1** (1996), No. 16, 319–393. MR1418952

[154] *Helices and vector bundles. Seminaire Rudakov*, London Mathematical Society Lecture Note Series, **148**. A.N. Rudakov, ed. Cambridge University Press, Cambridge, 1990.

[155] David J. Saltman, *Noether's problem over an algebraically closed field*, Invent. Math. **77** (1984), no. 1, 71–84, DOI 10.1007/BF01389135. MR751131

[156] Beniamino Segre, *Sur un problème de M. Zariski* (French), Colloques Internationaux du Centre National de la Recherche Scientifique, no. 24, Centre National de la Recherche Scientifique, Paris, 1950, pp. 135–138. MR0041480

[157] Beniamino Segre, *Sull'esistenza, sia nel campo razionale che nel campo reale, di involuzioni piane non birazionali* (Italian), Atti Accad. Naz. Lincei. Rend. Cl. Sci. Fis. Mat. Nat. (8) **10** (1951), 94–97. MR0042746

[158] Mingmin Shen and Charles Vial, *The Fourier transform for certain hyperkähler fourfolds*, Mem. Amer. Math. Soc. **240** (2016), no. 1139, vii+163, DOI 10.1090/memo/1139. MR3460114

[159] N. I. Shepherd-Barron, *Stably rational irrational varieties*, Univ. Torino, Turin, 2004, pp. 693–700. MR2112599

[160] V.V. Shokurov, *Prym varieties: theory and applications*, Math. USSR-Izv. **23** (1984), 83–147.

[161] P. Sosna, *Some remarks on phantom categories and motives*, preprint arXiv:1511.07711.

[162] Bhama Srinivasan and Judith D. Sally (eds.), *Emmy Noether in Bryn Mawr*, Springer-Verlag, New York-Berlin, 1983. MR713788

[163] Gonçalo Tabuada, *Noncommutative motives*, University Lecture Series, vol. 63, American Mathematical Society, Providence, RI, 2015. With a preface by Yuri I. Manin. MR3379910

[164] S. Tanimoto and A. Várilly-Alvarado, *Kodaira dimension of moduli of special cubic fourfolds*, preprint arXiv:1509.01562, 2015.

[165] Bertrand Toën, *The homotopy theory of dg-categories and derived Morita theory*, Invent. Math. **167** (2007), no. 3, 615–667, DOI 10.1007/s00222-006-0025-y. MR2276263

[166] Burt Totaro, *Hypersurfaces that are not stably rational*, J. Amer. Math. Soc. **29** (2016), no. 3, 883–891, DOI 10.1090/jams/840. MR3486175

[167] Burt Totaro, *The integral cohomology of the Hilbert scheme of two points*, Forum Math. Sigma **4** (2016), e8, 20, DOI 10.1017/fms.2016.5. MR3490758

[168] Charles Vial, *Exceptional collections, and the Néron-Severi lattice for surfaces*, Adv. Math. **305** (2017), 895–934, DOI 10.1016/j.aim.2016.10.012. MR3570150

[169] V. Voevodsky, *A nilpotence theorem for cycles algebraically equivalent to zero*, Internat. Math. Res. Notices **4** (1995), 187–198, DOI 10.1155/S1073792895000158. MR1326064

[170] Claire Voisin, *Remarks on zero-cycles of self-products of varieties*, in Moduli of vector bundles (Proceedings of the Taniguchi congress on vector bundles), Maruyama Ed., Decker, 1994, pp. 265–285.

[171] Claire Voisin, *Abel-Jacobi map, integral Hodge classes and decomposition of the diagonal*, J. Algebraic Geom. **22** (2013), no. 1, 141–174, DOI 10.1090/S1056-3911-2012-00597-9. MR2993050

[172] Claire Voisin, *Bloch's conjecture for Catanese and Barlow surfaces*, J. Differential Geom. **97** (2014), no. 1, 149–175. MR3229054

[173] Claire Voisin, *Degree 4 unramified cohomology with finite coefficients and torsion codimension 3 cycles* (English, with English and French summaries), EMS Ser. Congr. Rep., Eur. Math. Soc., Zürich, 2012, pp. 347–368, DOI 10.4171/119-1/20. MR2987669

[174] Claire Voisin, *Théorème de Torelli pour les cubiques de* $\mathbf{P}^5$ (French), Invent. Math. **86** (1986), no. 3, 577–601, DOI 10.1007/BF01389270. MR860684

[175] Claire Voisin, *Hodge theory and complex algebraic geometry. I*, Reprint of the 2002 English edition, Cambridge Studies in Advanced Mathematics, vol. 76, Cambridge University Press, Cambridge, 2007. Translated from the French by Leila Schneps. MR2451566

[176] Claire Voisin, *Some aspects of the Hodge conjecture*, Jpn. J. Math. **2** (2007), no. 2, 261–296, DOI 10.1007/s11537-007-0639-x. MR2342587

[177] Claire Voisin, *Unirational threefolds with no universal codimension* 2 *cycle*, Invent. Math. **201** (2015), no. 1, 207–237, DOI 10.1007/s00222-014-0551-y. MR3359052

[178] Claire Voisin, *On the universal* CH_0 *group of cubic hypersurfaces*, arXiv:1407.7261, to appear in Journal of the European Mathematical Society.

[179] Claire Voisin, *Chow rings, decomposition of the diagonal, and the topology of families*, Annals of Mathematics Studies, vol. 187, Princeton University Press, Princeton, NJ, 2014. MR3186044

[180] Claire Voisin, *Stable birational invariants and the Lüroth problem*, in Advances in geometry and mathematical physics (Lectures given at the Geometry and Topology conference, Harvard University, 2014), H.-D. Cau and S.-T. Yau, eds., Surveys in Differential Geometry **21**, International Press, 2016, pp. 313–332.

[181] Oscar Zariski, *On Castelnuovo's criterion of rationality* $p_a = P_2 = 0$ *of an algebraic surface*, Illinois J. Math. **2** (1958), 303–315. MR0099990

[182] Steven Zucker, *The Hodge conjecture for cubic fourfolds*, Compositio Math. **34** (1977), no. 2, 199–209. MR0453741

Department of Mathematics, Yale University, 10 Hillhouse Avenue, New Haven, Connecticut 06520-9128
E-mail address: asher.auel@yale.edu

Institut de Mathématiques de Toulouse, Université Paul Sabatier, 118 route de Narbonne, 31062 Toulouse Cedex 9, France
E-mail address: marcello.bernardara@math.univ-toulouse.fr

Degenerations of Hodge structure

Colleen Robles

ABSTRACT. Two interesting questions in algebraic geometry are: (i) how can a smooth projective variety degenerate? and (ii) given two such degenerations, when can we say that one is "more singular/degenerate" than the other? Schmid's Nilpotent Orbit Theorem yields Hodge-theoretic analogs of these questions, and the Hodge-theoretic answers in turn provide insight into the motivating algebro-geometric questions, sometimes with applications to the study of moduli. Recently the Hodge-theoretic questions have been completely answered. This is an expository survey of that work.

1. Introduction

Two motivating questions from algebraic geometry are:

QUESTION 1.1. How can a smooth projective variety degenerate?

More precisely, let

$$f : \mathcal{X} \to S \tag{1.2}$$

be a family of polarized algebraic manifolds. That is, there is a surjective algebraic mapping $f : \overline{\mathcal{X}} \to \overline{S}$ of complex projective varieties such that the generic fibre $X_s = f^{-1}(s)$ is smooth, $S \subset \overline{S}$ is the Zariski open subset over which f has smooth fibres and $\mathcal{X} = f^{-1}(S)$. Assuming that the family (1.2) is well-understood, the first question is what can we say about the X_s when $s \in \overline{S} \backslash S$?

QUESTION 1.3. What are the "relations" between two such degenerations?

The second question is roughly asking for a stratification of $\overline{S}\backslash S$ with the property that the family is equi-singular along the strata, and "closure relations" between the strata. For example, one might consider a case in which the X_s are curves of genus g with at worst nodal singularities, and take the strata to correspond to the number of nodes.

One way to gain insight into Questions 1.1 and 1.3 is to ask the analogous questions of invariants associated with the smooth projective varieties. In this case, the invariant that we have in mind is a polarized Hodge structure. The Hodge theoretic analogs of the motivating Questions 1.1 and 1.3 arise by considering the period map

$$\Phi : S \to \Gamma\backslash D,$$

2010 *Mathematics Subject Classification.* Primary 14D07, 32G20.

The author was partially supported by NSF grants DMS 02468621 and 1361120.

which (roughly) assigns to $s \in S$ the Hodge structure on the primitive cohomology on X_s. Here D is a period domain parameterizing Hodge structures, and the assignment is defined up to the action of a discrete automorphism group Γ of D. (See 2.1 and 2.4 for details and further references.) The Hodge theoretic analog of the first motivating question is: how does $\Phi(t)$ degenerate as t approaches a boundary point $s_0 \in \overline{S}\backslash S$? Very roughly Schmid's Nilpotent Orbit Theorem 3.2, suitably interpreted through results of Cattani, Kaplan and Schmid (Theorem 3.4), says that $\Phi(t)$ degenerates to a limit mixed Hodge structure, call it LMHS(s_0).[1] (Detailed analysis of degenerations of polarized Hodge structures can be used to better understand degeneration of smooth projective varieties, and moduli spaces and their compactifications, see the surveys [**8**, **21**] and the references therein.) On the other hand, each (not necessarily closed or smooth) algebraic variety carries a mixed Hodge structure by Deligne [**9**]. If DMHS(s_0) denotes Deligne's mixed Hodge structure on X_{s_0}, then it is natural to ask how are the two mixed Hodge structures LMHS(s_0) and DMHS(s_0) related? In the case that $\dim \overline{S} = 1$ and the family $f : \overline{\mathfrak{X}} \to \overline{S}$ is semistable, the two mixed Hodge structures LMHS(s_0) and DMHS(s_0) are related by the the Clemens–Schmid exact sequence [**7**].

A semisimple Lie group $G \supset \Gamma$ acts homogeneously on D, and there is a natural action of G on the set of limit mixed Hodge structures. (In practice, G is a symplectic $\mathrm{Sp}(2g, \mathbb{R})$ or orthogonal $\mathrm{O}(a, b)$ group.) This first goal of this survey is to describe a classification of the G–conjugacy classes of limit mixed Hodge structures (4.1). This answers the Hodge theoretic analog of Question 1.1.

For the Hodge theoretic interpretation of the second question, we consider the case that $\dim \overline{S} = 2$ and s_0 admits a neighborhood $U \subset \overline{S}$ biholomorphic to a product of unit discs $\Delta \times \Delta$ that identifies s_0 with $(0,0)$ and so that $U \cap S = \Delta^* \times \Delta^*$ is a product of punctured discs. An illustrative example to keep in mind here is

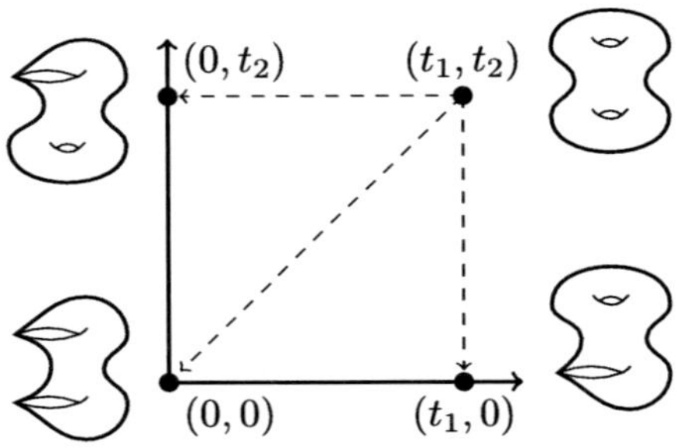

FIGURE 1.1. Degeneration of genus 2 curves

the case that $\Delta^* \times \Delta^*$ parameterizes smooth curves of genus 2, with one cycle degenerating to a node as $t_1 \to 0$, and another cycle degenerating to a node as $t_2 \to 0$, c.f. Figure 1.1. The idea is that each of the three 1–parameter degenerations $(t_1, t_2) \to (t_1, 0)$, $(t_1, t_2) \to (0, t_2)$ and $(t, t) \to (0, 0)$ will give limit mixed Hodge structures LMHS(t_1, 0), LMHS(0, t_2) and LMHS(0, 0), respectively. We regard LMHS(0, 0) as more degenerate/singular than LMHS(t_1, 0) and LMHS(0, t_2)

[1] The results of Cattani, Kaplan and Schmid hold in the more general setting of abstract variations of Hodge structure (2.4). Steenbrink [**25**] described the limit mixed Hodge structure in the geometric setting, with $\dim \overline{S} = 1$ and X_{s_0} a normal crossing divisor, in terms of the cohomology of certain intersections of components of X_{s_0}.

and declare "polarized relations" LMHS$(t_1, 0)$, LMHS$(0, t_2) \prec$ LMHS$(0, 0)$. The second goal of this survey is to classify the polarized relations between (representatives of) G–conjugacy classes of limit mixed Hodge structures (§4.2). This answers the Hodge theoretic analog of Question 1.3.

Both classifications we shall discuss are given by discrete combinatorial data in the form of "Hodge diamonds," weighted configurations of integer points in the pq–plane.

More generally, we can ask these questions in the setting of Mumford–Tate domains. The latter are generalizations of period domains: they are the classifying spaces of Hodge structures with (possibly) non-generic Hodge tensors [**12**]. As such they are realized as subdomains of period domains. Unfortunately, once we move to the more general setting of Mumford–Tate domains, the combinatorially simple Hodge diamonds do not suffice to classify the PMHS and the polarized relations amongst them. The general classifications are given by representation theoretic data (in the form of Weyl groups, Levi subgroups, and embeddings of SL(2) into the Mumford–Tate group) that is associated with the domain; see [**23**] and [**19**] for details. The goal of this article is to give an expository survey of that work in the relative simple setting of period domains. Related articles include [**13**] which studies a coarser notation of polarized relation that is defined in terms of the G–orbit structure of the topological boundary of D in the compact dual, and [**2**] which studies the representation theoretic structure of the nilpotent cones underlying a nilpotent orbit. (As will be discussed in §3, nilpotent orbits asymptotically approximate period maps near $s_0 \in \overline{S}\backslash S$.)

Acknowledgements. I learnt much of the material presented here from collaboration, correspondence and the work of several colleagues; I am especially indebted to M. Green, P. Griffiths, M. Kerr, R. Laza, W. McGovern and G. Pearlstein.

These notes were prepared for the Algebraic Geometry Summer Research Institute Graduate Student Bootcamp in Salt Lake City, Utah, July 06–10, 2015. I thank the Bootcamp organizers İ. Coşkun, T. de Fernex, A. Gibney and M. Leiblich for the opportunity to participate.

2. Hodge structures and their generalizations

We fix, once and for all, a rational vector space V, an integer n and a nondegenerate bilinear form $Q : V \times V \to \mathbb{Q}$ with the property $Q(u, v) = (-1)Q(v, u)$ for all $u, v \in V$. A brief review of Hodge theory follows; for more detail see [**3, 4, 22**] and the references therein.

2.1. Hodge structures. A (pure) *Hodge structure of weight* $0 \le n \in \mathbb{Z}$ on the rational vector space V is given by either of the following two equivalent objects:[2] A *Hodge decomposition*

$$V_{\mathbb{C}} = \bigoplus_{p+q=n} V^{p,q} \quad \text{such that} \quad \overline{V^{p,q}} = V^{q,p} . \tag{2.1}$$

[2]It is implicit in this definition that we are assuming that the Hodge structure is effective ($V^{p,q} = 0$ if either $p < 0$ or $q < 0$). Neither this nor the assumption that the weight is non-negative is necessary (or even desirable), but we restrict to this case for notational/expository clarity and convenience.

A (finite, decreasing) *Hodge filtration*

$$\begin{array}{c} 0 \subset F^n \subset F^{n-1} \subset \cdots \subset F^1 \subset F^0 = V_{\mathbb{C}} \\ \text{such that} \quad V_{\mathbb{C}} = F^k \oplus \overline{F^{n+1-k}}. \end{array} \tag{2.2}$$

The equivalence of the two definitions is given by

$$F^k = \bigoplus_{p \geq k} V^{p,n-p} \quad \text{and} \quad V^{p,q} = F^p \cap \overline{F^q}.$$

EXAMPLE 2.3. The Hodge Theorem asserts that the n-th cohomology group $V = H^n(X, \mathbb{Q})$ of a compact Kähler manifold admits a Hodge structure of weight n, with $V^{p,q} = H^{p,q}(X) \subset H^n(X, \mathbb{C})$ the cohomology classes in represented by (p,q)–forms.

The *Hodge numbers* $\mathbf{h} = (h^{p,q})$ and $\mathbf{f} = (f^p)$ are

$$h^{p,q} := \dim_{\mathbb{C}} V^{p,q} \quad \text{and} \quad f^p := \dim_{\mathbb{C}} F^p.$$

A weight n Hodge structure on V is *Q–polarized* if the *Hodge–Riemann bilinear relations* hold:

$$Q(V^{p,q}, V^{r,s}) = 0 \quad \text{if} \quad (p,q) \neq (s,r), \tag{2.4a}$$

$$\mathbf{i}^{p-q} Q(v, \bar{v}) > 0 \quad \text{for all} \quad 0 \neq v \in V^{p,q}. \tag{2.4b}$$

The *period domain* $D = D_{\mathbf{h},Q}$ is the set of all Q–polarized Hodge structures on V with Hodge numbers $\mathbf{h}$. It is a homogeneous space with respect to the action of the real automorphism group

$$G := \mathrm{Aut}(V_{\mathbb{R}}, Q),$$

and the isotropy group is compact. If n is odd, then $G \simeq \mathrm{Sp}(2g, \mathbb{R})$, where $\dim V = 2g$; if $n = 2k$ is even, then $G \simeq \mathrm{O}(a, b)$ where $a = \sum h^{k+2p,k-2p}$ and $b = \sum h^{k+1+2p,k-1-2p}$.

EXAMPLE 2.5. Let $X \subset \mathbb{P}^m$ be a projective algebraic manifold of dimension d with hyperplane class $\omega \in H^2(X, \mathbb{Z})$. Given $n \leq d$, the primitive cohomology

$$V = P^n(X, \mathbb{Q}) := \{\alpha \in H^n(X, \mathbb{Q}) \mid \omega^{d-n+1} \wedge \alpha = 0\}$$

inherits the weight n Hodge decomposition $V_{\mathbb{C}} = \oplus_{p+q=n} H^{p,q}(X) \cap V_{\mathbb{C}}$ from $H^n(X, \mathbb{Q})$. The Hodge–Riemann bilinear relations assert that this Hodge structure is polarized by $Q(\alpha, \beta) := (-1)^{n(n-1)} \int_X \alpha \wedge \beta \wedge \omega^{d-n}$.

With respect to the Hodge filtration (2.2), the first Hodge–Riemann bilinear relation (2.4a) asserts that $F = (F^p)$ is *Q–isotropic*

$$Q(F^p, F^q) = 0, \quad \text{for all} \quad p + q = n + 1. \tag{2.6}$$

Equivalently, the Hodge filtration defines a point in the rational homogeneous variety

$$\check{D} := \mathrm{Flag}^Q(\mathbf{f}, V_{\mathbb{C}})$$

of Q–isotropic filtrations $F^\bullet = (F^p)$ of $V_{\mathbb{C}}$; the variety $\check{D}$ is known as the *compact dual* (*of* D). The complex automorphism group

$$G_{\mathbb{C}} := \mathrm{Aut}(V_{\mathbb{C}}, Q)$$

acts transitively on $\check{D}$, and contains the period domain D as an open subset. In summary, the compact dual $\check{D}$ parameterizes filtrations F of $V_{\mathbb{C}}$ satisfying the first Hodge–Riemann bilinear relation, and the period domain D parameterizes filtrations satisfying both Hodge–Riemann bilinear relations.

2.2. Mixed Hodge structures.

2.2.1. *Definition and examples.* A *mixed Hodge structure* (MHS) on V is given an increasing filtration $W = (W_\ell)$ of V, and a decreasing filtration $F = (F^p)$ of $V_{\mathbb{C}}$ with the property that F induces a weight ℓ Hodge structure on the graded quotients

$$W_\ell^{\mathrm{gr}} := W_\ell / W_{\ell-1}\,.$$

EXAMPLE 2.7. If X is a Kähler manifold of dimension d and

$$V = H(X,\mathbb{Q}) := \bigoplus_n H^n(X,\mathbb{Q})\,,$$

then $W_\ell = \oplus_{n\le\ell} H^n(X,\mathbb{Q})$ and $F^k = \oplus_{p\ge k} H^{p,\bullet}(X)$ defines a mixed Hodge structure on V.

EXAMPLE 2.8. Alternatively, if X is a Kähler manifold of dimension d and $V = H(X,\mathbb{Q})$, then $W_\ell = \oplus_{n\ge 2d-\ell} H^n(X,\mathbb{Q})$ and $F^k = \oplus_{q\le d-k} H^{\bullet,q}(X)$ defines a mixed Hodge structure on V.

EXAMPLE 2.9. Deligne [**9**] has shown that the cohomology $H^n(X,\mathbb{Q})$ of an algebraic variety X admits a (functorial) mixed Hodge structure. Here X need not be smooth or closed. However, when X is smooth and closed, Deligne's MHS is the (usual) Hodge structure of Example 2.3. For an expository introduction to mixed Hodge structures on algebraic varieties see [**10**]; for a thorough treatment see [**22**].

2.2.2. *Deligne splitting.* Given a mixed Hodge structure (W,F) on V there exists a unique splitting

$$V_{\mathbb{C}} = \bigoplus I^{p,q} \tag{2.10a}$$

with the properties that

$$F^p = \bigoplus_{p\ge r} I^{r,\bullet}\,, \quad W_\ell = \bigoplus_{p+q\le\ell} I^{p,q} \tag{2.10b}$$

and

$$\overline{I^{p,q}} \equiv I^{q,p} \mod \bigoplus_{\substack{r<q\\ s<p}} I^{r,s}\,. \tag{2.10c}$$

The splitting is given by

$$\begin{aligned} I^{p,q} &= F^p \cap W_{p+q} \cap \left(\overline{F^q} \cap W_{p+q} + \overline{U^{q-1}_{p+q-2}}\right)\,, \quad \text{where}\\ U^a_b &:= \sum_{j\ge 0} F^{a-j} \cap W_{b-j}\,. \end{aligned}$$

Note that (2.10b) implies that the

$$\text{Hodge decomposition on } W_\ell^{\mathrm{gr}} \text{ induced by } F \text{ is } W_\ell^{\mathrm{gr}} \otimes \mathbb{C} \simeq \bigoplus_{p+q=\ell} I^{p,q}\,. \tag{2.11}$$

The MHS is $\mathbb{R}$*–split* if $\overline{I^{p,q}} = I^{q,p}$.

EXAMPLE 2.12. The Deligne splittings of the MHS in Examples 2.7 and 2.8 are given by $I^{p,q} = H^{p,q}(X)$ and $I^{p,q} = H^{d-q,d-p}(X)$, respectively. Both are $\mathbb{R}$–split.

Let

$$\Lambda_{\mathbb{C}}^{-1,-1} := \left\{ \xi \in \mathrm{End}(V_{\mathbb{C}}) \;\middle|\; \xi(I^{p,q}) \subset \bigoplus_{\substack{r<p\\ s<q}} I^{r,s} \ \forall\, p,q \right\}.$$

Then $\Lambda_{\mathbb{C}}^{-1,-1}$ is a nilpotent subalgebra of $\mathrm{End}(V_{\mathbb{C}})$ and is defined over $\mathbb{R}$. Deligne showed that given a MHS (W,F) there exists a unique $\delta \in \Lambda_{\mathbb{R}}^{-1,-1}$ so that $(W,\tilde{F})$, with $\tilde{F} = e^{\mathrm{i}\delta}F$, is an $\mathbb{R}$–split PMHS. An important property of this new $\tilde{F}$ is that it determines the same Hodge structure on W_ℓ^{gr} as the original F. In particular, if $V_{\mathbb{C}} = \oplus \tilde{I}^{p,q}$ is the Deligne splitting for $(W,\tilde{F})$, then

$$\dim_{\mathbb{C}} I^{p,q} \;=\; \dim_{\mathbb{C}} \tilde{I}^{p,q}\,, \tag{2.13}$$

for all p,q.

2.3. Polarized mixed Hodge structures.

2.3.1. *Jacobson–Morosov filtrations.* Every nilpotent endomorphism $N : V \to V$ determines a unique increasing filtration $W(N) = (W_\ell(N))$ of V with the two properties that

$$N(W_\ell(N)) \;\subset\; W_{\ell-2}(N) \tag{2.14a}$$

and

(2.14b) the induced $N^\ell : W_\ell^{\mathrm{gr}}(N) \to W_{-\ell}^{\mathrm{gr}}(N)$ is an isomorphism for all $\ell \geq 0$.

Moreover, if N lies in the Lie algebra

$$\mathfrak{g} \;:=\; \mathrm{End}(V,Q)$$

of $G = \mathrm{Aut}(V,Q)$, then the filtration $W(N)$ is Q–isotropic.

EXERCISE 2.15. Suppose that $N^k \neq 0$ and $N^{k+1} = 0$. Show that $W(N)$ is given inductively by

$$\begin{aligned} W_k \;&=\; \ker N^{k+1} \;=\; V \quad\text{and}\quad W_{-k-1} \;=\; \mathrm{im}\, N^{k+1} \;=\; 0\,,\\ W_{k-1} \;&=\; \ker N^k \quad\text{and}\quad W_{-k} \;=\; \mathrm{im}\, N^k\,, \end{aligned}$$

and for all $0 \leq \ell \leq k-2$,

$$W_\ell \;=\; \{v \in W_{\ell+1} \mid N^\ell v \subset W_{-\ell-2}\} \quad\text{and}\quad W_{-\ell-1} \;=\; N^{\ell+1}(W_\ell)\,.$$

Notice that the first nontrivial subspace $W_{-k}(N)$ in the Jacobson–Morosov filtration is indexed by a negative integer (if $N \neq 0$). The "shifted" filtration, with nontrivial subspaces indexed by nonnegative integers is denoted $W(N)[-k]$, and given by

$$W_\ell(N)[-k] \;:=\; W_{\ell-k}(N)\,.$$

REMARK 2.16. It is sometimes useful to describe the Jacobson–Morosov filtration in terms of the action of a three-dimensional subalgebra $\mathfrak{s} \subset \mathrm{End}(V)$ that is isomorphic to $\mathfrak{sl}(2)$ and contains N. Specifically, the Jacobson–Morosov Theorem asserts that there exist $Y, N^+ \in \mathrm{End}(V)$ so that

$$[Y,N] \;=\; -2N\,, \quad [N^+,N] \;=\; Y \quad\text{and}\quad [Y,N^+] \;=\; 2N^+\,. \tag{2.17}$$

When $N \in \mathfrak{g}$, we can choose $Y, N^+ \in \mathfrak{g}$. The relations (2.17) imply that $\{N, Y, N^+\}$ span a subalgebra of $\mathrm{End}(V)$ that is isomorphic to $\mathfrak{sl}(2)$. Moreover, the element Y acts on V by integer eigenvalues. If

$$V = \bigoplus_\ell V_\ell$$

is the Y–eigenspace decomposition of V, then

$$W_\ell(N) = \bigoplus_{m \le \ell} V_m .$$

The Jacobi identity and (2.17) imply $N(V_\ell) \subset N_{\ell-2}$; from this we see that (2.14a) holds. Note that $W_\ell^{\mathrm{gr}}(N) \simeq V_\ell$. It is a classical result from the representation theory of $\mathfrak{sl}(2)$ that $N^\ell : V_\ell \to V_{-\ell}$ is an isomorphism for all $\ell \ge 0$; from this we see that (2.14b) holds.

2.3.2. *Definition and examples.* A *Q–polarized mixed Hodge structure* (PMHS) on V is given by a mixed Hodge structure (W, F) and a set $\mathcal{N} \subset \mathfrak{g}_\mathbb{R}$ of nilpotent elements with the properties:

(i) For all $N \in \mathcal{N}$ we have $N^{n+1} = 0$ and $W = W(N)[-n]$.

(ii) The filtration F is Q–isotropic, and $N(F^p) \subset F^{p-1}$ for all $N \in \mathcal{N}$ and p.

(iii) The filtration F induces a weight $n+\ell$ Hodge structure on the *primitive space*

$$P(N)_\ell := \ker\{N^\ell : W^{\mathrm{gr}}_{n+\ell} \to W^{\mathrm{gr}}_{n-\ell-2}\}$$

that is polarized by

$$Q^N_\ell(\cdot,\cdot) := Q(\cdot, N^\ell \cdot),$$

for all $\ell \ge 0$.

We sometimes say that the mixed Hodge structure (W, F) is *polarized by $\mathcal{N}$*. From (i) we see that the filtration W is determined by $\mathcal{N}$, and we will often write $(F, \mathcal{N})$ for the PMHS $(W, F, \mathcal{N})$.

EXAMPLE 2.18. Let $X \subset \mathbb{P}^m$ be a projective algebraic manifold of dimension n with hyperplane class $\omega \in H^2(X, \mathbb{Z})$. Let $V = H(X, \mathbb{Q})$ and define $Q(\alpha, \beta) = (-1)^{k(k-1)/2} \int_X \alpha \wedge \beta$, with $k = \deg \alpha$. The ray $\mathcal{N} = \{t\omega \mid t > 0\}$ spanned by the Lefschetz operator $\omega : H^\bullet(X, \mathbb{Q}) \to H^{\bullet+2}(X, \mathbb{Q})$ polarizes the mixed Hodge structure defined of Example 2.8. (Alternatively, there is a canonical dual N^*_ω : $H^\bullet(X, \mathbb{Q}) \to H^{\bullet-2}(X, \mathbb{Q})$ to the Lefschetz operator, and the mixed Hodge structure of Example 2.7 is polarized by the ray $\mathcal{N} = \{t\, N^*_\omega \mid t > 0\}$.)

REMARK 2.19. Given a PMHS $(W, F, \mathcal{N})$, let $(W, \tilde{F} = e^{\mathrm{i}\delta} F)$ be the $\mathbb{R}$–split MHS of §2.2.2. Then $(W, \tilde{F}, \mathcal{N})$ is a PMHS, and $\tilde{F}$ determines the same Q^N_ℓ–polarized Hodge structures on $P(N)_\ell$ as F for all $\ell \ge 0$. Moreover, $\delta \in \Lambda^{-1,-1} \cap \mathfrak{g}_\mathbb{R}$ and $[\delta, N] = 0$ for all $N \in \mathcal{N}$. Finally, if $V_\mathbb{C} = \oplus I^{p,q}$ is the Deligne splitting for $(W, \tilde{F})$, then $N(I^{p,q}) \subset I^{p-1,q-1}$ for all $N \in \mathcal{N}$.

2.4. Variation of Hodge structure. Let S be a complex manifold with fundamental group $\pi_1(M)$ and universal cover $\tilde{S}$. Let $\rho : \pi_1(S) \to \mathrm{Aut}(V, Q)$ be a representation of the fundamental group. Then

$$\mathcal{V} := \tilde{S} \times_{\pi_1(S)} V$$

defines a flat vector bundle over S. Let ∇ denote the flat connection. The bilinear form Q induces a flat form $\mathcal{Q}$ on $\mathcal{V}$. A *(polarized) variation of Hodge structure*

(VHS) over S is given by a holomorphic filtration $\mathcal{F}^\bullet$ of $\mathcal{V}_\mathbb{C}$ that defines a $\mathcal{Q}_s$-polarized Hodge structure on each fibre $\mathcal{V}_s$, $s \in S$, and with the property that $\nabla \mathcal{F}^p \subset \Omega^1_S \otimes \mathcal{F}^{p-1}$. The variation of Hodge structure induces a *period map*

$$\Phi : S \to \Gamma\backslash D\,,$$

where $\Gamma = \rho(\pi_1(S)) \subset \mathrm{Aut}(V, Q)$. Geometrically, VHS arise when considering a family $\mathfrak{X} \to S$ of polarized algebraic manifolds: one obtains a VHS $\mathcal{V} \to S$ with fibres $\mathcal{V}_s$ isomorphic to the primitive cohomology $P^n(\mathfrak{X}_s, \mathbb{Q})$, and $\mathcal{F}_s$ the Hodge filtration, see [**15, 16**].

3. Nilpotent orbits

The significance of nilpotent orbits comes from Schmid's Nilpotent Orbit Theorem (§3.2) which asserts that that every (lifting of a) period map (§2.4) is well approximated by a nilpotent orbit. In particular, the asymptotic behavior of a period mapping is encoded by nilpotent orbits. Moreover, results of Cattani, Kaplan and Schmid imply that a nilpotent orbit is equivalent to a PMHS (Theorem 3.4); this is the sense in which

a PMHS arises from a degeneration of Hodge structure.

3.1. Definition. A *nilpotent orbit* is a map $\theta : \mathbb{C}^r \to \check{D}$ of the form

$$\theta(z) \;=\; \exp(\textstyle\sum z_j N_j) \cdot F\,,$$

with $F \in \check{D}$ and $\{N_1, \ldots, N_r\} \subset \mathfrak{g}_\mathbb{R}$ a set of commuting nilpotent elements, and having the properties that:

(i) $N_j(F^p) \subset F^{p-1}$ for all j, p, and
(ii) $\theta(z) \in D$ when $\mathrm{Im}(z_j) \gg 0$ for all j.

3.2. Schmid's Nilpotent Orbit Theorem. Fix a VHS $(\mathcal{V}, \mathcal{Q}, \mathcal{F})$ over S as in §2.4, and let $\Phi : S \to \Gamma\backslash D$ denote the associated period map. In practice one is interested in the case that S is a Zariski open subset of a compact analytic space $\overline{S}$, and wants to describe the singularities of Φ on the boundary $\overline{S}\backslash S$. Applying Hironaka's resolution of singularities [**17**], we may assume that $\overline{S}$ is smooth. If $\overline{S}\backslash S$ has codimension greater than two, then Φ extends holomorphically to $\overline{S}$ [**14**]. In the case that the boundary has codimension one, we may again apply Hironaka's resolution of singularities to assume that $\overline{S}\backslash S$ is locally a normal crossing divisor. That is, every point $s \in \overline{S}$ admits a neighborhood of the form Δ^m and with the property that $\Delta^m \cap S = (\Delta^*)^r \times \Delta^{m-r}$; here

$$\Delta \;:=\; \{t \in \mathbb{C} \;:\; |t| < 1\}$$

is the unit disc, and

$$\Delta^* \;:=\; \{t \in \mathbb{C} \;:\; 0 < |t| < 1\}$$

is the punctured unit disc.

The nilpotent orbit theorem is a local statement, describing the behavior of the period map $\Phi(t)$ as $t \to t_o \in \overline{S}\backslash S$, so we now restrict Φ to $(\Delta^*)^r \times \Delta^{m-r}$. For simplicity of exposition we will take $m = r$ and consider the period map

$$\Phi : (\Delta^*)^r \;\to\; \Gamma\backslash D\,, \tag{3.1}$$

with $\Gamma = \rho(\pi_1((\Delta^*)^r)) \subset \mathrm{Aut}(V, Q)$; for the general statement of the Nilpotent Orbit Theorem see [**24**]. The fundamental group $\pi_1((\Delta^*)^r)$ is generated by elements

$\{\gamma'_1, \ldots, \gamma'_r\}$ where γ'_j may be identified with the counter-clockwise generator of the fundamental group of the j–th copy of Δ^* in $(\Delta^*)^r$. The images $\gamma_j = \rho(\gamma'_j) \in \Gamma$ are the *monodromy transformations.* They pairwise commute and are known to be quasi-unipotent.[3] Quasi-unipotency implies there exist $0 \le m_j \in \mathbb{Z}$ and nilpotent $N_j \in \mathfrak{g}_\mathbb{R}$ so that

$$\gamma_j^{m_j} = \exp(m_j N_j) .$$

Let

$$\mathcal{H} := \{z \in \mathbb{C} \mid \operatorname{Im} z > 0\}$$

denote the upper-half plane, so that $\mathcal{H} \to \Delta^*$, sending $z \mapsto t = e^{2\pi \mathbf{i} z}$, is the universal covering map. Then γ'_j acts on $\mathcal{H}^r$ by

$$\gamma'_j \cdot (z_1, \ldots, z_r) = (z_1, \ldots, z_{j-1}, z_j + 1, z_{j+1}, \ldots, z_r) =: z + \varepsilon_j$$

by the translation replacing the j–th coordinate z_j with $z_j + 1$. Fixing a lift $\tilde{\Phi} : \mathcal{H}^r \to D$ of the period map (3.1), we have $\gamma_j \cdot \tilde{\Phi}(z) = \tilde{\Phi}(z + \varepsilon_j)$. In particular, the map

$$\widetilde{\Psi} : \mathcal{H}^r \to \check{D} \quad \text{sending} \quad z \mapsto \exp\left(-\textstyle\sum_j m_j z_j N_j\right) \cdot \tilde{\Phi}(z)$$

descends to a well-defined map $\Psi : (\Delta^*)^r \to \check{D}$.

THEOREM 3.2 (Schmid's Nilpotent Orbit Theorem [**24**]). *The map Ψ extends holomorphically to $\Delta^r \to \check{D}$. Setting $F := \Psi(0)$, the map $\theta(z) := \exp(\sum_j z_j N_j) \cdot F$ is a nilpotent orbit. Moreover, given any G–invariant distance d on D, there exist constants $0 \le \alpha, \beta, C$ such that $\theta(z) \in D$ and*

$$d\left(\tilde{\Phi}(z), \theta(z)\right) \le C \textstyle\sum_j (\operatorname{Im} z_j)^\beta \exp(-2\pi(\operatorname{Im} z_j)/m_j) \tag{3.3}$$

so long as $\operatorname{Im} z_j > \alpha$.

The bound (3.3)[4] is the precise sense in which the nilpotent orbit θ strongly approximates the lifted period map $\tilde{\Phi}$ as $\operatorname{Im} z_j \to \infty$. The constants α, β, C depend only on d, the m_j, the Hodge numbers $\mathbf{h} = (h^{p,q})$ and the weight n.

3.3. Relationship to PMHS. Fix $F \in \check{D}$ and pairwise commuting nilpotent $N_1, \ldots, N_r \in \mathfrak{g}_\mathbb{R}$. Let

$$\sigma := \left\{ \textstyle\sum_j x_j N_j \;\middle|\; x_j > 0 \right\} \subset \mathfrak{g}_\mathbb{R}$$

be the *nilpotent cone* spanned by the $\{N_j\}$.

THEOREM 3.4 (Cattani,Kaplan, Schmid). *The map $\theta : \mathbb{C}^r \to \check{D}$ sending*

$$z \mapsto \exp\left(\textstyle\sum_j z_j N_j\right) \cdot F$$

is a nilpotent orbit if and only if the Jacobson–Morosov filtration $W(N)$ is independent of $N \in \sigma$, so that $W(\sigma)$ is well-defined, and $(W(\sigma)[-n], F, \sigma)$ is a PMHS. Moreover, if $(W(N)[-n], \tilde{F})$ is the $\mathbb{R}$–split PMHS constructed by Deligne (§2.2.2), then the nilpotent orbits $\theta(z)$ and $\tilde{\theta}(z) = \exp(\sum z_j N_j) \cdot \tilde{F}$ agree to first order as $\operatorname{Im} z_j \to \infty$.

---

[3]In the geometric setting quasi-unipotency is due to Katz [**18**] and Landman [**20**]; the general statement is due to Borel [**24**, (4.5)].

[4]This bound is an improvement, due to Deligne, of the initial distance estimate given in [**24**].

REMARK 3.5. The common asymptotic limit

$$\Phi_\infty(\sigma, F) := \lim_{\substack{\mathrm{Im}\, z \to \infty \\ \mathrm{Re}\, z \text{ bdd}}} \exp(zN) \cdot F,$$

with $\mathrm{Re}\, z$ bounded and $z \in \mathbb{C}$, of the two nilpotent orbits is independent of our choice of $N \in \sigma$, [**6**]; as a filtration of $V_\mathbb{C}$, the point $\Phi_\infty(\sigma, F) \in \check{D}$ is given by the Deligne splitting (2.10) as

$$\Phi^p_\infty(\sigma, F) = \bigoplus_{q \le n-p} I^{\bullet,q}.$$

Moreover, note that $\exp(\zeta N) \cdot F \in D$ for all $\mathrm{Im}\, \zeta \gg 0$ implies $\Phi_\infty(\sigma, F)$ lies in the topological closure $\overline{D} \subset \check{D}$ of D.

PROOF. Given a one–variable nilpotent orbit $\exp(zN)F$, Schmid [**24**] proved that (F, N) is a PMHS. Given a several–variable nilpotent orbit, the independence of the Jacobson–Morosov filtration $W(N)$ of the choice of N in the underlying nilpotent cone was proven by Cattani and Kaplan [**5**]. From these two results it follows that a several–variable nilpotent orbit determines a PMHS. The converse was proved by Cattani, Kaplan and Schmid [**6**]. The asymptotic first-order agreement of θ and $\tilde{\theta}$ is also established in [**6**]. □

4. Classifications

4.1. Classification of $\mathbb{R}$–split PMHS. Notice that G acts on the set of PMHS: given $g \in G$ and a PMHS $(W, F, \mathcal{N})$ we have

$$g \cdot (W, F, \mathcal{N}) := (g \cdot W, g \cdot F, \mathrm{Ad}_g \mathcal{N}).$$

Moreover, $(W, F, \mathcal{N})$ is $\mathbb{R}$–split if and only if $g \cdot (W, F, \mathcal{N})$ is. In this section will classify the $\mathbb{R}$–split PMHS. The classification is given by Hodge diamonds, which depend only on the MHS (W, F), and it is a consequence of (2.13) that (W, F) and $(W, \tilde{F})$ have the same Hodge diamond.

Given a MHS (W, F), let $V_\mathbb{C} = \oplus I^{p,q}_{W,F}$ be the Deligne splitting (§2.2.2). The *Hodge diamond* of (W, F) is the function $\Diamond(W, F) : \mathbb{Z} \times \mathbb{Z} \to \mathbb{Z}$ given by

$$\Diamond(W, F)(p, q) := \dim_\mathbb{C} I^{p,q}.$$

LEMMA 4.1 ([**19**]). *The Hodge diamond $\Diamond = \Diamond(W, F, \mathcal{N})$ of a PMHS on a period domain D parameterizing weight n Hodge structures with Hodge numbers $\mathbf{h} = (h^{p,q})_{p+q=n}$ satisfies the following four properties: The columns of the Hodge diamond sum to the Hodge numbers*

$$\textstyle\sum_p \Diamond(p, q) = h^{n-q,q}. \tag{4.2a}$$

The Hodge diamond is symmetric about the diagonal $p = q$:

$$\Diamond(p, q) = \Diamond(q, p). \tag{4.2b}$$

The Hodge diamond is symmetric about $p + q = n$:

$$\Diamond(p, q) = \Diamond(n - q, n - p). \tag{4.2c}$$

The values $\Diamond(p, q)$ are non-increasing as one moves away from $p + q = n$ along a(n off) diagonal:

$$\Diamond(p, q) \ge \Diamond(p + 1, q + 1) \quad \textit{for all} \quad p + q \ge n. \tag{4.2d}$$

Note that the four conditions (4.2) imply that the Hodge diamond of a PMHS "lies in" the square $[0,n] \times [0,n]$; that is

$$\Diamond(p,q) \neq 0 \quad \text{implies} \quad 0 \leq p,q \leq n\,.$$

PROOF. The property (4.2a) follows from $F \in \check{D}$ and the first equation of (2.10b); property (4.2b) is due to (2.11); and properties (4.2c) and (4.2d) to (2.14b). □

Given a PMHS $(F,\mathcal{N})$, we will denote the Hodge diamond by $\Diamond(F,\mathcal{N})$. The following proposition asserts that (i) every non-negative function satisfying (4.2) may be realized as the Hodge diamond of an $\mathbb{R}$–split PMHS, and (ii) the $\mathbb{R}$–split PMHS on D are classified, up to the action of G, by their Hodge diamonds.

THEOREM 4.3 ([**19**]). *Any function $f : \mathbb{Z} \times \mathbb{Z} \to \mathbb{Z}_{\geq 0}$ satisfying* (4.2) *may be realized as the Hodge diamond $\Diamond(F,N)$ of an $\mathbb{R}$–split polarized mixed Hodge structure (F,N), $N \in \mathfrak{g}_\mathbb{R}$, on the period domain D. Moreover, $\Diamond(F_1,N_1) = \Diamond(F_2,N_2)$ if and only if $(F_2,N_2) = (g \cdot F_1, \mathrm{Ad}_g N_1)$ for some $g \in G$.*

The proof is essentially a consequence of the classification of nilpotent $N \in \mathfrak{g}_\mathbb{R}$ by "signed Young diagrams," and the fact that the latter are determined by the Hodge diamonds; see [**19**] for details.

REMARK 4.4. By virtue of the equivalence between $\mathbb{R}$–split PMHS and horizontal SL(2)–orbits on D, Theorem 4.3 is also a classification $G(\mathbb{R})$–conjugacy classes of horizontal SL(2)–orbits on period domains.

REMARK 4.5. There is an interesting relationship between PMHS and the topological boundary $\partial(D) \subset \check{D}$ of D. First, recall (Remark 3.5) that the asymptotic limit

$$\Phi_\infty(N,F) \;=\; \lim_{\substack{\mathrm{Im}\, z \,\to\, \infty \\ \mathrm{Re}\, z \text{ bdd}}} \exp(zN)\cdot F \;\in\; \partial(D)$$

of the nilpotent orbit lies in the boundary if $N \neq 0$. Note that $\partial(D)$ decomposes into a disjoint union of G orbits, and the map Φ_∞ is G–equivariant. Thus, from Theorem 4.3, we obtain an induced map Φ_∞ from the set $\Diamond(D)$ of Hodge diamonds to the set $\mathcal{O}(D)$ of G–orbits in $\overline{D}$. This map is injective [**19**]. In particular, the Hodge diamonds index the "polarizable" G–orbits in $\overline{D}$. In [**13**] the natural closure relations between these orbits are used to define the notion of an extremal degeneration of Hodge structures. Theorem 4.18 (and more generally the results of [**19**]) may be viewed as refining, or "filling-in", results of [**13**].

We finish this section by giving a number of examples illustrating Theorem 4.3. In each of the examples that follows we fix a period domain D (by specifying the Hodge numbers and fixing a polarization Q), and apply Theorem 4.3 to list the Hodge diamonds. The diamonds are represented by labeled configurations of points in the pq–plane: the node at (p,q) is labeled with the (nonzero) value of $\Diamond(p,q)$.

Notation. The following notation will be used to characterize the flags $F^\bullet \in \check{D}$ realizing a given Hodge diamond in the examples below. Observe that

$$Q^*(\cdot,\cdot) \;:=\; \mathbf{i}^n\, Q(\cdot,\bar{\cdot})$$

defines a nondegenerate Hermitian form on $V_\mathbb{C}$. Given a subspace $E \subset V_\mathbb{C}$, let

$$E_0 \;:=\; \{v \in E \mid Q^*(v,E) = 0\}\,.$$

Notice that Q^* induces a nondegenerate Hermitian form Q_0^* on E/E_0. We will write $Q_0^* > 0$ when this form is positive definite.

EXAMPLE 4.6 (Hodge numbers $\mathbf{h} = (g, g)$). There are $g+1$ Hodge diamonds, which we denote $\Diamond_r$, with $0 \le r \le g$, and picture as

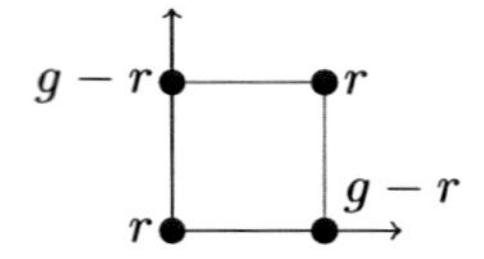

In this case Q is skew-symmetric, and $Q^*(\cdot,\cdot) = \mathbf{i}\, Q(\cdot, \bar{\cdot})$ is a nondegenerate Hermitian form on $V_\mathbb{C} \simeq \mathbb{C}^{2g}$. The flags $F^\bullet = (F^1) \in \check{D}$ consist of a single subspace and the compact dual $\check{D} = \mathrm{Gr}^Q(g, V_\mathbb{C})$ is the Lagrangian grassmannian. The subspaces $E \in \check{D}$ realizing the Hodge diamond $\Diamond_r$ form a G–orbit

$$\mathcal{O}_r \ := \ \{E \in \mathrm{Gr}^Q(g, V_\mathbb{C}) \mid E_0 = E \cap \overline{E}\,, \dim E_0 = r\,,\ Q_0^* > 0\}\,.$$

Note that $D = \mathcal{O}_0$ and $\mathcal{O}_s \subset \overline{\mathcal{O}}_r$ if and only if $r \le s$.

REMARK 4.7. Recall Figure 1.1 and the associated limit mixed Hodge structures (which are PMHS). Here we have $g = 2$, and the Hodge diamond for $\mathrm{LMHS}(t_1, 0)$ and $\mathrm{LMHS}(0, t_2)$, with $t_1 t_2 \neq 0$, is $\Diamond_1$; likewise, the Hodge diamond for $\mathrm{LMHS}(0,0)$ is $\Diamond_2$.

EXAMPLE 4.8 (Hodge numbers $\mathbf{h} = (1, a, 1)$). The Hodge diamonds are

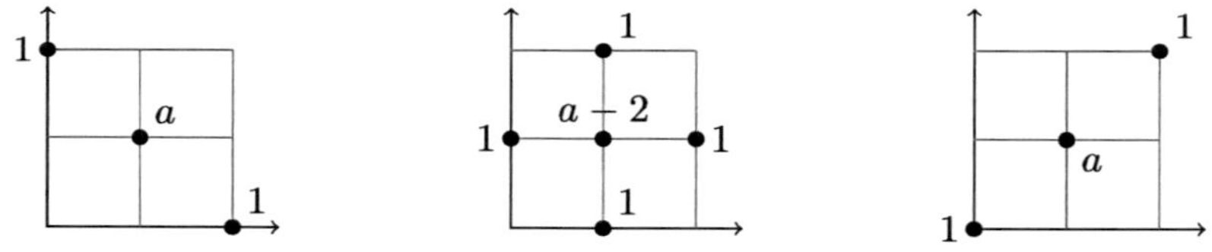

Here the second diamond arises only if $a \ge 2$, and the third only if $a \ge 1$.

EXAMPLE 4.9 (Hodge numbers $\mathbf{h} = (b, a, b)$). The Hodge diamonds $\Diamond_{r,s}$ are indexed by $0 \le r, s$ satisfying $r + s \le b$ and $r + 2s \le a$.

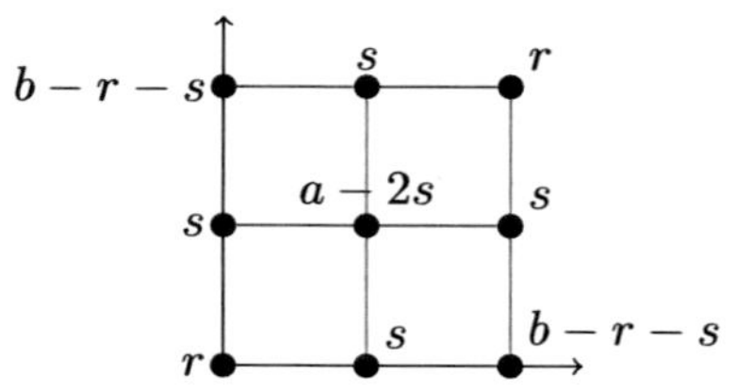

In this case Q is symmetric, and $Q^*(\cdot,\cdot) = -Q(\cdot,\bar{\cdot})$ defines a nondegenerate Hermitian form on $V_\mathbb{C} \simeq \mathbb{C}^{a+2b}$. The flags $(F^2 \subset F^1) \in \check{D} = \mathrm{Flag}^Q(b, a+b, V_\mathbb{C})$ satisfy $F^1 = (F^2)^\perp$; that is, the flag $F^\bullet$ is completely determined by the first subspace F^2, so that $\check{D} \simeq \mathrm{Gr}^Q(b, \mathbb{C}^{a+2b})$. The flags realizing the Hodge diamond $\Diamond_{r,s}$ form a G–orbit $\mathcal{O}_{r,s} \in \overline{D}$ that is characterized by

$$\mathcal{O}_{r,s} \ = \ \left\{E \in \mathrm{Gr}^Q(b, \mathbb{C}^{a+2b}) \ \middle|\ \dim E \cap \overline{E} = r\,,\ \dim E_0 = r+s\,,\ Q_0^* > 0\right\}$$

Note that $D = \mathcal{O}_{0,0}$ and $\mathcal{O}_{t,u} \subset \overline{\mathcal{O}}_{r,s}$ if and only if $r \le t$ and $r + s \le t + u$.

4.2. Degeneracy relations between $\mathbb{R}$–split PMHS.

4.2.1. *Polarized relations on Hodge diamonds.* Schmid's Nilpotent Orbit Theorem 3.2 provides the link between the geometry and the Hodge theory. Specifically, the lift $\tilde{\Phi} : \mathcal{H} \times \mathcal{H} \to D$ of any period map $\Phi : \Delta^* \times \Delta^* \to \Gamma\backslash D$ with unipotent monodromies is a approximated by a two-variable nilpotent orbit

$$\theta(z_1, z_2) = \exp(z_1 N_1 + z_2 N_2) \cdot F . \tag{4.10}$$

We may assume, without loss of generality, that the associated PMHS (F, σ), with $\sigma = \{x_1 N_1 + x_2 N_2 \mid x_j > 0\}$ the underlying nilpotent orbit, is $\mathbb{R}$–split (Theorem 3.4). Note that $z_1 \mapsto \theta(z_1, \mathbf{i}) = \exp(z_1 N_1) e^{\mathbf{i} N_2} \cdot F$ is a one-variable nilpotent orbit with corresponding PMHS $(e^{\mathbf{i}N_2} \cdot F, N_1)$. Fixing $z_2 = \mathbf{i} \in \mathcal{H}$ and letting $\mathrm{Im}\, z_1 \to \infty$ corresponds to fixing $t_2 \in \Delta^*$ and letting $t_1 \to 0$. So it is natural to regard the PMHS $(e^{\mathbf{i}N_2} \cdot F, N_1)$ as "less degenerate" than (F, N). (Cf. Figure 1.1 and the related discussion.) This motivates the following definition: given *any* $\mathbb{R}$–split two-variable nilpotent orbit (4.10), let (F_1, N_1) be the $\mathbb{R}$–split PMHS associated to $(e^{\mathbf{i}N_2} \cdot F, N_1)$ as in §2.2.2. Then we say the corresponding Hodge diamonds satisfy the *polarized relation*

$$\Diamond(F_1, N_1) \preceq \Diamond(F, N) .$$

The polarized relations are classified in Theorem 4.18. The classification requires the notion of a "primitive Hodge diamond."

REMARK 4.11. Recall the map $\Phi_\infty : \Diamond(D) \to \mathcal{O}(D)$ of Remark 4.5. Observe that the G–orbit $\mathcal{O}(F, N) = \Phi_\infty(\Diamond(F, N)) \subset \overline{D}$ is contained in the closure of $\mathcal{O}(F_1, N_1) = \Phi_\infty(\Diamond(F_1, N_1))$. In particular, if we define a partial order on the set $\mathcal{O}(D)$ of G–orbits in $\overline{D}$ by $\mathcal{O}_1 \le \mathcal{O}$ if $\mathcal{O} \subset \overline{\mathcal{O}}_1$, then the map Φ_∞ preserves the two relations. (However, beware that the polarized relation on Hodge diamonds is not, in general, a partial order: transitivity may fail. See Example 4.26.)

4.2.2. *Primitive subspaces.* Fix a $\mathbb{R}$–split PMHS (F, N), and let $V_\mathbb{C} = \oplus I^{p,q}$ be the Deligne splitting (§2.2.2). Set

$$P(N)^{p,q} := \ker\{N^{\ell+1} : I^{p,q} \to I^{-p-1,-q-1}\} , \tag{4.12a}$$

and define the *weight $n + \ell$ N-primitive subspace*

$$P(N)_{n+\ell,\mathbb{C}} := \bigoplus_{p+q=n+\ell} P(N)^{p,q} . \tag{4.12b}$$

Since (F, N) is $\mathbb{R}$–split we see that the $P(N)_{n+\ell}$ is defined over $\mathbb{R}$. Moreover, from the second equation of (2.10b) and Remark 2.16 it may be deduced that

$$V_\mathbb{R} = \bigoplus_{\substack{0 \le \ell \\ 0 \le a \le \ell}} N^a P(N)_{n+\ell} . \tag{4.13}$$

In particular, the decomposition (4.12) determines the Deligne bigrading $V_\mathbb{C} = \oplus I^{p,q}$ of $(F, W(N))$. Moreover, (4.12a) is a weight $n + \ell$ Hodge decomposition of $P(N)_{\ell,\mathbb{R}}$ polarized by

$$Q^N_\ell(\cdot, \cdot) := Q(\cdot, N^\ell \cdot) .$$

The *N-primitive Hodge–Deligne numbers* are the

$$j^{p,q} := \dim_\mathbb{C} P(N)^{p,q} .$$

The *weight $n+\ell$ primitive part* of $\Diamond(F,N)$ is the function

$$\Diamond^{\mathrm{prim}}_{n+\ell}(F,N) : \mathbb{Z}\times\mathbb{Z} \to \mathbb{Z}_{\ge 0} \quad \text{sending} \quad (p,q) \mapsto j^{p,q}\,.$$

Likewise, the *primitive part* of $\Diamond(F,N)$ is the sum

$$\Diamond^{\mathrm{prim}}(F,N) \ = \ \sum_{\ell=0}^{n} \Diamond^{\mathrm{prim}}_{n+\ell}(F,N)$$

of the weight k primitive Hodge sub-diamonds. Note that $\Diamond^{\mathrm{prim}}_{n+\ell}(F,N)$ not a Hodge diamond: (4.2c) and (4.2d) will fail whenever $N\neq 0$. We will call any such $\Diamond^{\mathrm{prim}}(F,N)$ a *primitive sub-diamond for the period domain D*. From (4.13) we see that

(4.14) $\Diamond^{\mathrm{prim}}(F,N)$ *determines* $\Diamond(F,N)$ *(and visa versa).*

To be more precise, given $f:\mathbb{Z}\times\mathbb{Z}\to\mathbb{Z}_{\ge 0}$ define $f[\ell]:\mathbb{Z}\times\mathbb{Z}\to\mathbb{Z}_{\ge 0}$ by $(p,q)\mapsto f(p+\ell,q+\ell)$. Then (4.13) implies

$$\Diamond(F,N) \ = \ \sum_{\substack{0\le\ell\\ 0\le a\le\ell}} \Diamond^{\mathrm{prim}}_{n+\ell}(F,N)[a]\,.$$

From Theorem 4.3 we then obtain

Corollary 4.15. *The G–conjugacy class of an $\mathbb{R}$–split PMHS (F,N) on D is determined by the primitive sub-diamond $\Diamond^{\mathrm{prim}}(F,N)$.*

Example 4.16 (Hodge numbers $\mathbf{h}=(g,g)$). The primitive Hodge diamond for the $\mathbb{R}$–split PMHS (F_r,N_r) with Hodge diamond $\Diamond_r$ of Example 4.6 is

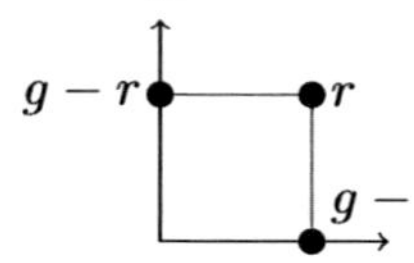

Example 4.17 (Hodge numbers $\mathbf{h}=(b,a,b)$). The primitive Hodge diamond for the $\mathbb{R}$–split PMHS $(F_{r,s},N_{r,s})$ with Hodge diamond $\Diamond_{r,s}$ of Example 4.9 is

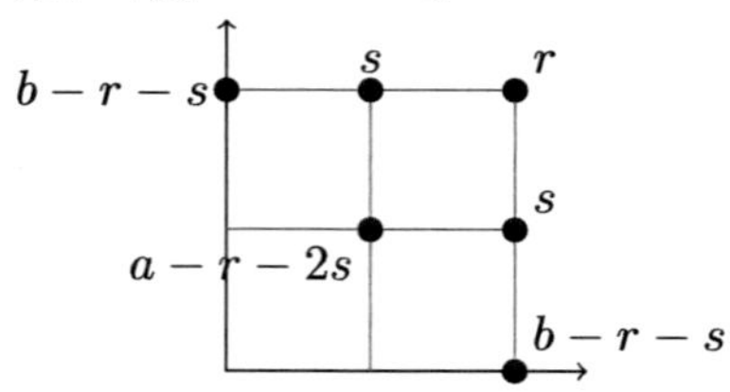

Theorem 4.18 ([**19**]). *Let D be a period domain parameterizing weight n, Q–polarized Hodge structures on $V_{\mathbb{R}}$. Let $[F_1,N_1],[F_2,N_2]\in\Psi_D$. Then $[F_1,N_1]\preceq[F_2,N_2]$ if and only if $\Diamond(F_2,N_2)$ can be expressed as a sum*

$$\Diamond(F_2,N_2) \ = \ \sum_{\substack{0\le k\\ 0\le a\le\ell}} \Diamond(F'_\ell,N'_\ell)[a]$$

for Hodge diamonds $\Diamond(F'_\ell,N'_\ell)$ on the period domains D_ℓ parameterizing weight $n+\ell$, $Q^{N_1}_\ell$–polarized Hodge structures on $P(N_1)_{n+\ell}$ with Hodge numbers $\{j_1^{p,q}\mid p+q=n+\ell\}$.

PROOF. The theorem is a consequence of the Cattani–Kaplan–Schmid Several Variable SL(2)–Orbit Theorem [**6**]. See [**19**] for details. □

We finish this section by giving a number of examples illustrating Theorem 4.18.

EXAMPLE 4.19 (Hodge numbers $\mathbf{h} = (g, g)$). The polarized relations among the Hodge diamonds in Example 4.6 are $\diamond_s \prec \diamond_r$ if and only if $s > r$. In particular, in this case $\prec$ is a linear order. Moreover, the map $\Phi_\infty : \diamond(D) \to \mathcal{O}(D)$ of Remark 4.5 sending $\diamond_r \mapsto \mathcal{O}_r$ is a bijection preserving the orders.

REMARK 4.20. Notice that each of the polarized relations in Example 4.19 can be realized geometrically. For example, if $g = 2$, then the polarized relations $\diamond_0 \prec \diamond_1 \prec \diamond_2$ can be realized by degenerating a curve of genus two so that it acquires one, and then two, nodes.

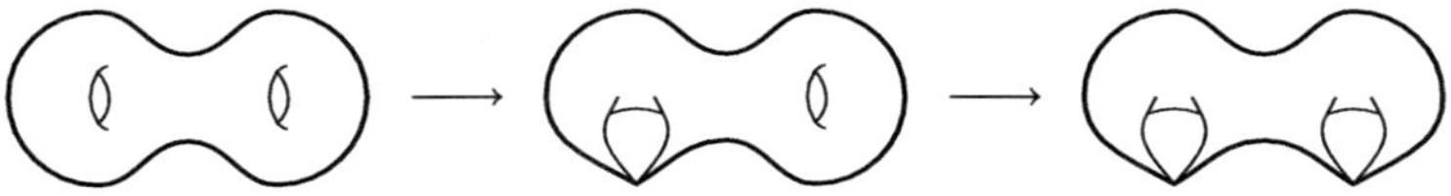

EXAMPLE 4.21 (Hodge numbers $\mathbf{h} = (b, a, b)$). The polarized relations amongst the Hodge diamonds in Example 4.9 are $\diamond_{r,s} \preceq \diamond_{t,u}$ if and only if $r \le t$ and $r + s \le t + u$. As in Example 4.19 the map $\Phi_\infty : \diamond(D) \to \mathcal{O}(D)$ of Remark 4.5 sending $\diamond_{r,s} \mapsto \mathcal{O}_{r,s}$ is a bijection preserving the relations. In particular, if $b = 1$, then $\prec$ is a linear order. However, if $b > 1$, then $\prec$ is a partial order, but not a linear order. For example, if $b = 2$ and $a \ge 4$, then the partial order is represented by the diagram

(4.22)

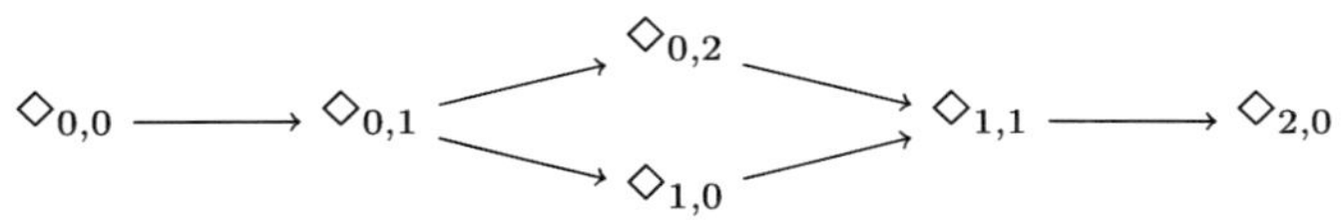

with, for example, $\diamond_{0,0} \to \diamond_{0,1}$ indicating $\diamond_{0,0} \prec \diamond_{0,1}$. Likewise for $b = 3$ and $a \ge 6$ we have

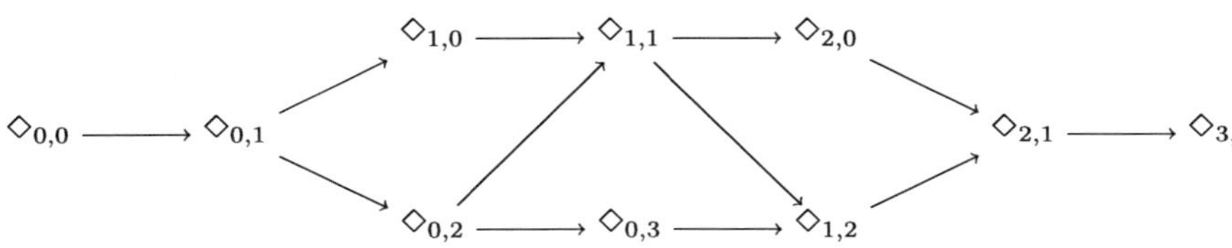

REMARK 4.23. The period domains for polarized Hodge structures with Hodge numbers $\mathbf{h} = (g, g)$ or $\mathbf{h} = (1, a, 1)$ are Hermitian symmetric period domains. More generally, if D is any Hermitian symmetric Mumford–Tate domain, then $\prec$ is a linear order [**19**].

REMARK 4.24 (Geometric realization of the polarized relations for $\mathbf{h} = (2, 27, 2)$). In analogy with Remark 4.20, each of the polarized relations of (4.22) may be realized by degenerations of Horikawa surfaces; see the forthcoming [**11**] for details.

PROBLEM 4.25. In analogy with Remarks 4.20 and 4.24, give geometric realizations of the polarized relations for other values of $\mathbf{h}$.

EXAMPLE 4.26 (Hodge numbers $\mathbf{h} = (1, 2, 2, 1)$). From Theorem 4.3 we see that the Hodge diamonds include the following three

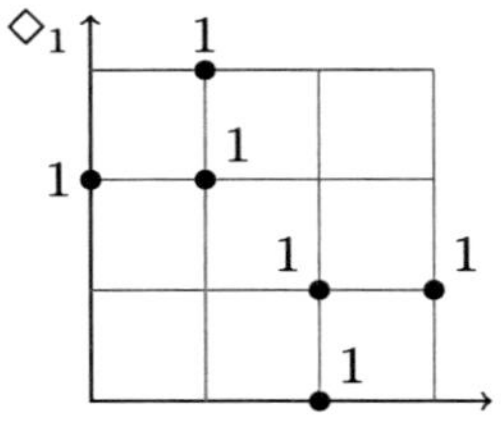

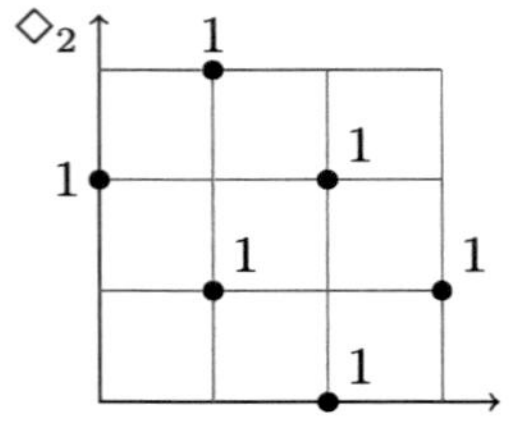

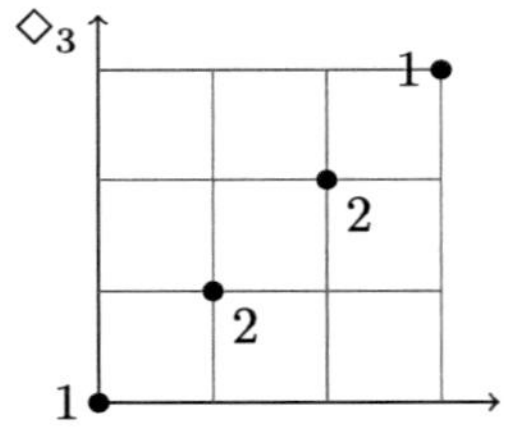

From Theorem 4.18 we see that $\diamond_1 \prec \diamond_2 \prec \diamond_3$, but $\diamond_1 \not\prec \diamond_3$. That is, *the polarized relation is not transitive*, and therefore can not define a partial order. There is a similar failure of transitivity for $\mathbf{h} = (1, a, a, 1)$, $a \geq 2$, cf. [**19**].

QUESTION 4.27. What are the geometric implications of the failure of transitivity in the Calabi–Yau 3-fold case (Example 4.26)?

PROBLEM 4.28. Identify the Hodge diamonds for $\mathbf{h} = (1, a, a, 1)$, and determine their polarized relations [**19**].

REMARK 4.29. Work of Bloch, Kerr and Vanhove [**1**] yields geometric realizations of the polarized relations on Calabi–Yau 3-folds, see [**19**].

REMARK 4.30. In the case that D is a period domain parameterizing polarized Hodge structures with Hodge numbers $\mathbf{h} = (1, \ldots, 1)$ or $\mathbf{h} = (1, \ldots, 1, 2, 1, \ldots, 1)$, then $\prec$ is a partial order. In both these cases, the isotropy group (the subgroup of G stabilizing $\varphi \in D$) is a compact torus (equivalently the complex parabolic subgroup of $G_\mathbb{C}$ stabilizing the Hodge filtration is a Borel). More generally, if D is a Mumford–Tate domain with isotropy a compact torus, then $\prec$ is a partial order [**19**].

References

[1] Spencer Bloch, Matt Kerr, and Pierre Vanhove. Local mirror symmetry and the sunset feynman integral. arxiv:601.08181, 2016.

[2] P. Brosnan, G. Pearlstein, and C. Robles. Nilpotent cones and their representation theory. In preparation; an expository chapter in a collection honoring Steve Zucker's work, 2015.

[3] James Carlson, Stefan Mü ller-Stach, and Chris Peters, *Period mappings and period domains*, Cambridge Studies in Advanced Mathematics, vol. 85, Cambridge University Press, Cambridge, 2003. MR2012297

[4] Eduardo Cattani, Fouad El Zein, Phillip A. Griffiths, and Lê Dũng Tráng (eds.), *Hodge theory*, Mathematical Notes, vol. 49, Princeton University Press, Princeton, NJ, 2014. MR3288678

[5] Eduardo Cattani and Aroldo Kaplan, *Polarized mixed Hodge structures and the local monodromy of a variation of Hodge structure*, Invent. Math. **67** (1982), no. 1, 101–115. MR664326

[6] Eduardo Cattani, Aroldo Kaplan, and Wilfried Schmid, *Degeneration of Hodge structures*, Ann. of Math. (2) **123** (1986), no. no. 3, 457–535. MR840721

[7] C. H. Clemens, *Degeneration of Kähler manifolds*, Duke Math. J. **44** (1977), no. 2, 215–290. MR0444662

[8] Olivier Debarre, *Periods and moduli*, Current developments in algebraic geometry, Math. Sci. Res. Inst. Publ., vol. 59, Cambridge Univ. Press, Cambridge, 2012, pp. 65–84. MR2931865

[9] Pierre Deligne, *Théorie de Hodge. III* (French), Inst. Hautes Études Sci. Publ. Math. **44** (1974), 5–77. MR0498552

[10] Alan H. Durfee, *A naive guide to mixed Hodge theory*, Singularities, Part 1 (Arcata, Calif., 1981), Proc. Sympos. Pure Math., vol. 40, Amer. Math. Soc., Providence, RI, 1983, pp. 313–320. MR713069

[11] M. Green, P. Griffiths, R. Laza, and C. Robles. Hodge theory and moduli of H-surfaces. In preparation.

[12] Mark Green, Phillip Griffiths, and Matt Kerr, *Mumford-Tate groups and domains*, Annals of Mathematics Studies, vol. 183, Princeton University Press, Princeton, NJ, 2012. Their geometry and arithmetic. MR2918237

[13] Mark Green, Phillip Griffiths, and Colleen Robles. Extremal degenerations of polarized Hodge structures. Preprint, arXiv:1403.0646, 2014.

[14] Phillip Griffiths and Wilfried Schmid, *Locally homogeneous complex manifolds*, Acta Math. **123** (1969), 253–302. MR0259958

[15] Phillip A. Griffiths, *Periods of integrals on algebraic manifolds. I. Construction and properties of the modular varieties*, Amer. J. Math. **90** (1968), 568–626. MR0229641

[16] Phillip A. Griffiths, *Periods of integrals on algebraic manifolds. II. Local study of the period mapping*, Amer. J. Math. **90** (1968), 805–865. MR0233825

[17] Heisuke Hironaka, *Resolution of singularities of an algebraic variety over a field of characteristic zero. I, II*, Ann. of Math. (2) **79** (1964), 109–203; ibid. (2) **79** (1964), 205–326. MR0199184

[18] Nicholas M. Katz, *Nilpotent connections and the monodromy theorem: Applications of a result of Turrittin*, Inst. Hautes Études Sci. Publ. Math. **39** (1970), 175–232. MR0291177

[19] Matt Kerr and Colleen Robles. Partial orders and polarized relations on limit mixed hodge structures. In prepration, 2015.

[20] Alan Landman, *On the Picard-Lefschetz transformation for algebraic manifolds acquiring general singularities*, Trans. Amer. Math. Soc. **181** (1973), 89–126. MR0344248

[21] Radu Laza, *Perspectives on the construction and compactification of moduli spaces*, Compactifying moduli spaces, Adv. Courses Math. CRM Barcelona, Birkhäuser/Springer, Basel, 2016, pp. 1–39. MR3495110

[22] Chris A. M. Peters and Joseph H. M. Steenbrink, *Mixed Hodge structures*, Ergebnisse der Mathematik und ihrer Grenzgebiete. 3. Folge. A Series of Modern Surveys in Mathematics [Results in Mathematics and Related Areas. 3rd Series. A Series of Modern Surveys in Mathematics], vol. 52, Springer-Verlag, Berlin, 2008. MR2393625

[23] Colleen Robles, *Classification of horizontal* SL(2)*s*, Compos. Math. **152** (2016), no. no. 5, 918–954. MR3505643

[24] Wilfried Schmid, *Variation of Hodge structure: the singularities of the period mapping*, Invent. Math. **22** (1973), 211–319. MR0382272

[25] Joseph Steenbrink, *Limits of Hodge structures*, Invent. Math. **31** (1975/76), no. 3, 229–257. MR0429885

Department of Mathematics, Duke University, Box 90320, Durham, North Carolina 27708-0320

E-mail address: `robles@math.duke.edu`

Questions about Boij–Söderberg theory

Daniel Erman and Steven V Sam

ABSTRACT. Boij-Söderberg theory focuses on the properties and duality relationship between two types of numerical invariants. One side involves the Betti table of a graded free resolution over the polynomial ring. The other side involves the cohomology table of a coherent sheaf on projective space. We discuss open questions and problems in Boij-Söderberg theory.

1. Background on Boij–Söderberg Theory

Boij–Söderberg theory focuses on the properties and duality relationship between two types of numerical invariants. One side involves the Betti table of a graded free resolution over the polynomial ring. The other side involves the cohomology table of a coherent sheaf on projective space. The theory began with a conjectural description of the cone of Betti tables of finite length modules, given in [**10**]. Those conjectures were proven in [**25**], which also described the cone of cohomology tables of vector bundles and illustrated a sort of duality between Betti tables and cohomology tables.

The theory itself has since expanded in many directions: allowing modules whose support has higher dimension, replacing vector bundles by coherent sheaves, working over rings other than the polynomial ring, and so on. But at its core, Boij–Söderberg theory involves:

(1) A classification, up to scalar multiple, of the possible Betti tables of some class of objects (for example, free resolutions of finitely generated modules of dimension $\leq c$).
(2) A classification, up to scalar multiple, of the cohomology tables of some class of objects (for examples, coherent sheaves of dimension $\leq n-c$).
(3) Intersection theory-style duality results between Betti tables and cohomology tables.

One motivation behind Boij and Söderberg's conjectures was the observation that it would yield an immediate proof of the Cohen–Macaulay version of the Multiplicity Conjectures of Herzog–Huneke–Srinivasan [**46**]. Eisenbud and Schreyer's [**25**] thus yielded an immediate proof of that conjecture, and the subsequent papers [**11, 26**] provided a proof of the Multiplicity Conjecture for non-Cohen–Macaulay modules. Other applications of the theory involve Horrocks'

2010 *Mathematics Subject Classification.* Primary 13D02.

The first author was partially supported by NSF DMS-1302057 and SS was partially supported by NSF DMS-1500069.

Conjecture [**32**], cohomology of tensor products of vector bundles [**29**], sparse determinantal ideals [**12**], concavity of Betti tables [**49**] and more. In fact, Boij–Söderberg theory has grown into an active area of research in commutative algebra and algebraic geometry. See §9 or [**36, 39**] for a summary of many of the related papers. In addition, many features from the theory have been implemented via Macaulay2 packages such as `BoijSoederberg.m2`, `BGG.m2`, and `TensorComplexes.m2` [**44**].

In this paper, we focus on discussing several open questions related to Boij–Söderberg theory. Of course, the choice of topics reflects our own bias and perspective. We also briefly review some of the major aspects of the theory, but those interested in a fuller expository treatment should refer to [**27**] or [**36**].

1.1. Betti tables. Let $S = \Bbbk[x_0, \dots, x_n]$ with the grading $\deg(x_i) = 1$ for all i, and with $\Bbbk$ any field. Let M be a finitely generated graded module over S. Since M is graded, it admits a minimal free resolution $\mathbf{F} = [F_0 \leftarrow F_1 \leftarrow \cdots \leftarrow F_p \leftarrow 0]$. The Betti table of M is a vector whose coordinates $\beta_{i,j}(M)$ encode the numerical data of the minimal free resolution of M. Namely, since each F_i is a graded free module, we can let $\beta_{i,j}(M)$ be the number of degree j generators of F_i; equivalently, we can write $F_i = \bigoplus_{j \in \mathbb{Z}} S(-j)^{\beta_{i,j}}$; also equivalently, the Betti numbers come from graded Tor groups with respect to the residue field: $\beta_{i,j}(M) := \dim \operatorname{Tor}_i(\mathbf{F}, \Bbbk)_j$.

For example, if $S = \Bbbk[x_0, x_1]$ and $M = S/(x_0, x_1^2)$ then the minimal free resolution of M is a Koszul complex

$$F = S \longleftarrow \begin{matrix} S^1(-1) \\ \oplus \\ S^1(-2) \end{matrix} \longleftarrow S^1(-3) \longleftarrow 0.$$

The Betti table of M is traditionally displayed as the following array or matrix:

$$\beta(M) = \begin{bmatrix} \beta_{0,0} & \beta_{1,1} & \cdots & \beta_{p,p} \\ \beta_{0,1} & \beta_{1,2} & \cdots & \beta_{p,p+1} \\ \vdots & \ddots & & \vdots \end{bmatrix}.$$

In the example above, we thus have

$$\beta(M) = \begin{bmatrix} 1 & 1 & - \\ - & 1 & 1 \end{bmatrix}.$$

There is a huge literature on the properties of Betti tables, and we refer the reader to [**18**] as a starting point. Yet there are also many fundamental open questions about Betti tables. The most notable question is Horrocks' Conjecture, due to Horrocks [**45**, Problem 24] and Buchsbaum-Eisenbud [**13**, p. 453], which proposes that the Koszul complex is the "smallest" free resolution. More precisely, one version of the conjecture proposes that if $c = \operatorname{codim} M$, then $\sum_{i,j} \beta_{i,j}(M) \geq 2^c$. The conjecture is known in five variables and other special cases [**3, 14**] but remains wide open in general.

Boij and Söderberg proposed taking the convex cone spanned by the Betti tables, and then focusing on this cone. This is like studying Betti tables "up to scalar multiple", which would remove the subtleties behind questions like Horrocks' Conjecture. Note that convex combinations are natural in this context, as $\beta(M \oplus M') = \beta(M) + \beta(M')$. We define $B^c(S)$ as the convex cone spanned by the Betti

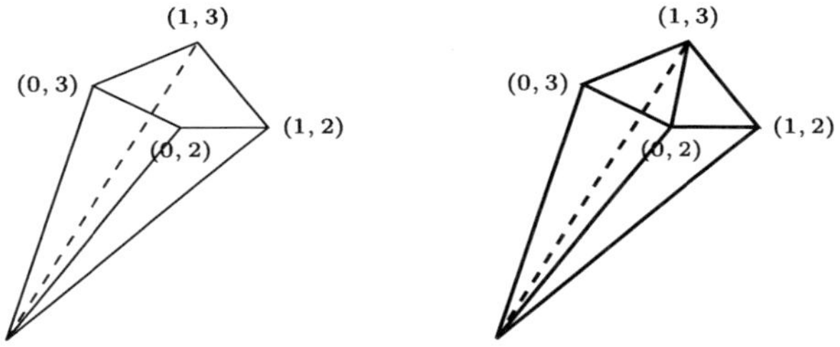

FIGURE 1. For a cone of Betti tables, the extremal rays come from Cohen–Macaulay modules with pure resolutions. The rays may thus be labelled by strictly increasing sequences of integers, as illustrated on the left. The cone has a simplicial fan structure, where simplices correspond to increasing chains of integers with respect to the partial order, as illustrated on the right.

tables of all S-modules of codimension $\geq c$:

$$B^c(S) := \mathbb{Q}_{\geq 0}\{\beta(M) \mid \operatorname{codim} M \geq c\} \subseteq \bigoplus_{i=0}^{n+1} \bigoplus_{j \in \mathbb{Z}} \mathbb{Q}.$$

Boij and Söderberg's original conjectures described the cone $B^{n+1}(S)$, which is the case of finite length S-modules [**10**]. This description was based on the notion of a **pure resolution of type** $d = (d_0, \ldots, d_{n+1}) \in \mathbb{Z}^{n+2}$, which is an acyclic complex where the ith term is generated entirely in degree d_i; in other words, it is a minimal free complex of the form:

$$S(-d_0)^{\beta_{0,d_0}} \longleftarrow S(-d_1)^{\beta_{1,d_1}} \longleftarrow S(-d_2)^{\beta_{2,d_2}} \longleftarrow \cdots \longleftarrow S(-d_p)^{\beta_{n+1,d_{n+1}}} \longleftarrow 0.$$

Any such resolution must satisfy $d_0 < d_1 < \cdots < d_{n+1}$, and Boij and Söderberg conjectured that for any strictly increasing vector $(d_0, \ldots, d_{n+1})$ there was such a pure resolution. It was known that if such a resolution existed, then the vector d determined a unique ray in $B^{n+1}(S)$ [**10**, §2.1], and Boij and Söderberg conjectured that these were precisely the extremal rays of $B^{n+1}(S)$. So the extremal rays of $B^{n+1}(S)$ should be in bijection with strictly increasing vectors (sometimes called **degree sequences**) $d = (d_0, \ldots, d_{n+1})$. With the advent of hindsight, this is a natural guess: pure resolutions will give the Betti tables with the fewest possible nonzero entries, and so they will always produce extremal rays.

There is a natural termwise partial order on such vectors, where $d \leq d'$ if $d_i \leq d'_i$ for all i, and Boij and Söderberg also conjectured that this partial order endowed $B^{n+1}(S)$ with the structure of a simplicial fan, where $B^{n+1}(S)$ is the union of simplicial cones and the simplices correspond to maximal chains of degree sequences with respect to the partial order. For instance, the cone in Figure 1 decomposes as the union of two simplicial cones; the two cones correspond to the chains $(0,2) < (1,2) < (1,3)$ and $(0,2) < (0,3) < (1,3)$.

The existence of pure resolutions was first proven in [**22**] in characteristic zero and in [**25**] in arbitrary characteristic; further generalizations appear in [**7, 38**]. The rest of Boij and Söderberg's conjectures were proven in [**25**] and the theory has since been extended from finite length modules to finitely generated modules [**11**] and to bounded complexes of modules [**19**].

One of the most striking corollaries of Boij–Söderberg theory is the resulting decomposition of Betti tables. The simplicial structure of the cone of Betti tables provides an algorithm for writing a Betti table $\beta(M)$ as a positive, rational sum of the Betti tables of pure resolutions. For instance, returning to our example of $M = \Bbbk[x,y]/(x,y^2)$, we have:

$$\beta(M) = \begin{bmatrix} 1 & 1 & - \\ - & 1 & 1 \end{bmatrix} = \frac{1}{3}\begin{bmatrix} 2 & 3 & - \\ - & - & 1 \end{bmatrix} + \frac{1}{3}\begin{bmatrix} 1 & - & - \\ - & 3 & 2 \end{bmatrix} \tag{1}$$

The decomposition of $\beta(M)$ will always be a finite sum, which is an immediate consequence of the fact that every Betti table has only finitely many nonzero entries, and there are thus a finite number of extremal rays which could potentially contribute to that Betti table. This decomposition algorithm is implemented in Macaulay2, and it is extremely efficient.

These decompositions provide a counterintuitive aspect of Boij–Söderberg theory.

EXAMPLE 1.1. Let $S = \Bbbk[x,y]$ and $M = S/(x,y^2)$. The pure resolutions appearing on the right-hand side of (1) are resolutions of actual modules. If we let $M' = S/(x,y)^2$ and $M'' := \operatorname{coker}\begin{pmatrix} x & y & 0 \\ 0 & x & y \end{pmatrix}$ then we have:

$$\beta(M) = \tfrac{1}{3}\beta(M') + \tfrac{1}{3}\beta(M'').$$

Yet the above equation is purely numerical: Boij and Söderberg did not address whether the free resolution of M could be built from the free resolutions of M' and M'', and Eisenbud and Schreyer's proof provides no such categorification. We discuss this in more detail in §2.

If we study finitely generated modules which are not necessarily of finite length, then the description of the cone of Betti tables is a bit more subtle. For the cone $B^c(S)$ of Betti tables of modules of modules of codimension $\geq c$, the extremal rays still correspond to pure resolutions of Cohen–Macaulay modules, but now the modules may have codimension between c and $n+1$. For instance, if we consider the non-Cohen–Macaulay, cyclic module $M = \Bbbk[x,y,z]/(x^2,xy,xz,yz)$, then the Boij–Söderberg decomposition will be:

$$\beta(M) = \begin{bmatrix} 1 & - & - & - \\ - & 4 & 4 & 1 \end{bmatrix} = \frac{1}{3}\begin{bmatrix} 1 & - & - & - \\ - & 6 & 8 & 3 \end{bmatrix} + \frac{2}{3}\begin{bmatrix} 1 & - & - \\ - & 3 & 2 \end{bmatrix}.$$

The Betti tables on the right can be realized as the Betti tables of $S/(x_1,x_2,x_3)^2$ and $S/(x_1,x_2)^2$, which are Cohen–Macaulay modules of codimension 3 and 2, respectively. We denote these as pure resolutions of type $(0,2,3,4)$ and type $(0,2,3,\infty)$ respectively, because under this convention, the termwise partial order still induces a simplicial fan structure on $B^c(S)$.

Even more generally, we could leave the world of modules and look instead at bounded complexes of free modules or equivalently at elements of $\mathrm{D}^b(S)$. The corresponding cone is yet again spanned by pure resolutions of Cohen–Macaulay modules, but where we also allow homological shifts. With the right conventions, the whole story carries over in great generality in this context, including the decomposition theorem. For instance, on $S = \Bbbk[x,y]$ we could take the complex

$$\mathbf{F} := \left[S^1 \xleftarrow{\begin{pmatrix} x & y \end{pmatrix}} S^2(-1) \xleftarrow{\begin{pmatrix} -y^2 & xy \\ xy & -x^2 \end{pmatrix}} S^2(-3) \xleftarrow{\begin{pmatrix} y \\ x \end{pmatrix}} S^1(-4) \right],$$

which has finite length homology $H_0\mathbf{F} = \Bbbk$, $H_1\mathbf{F} = \Bbbk(-2)$. We then have the decomposition:

$$\beta(\mathbf{F}) = \begin{bmatrix} 1 & 2 & - & - \\ - & - & 2 & 1 \end{bmatrix} = \frac{1}{2}\begin{bmatrix} - & 1 & - & - \\ - & - & 3 & 2 \end{bmatrix} + \frac{1}{2}\begin{bmatrix} 2 & 3 & - & - \\ - & - & 1 & - \end{bmatrix}.$$

See [**19**] for more on Boij–Söderberg theory for complexes.

1.2. Cohomology tables. On the sheaf cohomology side, we fix a coherent sheaf $\mathcal{E}$ on $\mathbb{P}^n_\Bbbk$. We define the **cohomology table** of $\mathcal{E}$ as a vector whose coordinates $\gamma_{i,j}(\mathcal{E})$ encode graded sheaf cohomology groups of $\mathcal{E}$. In particular, $\gamma_{i,j}(\mathcal{E}) := \dim_\Bbbk \mathrm{H}^i(\mathbb{P}^n, \mathcal{E}(j))$.

When displaying cohomology tables, we follow the tradition introduced in [**21**] and write:

$$\gamma(\mathcal{E}) = \begin{matrix} \cdots & \gamma_{n,-n-2} & \gamma_{n,-n-1} & \gamma_{n,-n} & \gamma_{n,-n+1} & \cdots \\ \cdots & \gamma_{n-1,-n-1} & \gamma_{n-1,-n} & \gamma_{n-1,-n+1} & \gamma_{n-1,-n+2} & \cdots \\ & & \ddots & & & \\ \cdots & \gamma_{1,-3} & \gamma_{1,-2} & \gamma_{1,-1} & \gamma_{1,0} & \cdots \\ \cdots & \gamma_{0,-2} & \gamma_{0,-1} & \gamma_{0,0} & \gamma_{0,1} & \cdots \end{matrix}$$

Thus for instance, if $\mathcal{E} = \mathcal{O}_{\mathbb{P}^2}$ then we have:

$$\gamma(\mathcal{O}_{\mathbb{P}^2}) = \begin{matrix} \cdots & 3 & 1 & - & - & - & - & \cdots \\ \cdots & - & - & - & - & - & - & \cdots \\ \cdots & - & - & 1 & 3 & 6 & 10 & \cdots \end{matrix}$$

where the 3 on the bottom row corresponds to $\gamma_{0,1}(\mathcal{O}_{\mathbb{P}^2}(1)) = \dim \mathrm{H}^0(\mathbb{P}^2, \mathcal{O}_{\mathbb{P}^2}(1)) = 3$.

While sheaf cohomology is central to all of modern algebraic geometry, the research on cohomology tables is nowhere near as extensive as the work on Betti tables. In fact, cohomology tables seem to have first appeared in [**21**], where the cohomology table of a sheaf $\mathcal{F}$ is written as the Betti table of the corresponding Tate resolution over the exterior algebra.

As with Betti tables, there are many open questions related to cohomology tables. One notable question is whether there exists a non-split rank 2 vector bundle on $\mathbb{P}^n$ for any $n \geq 5$. This question could be answered entirely with information in the cohomology table. A coherent sheaf on $\mathbb{P}^n$ is a vector bundle if and only if there are only finitely many zero entries in the cohomology table in the rows corresponding to H^i where $i = 1, 2, \ldots, n-1$; and it is a sum of line bundles if and only if all of those intermediate entries are zero. And since the cohomology table is a refinement of the Hilbert polynomial, the rank of the vector bundle can be read off from the cohomology table as well.

Inspired by the Boij–Söderberg conjectures, Eisenbud and Schreyer introduced the convex cone spanned by the cohomology tables. We start by focusing on vector bundles; define $\mathrm{C}_{\mathrm{vb}}(\mathbb{P}^n)$ as the convex cone spanned by the cohomology tables of all vector bundles on $\mathbb{P}^n$:

$$\mathrm{C}_{\mathrm{vb}}(\mathbb{P}^n) := \mathbb{Q}_{\geq 0}\{\gamma(\mathcal{E}) \mid \mathcal{E} \text{ is a vector bundle on } \mathbb{P}^n\} \subseteq \prod_{i=0}^{n+1} \prod_{j \in \mathbb{Z}} \mathbb{Q}.$$

Eisenbud and Schreyer give a complete description of this cone in [**25**]. A **supernatural bundle** is a vector bundle with as few nonzero cohomology groups as possible; more precisely, $\mathcal{E}$ is supernatural of type $f = (f_1, \dots, f_n) \in \mathbb{Z}^n$ if

$$\mathrm{H}^i(\mathbb{P}^n, \mathcal{E}(j)) \neq 0 \iff \begin{cases} i = n \text{ and } j < f_n \\ 1 \leq i \leq n-1 \text{ and } f_{i+1} < j < f_i \\ i = 0 \text{ and } j > f_1. \end{cases}$$

Any such bundle must satisfy $f_1 > f_2 > \cdots > f_n$ and such a strictly decreasing sequence of integers is called a **root sequence**. Eisenbud and Schreyer proved that there exists a supernatural bundle with any specified root sequence; as was the case with pure resolutions, the vector f thus determines a unique ray in $\mathrm{C_{vb}}(\mathbb{P}^n)$, and Eisenbud and Schreyer proved that these were precisely the extremal rays of that cone. Again with the advent of hindsight, this is a natural guess: supernatural bundles give the cohomology tables with the fewest possible nonzero entries, and so they will always produce extremal rays.

The termwise partial order on root sequences then induces a simplicial fan structure on $\mathrm{C_{vb}}(\mathbb{P}^n)$ in a manner entirely parallel to the story for the cone of Betti tables.

EXAMPLE 1.2. Let $\mathcal{E}$ be the cokernel of a generic map $\phi\colon \mathcal{O}_{\mathbb{P}^2}(-2)^2 \to \mathcal{O}_{\mathbb{P}^2}(-1)^5$. Then $\mathcal{E}$ is a rank 3 bundle on $\mathbb{P}^2$. A computation in Macaulay2 yields the decomposition

$$\begin{aligned}\gamma(\mathcal{E}) &= \begin{bmatrix} \dots & 20 & 10 & 3 & - & - & - & - & - & \dots \\ \dots & - & - & 1 & 2 & - & - & - & - & \dots \\ \dots & - & - & - & - & 5 & 13 & 24 & \dots \end{bmatrix} \\ &= \frac{1}{3}\begin{bmatrix} \dots & 15 & 6 & - & - & - & - & - & \dots \\ \dots & - & - & 3 & 3 & - & - & - & \dots \\ \dots & - & - & - & - & 6 & 15 & 27 & \dots \end{bmatrix} \\ &\quad + \begin{bmatrix} \dots & 15 & 8 & 3 & - & - & - & - & \dots \\ \dots & - & - & - & 1 & - & - & - & \dots \\ \dots & - & - & - & - & 3 & 8 & 15 & \dots \end{bmatrix},\end{aligned}$$

where the tables used in the decomposition can be realized as cohomology tables of supernatural bundles. In fact, if $\mathcal{E}' := (\mathrm{Sym}_2\, \Omega^1_{\mathbb{P}^2})(2)$ and $\mathcal{E}'' := \Omega^1_{\mathbb{P}^2}(1)$, then we have

$$= \tfrac{1}{3}\gamma(\mathcal{E}') + \tfrac{1}{3}\gamma(\mathcal{E}'').$$

We return to this example in §2.

All of the results about vector bundles can be extended to coherent sheaves as well. Let $C(\mathbb{P}^n)$ be the cone of cohomology tables of coherent sheaves $\mathcal{F}$ on $\mathbb{P}^n$. In [**26**], Eisenbud and Schreyer describe this cone in detail. Every extremal ray comes from a sheaf $\mathcal{E}$, which is supported on a linear subspace $\mathbb{P}^a \subseteq \mathbb{P}^n$ for some $a \leq n$, and where $\mathcal{E}$ is a supernatural vector bundle on $\mathbb{P}^a$. As in the case of vector bundles, the extremal rays thus correspond to root sequences, and an appropriate partial order induces a simplicial fan structure on $C(\mathbb{P}^n)$. But there is one key difference for coherent sheaves: the decomposition might involve an infinite number of summands [**26**, Example 0.3]. There are open questions even in this case.

QUESTION 1.3. *Under what conditions will an infinite sum of supernatural sheaves correspond to a cohomology table, at least up to scalar multiple?*

2. Categorification

The most natural and mysterious question raised by Boij–Söderberg theory has to do with categorifying the decompositions into pure/supernatural tables.

QUESTION 2.1. (1) *For a graded module M, does the decomposition of $\beta(M)$ into pure Betti tables "lift" to a corresponding decomposition of the module M itself?*

(2) *For a vector bundle $\mathcal{E}$, does the decomposition of $\gamma(\mathcal{E})$ into supernatural cohomology tables "lift" to a corresponding decomposition of the bundle $\mathcal{E}$ itself?*

Let us return to Examples 1.1 and 1.2, and continue with that notation. On the module side, we had the decomposition

$$\beta(M) = \tfrac{1}{3}\beta(M') + \tfrac{1}{3}\beta(M'')$$

and on the vector bundle side we had

$$\gamma(\mathcal{E}) = \tfrac{1}{3}\gamma(\mathcal{E}') + \gamma(\mathcal{E}'').$$

If we want to decompose M or $\mathcal{E}$ into pieces corresponding to the Boij–Söderberg decomposition, then there are three challenges:

(1) **Denominators:** Is there a natural way to systematically clear the denominators that arise in these sorts of decompositions?
(2) **Moduli:** Modules with pure resolutions and supernatural bundles can have nontrivial moduli. Namely, there can be a lot of modules/vector bundles with the same pure Betti table/supernatural cohomology table, or even the same table up to scalar multiple.[1] How do we select the appropriate module or bundle to represent the summands appearing in the decomposition?
(3) **Filtration:** Even if we overcome the first two obstacles, we still need to assemble these pieces, perhaps through a filtration. How do we build such a filtration?

On the module side, almost nothing is known. Eisenbud, Erman and Schreyer obtain categorifications under highly restrictive hypotheses in [**20**], but their techniques avoid the issues raised by denominators and moduli.

The example above is already challenging. We might clear denominators by replacing M by $M^{\oplus 3N}$ for some N. This would yield:

$$\beta(M^{\oplus 3N}) = \begin{pmatrix} 2N & 3N & - \\ - & - & N \end{pmatrix} + \begin{pmatrix} N & - & - \\ - & 3N & 2N \end{pmatrix}.$$

The diagrams on the righthand correspond to unique modules. Continuing with the notation of Example 1.1, the diagrams on the right must correspond to $(M')^{\oplus N}$ and $(M'')^{\oplus N}$. We would thus want to realize $M^{\oplus 3N}$ as an extension of $(M')^{\oplus N}$ and $(M'')^{\oplus N}$. Since there are no nonzero extensions of M' by M'', the only possibility is an extension

$$0 \to (M')^{\oplus N} \to M^{\oplus N} \to (M'')^{\oplus N} \to 0.$$

[1]For example, if ϕ is any sufficiently general 2×4 matrix of linear forms on $S = \Bbbk[x, y, z]$, then the cokernel of ϕ will define a pure resolution with degree sequence $(0, 1, 3, 4)$. Varying ϕ produces non-isomorphic pure resolutions. Similar constructions work for supernatural bundles [**29**, §6].

As originally observed by Sam and Weyman [**55**], this is also impossible: $M^{\oplus N}$ is annihilated by the linear form x whereas the submodule $(M')^{\oplus N}$ is not annihilated by any linear form.

However, on the vector bundle side (in characteristic 0), some recent progress was made. In fact, the first two issues now seem to have a satisfactory answer. The most naive idea for clearing denominators is to take direct sums of $\mathcal{E}$, which is equivalent to applying $\mathcal{E} \mapsto \mathcal{E} \otimes \mathcal{O}^{\oplus N}$. Instead, we introduce a collection of Fourier–Mukai transforms which can be thought of as "twisting" by something like an Ulrich sheaf. These transforms clear denominators in exactly the way one would hope for [**34**, Corollary 1.4]. In addition, these functors also resolve the moduli issue, as they send every supernatural bundle to a direct sum of the GL-equivariant supernatural bundles [**34**, Corollary 1.9].

In the case of Example 1.2, this is sufficient to categorify the decomposition. In fact, after applying an appropriate Fourier–Mukai transform Φ, we obtain

$$\Phi(\mathcal{E}) \cong \mathcal{E}' \oplus \mathcal{E}''^{\oplus 3}.$$

However, beyond the cases covered in [**34**, Corollary 1.9], there is no general procedure for producing a filtration from the various supernatural pieces.

CONJECTURE 2.2. *Fix a vector bundle $\mathcal{E}$ on $\mathbb{P}^n$. Let f be the root sequence corresponding to the first step of the Boij–Söderberg decomposition of $\mathcal{E}$ and let $\mathcal{F}_f$ be the equivariant supernatural bundle of type f constructed in* [**26**, Theorem 6.2].

Then there exists $r > 0$, a Fourier–Mukai transform Φ, and a short exact sequence

$$0 \to \mathcal{F}_f^{\oplus r} \to \Phi(\mathcal{E}) \to \mathcal{E}' \to 0$$

where:

(1) *The cohomology table of $\Phi(\mathcal{E})$ is a scalar multiple of the cohomology table of $\mathcal{E}$, and*
(2) $\gamma(\mathcal{E}) = \gamma(\mathcal{F}_f^{\oplus r}) + \gamma(\mathcal{E}')$.

In essence, the conjecture says that the Fourier–Mukai transforms constructed in [**34**] can be used to categorify the first step of the Boij–Söderberg decomposition of any vector bundle. If true, one could iterate this to categorify the entire decomposition.

We can ask for similar results for modules. Even finding a natural categorification in the single case of Example 1.1 would represent significant progress.

PROBLEM 2.3. *Find an operation, other than direct sums, which naturally clears denominators on the module side.*

3. Boij–Söderberg theory and the tails of infinite resolutions

The first major progress in extending Boij–Söderberg theory to other projective varieties was due to Eisenbud and Schreyer, who observed that the cone of cohomology tables depends only mildly on the projective scheme [**28**, Theorem 5]. In particular, they show that if $X \subseteq \mathbb{P}^n$ is a projective, d-dimensional subscheme which has an Ulrich sheaf, then the cone of cohomology tables on X (with respect to $\mathcal{O}_X(1)$) equals the cone of cohomology tables for $\mathbb{P}^d$. This provides further motivation for Eisenbud and Schreyer's question about the existence of Ulrich sheaves [**24**, p. 543].

QUESTION 3.1 (Eisenbud-Schreyer). *Does every $X \subseteq \mathbb{P}^n$ have an Ulrich sheaf?*

Eisenbud and Erman extended Boij–Söderberg theory to many other graded rings, but only if one restricts to Betti tables of perfect complexes with finite length homology [**19**, Theorem 0.8]. Their result is similar to the Eisenbud and Schreyer result, but with a linear Koszul complex replacing the Ulrich sheaf. Namely, they show that if R is a graded ring of dimension d which has d independent linear forms, then the cone of Betti tables of perfect complexes $\mathbf{F}$ where all homology modules of $\mathbf{F}$ have finite length lines up with the corresponding cone of Betti tables for $\Bbbk[x_1, \ldots, x_d]$.

Restricting to perfect complexes makes the situation much simpler, as one entirely avoids the complications involved in the study of infinite resolutions [**2, 52**]. While the cone of Betti tables of resolutions has been worked out for some other graded rings [**4, 42, 47**], these are all fairly simple rings which avoid many of the complexities of infinite resolutions since they have finite Cohen–Macaulay representation type (i.e., there are only finitely many isomorphism classes of indecomposable maximal Cohen–Macaulay modules). Some natural next cases here are other rings of finite Cohen–Macaulay representation type: quadric hypersurfaces and the Veronese surface.

In addition, recent work of Avramov, Gibbons, and Wiegand provides the first examples beyond the case of finite Cohen–Macaulay representation type. More than just the cone, they work out the full semigroup of Betti tables for graded, Gorenstein algebras with Hilbert series $1 + es + s^2$ for some e; when $e \geq 3$, these rings have wild representation type.

REMARK 3.2. In [**42, 47**], the extremal rays of the Boij–Söderberg cone are still pure free resolutions. These can be constructed for the quadric hypersurface $\Bbbk[x_0, \ldots, x_n]/(x_0^2)$ when $\Bbbk$ is a field of characteristic 0 by noting that this ring is the symmetric algebra on a super vector space with 1 odd variable x_0 and n even variables $x_1, \ldots, x_n$ and then applying [**22**, Theorem 0.1]. However, we have been unable to deform these resolutions to get pure free resolutions over higher rank quadrics.

An alternate approach, and one which is common in the study of infinite resolutions, would be to focus on the tails of the infinite resolutions. In other words, we could define a cone of Betti tables where we only focus on modules M which arise as "sufficiently high" syzygy modules. The notion of "sufficiently high" will depend on the context, though [**23**, §7] provides a notion in the case where R is a complete intersection ring, and this is a natural starting point.

PROBLEM 3.3. *Let R be a graded complete intersection $R = S/(f_1, \ldots, f_c)$ and let $\mathrm{B}^{tail}(R)$ be the cone of Betti tables $\beta(M)$ where M is a sufficiently high syzygy module. Describe $\mathrm{B}^{tail}(R)$.*

Again, there are many natural variants of this problem, such as where R is: a Golod ring, a ring defined by a toric or monomial ideal, or the homogeneous coordinate ring of a curve under a very positive embedding. In a different direction, see [**6**, Conjecture 1.6] for an open question about the local ring case.

4. Exact sequences

The long exact sequence in cohomology provides a central tool for studying the cohomology of a coherent sheaf, and it could thus be extremely fruitful to develop machinery that explores Boij–Söderberg theory in exact sequences. The main challenge is that cohomology tables (or Betti tables) are subadditive, but not necessarily additive, over short exact sequences.

Suppose we have a short exact sequence of coherent sheaves on $\mathbb{P}^n$

$$0 \to \mathcal{A} \to \mathcal{E} \to \mathcal{B} \to 0$$

where we fully understand the sheaf cohomology of $\mathcal{A}$ and $\mathcal{B}$. A good example to keep in mind is the case where $\mathcal{A}$ and $\mathcal{B}$ are supernatural bundles. After twisting by $\mathcal{O}(j)$ we get a long exact sequence in cohomology:

$$\cdots \to \mathrm{H}^i(\mathcal{E}(j)) \to \mathrm{H}^i(\mathcal{B}(j)) \to \mathrm{H}^{i+1}(\mathcal{A}(j)) \to \mathrm{H}^{i+1}(\mathcal{E}(j)) \to \cdots$$

If all connecting maps in cohomology are zero, then this maximizes all entries in the cohomology table of $\mathcal{E}$ in the sense that $\mathrm{H}^i(\mathcal{E}(j)) \cong \mathrm{H}^i(\mathcal{A}(j)) \oplus \mathrm{H}^i(\mathcal{B}(j))$. If this holds for all j then we have $\gamma(\mathcal{E}) = \gamma(\mathcal{A}) + \gamma(\mathcal{B})$. On the other hand, if some of the connecting maps were nonzero, then this reduces the cohomology table of $\gamma(\mathcal{E})$ via "consecutive cancellations", i.e., by simultaneously reducing $\mathrm{H}^i(\mathcal{E}(j))$ and $\mathrm{H}^{i+1}(\mathcal{E}(j))$ by the same amount.[2]

We thus expect some ambiguities in the cohomology table of an extension bundle like $\mathcal{E}$, and here is where Boij–Söderberg theory might help. Some patterns of consecutive cancellations will yield tables which lie inside the cone $C(\mathbb{P}^n)$ of genuine cohomology tables, whereas other patterns might lie outside of that cone. The following notation will be helpful.

NOTATION 4.1. *For a root sequence f we write σ_f for the supernatural cohomology table of rank one and with root sequence f. If f has one entry, so $f = (i)$, then we simply use σ_i.*

EXAMPLE 4.2. Consider a rank 10 extension bundle:

$$0 \to \mathcal{O}^5_{\mathbb{P}^1}(-2) \to \mathcal{E} \to \mathcal{O}^5_{\mathbb{P}^1}(2) \to 0.$$

The maximal possible cohomology table is if the extension splits:

$\cdots$	50	40	30	20	15	10	5	–	–	–	$\cdots$	
$\cdots$	–	–	–	5	10	15	20	30	40	50	$\cdots$	.

However, there might be some consecutive cancellations, and taking into account the symmetry imposed by Serre duality, the possibilities are:

$\cdots$	50	40	30	20	$15-a$	$10-b$	$5-a$	–	–	–	$\cdots$	
$\cdots$	–	–	–	$5-a$	$10-b$	$15-a$	20	30	40	50	$\cdots$	.

But which pairs a and b can actually arise? It takes a bit of work to answer this. If we want the cohomology table to remain inside the cone then we get:

$$10 - b \geq 2(5-a); \qquad \text{equivalently, } 2a \geq b.$$

This turns out to be the only constraint, and the range of possibilities for the cohomology table of $\mathcal{E}$ can thus be described by a triangle in the cone of cohomology tables, with corners corresponding to $(a, b) = (0,0), (5,5)$ and $(5, 10)$. The corners of that triangle correspond to sums of supernatural tables, as illustrated in Figure 2.

More generally, if $\mathcal{E}$ is an extension of two sheaves $\mathcal{A}$ and $\mathcal{B}$, then the possible cohomology tables of $\mathcal{E}$ is a polytope inside of $C(\mathbb{P}^n)$.

[2]The terminology of consecutive cancellations is common in the literature on Betti tables, but does not appear to have been used in discussion of sheaf cohomology.

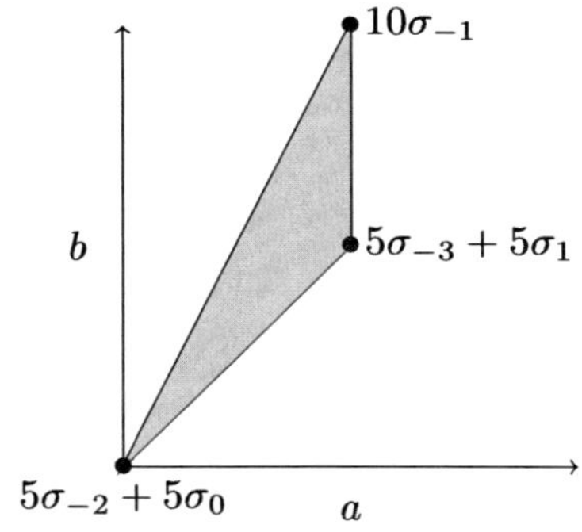

FIGURE 2. The potential cohomology tables for an extension bundle form a polytope in the cone of cohomology tables; with $\mathcal{E}$ as in Example 4.2, the polytope is a triangle.

PROBLEM 4.3. *Fix a short exact sequence of vector bundles*

$$0 \to \mathcal{A} \to \mathcal{E} \to \mathcal{B} \to 0.$$

(1) *Assume that $\mathcal{A}$ is supernatural of type f and $\mathcal{B}$ is supernatural of type f'. Describe the polytope of possible cohomology tables of $\mathcal{E}$ in terms of supernatural tables σ_g for various root sequences g.*
(2) *For more general bundles $\mathcal{A}$ and $\mathcal{B}$, describe the polytope of possible cohomology tables of $\mathcal{E}$ in terms of the Boij–Söderberg decompositions of $\mathcal{A}$ and $\mathcal{B}$.*

There are many natural variants of this question. For instance, one could aim to describe the potential cohomology tables of $\mathcal{A}$ in terms of $\mathcal{E}$ and $\mathcal{B}$; one could work with longer exact sequences; or one could pose analogous questions about Betti tables.

5. Boij–Söderberg theory over a DVR

Fix a coherent sheaf $\mathcal{F}$ on $\mathbb{P}^n_{\mathbb{Z}_p}$ that is flat over $\mathbb{Z}_p$. How much can the cohomology table of $\mathcal{F}$ "jump" when you pass from the generic point $\mathbb{Q}_p$ to the special fiber $\mathbb{F}_p$? Are there any restrictions other than having a constant Hilbert polynomial? A version of Boij–Söderberg theory with $\mathbb{Z}_p$-coefficients would offer insights into this question.

More generally, fix a DVR R and let $S := R[x_0, \ldots, x_n]$, where $\deg(x_i) = 1$ for all i. In place of finite length modules, consider graded S-modules which are finite rank, free R-modules.

PROBLEM 5.1. *Let $S = R[x_0, \ldots, x_n]$ with R a DVR. Describe:*

(1) *The cone of Betti tables of finitely generated, graded S-modules M which are flat and finitely generated as R-modules.*
(2) *The cone of cohomology tables of vector bundles on $\mathbb{P}^n_R$.*

The functorial approach in **[19]** offers a possible approach.

PROBLEM 5.2. *With notation as above:*

(1) *Extend the construction of the functor* Φ *from* [**19**, §2] *to a pairing of derived categories:*

$$\Phi_R\colon \mathrm{D}^b(R[x_0,\dots,x_n]) \times \mathrm{D}^b(\mathbb{P}^n_R) \to \mathrm{D}^b(R[t]).$$

(2) *Give explicit formulas, extending* [**19**, Theorem 2.3], *that relate the numerical invariants of* $\Phi_R(\mathbf{F},\mathcal{E})$ *to the numerical invariants of* $\mathbf{F}$ *and* $\mathcal{E}$.

Even more generally, one might allow R to be a regular local ring of dimension > 1. However when R is a DVR, the target of the functor Φ has global dimension 2. Since much more is known about Betti tables in this context, we expect that the case where R is a DVR is more tractable.

6. Non-commutative analogues

Consider the degree sequence $(0,1,3,4)$. The Herzog–Kühl equations state that any finite length module over a polynomial ring in 3 variables with a pure resolution of type $(0,1,3,4)$ is a multiple of the following table:

$$\begin{bmatrix} 1 & 2 & & \\ & & 2 & 1 \end{bmatrix}. \tag{2}$$

One can easily deduce that this Betti table is non-realizable: this would be the Betti table of $\Bbbk[x,y,z]/I$ where I is generated by two linear forms, which must then have a linear Koszul relation.

However, there is a way to realize this as the Betti table of a finite length module if we are willing to replace $\Bbbk[x,y,z]$ by another algebra. In particular, define a 3-dimensional Lie algebra $\mathfrak{H}$ (Heisenberg Lie algebra) with basis $\{x,y,z\}$ and the following multiplication

$$[x,y] = z, \qquad [x,z] = [y,z] = 0.$$

Recall that given any Lie algebra $\mathfrak{g}$, its universal enveloping algebra $\mathrm{U}(\mathfrak{g})$ is the tensor algebra on $\mathfrak{g}$ modulo the relations $x\otimes y - y\otimes x = [x,y]$ for all $x,y\in\mathfrak{g}$. In general, $\mathrm{U}(\mathfrak{g})$ is only a filtered algebra, and the associated graded algebra is the symmetric algebra $\mathrm{Sym}(\mathfrak{g})$ by the Poincaré–Birkhoff–Witt theorem. However, if $\mathfrak{g}$ is graded, then the same is true for $\mathrm{U}(\mathfrak{g})$. In our case, $\mathfrak{H}$ is graded via $\deg(x) = \deg(y) = 1$ and $\deg(z) = 2$. In this case, minimal free resolutions of graded modules are well-defined.

The Chevalley–Eilenberg complex (which becomes the usual Koszul complex under the PBW degeneration mentioned above) always gives a free resolution of the residue field, but is not necessarily minimal. For $\mathfrak{H}$, the Chevalley–Eilenberg complex looks like:

$$0 \to \mathrm{U}(\mathfrak{H})(-4) \to \begin{matrix} \mathrm{U}(\mathfrak{H})(-2) \\ \oplus \\ \mathrm{U}(\mathfrak{H})(-3)^{\oplus 2} \end{matrix} \to \begin{matrix} \mathrm{U}(\mathfrak{H})(-1)^{\oplus 2} \\ \oplus \\ \mathrm{U}(\mathfrak{H})(-2) \end{matrix} \to \mathrm{U}(\mathfrak{H}).$$

The two terms of degree 2 cancel since it corresponds to the redundancy $xy-yx=z$. So we get the following minimal free resolution:

$$0 \to \mathrm{U}(\mathfrak{H})(-4) \to \mathrm{U}(\mathfrak{H})(-3)^{\oplus 2} \to \mathrm{U}(\mathfrak{H})(-1)^{\oplus 2} \to \mathrm{U}(\mathfrak{H})$$

and hence we can realize (2).

This suggests that non-realizable integral points in the Boij–Söderberg cone may be realizable over $\mathrm{U}(\mathfrak{g})$ for a $\mathbb{Z}_{>0}$-graded Lie algebra $\mathfrak{g}$. Note that a finite-dimensional $\mathbb{Z}_{>0}$-graded Lie algebra $\mathfrak{g}$ is necessarily nilpotent since $[\mathfrak{g}, \mathfrak{g}_i] \subset \mathfrak{g}_{i+1}$. Also, a standard graded polynomial ring in n variables is $\mathrm{U}(\mathbb{k}^n)$ where $\mathbb{k}^n$ is the abelian Lie algebra of dimension n concentrated in degree 1.

In dimension 2, every nilpotent Lie algebra is abelian, and in dimension 3, the only possibilities are $\mathbb{k}^3$ and $\mathfrak{H}$ (to normalize, we insist that the generators of the Lie algebra consist of degree 1 elements). It is easy to see that every integral point on a pure ray in the Boij–Söderberg cone in 2 variables is realizable, so we offer the following question:

QUESTION 6.1. *For every degree sequence* (d_0, d_1, d_2, d_3), *and every integral point on the corresponding ray in the Boij–Söderberg cone, is there a finite length module either over* $\mathbb{k}[x, y, z]$ *or* $\mathrm{U}(\mathfrak{H})$ *whose Betti table is that integral point?*

Of course, there is a natural extension for any number of variables:

QUESTION 6.2. *For every degree sequence* $(d_0, \ldots, d_n)$, *and every integral point on the corresponding ray in the Boij–Söderberg cone, does there exist an* n*-dimensional* $\mathbb{Z}_{>0}$*-graded Lie algebra* $\mathfrak{g}$ *generated in degree 1, and a finite length module over* $\mathrm{U}(\mathfrak{g})$ *whose Betti table is that integral point?*

One source of examples for $\mathbb{Z}_{>0}$-graded Lie algebras are the nilpotent radicals of parabolic subalgebras of (split) reductive Lie algebras. For example, parabolic subalgebras of $\mathfrak{gl}_n$ are subalgebras of block upper-triangular matrices (where the block sizes are fixed) the nilpotent radical is the subalgebra where the diagonal blocks are identically 0. Over a field of characteristic 0, Kostant's version of the Borel–Weil–Bott theorem (see [**54**, §2] for a convenient reference and combinatorial details for the symplectic Lie algebra which will be mentioned below) calculates the Tor groups of the restriction of an irreducible representation from the reductive Lie algebra to the nilpotent one. In fact, the EFW complexes are a special case of Kostant's calculation where the reductive Lie algebra is $\mathfrak{gl}_{n+1}$ and we take the nilpotent radical of the subalgebra of block upper-triangular matrices with block sizes 1 and n (in this case, the nilpotent radical is abelian).

Furthermore, one can construct many kinds of pure resolutions using this construction, though not necessarily all degree sequences are realizable (in the EFW case, they are). We point out that $\mathfrak{H}$ can be realized as the nilpotent radical for a parabolic subalgebra of the symplectic Lie algebra $\mathfrak{sp}_4$ (see [**54**, §2.2]), and using this representation-theoretic perspective, one can realize more integral points which are not realizable over $\mathbb{k}[x, y, z]$ than the example presented above. Details will appear in forthcoming work of the second author.

REMARK 6.3. We are appealing to the fact that $\mathrm{U}(\mathfrak{g})$ "looks like" a graded polynomial algebra, which is made precise by the Poincaré–Birkhoff–Witt theorem. One could also study other non-commutative algebras that look like graded polynomial algebras. One desirable property is finite global dimension. As a starting point, one may consider Artin–Schelter algebras, see [**1**].

7. Connection with Stillman's Conjecture

Let $\widehat{S} := \mathbb{k}[x_0, x_1, x_2, \ldots]$ and fix positive integers $e_1, \ldots, e_r$. Let I be an ideal with generators in degrees $e_1, \ldots, e_r$. Stillman's Conjecture asks whether there is an upper bound for the projective dimension of $\widehat{S}/I$ which depends only on $e_1, \dots, e_r$; in particular, the bound should not depend on the ideal I [**53**, Problem 3.14].

Caviglia has shown that a positive answer to Stillman's Conjecture is equivalent to the existence of an upper bound on the regularity of $\widehat{S}/I$ which depends only on $e_1, \dots, e_r$ [**52**, Theorem 29.5]. This allows us to rephrase Stillman's conjecture as a finiteness statement about Betti tables. For simplicity, we focus on the case where all of the e_i have the same value e. Then Stillman's conjecture is equivalent to the following:

CONJECTURE 7.1 (Alternate Stillman). *There are only finitely many Betti tables $\beta(M)$ of $\widehat{S}$-modules M satisfying:*

$$\beta_{0,j}(M) = \begin{cases} 1 & j = 0 \\ 0 & j \neq 0 \end{cases} \quad \text{and} \quad \beta_{1,j}(M) = \begin{cases} r & j = e \\ 0 & j \neq e \end{cases}.$$

Note that any finitely presented $\widehat{S}$-module has a presentation matrix that involves only finitely many variables. Hence the Betti table of every $\widehat{S}$-module can be decomposed as a sum of the Betti tables of Cohen–Macaulay modules with pure resolutions. Thus, by taking the union of the cones of Betti tables of resolutions over $\Bbbk[x_0, x_1, \dots, x_n]$ as $n \to \infty$, we will obtain the cone $\mathrm{B}(\widehat{S})$ of resolutions of Betti tables of finitely presented $\widehat{S}$-modules.

One could then imagine trying to prove Stillman's Conjecture via Boij–Söderberg theory by showing that there are only finitely many *lattice points* in $\mathrm{B}(\widehat{S})$ that satisfy the conditions of the conjecture. However, this turns out to be false.

PROPOSITION 7.2. *Fix any $e \geq 1$ and $r \geq 2$. Then there exist infinitely many lattice points $D \in \mathrm{B}(\widehat{S})$ which satisfy*

$$\beta_{0,j}(D) = \begin{cases} 1 & j = 0 \\ 0 & j \neq 0 \end{cases} \quad \text{and} \quad \beta_{1,j}(D) = \begin{cases} r & j = e \\ 0 & j \neq e \end{cases}.$$

In other words, there exist infinitely many lattice points in $D \in \mathrm{B}(\widehat{S})$ that look like the Betti table of an algebra generated by r forms of degree e. In fact, there exist infinitely many D satisfying the above and which also look like pure resolutions.

Yet this may not be the end of the story. Remark 7.4 shows that the "virtual" Betti tables constructed in the proof of the proposition cannot arise as the Betti table of an actual module, and we know of no other such infinite families.

QUESTION 7.3. *Ignoring pure Betti table counterexamples, like those constructed in Proposition 7.2, do there exist finitely many lattice points in $D \in \mathrm{B}(\widehat{S})$ that look like the Betti table of an algebra generated by r forms of degree e?*

A positive answer to this question would certainly be interesting as it would imply Stillman's Conjecture. But a negative answer would also be interesting for a different reason: assuming that we also believe Stillman's Conjecture, a negative answer to Question 7.3 would suggest that almost all lattice points in the cone of Betti tables that "look like" the Betti table of a cyclic module do not come from an actual Betti table. In other words, a negative answer would suggest that the "noise" of the fake Betti tables inside the cone overwhelms the "signal" of the actual Betti tables, at least for some questions.

PROOF OF PROPOSITION 7.2 (SKETCH). Fix $p \geq 0$ and let $n := r + p(r-1)$. For each such p, we will define a degree sequence $d^{(p)} \in \mathbb{Z}^{n+1}$, and we will show that for infinitely many of these degree sequences, the smallest integral pure diagram of type $d^{(p)}$ satisfies the conditions of the proposition. We set $d_0^{(p)} = 0$ and $d_1^{(p)} = e$. For $i = 2, \dots, n$ we then set $d_i^{(p)} := e(p+i)$.

We let π be the normalized pure diagram of type $d^{(p)}$, so that $\beta_{0,0}(\pi) = 1$ by assumption. It follows from [**10**, §2.1] that

$$\beta_{i,d_i^{(p)}} = \frac{\prod_{j\neq 0}^{n} d_j^{(p)}}{\prod_{j\neq i}^{n} |d_i^{(p)} - d_j^{(p)}|}.$$

To complete the proof we must show that each of these expressions is an integer. Since every entry of $d^{(p)}$ is divisible by e, we can reduce to the case where $e = 1$. A detailed but elementary computation then confirms that each of these expressions is an integer. □

REMARK 7.4. For a given $r \geq 2$ and $p > 0$, the virtual Betti table constructed in the proof of Proposition 7.2 would correspond to a module with codimension $r + p(r-1)$. But this would be the Betti table of a module $\beta(S/I)$ where I is an ideal with r generators. In particular, the codimension of S/I is bounded above by r, which is a contradiction. Thus, the pure Betti tables constructed in the proof of Proposition 7.2 cannot correspond to an actual Betti table. Of course, some scalar multiple of each table does come from an actual Betti table.

EXAMPLE 7.5. We consider the examples from the proof of Proposition 7.2 in the case $e = 2$ and $r = 3$. These would correspond to algebras S/I where I is generated by three quadrics. For any prime p, there is a pure diagram with degree sequence

$$d^{(p)} := (0, 2, 4+2p, 6+2p, \dots, 6+6p) \in \mathbb{Z}^{2p+4}$$

that looks like the Betti table of such an algebra. With $p = 3$, this yields

$$\begin{pmatrix} 1 & & & & & & & \\ & 3 & & & & & & \\ & & & & & & & \\ & & & & & & & \\ & & 42 & & & & & \\ & & & 126 & & & & \\ & & & & 168 & & & \\ & & & & & 120 & & \\ & & & & & & 45 & \\ & & & & & & & 7 \end{pmatrix}.$$

Yet a result of Eisenbud and Huneke implies that such an algebra has projective dimension at most 4 [**50**, Theorem 3.1], and so Stillman's Conjecture is true in this case, despite the presence of these "fake" Betti tables.

8. Extremal rays

A number of questions remain about the extremal rays. To begin with, Eisenbud, Fløystad, and Weyman conjecture that every sufficiently large integral point on an extremal ray comes from an actual Betti table [**22**, Conjecture 6.1]. This conjecture remains open, though it is known to be false for interior rays in the cone [**20**, Example 1.7]. One could ask a similar question for cohomology tables.

In addition, GL-equivariance plays a serendipitous role in Boij–Söderberg theory. We restrict to characteristic zero and to the finite length cone $\mathrm{B}^{n+1}(S)$ and the vector bundle cone $\mathrm{C}_{\mathrm{vb}}(\mathbb{P}^n)$. One can construct each extremal ray of these two cones by a GL_{n+1}-equivariant object: see [**22**, §3] for the equivariant pure resolutions and [**25**, Theorem 6.2] for the equivariant supernatural bundles. An immediate corollary is that *every* ray in $\mathrm{B}^{n+1}(S)$ or $\mathrm{C}_{\mathrm{vb}}(\mathbb{P}^n)$ can be realized by an equivariant object:

COROLLARY 8.1. *Let* $\Bbbk$ *be a field of characteristic* 0 *and let* $S = \Bbbk[x_0, \ldots, x_n]$.

(1) *For any finite length, graded,* S*-module* M*, there exists a finite length, graded, and* GL_{n+1}*-equivariant module* N *such that* $\beta(N)$ *is a scalar multiple of* $\beta(M)$.
(2) *For any vector bundle* $\mathcal{E}$ *on* $\mathbb{P}^n_k$*, there is a* GL_{n+1}*-equivariant vector bundle* $\mathcal{F}$ *on* $\mathbb{P}^n$ *such that* $\gamma(\mathcal{E})$ *is a scalar multiple of* $\gamma(\mathcal{F})$.

QUESTION 8.2. *Is there an intrinsic reason why every Betti table of a finite length module, and every cohomology table of a vector bundle, can be realized (up to scalar multiple) by an equivariant object? Is there a simpler proof of this fact?*

9. More topics

9.1. Toric Boij–Söderberg theory. Many researchers have observed that it would be natural to try to extend Boij–Söderberg theory to toric varieties and free complexes over their multigraded Cox rings. Eisenbud and Schreyer have made a conjecture about the extremal rays of the cone of cohomology on $\mathbb{P}^1 \times \mathbb{P}^1$ [**28**]; several authors have partial results for cones of Betti tables over a polynomial ring with $\mathbb{Z}^2$ or $\mathbb{Z}^s$ grading [**5**, **8**, **35**]; and [**19**, §11] develops some of the duality aspects for toric varieties. Yet even taken together, these results are far from providing a complete picture for any toric variety, and it appears to be quite challenging to develop such a picture even for $\mathbb{P}^1 \times \mathbb{P}^1$. A different possible direction is to extend Boij and Smith's work on cones of Hilbert functions to the multigraded case; see their remarks in [**9**, p. 10,317].

9.2. Equivariant Boij–Söderberg theory. Since pure free resolutions over a polynomial ring have equivariant realizations over a field of characteristic 0, one can ask whether there is a meaningful description of the cone of "equivariant Betti tables". See [**55**, §4] for the setup and partial progress. A natural variant is to work on a Grassmannian and to focus on GL-equivariant free resolutions and GL-equivariant cohomology tables. There is work in this direction due to Ford–Levinson [**40**] and Ford–Levinson–Sam [**41**].

9.3. Monomial ideals and combinatorics. A number of authors have explored the boundary between Boij–Söderberg theory and combinatorial commutative algebra. One remarkable recent result is Mayes-Tang's proof [**48**] of Engström's Conjecture on stabilization of decompositions of powers of monomial ideals [**30**]. This raises the question of whether similar asymptotic stabilization results can be expected in other contexts.

In another direction, Fløystad [**38**] has proposed a vast generalization of Boij–Söderberg cones by moving from Betti tables to certain triples of homological data which are defined on the category of squarefree monomial modules. This in turn has relations to the cone of hypercohomology tables of complexes of coherent sheaves and Tate resolutions [**37**], and Fløystad formulates many fascinating questions in those papers.

See [**15**, **31**, **43**, **51**] for some other work that mixes Boij-Söderberg theory and combinatorics.

9.4. Asymptotic Boij–Söderberg decompositions of Veroneses. Let X be a smooth projective variety with a very ample divisor A. For any $d > 0$ we can embed $X \subseteq \mathbb{P}^{r_d}$ by the complete linear series $|dA|$ and study the syzygies of X. Ein and Lazarsfeld ask: as $d \to \infty$, which Betti numbers of X will be nonzero? They show that in the limit, the answer only depends on the dimension of the variety of X, and not on any more refined geometric properties about X [**17**]. A natural followup is to ask about the asymptotic behavior of the Boij–Söderberg decompositions of these Betti tables. See [**16**, Problem 3.6] and [**33**].

Acknowledgments

We would like to thank the participants of the Boij–Söderberg Theory group at the Bootcamp for the 2015 Algebraic Geometry Summer Research Institute, whose ideas and questions informed this note. Several of these questions were developed in speculative conversations with colleagues, and we also thank those individuals, including: Christine Berkesch Zamaere, Mats Boij, David Eisenbud, Courtney Gibbons, Jason McCullough, Frank-Olaf Schreyer, Greg Smith, Jerzy Weyman, and many more.

References

[1] Michael Artin and William F. Schelter, *Graded algebras of global dimension* 3, Adv. in Math. **66** (1987), no. 2, 171–216. MR917738

[2] Luchezar L. Avramov, *Infinite free resolutions*, Six lectures on commutative algebra (Bellaterra, 1996), Progr. Math., vol. 166, Birkhäuser, Basel, 1998, pp. 1–118. MR1648664

[3] Luchezar L. Avramov and Ragnar-Olaf Buchweitz, *Lower bounds for Betti numbers*, Compositio Math. **86** (1993), no. 2, 147–158. MR1214454

[4] Christine Berkesch, Jesse Burke, Daniel Erman, and Courtney Gibbons, *The cone of Betti diagrams over a hypersurface ring of low embedding dimension*, J. Pure Appl. Algebra **216** (2012), no. 10, 2256–2268. MR2925819

[5] Christine Berkesch, Daniel Erman, and Manoj Kummini, *Three flavors of extremal Betti tables*, Commutative algebra, Springer, New York, 2013, pp. 99–121. MR3051372

[6] Christine Berkesch, Daniel Erman, Manoj Kummini, and Steven V. Sam, *Shapes of free resolutions over a local ring*, Math. Ann. **354** (2012), no. 3, 939–954. MR2983074

[7] Christine Berkesch Zamaere, Daniel Erman, Manoj Kummini, and Steven V. Sam, *Tensor complexes: multilinear free resolutions constructed from higher tensors*, J. Eur. Math. Soc. (JEMS) **15** (2013), no. 6, 2257–2295. MR3120743

[8] Mats Boij and Gunnar Fløystad, *The cone of Betti diagrams of bigraded Artinian modules of codimension two*, Combinatorial aspects of commutative algebra and algebraic geometry, Abel Symp., vol. 6, Springer, Berlin, 2011, pp. 1–16. MR2810421

[9] Mats Boij and Gregory G. Smith, *Cones of Hilbert functions*, Int. Math. Res. Not. IMRN **20** (2015), 10314–10338. MR3455868

[10] Mats Boij and Jonas Söderberg, *Graded Betti numbers of Cohen-Macaulay modules and the multiplicity conjecture*, J. Lond. Math. Soc. (2) **78** (2008), no. 1, 85–106. MR2427053

[11] Mats Boij and Jonas Söderberg, *Betti numbers of graded modules and the multiplicity conjecture in the non-Cohen-Macaulay case*, Algebra Number Theory **6** (2012), no. 3, 437–454. MR2966705

[12] Adam Boocher, *Free resolutions and sparse determinantal ideals*, Math. Res. Lett. **19** (2012), no. 4, 805–821. MR3008416

[13] David A. Buchsbaum and David Eisenbud, *Algebra structures for finite free resolutions, and some structure theorems for ideals of codimension* 3, Amer. J. Math. **99** (1977), no. 3, 447–485. MR0453723

[14] H. Charalambous and E. G. Evans Jr., *Problems on Betti numbers of finite length modules*, Free resolutions in commutative algebra and algebraic geometry (Sundance, UT, 1990), Res. Notes Math., vol. 2, Jones and Bartlett, Boston, MA, 1992, pp. 25–33. MR1165315

[15] David Cook II, *The structure of the Boij-Söderberg posets*, Proc. Amer. Math. Soc. **139** (2011), no. 6, 2009–2015. MR2775378

[16] Lawrence Ein, Daniel Erman, and Robert Lazarsfeld, *Asymptotics of random Betti tables*, J. Reine Angew. Math. **702** (2015), 55–75. MR3341466

[17] Lawrence Ein and Robert Lazarsfeld, *Asymptotic syzygies of algebraic varieties*, Invent. Math. **190** (2012), no. 3, 603–646. MR2995182

[18] David Eisenbud, *The geometry of syzygies*, Graduate Texts in Mathematics, vol. 229, Springer-Verlag, New York, 2005. A second course in commutative algebra and algebraic geometry. MR2103875

[19] David Eisenbud and Daniel Erman, *Categorified duality in Boij–Söderberg theory and invariants of free complexes* (2012). `arXiv:1205.0449v2`.

[20] David Eisenbud, Daniel Erman, and Frank-Olaf Schreyer, *Filtering free resolutions*, Compos. Math. **149** (2013), no. 5, 754–772. MR3069361

[21] David Eisenbud, Gunnar Fløystad, and Frank-Olaf Schreyer, *Sheaf cohomology and free resolutions over exterior algebras*, Trans. Amer. Math. Soc. **355** (2003), no. 11, 4397–4426. MR1990756

[22] David Eisenbud, Gunnar Fløystad, and Jerzy Weyman, *The existence of equivariant pure free resolutions* (English, with English and French summaries), Ann. Inst. Fourier (Grenoble) **61** (2011), no. 3, 905–926. MR2918721

[23] David Eisenbud and Irena Peeva, *Matrix Factorizations for Complete Intersections and Minimal Free Resolutions* (2013). `arXiv:1306.2615v4`.

[24] David Eisenbud, Frank-Olaf Schreyer, and Jerzy Weyman, *Resultants and Chow forms via exterior syzygies*, J. Amer. Math. Soc. **16** (2003), no. 3, 537–579. MR1969204

[25] David Eisenbud and Frank-Olaf Schreyer, *Betti numbers of graded modules and cohomology of vector bundles*, J. Amer. Math. Soc. **22** (2009), no. 3, 859–888. MR2505303

[26] David Eisenbud and Frank-Olaf Schreyer, *Cohomology of coherent sheaves and series of supernatural bundles*, J. Eur. Math. Soc. (JEMS) **12** (2010), no. 3, 703–722. MR2639316

[27] Frank-Olaf Schreyer and David Eisenbud, *Betti numbers of syzygies and cohomology of coherent sheaves*, Proceedings of the International Congress of Mathematicians. Volume II, Hindustan Book Agency, New Delhi, 2010, pp. 586–602. MR2827810

[28] David Eisenbud and Frank-Olaf Schreyer, *Boij-Söderberg theory*, Combinatorial aspects of commutative algebra and algebraic geometry, Abel Symp., vol. 6, Springer, Berlin, 2011, pp. 35–48. MR2810424

[29] David Eisenbud and Frank-Olaf Schreyer, *The banks of the cohomology river*, Kyoto J. Math. **53** (2013), no. 1, 131–144. MR3049309

[30] Alexander Engström, *Decompositions of Betti diagrams of powers of monomial ideals: a stability conjecture*, Combinatorial methods in topology and algebra, Springer INdAM Ser., vol. 12, Springer, Cham, 2015, pp. 37–40. MR3467322

[31] Alexander Engström and Matthew T. Stamps, *Betti diagrams from graphs*, Algebra Number Theory **7** (2013), no. 7, 1725–1742. MR3117505

[32] Daniel Erman, *A special case of the Buchsbaum-Eisenbud-Horrocks rank conjecture*, Math. Res. Lett. **17** (2010), no. 6, 1079–1089. MR2729632

[33] D. Erman, *The Betti table of a high-degree curve is asymptotically pure*, Recent advances in algebraic geometry, London Math. Soc. Lecture Note Ser., vol. 417, Cambridge Univ. Press, Cambridge, 2015, pp. 200–206. MR3380450

[34] Daniel Erman and Steven V Sam, *Supernatural analogues of Beilinson monads*, Compos. Math. `arXiv:1506.07558v2`.

[35] Gunnar Fløystad, *The linear space of Betti diagrams of multigraded Artinian modules*, Math. Res. Lett. **17** (2010), no. 5, 943–958. MR2727620

[36] Gunnar Fløystad, *Boij-Söderberg theory: introduction and survey*, Progress in commutative algebra 1, de Gruyter, Berlin, 2012, pp. 1–54. MR2932580

[37] Gunnar Fløystad, *Zipping Tate resolutions and exterior coalgebras*, J. Algebra **437** (2015), 249–307. MR3351965

[38] Gunnar Fløystad, *Triplets of pure free squarefree complexes*, J. Commut. Algebra **5** (2013), no. 1, 101–139. MR3084124

[39] Gunnar Fløystad, Jason McCullough, and Irena Peeva, *Three themes of syzygies*, Bull. Amer. Math. Soc. (N.S.) **53** (2016), no. 3, 415–435. MR3501795

[40] Nic Ford and Jake Levinson, *Foundations of Boij–Söderberg Theory for Grassmannians*. `arXiv:1609.03446`.

[41] Nic Ford, Jake Levinson, and Steven V Sam, *Towards Boij–Söderberg theory for Grassmannians: the case of square matrices*. `arXiv:1608.04058`.

[42] Iulia Gheorghita and Steven V Sam, *The cone of Betti tables over three non-collinear points in the plane*, J. Commut. Algebra **8** (2016), no. 4, 537–548. MR3566529

[43] Courtney Gibbons, Jack Jeffries, Sarah Mayes, Claudiu Raicu, Branden Stone, and Bryan White, *Non-simplicial decompositions of Betti diagrams of complete intersections*, J. Commut. Algebra **7** (2015), no. 2, 189–206. MR3370483

[44] Daniel R. Grayson and Michael E. Stillman, *Macaulay 2, a software system for research in algebraic geometry*. Available at `http://www.math.uiuc.edu/Macaulay2/`.

[45] Robin Hartshorne, *Algebraic vector bundles on projective spaces: a problem list*, Topology **18** (1979), no. 2, 117–128. MR544153

[46] Jürgen Herzog and Hema Srinivasan, *Bounds for multiplicities*, Trans. Amer. Math. Soc. **350** (1998), no. 7, 2879–2902. MR1458304

[47] Manoj Kummini and Steven V. Sam, *The cone of Betti tables over a rational normal curve*, Commutative algebra and noncommutative algebraic geometry. Vol. II, Math. Sci. Res. Inst. Publ., vol. 68, Cambridge Univ. Press, New York, 2015, pp. 251–264. MR3496870

[48] Sarah Mayes-Tang, *Stabilization of Boij-Söderberg decompositions of ideal powers*. `arXiv:1509.08544v1`.

[49] Jason McCullough, *A polynomial bound on the regularity of an ideal in terms of half of the syzygies*, Math. Res. Lett. **19** (2012), no. 3, 555–565. MR2998139

[50] Jason McCullough and Alexandra Seceleanu, *Bounding projective dimension*, Commutative algebra, Springer, New York, 2013, pp. 551–576. MR3051385

[51] Uwe Nagel and Stephen Sturgeon, *Combinatorial interpretations of some Boij-Söderberg decompositions*, J. Algebra **381** (2013), 54–72. MR3030509

[52] Irena Peeva, *Graded syzygies*, Algebra and Applications, vol. 14, Springer-Verlag London, Ltd., London, 2011. MR2560561

[53] Irena Peeva and Mike Stillman, *Open problems on syzygies and Hilbert functions*, J. Commut. Algebra **1** (2009), no. 1, 159–195. MR2462384

[54] Steven V. Sam, *Homology of analogues of Heisenberg Lie algebras*, Math. Res. Lett. **22** (2015), no. 4, 1223–1241. MR3391884

[55] Steven V. Sam and Jerzy Weyman, *Pieri resolutions for classical groups*, J. Algebra **329** (2011), 222–259. MR2769324

Department of Mathematics, University of Wisconsin, Madison, Wisconsin 53706
E-mail address: derman@math.wisc.edu
URL: http://www.math.wisc.edu/~derman/

Department of Mathematics, University of Wisconsin, Madison, Wisconsin 53706
E-mail address: svs@math.wisc.edu
URL: http://www.math.wisc.edu/~svs/

A primer for unstable motivic homotopy theory

Benjamin Antieau and Elden Elmanto

Abstract. In this expository article, we give the foundations, basic facts, and first examples of unstable motivic homotopy theory with a view towards the approach of Asok-Fasel to the classification of vector bundles on smooth complex affine varieties. Our focus is on making these techniques more accessible to algebraic geometers.

Contents

1. Introduction

This primer is intended to serve as an introduction to the basic facts about Morel and Voevodsky's motivic, or $\mathbb{A}^1$, homotopy theory [**MV99**], [**Voe98**], with a focus on the unstable part of the theory. It was written following a week-long summer school session on this topic led by Antieau at the University of Utah in July 2015. The choice of topics reflects what we think might be useful for algebraic geometers interested in learning the subject.

In our view, the starting point of the development of unstable motivic homotopy theory is the resolution of Serre's conjecture by Quillen [**Qui76**] and Suslin [**Sus76**].

2010 *Mathematics Subject Classification.* Primary 13C10, 14F42, 55R50, 55S45.

Key words and phrases. Vector bundles, projective modules, motivic homotopy theory, Postnikov systems, algebraic K-theory.

The first author was supported by NSF Grant DMS-1461847.

The second author was supported by NSF Grant DMS-1508040.

Serre asked in [**Ser55**] whether every finitely generated projective module over $k[x_1, \dots, x_n]$ is free when k is a field. Put another way, the question is whether

$$\mathrm{Vect}_r(\operatorname{Spec} k) \to \mathrm{Vect}_r(\mathbb{A}^n_k)$$

is a bijection for all $n \geq 1$, where $\mathrm{Vect}_r(X)$ denotes the set of isomorphism classes of rank r vector bundles on X. Quillen and Suslin showed that this is true and in fact proved the analogous statement when k is replaced by a Dedekind domain. This suggested the following conjecture.

CONJECTURE 1.1 (Bass-Quillen). *Let X be a regular noetherian affine scheme of finite Krull dimension. Then, the pullback map* $\mathrm{Vect}_r(X) \to \mathrm{Vect}_r(X \times \mathbb{A}^n)$ *is a bijection for all $r \geq 1$ and all $n \geq 1$.*

The Bass-Quillen conjecture has been proved in many cases, but not yet in full generality. Lindel [**Lin81**] prove the conjecture when X is essentially of finite type over a field, and Popescu proved it when X is the spectrum of an unramified regular local ring (see [**Swa98**, Theorem 2.2]). Piecing these results together one can, for example, allow X to be the spectrum of a ring with the property that all its localizations at maximal ideals are smooth over a Dedekind ring with a perfect residue field see [**AHW15a**, Theorem 5.2.1]. For a survey of other results in this direction, see [**Lam06**, Section VIII.6].

If X is a reasonable topological space, such as a manifold, simplicial complex, or CW complex, then there are also bijections $\mathrm{Vect}_r^{\mathrm{top}}(X) \to \mathrm{Vect}_r^{\mathrm{top}}(X \times I^1)$, where I^1 is the unit interval and $\mathrm{Vect}_r^{\mathrm{top}}$ denotes the set of isomorphism classes of rank r topological complex vector bundles on X. Thus, the Quillen-Suslin theorem and the Bass-Quillen conjecture suggest that there might be a homotopy theory for schemes in which the affine line $\mathbb{A}^1$ plays the role of the contractible unit interval.

Additional evidence for this hypothesis is provided by the fact that many important cohomology theories for smooth schemes are $\mathbb{A}^1$-invariant. For example, the pullback maps in Chow groups

$$\mathrm{CH}^*(X) \to \mathrm{CH}^*(X \times_k \mathbb{A}^1),$$

Grothendieck groups

$$\mathrm{K}_0(X) \to \mathrm{K}_0(X \times_k \mathbb{A}^1),$$

and étale cohomology groups

$$\mathrm{H}^*_{\text{ét}}(X, \mu_\ell) \to \mathrm{H}^*_{\text{ét}}(X \times_k \mathbb{A}^1, \mu_\ell)$$

are isomorphisms when X is smooth over a field k and ℓ is invertible in k.

Now, we should note immediately, that the functor $\mathrm{Vect}_r : \mathrm{Sm}_k^{\mathrm{op}} \to \mathrm{Sets}$ is *not* itself $\mathbb{A}^1$-invariant. Indeed, there are vector bundles on $\mathbb{P}^1 \times_k \mathbb{A}^1$ that are not pulled back from $\mathbb{P}^1$. The reader can construct a vector bundle on the product such that the restriction to $\mathbb{P}^1 \times \{1\}$ is a non-trivial extension E of $\mathcal{O}(1)$ by $\mathcal{O}(-1)$ while the restriction to $\mathbb{P}^1 \times \{0\}$ is $\mathcal{O}(-1) \oplus \mathcal{O}(1)$. Surprisingly, for affine schemes, this proves not to be a problem: forcing Vect_r to be $\mathbb{A}^1$-invariant produces an object which still has the correct values on smooth affine schemes.

The construction of (unstable) motivic homotopy theory over a quasi-compact and quasi-separated scheme S takes three steps. The first stage is a homotopical version of the process of passing from a category of schemes to the topos of presheaves on the category. Specifically, one enlarges the class of spaces from Sm_S, the category of smooth schemes over S, to the category of presheaves of simplicial sets sPre(Sm$_S$) on Sm$_S$. An object X of sPre(Sm$_S$) is a functor $X : \mathrm{Sm}_S^{\mathrm{op}} \to \mathrm{sSets}$, where sSets is the category of simplicial sets, one model for the homotopy theory of CW complexes. Presheaves of sets give examples of simplicial presheaves by viewing a set as a discrete space. There is a Yoneda embedding $\mathrm{Sm}_S \to \mathrm{sPre}(\mathrm{Sm}_S)$ as usual. In the next stage, one imposes a descent condition, namely focusing on those presheaves that satisfy the appropriate homotopical version of Nisnevich descent. We note that one can construct motivic homotopy theory with other topologies, as we will do later in Section 5. The choice of Nisnevich topology is motivated by the fact that it is the coarsest topology where we can prove the purity theorem (Section 7) and the finest where we can prove representability of K-theory (Section 6.1). The result is a homotopy theory enlarging the category of smooth schemes over S but which does not carry any information about the special role $\mathbb{A}^1$ is to play. In the third and final stage, the projection maps $X \times_S \mathbb{A}^1 \to X$ are formally inverted.

In practice, care must be taken in each stage; the technical framework we use in this paper is that of model categories, although one could equally use ∞-categories instead, as has been done recently by Robalo [**Rob15**]. Model categories, Quillen functors (the homotopical version of adjoint pairs of functors), homotopy limits and colimits, and Bousfield's theory of localization are all explained in the lead up to the construction of the motivic homotopy category.

When S is regular and noetherian, algebraic K-theory turns out to be representable in $\mathrm{Spc}_S^{\mathbb{A}^1}$, as are many of its variants. A pleasant surprise however is that despite the fact that Vect_r is not $\mathbb{A}^1$-invariant on all of Sm_S, its $\mathbb{A}^1$-localization still has the correct values on smooth affine schemes over k. This is a crucial result of Morel [**Mor12**, Chapter 8], which was simplified in the Zariski topology by Schlichting [**Sch15**], and simplified and generalized by Asok, Hoyois, and Wendt [**AHW15a**]. This fact is at the heart of applications of motivic homotopy theory to the classification of vector bundles on smooth affine complex varieties by Asok and Fasel.

We describe now the contents of the paper. As mentioned above, the document below reflects topics the authors decided should belong in a first introduction to $\mathbb{A}^1$-homotopy theory, especially for people coming from algebraic geometry. Other surveys in the field which focus on different aspects of the theory include [**DLØ$^+$07**], [**Lev08**], [**Lev16**], [**Mor06**]; a textbook reference for the ideas covered in this survey is [**Mor12**]. Voevodsky's ICM address pays special attention to the topological motivation for the theory in [**Voe98**].

Some of these topics were the focus of Antieau's summer school course at the AMS Summer Institute in Algebraic Geometry at the University of Utah in July 2015. This includes the material in Section 2 on topological vector bundles. This section is meant to explain the power of the Postnikov tower approach to classification problems and entice the reader to dream of the possibilities were this possible in algebraic geometry.

The construction of the motivic homotopy category is given in Section 3 after an extensive introduction to model categories, simplicial presheaves, and the Nisnevich topology. Other topics are meant to fill gaps in the literature, while simultaneously illustrating the techniques common to the field. Section 4 establishes the basic properties of motivic homotopy theory over S. It is meant to be a kind of cookbook and contains many examples, exercises, and computations. In Section 5 we define and give examples of classifying spaces BG for linear algebraic groups G, and perform some calculations of their homotopy sheaves. The answers will involve algebraic K-theory which is discussed in Section 6. Following [**MV99**], with some modifications, we give a self-contained proof that algebraic K-theory is representable in the $\mathbb{A}^1$-homotopy category, and we identify its representing object as the $\mathbb{A}^1$-homotopy type of a classifying space BGL_∞. In Section 7, we prove the critical purity theorem which is the source of Gysin sequences. A brief vista at the end of the paper, in Section 8, illustrates how all of this comes together to classify vector bundles on smooth affine schemes. Finally, in Section 9, we gather some miscellaneous additional exercises.

Many things are not in this paper. We view the biggest omission as the exclusion of a presentation of the first non-zero homotopy sheaves of punctured affine spaces. Morel proved that

$$\pi_n^{\mathbb{A}^1}(\mathbb{A}^{n+1} - \{0\}) \cong \mathbf{K}_{n+1}^{\mathrm{MW}},$$

the $(n+1)$st unramified Milnor-Witt K-theory sheaf where $n \geq 1$. A proof may be found in [**Mor12**, Chapter 6].

Other topics we would include granted unlimited time include stable motivic homotopy theory, and in particular the motivic spectral sequence, the stable connectivity theorem of Morel, the theory of algebraic cobordism due to Levine-Morel [**LM07**], motivic cohomology and the work of Voevodsky and Rost on the Bloch-Kato conjecture, and the work [**DI10**] of Dugger and Isaksen on the motivic Adams spectral sequence.

Acknowledgements. These notes were commissioned by the organizers of the Graduate Student Bootcamp for the 2015 Algebraic Geometry Summer Research Institute. Each mentor at the bootcamp gave an hour-long lecture and then worked with a small group of graduate students and postdocs over the course of the week to understand their topic in greater detail. We would like to the thank the organizers, İzzet Coşkun, Tommaso de Fernex, Angela Gibney, and Max Lieblich, for creating a wonderful atmosphere in which to do this.

It was a pleasure to have the following students and postdocs in Antieau's group: John Calabrese (Rice), Chang-Yeon Cho (Berkeley), Ed Dewey (Wisconsin), Elden Elmanto (Northwestern), Márton Hablicsek (Penn), Patrick McFaddin (Georgia), Emad Nasrollahpoursamami (Caltech), Yehonatan Sella (UCLA), Emra Sertöz (Berlin), Arne Smeets (Imperial), Arnav Tripathi (Stanford), and Fei Xie (UCLA).

We would also like to thank John Calabrese, Gabriela Guzman, Marc Hoyois, Kirsten Wickelgren, and Benedict Williams for comments and corrections on earlier drafts of this paper.

Aravind Asok deserves special thanks for several useful conversations about material to include in the paper.

Finally, we thank the anonymous referee for their careful reading of the paper; they caught several mistakes, both trivial and non-trivial, that are corrected in this version. They also supplied many, many additional references.

2. Classification of topological vector bundles

We introduce the language of Postnikov towers and illustrate their use through several examples involving the classification of topological vector bundles. The point is to tempt the reader to imagine the power these tools would possess in algebraic geometry if they existed. General references for the material here include Hatcher and Husemoller's books [**Hat02, Hus75**].

2.1. Postnikov towers and Eilenberg-MacLane spaces. Let S^i denote the i-sphere, embedded in $\mathbb{R}^{i+1}$ as the unit sphere, and let $s = (1, 0, \ldots, 0)$ be the basepoint. Recall that if (X, x) is a pointed space, then

$$\pi_i(X, x) = [(S^i, s), (X, x)]_*,$$

the set of homotopy classes of pointed maps from the i-sphere to X. The set of path-components $\pi_0(X, x)$ is simply a set pointed by the component containing x. The fundamental group is $\pi_1(X, x)$, a not-necessarily-abelian group. The groups $\pi_i(X, x)$ are abelian for $i \geq 2$. For a path-connected space, $\pi_i(X, x)$ does not depend on x, so we will often omit x from our notation and write $\pi_i(X)$ or $\pi_i X$.

DEFINITION 2.1. A map of spaces $f : X \to Y$ is an **n-equivalence** if $\pi_0(f) : \pi_0(X) \to \pi_0(Y)$ is a bijection and if for each choice of a basepoint $x \in X$, the induced map

$$\pi_i(f) : \pi_i(X, x) \to \pi_i(Y, f(x))$$

is an isomorphism for each $i < n$ and a surjection for $i = n$. The map f is a **weak homotopy equivalence** if it is an ∞-equivalence.

Typically we are interested in working with spaces up to weak homotopy equivalence. The correct notion of a fibration in this setting is a Serre fibration. Let D^n denote the n-disk and I^1 the unit interval. A map $p : E \to B$ is a **Serre fibration** (or simply a **fibration** as we will not use any other notion of fibration for maps of topological spaces) if for every diagram

$$\begin{array}{ccc} D^n \times \{0\} & \longrightarrow & E \\ \big\downarrow{\scriptstyle i} & \nearrow & \big\downarrow{\scriptstyle p} \\ D^n \times I^1 & \longrightarrow & B \end{array}$$

of solid arrows, there exists a dotted lift making both triangles commute. In other words, p has the **right lifting property** with respect to the maps $D^n \times \{0\} \to D^n \times I^1$. This property is equivalent to having the right lifting property with respect to all maps $A \times I^1 \cup X \times \{0\} \to X$ for all CW pairs (X, A).

There is a functorial way of replacing an arbitrary map $f : X \to Y$ by a Serre fibration. Let P_f be the space consisting of pairs (x, ω) where $x \in X$ and $\omega : I^1 \to Y$ such that $\omega(0) = f(x)$. There is a natural inclusion $X \to P_f$ sending x to (x, c_x), where c_x is the constant path at $f(x)$, and there is a natural map $P_f \to Y$ sending (x, ω) to $\omega(1)$.

EXERCISE 2.2. Show that the map $X \to P_f$ is a homotopy equivalence and that $P_f \to Y$ is a fibration.

Given a fibration $p : E \to B$ and a basepoint $e \in E$, the subspace $F = p^{-1}(p(e))$ is the **fiber** of p at $p(e)$. The point e is inside F. The crucial fact about Serre fibrations is that the sequence

$$(F, e) \to (E, e) \to (B, p(e))$$

gives rise to a long exact sequence

$$\cdots \to \pi_n(F,e) \to \pi_n(E,e) \to \pi_n(B,p(e)) \to \pi_{n-1}(F,e)$$
$$\to \cdots \to \pi_0(F,e) \to \pi_0(E,e) \to \pi_0(B,p(e))$$

of homotopy groups. Some explanation of 'exactness' is required in low-degrees as they are only groups or pointed sets. We refer to [**BK72**, Section IX.4].

DEFINITION 2.3. The **homotopy fiber** $F_f(y)$ of a map $f : X \to Y$ over a point $y \in Y$ is the fiber of $P_f \to Y$ over y. A sequence $(F,x) \to (X,x) \xrightarrow{f} (Y,y)$ of pointed spaces is a **homotopy fiber sequence** if

(1) Y is path-connected,
(2) $f(F) = y$, and
(3) the natural map $F \to F_f(y)$ is a weak homotopy equivalence.

The homotopy fiber sequences are those which behave just as well as fiber sequences from the point of view of their homotopy groups.

EXERCISE 2.4. Given a space X and a point $x \in X$, the based loop space $\Omega_x X$ is the homotopy fiber of $x \to X$. When X is path-connected or the basepoint is implicit, we will write ΩX for $\Omega_x X$.

THEOREM 2.5. *Let X be a path-connected space, so that $\pi_0(X,x) = *$. There exists a commutative diagram of pointed spaces*

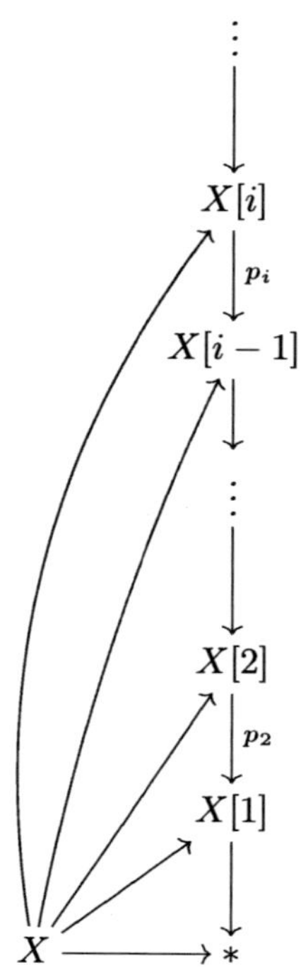

such that

(1)
$$\pi_j(X[i]) \cong \begin{cases} \pi_j X & j \leq i, \\ 0 & j > i; \end{cases}$$

(2) $X \to X[i]$ *is an* $(i+1)$*-equivalence;*
(3) *each map* $X[i+1] \to X[i]$ *is a Serre fibration;*
(4) *the natural map* $X \to \lim_i X[i]$ *is a weak homotopy equivalence.*

The space $X[i]$ *is the* i**th Postnikov section of** X*, and the diagram is called the* **Postnikov tower** *of* X.

PROOF. See Hatcher [**Hat02**, Chapter 4]. The basic idea is that one builds $X[i]$ from X by first attaching cells to X to kill π_{i+1}. Then, attaching cells to the result to kill π_{i+2}, and so on. □

DEFINITION 2.6. Let $i > 0$ and let G be a group, abelian if $i > 1$. A $K(G, i)$**-space** is a connected space Y such that $\pi_i(Y) \cong G$ and $\pi_j(Y) = 0$ for $j \neq i$. As a class, these are referred to as **Eilenberg-MacLane spaces**.

EXERCISE 2.7. The homotopy fiber of $X[i] \to X[i-1]$ is a $K(\pi_i X, i)$-space.

Suppose that Y is another space, and we want to construct a map $Y \to X$. We can hope to start with a map $Y \to X[1]$, lift it to a map $Y \to X[2]$, and on up the Postnikov tower. Using the fact that X is the limit of the tower, we would have constructed a map $Y \to X$.

What we need is a way of knowing when a map $Y \to X[i]$ lifts to a map $Y \to X[i+1]$. We need an **obstruction theory** for such lifts. Before getting into the details in the topological setting, we consider an example from algebra. Let

$$0 \to E \to F \to G \to 0$$

be an exact sequence of abelian groups. Let $g : H \to G$ be a homomorphism. When can we lift g to a map $f : H \to F$? The extension is classified by a class $p \in \mathrm{Ext}^1(G, E)$. This can be viewed as a map $G \to E[1]$ in the derived category $\mathrm{D}(\mathbb{Z})$. Composing with f, we get the pulled back extension $f^*p \in \mathrm{Ext}^1(H, E)$, viewed either as the composition $H \to G \to E[1]$ in $\mathrm{D}(\mathbb{Z})$, or as an induced extension

$$0 \to E \to F' \to H \to 0.$$

Now, we know that g lifts if and only if the extension F' splits if and only if $f^*p = 0 \in \mathrm{Ext}^1(H, E)$. The theory we explain now is a **nonabelian** version of this example.

DEFINITION 2.8. A homotopy fiber sequence $F \to X \xrightarrow{p} Y$ is **principal** if there is a delooping B of F (so that $\Omega B \simeq F$) and a map $k : Y \to B$ such that p is homotopy equivalent to the homotopy fiber of k. We will call k the **classifying map** of the principal fiber sequence.

EXAMPLE 2.9. The reduced cohomology $\widetilde{\mathrm{H}}^{i+1}(X, A)$ of a pointed space (X, x) with coefficients in an abelian group A is the kernel of the restriction map $\mathrm{H}^{i+1}(X, A) \to \mathrm{H}^{i+1}(\{x\}, A)$. If $i + 1 > 0$, then the reduced cohomology is isomorphic to the unreduced cohomology. Recall that the reduced cohomology of X can be represented as $\widetilde{\mathrm{H}}^{i+1}(X, A) = [(X, x), K(A, i+1)]_*$. That is, Eilenberg-MacLane spaces represent cohomology classes. Given a cohomology class $k \in \widetilde{\mathrm{H}}^{i+1}(X, A)$ viewed as a map $X \to K(A, i+1)$, the homotopy fiber of k is a space Y with $Y \to X$ having homotopy fiber $K(A, i)$.

Suppose that $k \in \widetilde{\mathrm{H}}^2(X, \mathbb{Z})$. The homotopy fiber sequence one gets is $K(\mathbb{Z}, 1) \to Y \to X$. Since $S^1 \simeq \mathbb{C}^* \simeq K(\mathbb{Z}, 1)$, we see that k corresponds to a topological complex line bundle (up to homotopy) $Y \to X$, as expected.

LEMMA 2.10. *Let $F \to X \to Y$ be a principal fibration classified by $k : Y \to B$ (so that $F \simeq \Omega B$). Let Z be a CW complex. Then, a map $Z \to Y$ lifts to $Z \to X$ if and only if the composition $Z \to Y \to B$ is nullhomotopic.*

PROOF. This follows from the definition of a fibration. Indeed, we can assume that $Y \to B$ is a fibration (by replacing the map by a fibration using P_k if necessary) and that X is the fiber over the basepoint. Applying the right lifting property to the case at hand, where $Z \times I^1 \to B$ is a nullhomotopy from the map $Z \to B$ to the map $Z \to \{b\} \subseteq B$, we see that the initial map $Z \to Y$ is homotopic to a map landing in the actual fiber of $Y \to B$. This fiber is X. □

THEOREM 2.11. *If X is simply connected (X is path-connected and $\pi_1(X) = 0$), then the Postnikov tower of X is a tower of principal fibrations. In particular, for each $i \geq 1$ there is a ladder of homotopy fiber sequences*

$$\begin{array}{ccc} K(\pi_i X, i) & \longrightarrow & X[i] \\ & & \downarrow \\ & & X[i-1] \xrightarrow{k_{i-1}} K(\pi_i X, i+1). \end{array}$$

*Specifically, $p_i : X[i] \to X[i-1]$ is the homotopy fiber of k_{i-1} and $K(\pi_i X, i) \to X[i]$ is the homotopy fiber of p_i. The map k_{i-1} is the $(i-1)$st **k-invariant** of X. It represents a cohomology class in $\mathrm{H}^{i+1}(X[i-1], \pi_i X)$.*

COROLLARY 2.12. *Let X be a simply connected space. For each $f : Y \to X[i-1]$, there is a uniquely determined class $f^* k_{i-1} \in \mathrm{H}^{i+1}(Y, \pi_i X)$. The map f lifts if and only if $f^* k_{i-1} = 0$.*

PROOF. This follows from Theorem 2.11 and Lemma 2.10, together with the fact that Eilenberg-MacLane spaces represent cohomology classes. □

A more complicated, equivariant version of the theory applies in the non-principal case (so that X is in particular not simply connected), but we ignore that for now as it is unnecessary in the topological applications we have in mind below. It is explained briefly in Section 8.

2.2. Representability of topological vector bundles.

DEFINITION 2.13. Let G be a topological group. Recall that a **G-torsor** on a space X is a space $p : Y \to X$ over X together with a (left) group action $a : G \times Y \to Y$ such that

(1) $p(a(g, y)) = p(y)$ (the action preserves fibers), and
(2) the natural map $G \times Y \to Y \times_X Y$ given by $(g, y) \mapsto (a(g, y), y)$ is an isomorphism.

Torsors for G are also called **principal G-bundles**.

EXAMPLE 2.14. The **trivial** G-torsor on X is $G \times X \to X$, with the projection map as the structure map.

Given G-torsors $p : Y \to X$ and $p' : Y' \to X'$, a morphism of G-torsors $(f, g) : (Y, p, X) \to (Y', p', X')$ is a map $f : Y \to Y'$ of G-spaces (i.e., compatible with the G-action) together with a map $g : X \to X'$ such that $g(p(y)) = p'(f(y))$. Given a map $g : X \to X'$ and a G-torsor $p' : Y' \to X'$, there is a uniquely determined G-torsor structure on $Y = X \times_{X'} Y'$. The projection map $f : Y \to X$ makes (f, g) into a morphism of G-torsors. We will write g^*Y' for the pull back bundle.

DEFINITION 2.15. A G-torsor $Y \to X$ is **locally trivial** if there is an open cover $\{g_i : U_i \to X\}_{i \in I}$ of X such that g_i^*Y is isomorphic as a G-torsor to $G \times U_i$ for each i. The subcategory of G-torsors on a fixed base X is naturally a groupoid. We write $\mathbf{Bun}_G(X)$ for the full subcategory of locally trivial G-torsors on X. The set of isomorphism classes of $\mathbf{Bun}_G(X)$ will be denoted $\mathrm{Tors}_G(X)$.

EXAMPLE 2.16. Let $L \to X$ be a complex line bundle. The fibers are in particular 1-dimensional complex vector spaces. Let $Y = L - s(X)$, where $S : X \to L$ is the 0-section. There is a natural action of the topological (abelian) group $\mathbb{C}^*$ on Y given simply by scalar multiplication in the fibers. In this case, Y becomes a principal $\mathbb{C}^*$-bundle on X.

THEOREM 2.17 (Steenrod). *Let G be a topological group. Then, there is a connected space $\mathrm{B}G$ with a G-torsor γ_G such that the natural pullback map $[X, \mathrm{B}G] \to \mathrm{Tors}_G(X)$ sending $f : X \to \mathrm{B}G$ to $f^*\gamma_G$ is an isomorphism for all paracompact Hausdorff spaces X. Moreover, $\Omega \mathrm{B}G \simeq G$.*

PROOF. See Husemoller [**Hus75**, Theorem 4.12.2]. The existence of $\mathrm{B}G$ can be proved in greater generality as the representability of certain functors satisfying Mayer-Vietoris and homotopy invariance properties. The total space of γ_G is a contractible space with a free G-action. Hence, $G \to \gamma_G \to \mathrm{B}G$ is a fiber sequence. It follows that $\Omega \mathrm{B}G \simeq G$. □

REMARK 2.18. Any CW complex is paracompact Hausdorff. Any differentiable manifold is paracompact and Hausdorff, as is the underlying topological space associated to a separated complex algebraic variety.

EXAMPLE 2.19. If A is an abelian group, then $K(A, n)$ can be given the structure of a topological abelian group. In this case, $\mathrm{B}K(A, n)$ is a $K(A, n+1)$-space. Indeed, $\Omega \mathrm{B}K(A, n) \simeq K(A, n)$.

DEFINITION 2.20. Let $p : Y \to X$ be a G-torsor, and let F be a space with a (left) G-action. There is then a left G-action on $Y \times F$, the diagonal action. Let F_Y denote the quotient $(Y \times F)/G$. There is a natural map $F_Y \to Y/G \cong X$. The fibers of this map are all isomorphic to F. The space $F_Y \to X$ is called the ***F*-bundle** associated to Y.

EXAMPLE 2.21. Let $p : Y \to X$ be a locally trivial $\mathrm{GL}_n(\mathbb{C})$-torsor. Let $\mathrm{GL}_n(\mathbb{C})$ act on $\mathbb{C}^n$ by matrix multiplication. Then, $\mathbb{C}^n_Y \to X$ is a vector bundle. In fact, this association gives a natural bijection

$$\mathrm{Tors}_{\mathrm{GL}_n(\mathbb{C})}(X) \cong \mathrm{Vect}_n^{\mathrm{top}}(X).$$

COROLLARY 2.22. *If X is a paracompact Hausdorff space, then there is a natural bijection*

$$[X, \mathrm{BGL}_n(\mathbb{C})] \cong \mathrm{Vect}_n^{\mathrm{top}}(X).$$

In the case of $\mathrm{GL}_n(\mathbb{C})$ we can construct a more explicit version of $\mathrm{BGL}_n(\mathbb{C})$ by using Grassmannians. Let $\mathrm{Gr}_n(\mathbb{C}^{n+k})$ denote the Grassmannian of n-plane bundles in $\mathbb{C}^{n+k}$, and let $\mathrm{Gr}_n = \mathrm{colim}_k \mathrm{Gr}_n(\mathbb{C}^{n+k})$ denote the colimit. Over each $\mathrm{Gr}_n(\mathbb{C}^{n+k})$ there is a canonical $\mathrm{GL}_n(\mathbb{C})$-bundle given by the Stiefel manifold $V_n(\mathbb{C}^{n+k})$, the space of n linearly independent vectors in $\mathbb{C}^{n+k}$. The map sending a set of linearly independent vectors to the subspace they span gives a surjective map

$$V_n(\mathbb{C}^{n+k}) \to \mathrm{Gr}_n(\mathbb{C}^{n+k}).$$

There is a natural free action of $\mathrm{GL}_n(\mathbb{C})$ on $V_n(\mathbb{C}^{n+k})$, and $V_n(\mathbb{C}^{n+k}) \to \mathrm{Gr}_n(\mathbb{C}^{n+k})$ is a locally trivial $\mathrm{GL}_n(\mathbb{C})$-torsor with this action.

LEMMA 2.23. *The space $V_n(\mathbb{C}^{n+k})$ is $2k$-connected for $n \geq 1$.*

PROOF. Consider the map $V_n(\mathbb{C}^{n+k}) \to \mathbb{C}^{n+k} - \{0\}$ sending a set of linearly independent vectors to the last vector. The fibers are all isomorphic to $V_{n-1}(\mathbb{C}^{n+k-1})$. If $n = 2$, the fiber is thus $V_1(\mathbb{C}^{1+k}) \cong \mathbb{C}^{1+k} - \{0\} \simeq S^{1+2k}$, which is certainly $2k$-connected. Therefore, by induction, we can assume that $V_{n-1}(\mathbb{C}^{n+k-1})$ is $2k$-connected, and since $\mathbb{C}^{n+k} - \{0\} \simeq S^{2n+2k-1}$ is $2n + 2k - 2$-connected, it follows that $V_n(\mathbb{C}^{n+k})$ is $2k$-connected for $n \geq 1$. □

As a result, the colimit $V_n = \mathrm{colim}_k V_n(\mathbb{C}^{n+k})$ is a contractible $\mathrm{GL}_n(\mathbb{C})$-torsor over Gr_n. Hence, $\mathrm{Gr}_n \simeq \mathrm{BGL}_n(\mathbb{C})$. There is a fairly easy way to see why Grassmannians should control $\mathrm{GL}_n(\mathbb{C})$-torsors on X, or equivalently vector bundles. Let $p : E \to X$ be a complex vector bundle of rank n. Suppose that E is a trivial on a finite cover $\{U_i\}_{i=1}^m$ of X. Let $s_i : X \to [0, 1]$ be a partition of unity subordinate to $\{U_i\}$, so that the support of each s_i is contained in U_i, and $s_1 + \cdots + s_m = 1_X$. Choose trivializations $t_i : \mathbb{C}^n \times U_i \to E|_{U_i}$. Now, we can define $g : E \to \mathbb{C}^{mn}$ by $g = \oplus_{i=1}^m g_i$, where $g_i = (s_i \circ p) \cdot (p_1 \circ t_i^{-1})$; outside U_i, $g_i = 0$. Now, the **Gauss map** g clearly defines a map $X \to \mathrm{Gr}_n(\mathbb{C}^{nm})$. The entire formalism of vector bundles can be based on these Gauss maps. See Husemoller [**Hus75**, Chapter 3].

We conclude the section with two remarks. First, the classifying space construction is functorial in homomorphisms of topological groups. That is, if there is a map of topological groups $G \to H$, then there is an induced map $\mathrm{B}G \to \mathrm{B}H$. The corresponding map $\mathrm{Tors}_G(X) \to \mathrm{Tors}_H(X)$ is an example of the fiber bundle construction. Indeed, since G acts on H, we can apply this construction to produce an H-torsor from a G-torsor.

EXAMPLE 2.24. The most important example for us will be the determinant map $\mathrm{BGL}_n(\mathbb{C}) \to \mathrm{BGL}_1(\mathbb{C})$, which gives the determinant map $\mathrm{Vect}_n^{\mathrm{top}}(X) \to \mathrm{Vect}_1^{\mathrm{top}}(X)$. The fiber of this map is just as important. Indeed, because

$$1 \to \mathrm{SL}_n(\mathbb{C}) \to \mathrm{GL}_n(\mathbb{C}) \to \mathrm{GL}_1(\mathbb{C}) \to 1$$

is an exact sequence of topological groups, the sequence $\mathrm{BSL}_n(\mathbb{C}) \to \mathrm{BGL}_n(\mathbb{C}) \to \mathrm{BGL}_1(\mathbb{C})$ turns out to be a homotopy fiber sequence. Hence, $[X, \mathrm{BSL}_n(\mathbb{C})]$ classifies topological complex vector bundles on X with trivial determinant.

The second remark is that one often works with BU_n rather than $\mathrm{BGL}_n(\mathbb{C})$. The natural inclusion $\mathrm{U}_n \to \mathrm{GL}_n(\mathbb{C})$ of the unitary matrices into all invertible complex matrices is a homotopy equivalence (using polar decomposition). Hence, $\mathrm{BU}_n \to \mathrm{BGL}_n(\mathbb{C})$ is also a homotopy equivalence. Philosophically, this corresponds to the fact that any complex vector bundle on a paracompact Hausdorff space admits a Hermitian metric.

2.3. Topological line bundles. Using the Grassmannian description of $\mathrm{BGL}_1(\mathbb{C})$, we find that $\mathrm{colim}_n\, \mathrm{Gr}_1(\mathbb{C}^n) \simeq \mathrm{BGL}_1(\mathbb{C})$. Of course, $\mathrm{Gr}_1(\mathbb{C}^n) \simeq \mathbb{CP}^{n-1}$. Hence, $\mathbb{CP}^\infty \simeq \mathrm{BGL}_1(\mathbb{C})$.

LEMMA 2.25. *The infinite complex projective space* $\mathbb{CP}^\infty$ *is a* $K(\mathbb{Z}, 2)$.

PROOF. Indeed, since $\mathrm{GL}_1(\mathbb{C}) \cong \mathbb{C}^* \simeq S^1$ and $\Omega\mathbb{CP}^\infty \simeq \mathrm{GL}_1(\mathbb{C})$, we find that $\mathbb{CP}^\infty \simeq K(\mathbb{Z}, 2)$. $\square$

As a result we may describe the set of line bundles on X in terms of a cohomology group:

COROLLARY 2.26. *The natural map* $\mathrm{Vect}_1^{\mathrm{top}}(X) \xrightarrow{c_1} \mathrm{H}^2(X, \mathbb{Z})$ *is a bijection for any paracompact Hausdorff space.*

2.4. Rank 2 bundles in low dimension. Recall that the cohomology of the infinite Grassmannian is

$$\mathrm{H}^*(\mathrm{Gr}_n, \mathbb{Z}) \cong \mathrm{H}^*(\mathrm{BGL}_n(\mathbb{C}), \mathbb{Z}) \cong \mathbb{Z}[c_1, \ldots, c_n],$$

where $|c_i| = 2i$. We will also need Bott's computation [**Bot58**, Theorem 5] of the homotopy groups of Gr_n in the stable range. If $i \leq 2n+1$, then

$$\pi_i \mathrm{Gr}_n \cong \begin{cases} \mathbb{Z} & \text{if } i \leq 2n \text{ is even,} \\ 0 & \text{if } i \leq 2n \text{ is odd,} \\ \mathbb{Z}/n! & \text{if } i = 2n+1. \end{cases}$$

When $i > 2n+1$ much less is known about the homotopy groups of Gr_n (except when $n = 1$). Playing the computation of the cohomology rings off of these homotopy groups gives a great deal of insight into the low stages of the Postnikov tower of Gr_n.

LEMMA 2.27. *The map* $\mathrm{Gr}_n \to \mathrm{Gr}_n[3] \simeq \mathrm{Gr}_n[2] \simeq K(\mathbb{Z}, 2)$ *is precisely* $c_1 \in \mathrm{H}^2(\mathrm{Gr}_n, \mathbb{Z})$.

PROOF. Indeed, this map is induced by the determinant map $\mathrm{GL}_n(\mathbb{C}) \to \mathrm{GL}_1(\mathbb{C})$. $\square$

We are in particular interested in the following part of the Postnikov tower of Gr_2:

$$\begin{array}{ccccc} K(\mathbb{Z}/2, 5) & \longrightarrow & \mathrm{Gr}_2[5] & & \\ & & \downarrow & & \\ K(\mathbb{Z}, 4) & \longrightarrow & \mathrm{Gr}_2[4] & \xrightarrow{k_4} & K(\mathbb{Z}/2, 6) \\ & & \downarrow & & \\ & & \mathrm{Gr}_2[3] \simeq K(\mathbb{Z}, 2) & \xrightarrow{k_3} & K(\mathbb{Z}, 5) \end{array}$$

Note that because $\mathrm{Gr}_2[3] \simeq \mathrm{Gr}_2[2]$ there is no obstruction to lifting a map $X \to \mathrm{Gr}_2[2]$ to a map $X \to \mathrm{Gr}_2[3]$ if X is a CW complex.

LEMMA 2.28. *The k-invariant k_3 is nullhomotopic. Hence,*

$$\mathrm{Gr}_2[4] \simeq K(\mathbb{Z}, 2) \times K(\mathbb{Z}, 4).$$

Moreover, this equivalence may be chosen so that the composition $\mathrm{Gr}_2 \to \mathrm{Gr}_2[4] \to K(\mathbb{Z}, 4)$ is $c_2 \in \mathrm{H}^4(\mathrm{Gr}_2, \mathbb{Z})$.

PROOF. The class $k_3 \in \mathrm{H}^5(K(\mathbb{Z}, 2), \mathbb{Z})$ vanishes simply because the cohomology of $K(\mathbb{Z}, 2) \simeq \mathbb{CP}^\infty$ is concentrate in even degrees. This gives the splitting claimed (we lifting the identity map $K(\mathbb{Z}, 2) \to Gr_2[2]$ up the Postnikov tower). Consider the map $\mathrm{Gr}_2 \to K(\mathbb{Z}, 2) \times K(\mathbb{Z}, 4)$ classified by the pair (c_1, c_2) in the cohomology of Gr_2. By definition, (c_1, c_2) factors through the functorial Postnikov section $\mathrm{Gr}_2 \to \mathrm{Gr}_2[4]$. It is enough to check that the induced map $\mathrm{Gr}_2[4] \to K(\mathbb{Z}, 2) \times K(\mathbb{Z}, 4)$ is a weak equivalence. We have already seen that it is an isomorphism on π_2. We have a map of fiber sequences

$$\begin{array}{ccccc} \mathrm{BSL}_2[4] & \longrightarrow & \mathrm{Gr}_2[4] & \longrightarrow & \mathrm{Gr}_1 \\ \downarrow & & \downarrow & & \downarrow \\ K(\mathbb{Z}, 4) & \longrightarrow & K(\mathbb{Z}, 2) \times K(\mathbb{Z}, 4) & \longrightarrow & K(\mathbb{Z}, 2), \end{array}$$

and the outside vertical arrows are weak equivalences by the Hurewicz isomorphism theorem. This proves the lemma. □

We can now classify rank 2 vector bundles on 4-dimensional spaces.

PROPOSITION 2.29. *Let X be a 4-dimensional space having the homotopy type of a CW complex. Then, the natural map*

$$\mathrm{Vect}_2^{\mathrm{top}}(X) \to \mathrm{H}^2(X, \mathbb{Z}) \times \mathrm{H}^4(X, \mathbb{Z})$$

is a bijection.

PROOF. The previous lemma shows that $[X, \mathrm{Gr}_2[4]] \to \mathrm{H}^2(X, \mathbb{Z}) \times \mathrm{H}^4(X, \mathbb{Z})$ is a bijection. The obstruction to lifting a given map $f : X \to \mathrm{Gr}_2[4]$ to $\mathrm{Gr}_2[5]$ is a class $f^*k_4 \in \mathrm{H}^6(X, \mathbb{Z}/2) = 0$. Similarly, the choice of lifts is bijective to a quotient of $\mathrm{H}^5(X, \mathbb{Z}/2)$, and this group is 0. Hence, for every such f there is a unique lift to $\mathrm{Gr}_2[5]$, and then the same reasoning gives a unique lift to $\mathrm{Gr}_2[m]$ for all $m \geq 5$. Since Gr_2 is the limit of its Postnikov tower, the proposition follows. □

If $\dim X = 5$, the situation is similar but more complicated. To state the theorem let us recall that a cohomology operation is a natural transformation of functors $H^i(-, R) \to H^j(-, R')$; by Yoneda such a map is classified by an element of $[K(i, R), K(j, R')] \simeq H^j(K(i, R), R')$.

PROPOSITION 2.30. *If X is a 5-dimensional space having the homotopy type of a CW complex, then the map $\mathrm{Vect}_2^{\mathrm{top}}(X) \to \mathrm{H}^2(X, \mathbb{Z}) \times \mathrm{H}^4(X, \mathbb{Z})$ is surjective, and the choice of lifts is parametrized by $\mathrm{H}^5(X, \mathbb{Z}/2)/\mathrm{im}(\mathrm{H}^3(X, \mathbb{Z}) \to \mathrm{H}^5(X, \mathbb{Z}/2))$, where the map $\mathrm{H}^3(X, \mathbb{Z}) \to \mathrm{H}^5(X, \mathbb{Z}/2)$ is a certain non-zero cohomology operation.*

PROOF. Consider the fiber sequence $K(\mathbb{Z}/2, 5) \to \mathrm{Gr}_2[5] \to \mathrm{Gr}_2[4]$. As above, $[X, \mathrm{Gr}_2[4]]$ is classified by the 1st and 2nd Chern classes. On a 5-dimensional space, once a lift to $\mathrm{Gr}_2[5]$ is specified, there is a unique lift all the way to Gr_2, just as in the proof of the previous proposition. The obstructions to finding a lift from $\mathrm{Gr}_2[4]$ to $\mathrm{Gr}_2[5]$ are in $\mathrm{H}^6(X, \mathbb{Z}/2)$, and hence all lift. Recall that to any fiber sequence there is an associated long exact sequence of fibrations. See [**Hat02**, Section 4.3]. Extending to the left a little bit, in our cases this is

$$\Omega\mathrm{Gr}_2[4] \to K(\mathbb{Z}/2,5) \to \mathrm{Gr}_2[5] \to \mathrm{Gr}_2[4].$$

However, $K(\mathbb{Z}/2,5) \to \mathrm{Gr}_2[5] \to \mathrm{Gr}_2[4]$ is principal, so it extends to the right one term as well:

$$\Omega\mathrm{Gr}_2[4] \to \Omega K(\mathbb{Z}/2,6) \to \mathrm{Gr}_2[5] \to \mathrm{Gr}_2[4] \to K(\mathbb{Z}/2,6),$$

where $\Omega K(\mathbb{Z}/2,6) \simeq K(\mathbb{Z}/2,5)$. It follows that there is an exact sequence of pointed sets

$$\mathrm{H}^1(X,\mathbb{Z}) \times \mathrm{H}^3(X,\mathbb{Z}) \to \mathrm{H}^5(X,\mathbb{Z}/2) \to \mathrm{Vect}_2^{\mathrm{top}}(X) \to \mathrm{H}^2(X,\mathbb{Z}) \times \mathrm{H}^4(X,\mathbb{Z}),$$

which is surjective on the right. Moreover, the map $\mathrm{H}^1(X,\mathbb{Z}) \times \mathrm{H}^3(X,\mathbb{Z}) \to \mathrm{H}^5(X,\mathbb{Z}/2)$ is a group homomorphism because it is induced by taking loops of a map. There is an action of $\mathrm{H}^5(X,\mathbb{Z}/2)$ on $\mathrm{Vect}_2^{\mathrm{top}}(X)$ such that two rank 2 vector bundles on X have the same Chern classes if and only if they are in the same orbit of $\mathrm{H}^5(X,\mathbb{Z}/2)$. There are no cohomology operations $\mathrm{H}^1(X,\mathbb{Z}) \to \mathrm{H}^5(X,\mathbb{Z}/2)$, since $\mathrm{H}^5(S^1,\mathbb{Z}/2) = 0$. However, there is a cohomology operation $\mathrm{H}^3(X,\mathbb{Z}) \to \mathrm{H}^5(X,\mathbb{Z}/2)$, often denoted $\mathrm{Sq}^2_{\mathbb{Z}}$. Note that this class is precisely Ωk_4. That is, since we have $k_4 : K(\mathbb{Z},2) \times K(\mathbb{Z},4) \to K(\mathbb{Z}/2,6)$, the loop space is $K(\mathbb{Z},1) \times K(\mathbb{Z},3) \to K(\mathbb{Z}/2,5)$. One can check, using the Postnikov tower and cohomology of $\mathrm{BSL}_2(\mathbb{C})$ that this class Ωk_4 is precisely the unique non-zero element of $\mathrm{H}^5(K(\mathbb{Z},3),\mathbb{Z}/2) \cong \mathbb{Z}/2$ by the next lemma. □

LEMMA 2.31. *The k-invariant $k_4 : \mathrm{Gr}_2[4] \to K(\mathbb{Z}/2,6)$ is non-trivial.*

PROOF. It is enough to show that the corresponding k-invariant $\mathrm{BSL}_2(\mathbb{C})[4] \to K(\mathbb{Z}/2,6)$ is non-trivial. Note that $\mathrm{BSL}_2(\mathbb{C}) \to \mathrm{BSL}_2(\mathbb{C})[4] \simeq K(\mathbb{Z},4)$ is a 5-equivalence and that $\mathrm{BSL}_2(\mathbb{C}) \to \mathrm{BSL}_2(\mathbb{C})[5]$ is a 6-equivalence. It follows that $\mathrm{H}^6(\mathrm{BSL}_2(\mathbb{C})[5],\mathbb{Z}/2) = 0$ since $\mathrm{H}^*(\mathrm{BSL}_2(\mathbb{C}),\mathbb{Z}/2) \cong \mathbb{Z}/2[c_2]$. On the other hand, $\mathrm{H}^6(\mathrm{BSL}_2(\mathbb{C})[4],\mathbb{Z}/2) \cong \mathrm{H}^6(K(\mathbb{Z}/4),\mathbb{Z}/2) \cong \mathbb{Z}/2$. If the extension $K(\mathbb{Z}/2,5) \to \mathrm{BSL}_2(\mathbb{C})[5] \to \mathrm{BSL}_2(\mathbb{C})[4]$ were split, the cohomology of $\mathrm{BSL}_2(\mathbb{C})[4]$ would inject into the cohomology of $\mathrm{BSL}_2(\mathbb{C})[5]$. Since this does not happen, we see that k_4 is non-zero. □

EXERCISE 2.32. Describe the obstruction class $k_4 \in \mathrm{H}^6(X,\mathbb{Z}/2)$ by computing the cohomology of $K(\mathbb{Z},2) \times K(\mathbb{Z},4)$ and finding k_4.

Finally, if $\dim X = 6$, there is a similar picture, except that there is an obstruction to realizing a given pair of Chern classes, and there is an additional choice of lift.

EXAMPLE 2.33. Let X be the 6-skeleton of $\mathrm{BGL}_3(\mathbb{C})$. There is a universal rank 3 vector bundle E on X with Chern classes $c_i(E) = c_i \in \mathrm{H}^{2i}(X,\mathbb{Z})$ for $i = 1,2,3$. On the other hand, we can ask if there is a rank 2 bundle F on X with Chern classes $c_i(F) = c_i$ for $i = 1,2$. This is the universal example where the obstruction above is nonzero and demonstrates the *incompressibility* of Grassmannians.

One can use the fact that $\mathrm{SU}_2 \cong \mathrm{SO}_3$, which is itself isomorphic to the 3-sphere, to find that $\pi_6 \mathrm{Gr}_2 \cong \pi_5 S^3 = \mathbb{Z}/2$. This leads to the following description of rank 2-bundles on a 6-dimensional CW complex.

PROPOSITION 2.34. *Let X be a 6-dimensional space with the homotopy type of a CW complex. The map $\mathrm{Vect}_2^{\mathrm{top}}(X) \to \mathrm{H}^2(X, \mathbb{Z}) \times \mathrm{H}^4(X, \mathbb{Z})$ has image precisely those pairs (c_1, c_2) such that $k_4(c_1, c_2) = 0$ in $\mathrm{H}^6(X, \mathbb{Z})$. If $k_4(c_1, c_2) = 0$, the set of lifts to $\mathrm{Gr}_2[5]$ is parameterized by a quotient of $\mathrm{H}^5(X, \mathbb{Z}/2)$ as above. Each lift then lifts to Gr_2, and the set of lifts from $\mathrm{Gr}_2[5]$ to Gr_2 is parametrized by a quotient of $\mathrm{H}^6(X, \mathbb{Z}/2)$.*

2.5. Rank 3 bundles in low dimension.

PROPOSITION 2.35. *Suppose that X is a 5-dimensional CW complex. Then, the natural map*

$$(c_1, c_2) : \mathrm{Vect}_3^{\mathrm{top}}(X) \to \mathrm{H}^2(X, \mathbb{Z}) \times \mathrm{H}^4(X, \mathbb{Z})$$

is an isomorphism.

PROOF. Indeed, $\mathrm{Gr}_3[4] \simeq K(\mathbb{Z}, 2) \times K(\mathbb{Z}, 4)$, just as for $\mathrm{Gr}_2[4]$. But, this time, $\pi_5 \mathrm{Gr}_3 = 0$. Hence, the next interesting problem is to lift from $\mathrm{Gr}_2[4]$ to $\mathrm{Gr}_2[6]$. The obstructions are in $\mathrm{H}^7(X, \mathbb{Z})$, and hence vanish. The lifts of a given map to $\mathrm{Gr}_3[4]$ are a quotient of $\mathrm{H}^6(X, \mathbb{Z}) = 0$. □

As a consequence, one sees immediately from the last section that every rank 3 vector bundle E on a 5-dimensional CW complex splits as $E_0 \oplus \mathbb{C}$ for some rank 2 vector bundle E_0. This is one example of a more general phenomenon we leave to the reader to discover.

PROPOSITION 2.36. *If X is a 6-dimensional closed real orientable manifold, then*

$$(c_1, c_2, c_3) : \mathrm{Vect}_3^{\mathrm{top}}(X) \to \mathrm{H}^2(X, \mathbb{Z}) \times \mathrm{H}^4(X, \mathbb{Z}) \times \mathrm{H}^6(X, \mathbb{Z})$$

is a injection with image the triples with c_3 an even multiple of a generator of $\mathrm{H}^6(X, \mathbb{Z}) \cong \mathbb{Z}$.

PROOF. As above, once we have constructed a map $X \to \mathrm{Gr}_3[6]$, there is a unique lift to Gr_3. Given a map $X \to \mathrm{Gr}_3[4] \simeq K(\mathbb{Z}, 2) \times K(\mathbb{Z}, 4)$, the obstruction to lifting to $\mathrm{Gr}_3[6]$ is a class in $\mathrm{H}^7(X, \mathbb{Z}) = 0$ since X is 6-dimensional. We have again an exact sequence of pointed sets

$$\mathrm{H}^1(X, \mathbb{Z}) \times \mathrm{H}^3(X, \mathbb{Z}) \to \mathrm{H}^6(X, \mathbb{Z}) \to \mathrm{Vect}_3^{\mathrm{top}}(X) \to \mathrm{H}^2(X, \mathbb{Z}) \times \mathrm{H}^4(X, \mathbb{Z}).$$

The map on the left is induced from a map $K(\mathbb{Z}, 1) \times K(\mathbb{Z}, 3) \to K(\mathbb{Z}, 6)$ which is Ωk_5, where k_5 is the k-invariant $K(\mathbb{Z}, 2) \times K(\mathbb{Z}, 4) \to K(\mathbb{Z}, 7)$. In particular, the image in $\mathrm{H}^6(X, \mathbb{Z})$ consists of torsion classes. But, $\mathrm{H}^6(X, \mathbb{Z}) \cong \mathbb{Z}$ by hypothesis. One can check that the composition $K(\mathbb{Z}, 6) \to \mathrm{Gr}_3[6] \xrightarrow{c_3} K(\mathbb{Z}, 6)$ is multiplication by 2. This completes the proof. □

In general, understanding vector bundles of a fixed dimension becomes more and more difficult as the dimension of the base space increases. The systematic approach to this kind of problem uses cohomology and Serre spectral sequences to determine Postnikov extensions one step a time. For an overview, see [**Tho66**].

3. The construction of the $\mathbb{A}^1$-homotopy category

The first definitions of $\mathbb{A}^1$-homotopy theory were given in [**MV99**] when the base scheme S is noetherian of finite Krull dimension. An equivalent homotopy theory was constructed by Dugger [**Dug01a**], and we will follow Dugger's definition, but with the added generality of allowing S to be quasi-compact and quasi-separated using Lurie's Nisnevich topology [**Lur16**, Section A.2.4]. We use model categories for the construction, but in the Section 4, where we give many properties of the homotopy theory, we emphasize the model-independence of the proofs.

3.1. Model categories. Model categories are a technical framework for working up to homotopy. The axioms guarantee that certain category-theoretic localizations exist without enlarging the ambient set-theoretic universe and that it is possible in some sense to compute the hom-sets in the localization. The theory generalizes the use of projective or injective resolutions in the construction of derived categories of rings or schemes.

References for this material include Quillen's original book on the theory [**Qui67**], Dwyer-Spalinski [**DS95**], Goerss-Jardine [**GJ99**], and Goerss-Schemmerhorn [**GS07**]. For consistency, we refer the reader where possible to [**GJ99**]. However, unlike some of these references, we assume that the category underlying M has all small limits and colimits. This is satisfied immediately in all cases of interest to us.

DEFINITION 3.1. Let M be a category with all small limits and colimits. A model category structure on M consists of three classes W, C, F of morphisms in M, called **weak equivalences**, **cofibrations**, and **fibrations**, subject to the following set of axioms.

M1 Given $X \xrightarrow{f} Y \xrightarrow{g} Z$ two composable morphisms in M, if any two of $g \circ f$, f, and g are weak equivalences, then so is the third.

M2 Each class W, C, F is closed under retracts.

M3 Given a diagram

$$\begin{array}{ccc} Z & \longrightarrow & E \\ {\scriptstyle i}\downarrow & \nearrow & \downarrow{\scriptstyle p} \\ X & \longrightarrow & B \end{array}$$

of solid arrows, a dotted arrow can be found making the diagram commutative if either

(a) p is an **acyclic fibration** ($p \in W \cap F$) and i is a cofibration, or

(b) i is an **acyclic cofibration** ($i \in W \cap C$) and p is a fibration.

(In particular, cofibrations i have the **left lifting property** with respect to acyclic fibrations, while fibrations p have the **right lifting property** with respect to acyclic cofibrations.)

M4 Any map $X \to Z$ in M admits two factorizations $X \xrightarrow{f} E \xrightarrow{p} Z$ and $X \xrightarrow{i} Y \xrightarrow{g} Z$, such that f is an acyclic cofibration, p is a fibration, i is a cofibration, and g is an acyclic fibration.

REMARK 3.2. In practice, a model category is determined by only W and either C or F. Indeed, C is precisely the class of maps in M having the left lifting property with respect to acyclic fibrations. Similarly, F consists of exactly those maps in M having the right lifting property with respect to acyclic cofibrations. The reader can prove this fact using the axioms or refer to [**DS95**, Proposition 3.13]. However, some caution is required. While one often sees model categories specified in the literature by just fixing W and either C or F, it usually has to be checked that these really do give M a model category structure.

REMARK 3.3. Many authors strengthen **M4** to assume the existence of *functorial* factorizations. This is satisfied in all model categories of relevance for this paper by [**Hov99**, Section 2.1] as they are all cofibrantly generated.

EXERCISE 3.4. Let A be an associative ring. Consider $\mathrm{Ch}_{\geq 0}(A)$, the category of non-negatively graded chain complexes of right A-modules. Since limits and colimits of chain complexes are computed degree-wise, $\mathrm{Ch}_{\geq 0}(A)$ is closed under all small limits and colimits. Let W be the class of quasi-isomorphisms, i.e., those maps $f : M_\bullet \to N_\bullet$ of chain complexes such that $\mathrm{H}_n(f) : \mathrm{H}_n(M_\bullet) \to \mathrm{H}_n(N_\bullet)$ is an isomorphism for all $n \geq 0$. Let F be the class of maps of chain complexes which are surjections in positive degrees. Describe the class C of maps satisfying the left lifting property with respect to $F \cap W$. Prove that W, C, F is a model category structure on $\mathrm{Ch}_{\geq 0}(A)$.

DEFINITION 3.5. A model category M has an initial object $\emptyset$ and a final object $*$, since it is closed under colimits and limits. An object X of M is **fibrant** if $X \to *$ is a fibration, and X is **cofibrant** if $\emptyset \to X$ is a cofibration. Given an object X of M, an acyclic fibration $QX \to X$ such that QX is cofibrant is called a **cofibrant replacement**. Similarly, if $X \to RX$ is an acyclic fibration with RX fibrant, then RX is called a **fibrant replacement** of X. These replacements always exist, by applying **M4** to $\emptyset \to X$ or $X \to *$.

EXAMPLE 3.6. In $\mathrm{Ch}_{\geq 0}(A)$, let M be a right A-module (viewed as a chain complex concentrated in degree zero). A projective resolution $P_\bullet \to M$ is an example of a cofibrant replacement. Indeed, such a resolution is an acyclic fibration. Moreover, the map $0 \to P_\bullet$ is a cofibration, since the cokernel is projective in each degree.

EXAMPLE 3.7. Let sSets be the category of simplicial sets. This is the category of functors $\Delta^{\mathrm{op}} \to \mathrm{Sets}$, where Δ is the category of finite non-empty ordered sets. (For details, see [**GJ99**].) There is a geometric realization functor $\mathrm{sSets} \to \mathrm{Spc}$, which sends a simplicial set $X_\bullet$ to a space $|X_\bullet|$. Let W denote the class of weak homotopy equivalences in sSets, i.e., those maps $f : X_\bullet \to Y_\bullet$ such that $|f| : |X_\bullet| \to |Y_\bullet|$ is a weak homotopy equivalence. Let C denote the class of level-wise monomorphisms. If F is the class of maps having the right lifting property with respect to acyclic cofibrations, then sSets together with W, C, F is a model category. In sSets, every object is cofibrant. The fibrant objects are the **Kan complexes**, namely those simplicial sets having a filling property for all horns. See [**GJ99**, Section I.3].

DEFINITION 3.8. A model category M is **pointed** if the natural map $\emptyset \to *$ is an isomorphism. Examples of pointed model categories include $\mathrm{Ch}_A^{\geq 0}$, which is pointed by the 0 object, and $\mathrm{sSets}_\star$, the category of *pointed* simplicial sets.

Now, we come to the main reason why model categories have been so successful in encoding homotopical ideas: the homotopy category of a model category.

DEFINITION 3.9. Let M be a category and W a class of morphisms in M. The localization of M by W, if it exists, is a category $M[W^{-1}]$ with a functor $L : M \to M[W^{-1}]$ such that

(1) $L(w)$ is an isomorphism for every $w \in W$,
(2) every functor $F : M \to N$ having the property that $F(w)$ is an isomorphism for all $w \in W$ factors uniquely through L in the sense that there is a functor $G : M[W^{-1}] \to N$ and a natural isomorphism of functors $G \circ L \simeq F$, and
(3) for any category N, the functor $\mathrm{Fun}(M[W^{-1}], N) \to \mathrm{Fun}(M, N)$ induced by composition with $L : M \to M[W^{-1}]$ is fully faithful.

The localization of M by W, if it exists, is unique up to categorical equivalence.

In general, there is no reason that a localization of M by W should exist much less be useful. The fundamental problem is that in attempting to concretely construct the morphisms in $M[W^{-1}]$, for example by hammock localization (hat piling), one discovers size issues, where it might be necessary to enlarge the universe in order to obtain a category: the morphisms sets in a category must be actual sets, not proper classes.

THEOREM 3.10 ([**Qui67**]). *Let M be a model category with class of weak equivalences W. Then, the localization $M[W^{-1}]$ exists. It is called the homotopy category of M, and we will denote it by* $\mathrm{Ho}(M)$.

RECIPE 3.11. It is generally difficult to compute $[X,Y] = \mathrm{Hom}_{\mathrm{Ho}(M)}(X,Y)$ given two objects $X, Y \in M$. We give a recipe. Replace X by a weakly equivalent cofibrant object QX, and Y by a weakly equivalent fibrant object RY. Then, $[X,Y] = \mathrm{Hom}_M(QX, RY)/\sim$, where $\sim$ is an equivalence relation on $\mathrm{Hom}_M(QX, RY)$ generalizing homotopy equivalence (see [**GJ99**, Section II.1]). See [**DS95**, Proposition 5.11] for a proof that this construction does indeed compute the set of maps in the homotopy category.

REMARK 3.12. In many cases, every object of M might be cofibrant, in which case one just needs to replace Y by RY and compute the homotopy classes of maps. This is for example the case in sSets.

REMARK 3.13. In Goerss-Jardine [**GJ99**, Section II.1], the homotopy category $\mathrm{Ho}(M)$ is itself defined to be the category of objects of M that are both fibrant and cofibrant, with maps given by $\mathrm{Hom}_{\mathrm{Ho}(M)}(A,B) = \mathrm{Hom}(A,B)/\sim$. Given an arbitrary X in M it is possible to assign to X a fibrant-cofibrant object RQX as follows. First, take, via **M4**, a factorization $\emptyset \to QX \to X$ where QX is cofibrant $QX \to X$ is a weak equivalence. Now, take a factorization $QX \to RQX \to *$ of the canonical map $QX \to *$ in which $QX \to RQX$ is an acyclic cofibration and $RQX \to *$ is a fibration. In particular, RQX is fibrant. Since compositions of cofibrations are cofibrations, RQX is also cofibrant. Moreover, if $f : X \to Y$ is a morphism, then it is possible using **M3** to (non-uniquely) assign to f a morphism $RQf : RQX \to RQY$ such that one gets a well-defined functor $M \to \mathrm{Ho}(M)$ (i.e., after enforcing $\sim$).

REMARK 3.14. In practice, we will work with simplicial model category structures, for which there exist objects $QX \times \Delta^1$, where Δ^1 is the standard 1-simplex (so that $|\Delta^1| = I^1$). In this case, the equivalence relation $\sim$ is precisely that of (left) homotopy classes of maps. See Definition 3.16.

EXERCISE 3.15. For chain complexes, the equivalence relation $\sim$ is precisely that of chain homotopy equivalence. (See [**Wei94**, Section 1.4].) Using the recipe above, compute

$$\mathrm{Hom}_{\mathrm{Ho}(\mathrm{Ch}_{\geq 0}(\mathbb{Z}))}(\mathbb{Z}/p, \mathbb{Z}[1]),$$

where $\mathbb{Z}/p[1]$ denotes the chain complex with $\mathbb{Z}/p$ placed in degree 1 and zeros elsewhere.

3.2. Mapping spaces. We will now explain simplicial model categories since we will need to discuss mapping spaces. For details, we refer the reader to [**GJ99**, II.2-3]. If X and Y are simplicial sets, then we may define the **simplicial mapping space** $\mathrm{map}_{\mathrm{sSets}}(X, Y)$ as the simplicial set with n-simplices given by

$$\mathrm{map}_{\mathrm{sSets}}(X, Y)_n := \mathrm{Hom}_{\mathrm{sSets}}(X \times \Delta^n, Y).$$

This simplicial set fits into a tensor-hom adjunction given by

$$\mathrm{Hom}_{\mathrm{sSets}}(Z \times X, Y) \cong \mathrm{Hom}_{\mathrm{sSets}}(Z, \mathrm{map}_{\mathrm{sSets}}(X, Y)).$$

Indeed, from this adjunction we may deduce the formula for $\mathrm{map}(X, Y)_n$ by evaluating at $Z = \Delta^n$.

Abstracting these formulas, one arrives at the axioms for a **simplicial category** [**GJ99**, II Definition 2.1]. A simplicial category is a category M equipped with

(1) a **mapping space functor**: $\mathrm{map} : M^{op} \times M \to \mathrm{sSets}$, written $\mathrm{map}_M(X, Y)$,
(2) an **action** of sSets, $M \times \mathrm{sSets} \to M$, written $X \otimes S$, and
(3) an **exponential**, $\mathrm{sSets}^{\mathrm{op}} \times M \to M$, written X^S for an object $X \in M$ and a simplicial set S

subject to certain compatibilities. The most important are that

$$- \otimes X : \mathrm{sSets} \leftrightarrows C : \mathrm{map}_M(X, -)$$

should be an adjoint pair of functors and that $\mathrm{Hom}_M(X, Y) \cong \mathrm{map}(X, Y)_0$ for all $X, Y \in M$.

Suppose that M is a simplicial category simultaneously equipped with a model structure. We would like the simplicial structure above to play well with the model structure. For example, if $i : A \to X$ is a cofibration, we expect $\mathrm{map}_M(Y, A) \to \mathrm{map}_M(Y, X)$ to be a fibration (and hence induce long exact sequences in homotopy groups) for any object Y as is the case in simplicial sets.

DEFINITION 3.16. Suppose that M is a model category which is also a simplicial category. Then M satisfies **SM7**, and is called a **simplicial model category**, if for any cofibration $i : A \to X$ and any fibration: $p : E \to B$ the map of simplicial sets (induced by the functoriality of map)

$$\mathrm{map}_M(X, E) \to \mathrm{map}_M(A, E) \times_{\mathrm{map}_M(A,B)} \mathrm{map}_M(X, B)$$

is a fibration of simplicial sets which is moreover a weak equivalence if either i or p is.

EXERCISE 3.17. Show that in a simplicial model category M, if $A \to X$ is a cofibration, then for any object Y, the natural map $\mathrm{map}_M(Y, A) \to \mathrm{map}_M(Y, X)$ is a fibration of simplicial sets.

Another feature of simplicial model categories is the fact that one may define a concept of homotopy that is more transparent than in an ordinary model category (where one defines left and right homotopies, see [**DS95**]). Suppose that $A \in M$ is a cofibrant object, then we say that two morphisms $f, g : A \to X$ are homotopic if there is a morphism: $H : A \otimes \Delta^1 \to X$ such that

$$\begin{array}{ccc} A \coprod A & \xrightarrow{d_1 \coprod d_0} & A \otimes \Delta^1 \\ {\scriptstyle f \coprod g} \downarrow & \swarrow_{H} & \\ X & & \end{array}$$

commutes. Write $f \sim g$ if f and g are homotopic.

EXERCISE 3.18. Prove that $\sim$ is an equivalence relation on $\mathrm{Hom}_M(A, X)$ when A is cofibrant.

In 3.11 we stated a recipe for calculating $[X, Y]$, the hom-sets in $\mathrm{Ho}(M)$. We replace X by a weakly equivalent cofibrant object QX, and Y by a weakly equivalent fibrant object RY. Then, we claimed that $[X, Y] = \mathrm{Hom}_M(QX, RY)/\sim$ where $\sim$ was an unspecified equivalence relation. For a simplicial model category, this equivalence relation can be taken to be the one just given. The fact the this is well defined is checked in [**GJ99**, Proposition 3.8].

3.3. Bousfield localization of model categories. One way of creating new model categories from old is via Bousfield localization. The underlying category remains the same, while the class of weak equivalences is enlarged. To describe these localizations, we first need to consider a class of functors between model categories that are well-adapted to their homotopical nature.

DEFINITION 3.19. Consider a pair of adjoint functors

$$F : M \rightleftarrows N : G$$

between model categories M and N. The pair is called a **Quillen pair**, or a pair of Quillen functors, if one of the following equivalent conditions is satisfied:

- F preserves cofibrations and acyclic cofibrations;
- G preserves fibrations and acyclic fibrations.

In this case, F is also called a **left Quillen functor**, and G a **right Quillen functor**.

Quillen pairs provide a sufficient framework for a pair of adjoint functors on model categories to descend to a pair of adjoint functors on the homotopy categories.

PROPOSITION 3.20. *Suppose that $F : M \rightleftarrows N : G$ is a pair of Quillen functors. Then, there are functors $\mathbf{L}F : M \to \mathrm{Ho}(N)$ and $\mathbf{R}G : N \to \mathrm{Ho}(M)$, each of which takes weak equivalences to isomorphisms, such that there is an induced adjunction $\mathbf{L}F : \mathrm{Ho}(M) \rightleftarrows \mathrm{Ho}(N) : \mathbf{R}G$ between homotopy categories.*

PROOF. See [**DS95**, Theorem 9.7]. □

REMARK 3.21. The familiar functors from homological algebra all arise in this way, so $\mathbf{L}F$ is called the left derived functor of F, while $\mathbf{R}G$ is the right derived functor of G. There is a recipe for computing the value of the derived functors on an arbitrary object X of M and Y of N. Specifically, $\mathbf{L}F(X)$ is weakly equivalent to $F(QX)$ where QX a cofibrant replacement of X. Similarly, $\mathbf{R}G(Y)$ is weakly equivalent to $G(RY)$ where RY is a fibrant replacement of Y.

REMARK 3.22. It follows from the previous remark that when a functorial cofibrant replacement functor $Q : M \to M$ exists, then we can factor $\mathbf{L}F : M \to \mathrm{Ho}(N)$ through $M \xrightarrow{Q} M \xrightarrow{F} N \to \mathrm{Ho}(N)$. As mentioned above, this is the case for all model categories in this paper. As such, we will often abuse notation and write $\mathbf{L}F$ for the functor $F \circ Q : M \to N$.

DEFINITION 3.23. A **Quillen equivalence** is a Quillen pair $F : M \rightleftarrows N : G$ such that $\mathbf{L}F : \mathrm{Ho}(M) \rightleftarrows \mathrm{Ho}(N) : \mathbf{R}G$ is an inverse equivalence.

DEFINITION 3.24. Let M be a simplicial model category with class of weak equivalences W. Suppose that I is a *set* of maps in M. An object X of M is **I-local** if it is fibrant and if for all $i : A \to B$ with $i \in I$, the induced morphism on mapping spaces $i^* : \mathrm{map}_M(B, X) \to \mathrm{map}_M(A, X)$ is a weak equivalence (of simplicial sets). A morphism $f : A \to B$ is an **I-local weak equivalence** if for every I-local object X, the induced morphism on mapping spaces $f^* : \mathrm{map}_M(B, X) \to \mathrm{map}_M(A, X)$ is a weak equivalence. Let W_I be the class of all I-local weak equivalences. By using **SM7**, $W \subseteq I$.

Let F_I denote the class of maps satisfying the right lifting property with respect to W_I-acyclic cofibrations ($W_I \cap C$). If (W_I, C, F_I) is a model category structure on M, we call this the **left Bousfield localization** of M with respect to I.

To distinguish between the model category structures on M, we will write $\mathrm{L}_I M$ for the left Bousfield model category structure on M. We will only write $\mathrm{L}_I M$ when the classes of morphisms defined above do define a model category structure.

When it exists, the Bousfield localization of M with respect to I is universal with respect to Quillen pairs $F : M \rightleftarrows N : G$ such that $\mathbf{L}F(i)$ is a weak equivalence in N for all $i \in I$.

EXERCISE 3.25. Show that if it exists, then the identity functors $\mathrm{id}_M : M \rightleftarrows M : \mathrm{id}_M$ induce a Quillen pair between M (on the left) and $\mathrm{L}_I M$.

We want to quote an important theorem asserting that in good cases the left Bousfield localization of a model category with respect to a set of morphisms exists. Some conditions, which we now define, are needed on the model category.

DEFINITION 3.26. A model category M is **left proper** if pushouts of weak equivalences along cofibrations are weak equivalences.

Note that this is a condition about how weak equivalences and cofibrations behave with respect to ordinary categorical pushouts. Model categories in which all objects are cofibrant are left proper [**Lur09**, Proposition A.2.4.2].

The next condition we need is for M to be **combinatorial**. This definition, due to Jeff Smith, is rather technical, so we leave it to the interested reader to refer to [**Lur09**, Definition A.2.6.1]. Recall that a category is **presentable** if it has all small colimits and is κ-compactly generated for some regular cardinal κ. For details, see the book of Adámek-Rosicky [**AR94**], although note that they call this condition *locally presentable*. We keep Lurie's terminology for the sake of consistency. The most important thing to know about combinatorial model categories for the purposes of this paper is that they are presentable as categories.

EXERCISE 3.27. Show that the model category structure on $\mathrm{Ch}_{\geq 0}(A)$ of Exercise 3.4 is left proper.

THEOREM 3.28. *If M is a left proper and combinatorial simplicial model category and I is a set of morphisms in M, then the left Bousfield localization $\mathrm{L}_I M$ exists and inherits a simplicial model category structure from M.*

PROOF. This is [**Lur09**, Proposition A.3.7.3]. □

We refer to [**Hir03**, Proposition 3.4.1] for the next result, which identifies the fibrant objects in the Bousfield localization.

PROPOSITION 3.29. *If M is a left proper simplicial model category and I is a set of maps such that $\mathrm{L}_I M$ exists as a model category, then the fibrant objects of $\mathrm{L}_I M$ are precisely the I-local objects of M.*

EXERCISE 3.30. Consider the model category structure given in Exercise 3.4 on $\mathrm{Ch}_{\geq 0}(\mathbb{Z})$. It is not hard to show that this is a simplicial model category using the Dold-Kan correspondence (see [**GJ99**]). Let I be the set of all morphisms between chain complexes of finitely generated abelian groups inducing isomorphisms on *rational* homology groups. Then, $\mathrm{L}_I\mathrm{Ch}_{\geq 0}(\mathbb{Z})$ is Quillen equivalent to $\mathrm{Ch}_{\geq 0}(\mathbb{Q})$ with the model category structure of Exercise 3.4. Show that every rational homology equivalence is an isomorphism in $\mathrm{Ho}(\mathrm{L}_I\mathrm{Ch}_{\geq 0}(\mathbb{Z}))$.

EXERCISE 3.31. Construct a category of $\mathbb{Q}$-local spaces, by letting I be a set of maps $f : X \to Y$ of simplicial sets such that $\mathrm{H}_*(f, \mathbb{Q})$ is an isomorphism.

3.4. Simplicial presheaves with descent. Let C be an essentially small category. Let $\mathrm{sPre}(C)$ denote the category of functors $X : C^{\mathrm{op}} \to \mathrm{sSets}$. This is the category of **simplicial presheaves** on C, and there is a Yoneda functor $h : C \to \mathrm{sPre}(C)$. Bousfield and Kan [**BK72**] defined a model category structure on $\mathrm{sPre}(C)$, the **projective model category structure**, which has a special universal property highlighted by Dugger [**Dug01a**]: it is the initial model category into which C embeds. Consider the following classes of morphisms in $\mathrm{sPre}(C)$:

- objectwise weak equivalences: those maps $w : X \to Y$ such that $w(V) : X(V) \to Y(V)$ is a weak equivalence of simplicial sets for all objects V of C,
- objectwise fibrations, and
- projective cofibrations, those maps having the left lifting property with respect to acyclic objectwise fibrations.

PROPOSITION 3.32. *The category of simplicial presheaves with the weak equivalences, fibrations, and cofibrations as above is a left proper combinatorial simplicial model category.*

PROOF. The reader can find a proof in [**Lur09**, Proposition A.2.8.2]. See [**Lur09**, Remark A.2.8.4] for left properness. □

Suppose now that C has a Grothendieck topology τ. Let $U_\bullet$ be an object of sPre(C) (so that each U_n is a presheaf of *sets* on C), and suppose that there is a map $U \to V$, where V is a representable object. We call $U \to V$ a **hypercover** if each U_n is a coproduct of representables, the induced map $U_0 \to V$ is a τ-cover, and each $U^{\Delta^n} \to U^{\partial\Delta^n}$ is a τ-cover in degree 0. For details about hypercovers, see [**AM69**, Section 8]. Except for the definition of the τ-local category below, we will not need hypercovers in the rest of the paper. The reader may safely just imagine these to be Čech complexes.

The standard example of a hypercover is the **Čech** complex $\check{U} \to V$ associated to a τ-cover $U \to V$. So, $\check{U}_n = U \times_V \cdots \times_V U$, the product of U with itself $n+1$ times over V. Roughly speaking, a hypercover looks just like a Čech complex, except that one is allowed to refine the Čech simplicial object by iteratively taking covers of the fiber products.

THEOREM 3.33. *The Bousfield localization of* sPre(C) *with respect to the class of hypercovers*

$$\check{U}_\bullet \to V$$

exists. We will denote this model category throughout the paper by L_τsPre(C).

PROOF. By Theorem 3.28, we only have to remark that there is up to isomorphism only a set of τ-hypercovers since C is small. □

REMARK 3.34. We will refer to τ-local objects and τ-local weak equivalences for the I-local notions when I is the class of morphisms in the theorem. In the τ-local model category L_τsPre(C), an object V of C (viewed as the functor it represents) is equivalent to the Čech complex of any τ-covering. Since sPre(C) with its projective model category structure is left proper, the fibrant objects of L_τsPre(C) are precisely the τ-local objects by Proposition 3.29. Hence, the fibrant objects of L_τsPre(C) are precisely the presheaves of Kan complexes X such that

$$X(V) \to \operatorname*{holim}_{\Delta} X(U) \tag{1}$$

is a weak equivalence for every τ-hypercover $U \to A$. In other words, the fibrant objects are the **homotopy sheaves of spaces**.

There is another, older definition of the homotopy theory of τ-homotopy sheaves due to Joyal and Jardine. It is useful to know that it is Quillen equivalent to the one given above.

DEFINITION 3.35. Let X be an object of sPre(C), V an object of C, and $x \in X(V)$ a basepoint. We can define a presheaf of sets (or groups or abelian groups) $\pi_n(X, x)$ on $C_{/V}$, the category of objects in C over V, by letting

$$\pi_n(X, x)(U) = \pi_n(X(U), f^*(x))$$

for $g : U \to V$ an object of $C_{/V}$. Let $\pi_n^\tau(X, x)$ be the sheafification of $\pi_n(X, x)$ in the τ-topology restricted to $C_{/V}$. These are the **τ-homotopy sheaves** of X.

Let W_τ denote the class of maps $s : X \to Y$ in sPre(C) such that $s_* : \pi_n^\tau(X, x) \to \pi_n^\tau(Y, s(x))$ is an isomorphism for all V and all basepoints $x \in X(V)$. Jardine proved that together with W_τ, the class of objectwise cofibrations determines a model category structure $\mathrm{sPre}_J(C)$ on sPre(C).

THEOREM 3.36 (Dugger-Hollander-Isaksen [**DHI04**]). *The identity functor*

$$\mathrm{L}_\tau \mathrm{sPre}(C) \to \mathrm{sPre}_J(C)$$

is a Quillen equivalence.

EXAMPLE 3.37. A τ-sheaf of sets on C, when viewed as a presheaf of simplicial sets, is in particular fibrant. It follows that when τ is subcanonical (i.e., every representable presheaf is in fact a sheaf) the Yoneda embedding $C \to \mathrm{sPre}(C)$ factors through the category of fibrant objects for the τ-local model category on $\mathrm{sPre}(C)$. Thus, there is a fully faithful Yoneda embedding $C \to \mathrm{Ho}(\mathrm{L}_\tau \mathrm{sPre}(C))$.

3.5. The Nisnevich topology. In this section S denotes a quasi-compact and quasi-separated scheme. We denote by Sm_S the category of finitely presented smooth schemes over S. Recall that while all smooth schemes U over S are locally of finite presentation by definition, saying that $U \to S$ is finitely presented means in addition to local finite presentation that the morphism is quasi-compact and quasi-separated. Note that Sm_S is an essentially small category because smooth implies locally of finite presentation and because S is quasi-compact and quasi-separated.

DEFINITION 3.38 (Lurie [**Lur16**, Section A.2.4]). The Nisnevich topology on Sm_S is the topology generated by those finite families of étale morphisms $\{p_i : U_i \to X\}_{i \in I}$ such that there is a finite sequence $\emptyset \subseteq Z_n \subseteq Z_{n-1} \subseteq \cdots \subseteq Z_1 \subseteq Z_0 = X$ of finitely presented closed subschemes of X such that

$$\coprod_{i \in I} p_i^{-1}(Z_m - Z_{m+1}) \to Z_m - Z_{m+1}$$

admits a section for $0 \leq m \leq n-1$.

REMARK 3.39. The referee pointed out that Hoyois has proved in a preprint [**Hoy16**] that this definition is equivalent (for S quasi-compact and quasi-separated) to the original definition of Nisnevich [**Nis89**], which says that an étale cover $U \to X$ is Nisnevich if it is surjective on k-points for all fields k.

EXERCISE 3.40. Show that when S is noetherian of finite Krull dimension, then a finite family of étale morphisms $\{p_i : U_i \to X\}_{i \in I}$ is a Nisnevich cover if and only if for each point $x \in X$ there is an index $i \in I$ and a point $y \in U_i$ over x such that the induced map $k(x) \to k(y)$ is an isomorphism. This is the usual definition of a Nisnevich cover, as used for example by [**MV99**].

EXAMPLE 3.41. Let k be a field of characteristic different than 2 and $a \in k$ a non-zero element. We cover $\mathbb{A}^1$ by the Zariski open immersion $\mathbb{A}^1 - \{a\} \to \mathbb{A}^1$ and the étale map $\mathbb{A}^1 - \{0\} \to \mathbb{A}^1$ given by $x \mapsto x^2$. This étale cover is Nisnevich if and only if a is a square in k.

EXERCISE 3.42. Zariski covers are in particular Nisnevich covers. For example, we will use later the standard cover of $\mathbb{P}^1$ by two copies of $\mathbb{A}^1$.

Of particular importance in the Nisnevich topology are the so-called elementary distinguished squares.

DEFINITION 3.43. A pullback diagram

$$\begin{array}{ccc} U \times_X V & \longrightarrow & V \\ \big\downarrow & & \big\downarrow{\scriptstyle p} \\ U & \xrightarrow{\ i\ } & X \end{array}$$

of S-schemes in Sm_S is an **elementary distinguished (Nisnevich) square** if i is a Zariski open immersion, p is étale, and $p^{-1}(X-U) \to (X-U)$ is an isomorphism of schemes where $X-U$ is equipped with the reduced induced scheme structure.

The proof of the following lemma is left as an easy exercise for the reader.

LEMMA 3.44. *In the notation above, $\{i : U \to X, p : V \to X\}$ is a Nisnevich cover of X.*

EXAMPLE 3.45. If a is a square in Example 3.41, then we obtain a Nisnevich cover which does not come from an elementary distinguished square. However, if we remove one of the square roots of a from $\mathbb{A}^1 - \{0\}$, then we do obtain an elementary distinguished square.

EXERCISE 3.46. Let p be a prime, let $X = \operatorname{Spec} \mathbb{Z}_{(p)}$, and let $V = \operatorname{Spec} \mathbb{Z}_{(p)}[i] \to X$, where $i^2 + 1 = 0$. Let $U = \operatorname{Spec} \mathbb{Q} \to X$. Then, $\{U, V\}$ is an étale cover of $\operatorname{Spec} \mathbb{Z}_{(p)}$ for all odd p. It is Nisnevich if and only if in addition $p \equiv 1 \mod 4$.

EXAMPLE 3.47. Let $X = \operatorname{Spec} R$, where R is a discrete valuation ring with field of fractions K. Suppose that $p : V \to X$ is an étale map where V is the spectrum of another discrete valuation ring S. Then, the square

$$\begin{array}{ccc} U \times_X V & \longrightarrow & V \\ \downarrow & & \downarrow p \\ U & \xrightarrow{i} & X \end{array}$$

with $i : U = \operatorname{Spec} K \to X$ is an elementary distinguished square if and only if the inertial degree of $R \to S$ is 1.

DEFINITION 3.48. The **Nisnevich-local model category** $\mathrm{L}_{\mathrm{Nis}}\mathrm{sPre}(\mathrm{Sm}_S)$ will be denoted simply by Spc_S, and the fibrant objects of Spc_S will be called **spaces**. So, a space is a presheaf of Kan complexes on Sm_S satisfying Nisnevich hyperdescent in the sense that the arrows (1) are weak equivalences for Nisnevich hypercovers.

WARNING 3.49. There are three candidates for the $\mathbb{A}^1$-homotopy theory over S. One is the $\mathbb{A}^1$-localization of the Joyal-Jardine Nisnevich-local model structure [**Jar87**]. The other is that used by [**AHW15a**], which imposes descent only for covers. Finally, we impose descent for all hypercovers. When S is noetherian of finite Krull dimension, all three definitions are Quillen equivalent. In all cases, our definition is equivalent to the Joyal-Jardine definition, by the main result of [**DHI04**].

NOTATION 3.50. If X and Y are presheaves of simplicial sets on Sm_S, we will write $[X, Y]_{\mathrm{Nis}}$ for the set of Nisnevich homotopy classes of maps from X to Y, which is the hom-set from X to Y in the homotopy category of $\mathrm{L}_{\mathrm{Nis}}\mathrm{sPre}(\mathrm{Sm}_S)$. The pointed version is written $[X, Y]_{\mathrm{Nis},\star}$. When necessary, we will write $[X, Y]_s$ for the homotopy classes of maps from X to Y in $\mathrm{sPre}(\mathrm{Sm}_S)$, and similarly we write $[X, Y]_{s,\star}$ for the homotopy classes of pointed maps.

NOTATION 3.51. We will write $\mathrm{L}_{\mathrm{Nis}}$ for the left derived functor of the identity functor $\mathrm{sPre}(\mathrm{Sm}_S) \to \mathrm{L}_{\mathrm{Nis}}\mathrm{sPre}(\mathrm{Sm}_S)$. Thus, $\mathrm{L}_{\mathrm{Nis}}$ is computed by taking a cofibrant replacement functor with respect to the Nisnevich-local model category structure on $\mathrm{sPre}(\mathrm{Sm}_S)$.

EXAMPLE 3.52. Given a scheme X essentially of finite presentation over S, we abuse notation and also view X as the presheaf it represents on Sm_S. So, if Y is a finitely presented smooth S-scheme, then $X(Y) = \mathrm{Hom}_S(Y, X)$. Since the Nisnevich topology is subcanonical, it is an easy exercise to see that X is fibrant. Indeed, homotopy limits of discrete spaces are just computed as limits of the underlying sets of components. We will discuss homotopy limits and colimits further in Section 4.

The next proposition is a key tool for practically verifying Nisnevich fibrancy for a given presheaf of simplicial sets on Sm_S.

PROPOSITION 3.53. *Suppose that S is a noetherian scheme of finite Krull dimension. A simplicial presheaf F on Sm_S is Nisnevich-fibrant if and only if for every elementary distinguished square*

$$\begin{array}{ccc} U \times_X V & \longrightarrow & V \\ \downarrow & & \downarrow p \\ U & \xrightarrow{i} & X \end{array}$$

the natural map

$$F(X) \to F(V) \times_{F(U \times_X V)} F(U)$$

is a weak equivalence of simplicial sets and $F(\emptyset)$ is a final object.

Presheaves possessing the property in the proposition are said to satisfy the **Brown-Gersten property** [**BG73**] or the **excision** property, although Brown and Gersten studied the Zariski analog.

PROOF. Let us first indicate the references for this theorem. The proof in [**MV99**, Section 3.1] applies to sheaves of sets (i.e. sheaves valued in discrete simplicial sets); to deduce the simplicial version, one uses the techniques in [**BG73**].

One way the reader can get straight to the case of simplicial presheaves is via the following argument. The Nisnevich topology is generated by a **cd-structure**, a collection of squares in Sm_S stable under isomorphism; the cd-structure corresponding to the Nisnevich topology is given by the elementary distinguished squares. This observation amounts to [**Voe10b**, Proposition 2.17, Remark 2.18]. In [**Voe10a**, Section 2], Voevodsky gives conditions on a category C equipped with a cd-structure for when the sheaf condition on a presheaf of sets on C coincides with the excision condition with respect to the cd-structure. For a proof of the corresponding claim for simplicial presheaves (and hyperdescent), one can refer to [**AHW15a**, Theorem 3.2.5].

For another, direct approach see [**Dug01b**]. □

Let $\mathrm{Sm}_S^{\mathrm{Aff}}$ denote the full subcategory category of Sm_S consisting of (absolutely) affine schemes. A presheaf X on $\mathrm{Sm}_S^{\mathrm{Aff}}$ satisfies **affine Nisnevich excision** if it satisfies excision for the cd-structure on $\mathrm{Sm}_S^{\mathrm{Aff}}$ consisting of cartesian squares

$$\begin{array}{ccc} \mathrm{Spec}\, R'_f & \longrightarrow & \mathrm{Spec}\, R' \\ \downarrow & & \downarrow \\ \mathrm{Spec}\, R_f & \longrightarrow & \mathrm{Spec}\, R, \end{array}$$

where $\operatorname{Spec} R' \to \operatorname{Spec} R$ is étale, $f \in R$, and $R/(f) \cong R'/(f)$. An important result of [**AHW15a**] says that the topology generated by the affine Nisnevich cd-structure is the same as the Nisnevich topology restricted to $\mathrm{Sm}_S^{\mathrm{Aff}}$.

3.6. The $\mathbb{A}^1$-homotopy category. To define the $\mathbb{A}^1$-homotopy category, we perform a further left Bousfield localization of $\mathrm{L}_{\mathrm{Nis}}\mathrm{sPre}(\mathrm{Sm}_S)$. As above, S denotes a quasi-compact and quasi-separated scheme.

DEFINITION 3.54. Let I be the class of maps $\mathbb{A}^1 \times_S X \to X$ in $\mathrm{L}_{\mathrm{Nis}}\mathrm{sPre}(\mathrm{Sm}_S)$ as X ranges over all objects of Sm_S. Since Sm_S is essentially small, we can choose a subset $J \subseteq I$ containing maps $\mathbb{A}^1 \times_S X \to X$ as X ranges over a representative of each isomorphism class of Sm_S.

The **$\mathbb{A}^1$-homotopy theory** of S is the left Bousfield localization

$$\mathrm{L}_{\mathbb{A}^1}\mathrm{L}_{\mathrm{Nis}}\mathrm{sPre}(\mathrm{Sm}_S)$$

of $\mathrm{L}_{\mathrm{Nis}}\mathrm{sPre}(\mathrm{Sm}_S)$ with respect to J. Its homotopy category will be called the **$\mathbb{A}^1$-homotopy category** of S. Let $\mathrm{Spc}_S^{\mathbb{A}^1}$ be $\mathrm{L}_{\mathbb{A}^1}\mathrm{L}_{\mathrm{Nis}}\mathrm{sPre}(\mathrm{Sm}_S)$. Fibrant objects of $\mathrm{Spc}_S^{\mathbb{A}^1}$ will be called **$\mathbb{A}^1$-spaces** or **$\mathbb{A}^1$-local spaces**. Note that the simplicial presheaf underlying any $\mathbb{A}^1$-space is in particular a space in the sense that it is fibrant in Spc_S. The homotopy category of $\mathrm{Spc}_S^{\mathbb{A}^1}$ will always be written as $\mathrm{Ho}(\mathrm{Spc}_S^{\mathbb{A}^1})$, and usually functoriality or naturality statements will be made with respect to the homotopy category.

PROPOSITION 3.55. *The Bousfield localization* $\mathrm{Spc}_S^{\mathbb{A}^1} = \mathrm{L}_{\mathbb{A}^1}\mathrm{L}_{\mathrm{Nis}}\mathrm{sPre}(\mathrm{Sm}_S)$ *exists.*

PROOF. The simplicial structure, left properness, and combinatoriality are inherited by Spc_S from $\mathrm{sPre}(\mathrm{Sm}_S)$, and hence by Theorem 3.28 the Bousfield localization exists. □

NOTATION 3.56. If X and Y are presheaves of simplicial sets on Sm_S, we will write $[X, Y]_{\mathbb{A}^1}$ for the set of $\mathbb{A}^1$-homotopy classes of maps from X to Y, which is the hom-set from X to Y in the homotopy category of $\mathrm{L}_{\mathbb{A}^1}\mathrm{L}_{\mathrm{Nis}}\mathrm{sPre}(\mathrm{Sm}_S)$. The pointed version is written $[X, Y]_{\mathbb{A}^1,\star}$.

NOTATION 3.57. We will write $\mathrm{L}_{\mathbb{A}^1}$ for the left derived functor of the identity functor $\mathrm{L}_{\mathrm{Nis}}\mathrm{sPre}(\mathrm{Sm}_S) \to \mathrm{L}_{\mathbb{A}^1}\mathrm{L}_{\mathrm{Nis}}\mathrm{sPre}(\mathrm{Sm}_S)$. Thus, $\mathrm{L}_{\mathbb{A}^1}\mathrm{L}_{\mathrm{Nis}}$ is computed by taking a cofibrant replacement functor with respect to the $\mathbb{A}^1$-local model category structure on $\mathrm{sPre}(\mathrm{Sm}_S)$.

REMARK 3.58. It is common to call an $\mathbb{A}^1$-space, an $\mathbb{A}^1$-local space, and indeed the fibrant objects of $\mathrm{Spc}_S^{\mathbb{A}^1}$ are $\mathbb{A}^1$-local. In fact, a simplicial presheaf X in $\mathrm{sPre}(\mathrm{Sm}_S)$ is $\mathbb{A}^1$-local, i.e., a fibrant object of $\mathrm{Spc}_S^{\mathbb{A}^1}$, if it

(1) takes values in Kan complexes (so that it is fibrant in $\mathrm{sPre}(\mathrm{Sm}_S)$),
(2) satisfies Nisnevich hyperdescent (so that it is fibrant in Spc_S), and
(3) if $X(U) \to X(\mathbb{A}^1 \times_S U)$ is a weak equivalence of simplicial sets for all U in Sm_S.

EXERCISE 3.59. Construct the pointed version $\mathrm{Spc}_{S,\star}^{\mathbb{A}^1}$ of $\mathrm{Spc}_S^{\mathbb{A}^1}$, the homotopy theory of pointed $\mathbb{A}^1$-spaces. We will have occasion to use this pointed version as well as the Quillen adjunction

$$\mathrm{Spc}_S^{\mathbb{A}^1} \rightleftarrows \mathrm{Spc}_{S,\star}^{\mathbb{A}^1},$$

which sends a presheaf of spaces X to the pointed presheaf of spaces X_+ obtained by adding a disjoint basepoint.

DEFINITION 3.60. The weak equivalences in $\mathrm{Spc}_S^{\mathbb{A}^1}$ are called **$\mathbb{A}^1$-weak equivalences** or **$\mathbb{A}^1$-local weak equivalences**.

Here is an expected class of $\mathbb{A}^1$-weak equivalences.

DEFINITION 3.61. Let $f, g : X \to Y$ be maps of simplicial presheaves. We say that f and g are $\mathbb{A}^1$-homotopic if there exists a map $H : F \times \mathbb{A}^1 \to G$ such that $H \circ (\mathrm{id}_F \times i_0) = f$ and $H \circ (\mathrm{id}_F \times i_1) = g$. A map $g : F \to G$ is an $\mathbb{A}^1$-homotopy equivalence if there exists morphisms $h : G \to F$ and that $h \circ g$ and $g \circ h$ are $\mathbb{A}^1$-homotopic to id_F and id_G respectively.

EXERCISE 3.62. Show that if $p : E \to X$ is a vector bundle in Sm_S, then p is an $\mathbb{A}^1$-homotopy equivalence.

EXERCISE 3.63. Show that any $\mathbb{A}^1$-homotopy equivalence $f : F \to G$ is an $\mathbb{A}^1$-weak equivalence. Note that there are many more $\mathbb{A}^1$-weak equivalences.

4. Basic properties of $\mathbb{A}^1$-algebraic topology

This long section is dedicated to outlining the basic facts that form the substrate of the unstable motivic homotopy theorists' work. Examples and basic theorems abound, and we hope that it provides a helpful user's manual. Most non-model category theoretic results below are due to Morel and Voevodsky [**MV99**].

Throughout this section we fix a quasi-compact and quasi-separated base scheme S, and we study the model category

$$\mathrm{Spc}_S^{\mathbb{A}^1} = \mathrm{L}_{\mathbb{A}^1}\mathrm{L}_{\mathrm{Nis}}\mathrm{sPre}(\mathrm{Sm}_S).$$

4.1. Computing homotopy limits and colimits through examples. An excellent source for the construction of homotopy limits or colimits is the exposition of Dwyer and Spalinski [**DS95**]. We start with an example from ordinary homotopy theory. Consider the following morphism of pullback diagrams of topological spaces:

$$\left(\star \to S^1 \leftarrow \star\right) \to \left(\star \to S^1 \leftarrow P_\star S^1\right),$$

where $P_\star S^1$ is the path space of S^1 consisting of paths beginning at the basepoint of S^1. This diagram is a homotopy equivalence in each spot. However, the pullback of the first is just a point, while the pullback of the second is the loop space ΩS^1, which is homotopy equivalent to the discrete space $\mathbb{Z}$. This example illustrates that some care is needed when forming the homotopically correct notion of pullback.

Similarly, consider the maps of pushout diagrams

$$\left(* \leftarrow S^0 \to D^1\right) \to \left(* \leftarrow S^0 \to *\right),$$

where D^1 is the 1-disk. Again, this map is a homotopy equivalence in each place. But, the pushout in the first case is S^1, and in the second case it is just a point. Again, care is required in order to compute the *correct* pushout.

The key in these examples is that $P_\star S^1 \to S^1$ is a fibration, while $S^0 \to D^1$ is a cofibration. By uniformly replacing pullback diagrams with pullback diagrams where the maps are fibrations, and then taking the pullback, one obtains a *homotopy-invariant* notion of pullback, the homotopy pullback. Similarly, by replacing pushout diagrams with homotopy equivalent diagrams in which the morphisms are cofibrations, one obtains homotopy pushouts.

DEFINITION 4.1. A **homotopy pullback** diagram in a model category M is a pullback diagram

$$\begin{array}{ccc} c & \longrightarrow & d \\ \downarrow & & \downarrow \\ e & \longrightarrow & f \end{array}$$

in M where at least one of $e \to f$ or $d \to f$ is a fibration in M. Given a pullback diagram $e \to f \leftarrow d$, the **homotopy pullback** is the pullback of either the diagram $e' \to f \leftarrow d$ or $e \to f \leftarrow d'$ where $e' \to f$ (resp $d' \to f$) is the fibrant replacement via **M4** of $e \to f$ (resp. $d \to f$).

EXERCISE 4.2. Show that homotopy pullbacks are independent up to weak equivalence of any choices made.

To put this notion on a more precise footing, we make the following construction.

PROPOSITION 4.3 ([**Lur09**, Proposition A.2.8.2]). *Let M be a combinatorial model category and I a small category. The pointwise weak equivalences and pointwise fibrations determine a model category structure on M^I called the* **projective model category structure**, *which we will denote by M^I_{proj}. The pointwise weak equivalences and pointwise cofibrations determine a model category structure on M^I called the* **injective model category structure**, *which we will denote by M^I_{inj}.*

We have already seen the projective model category structure in our discussion of presheaves of spaces on a small category. These two model categories on M^I can be used to compute homotopy limits and homotopy colimits.

LEMMA 4.4. *The functor $\Delta : M \to M^I$ taking $m \in M$ to the constant functor $I \to M$ on m admits both a left and a right adjoint.*

PROOF. Note that the category M is presentable by the definition of a combinatorial model category. This means that M has all small colimits and is λ-compactly generated for some regular cardinal λ. By the adjoint functor theorem [**AR94**, Theorem 1.66][1], it suffices to prove that Δ is accessible, preserves limits, and preserves small colimits. However, accessibility of Δ simply means that it commutes with κ-filtered colimits for some regular cardinal κ. Since we will show that it commutes with all small colimits, accessibility is an immediate consequence.

[1] This gives the criterion for the existence of a left adjoint for a functor between locally presentable categories; it is somewhat easier to prove that a functor between locally presentable categories which preserves small colimits is a left adjoint. There is a good discussion of these issues on the nLab.

To prove that Δ commutes with small limits, let $y \cong \lim_k y_k$ be a limit in M. Consider an object $x : I \to M$ of M^I. Then,

$$\begin{aligned}
\hom_{M^I}(x, \Delta(y)) &\cong \mathrm{eq}\left(\prod_{i\in I} \hom_M(x(i), y(i)) \rightrightarrows \prod_{f\in \mathrm{Ar}(I)} \hom_M(x(i), y(j))\right) \\
&\cong \mathrm{eq}\left(\prod_{i\in I} \lim_k \hom_M(x(i), y_k(j)) \rightrightarrows \prod_{f\in \mathrm{Ar}(I)} \lim_k \hom_M(x(i), y_k(j))\right) \\
&\cong \lim_k \mathrm{eq}\left(\prod_{i\in I} \hom_M(x(i), y_k(j)) \rightrightarrows \prod_{f\in \mathrm{Ar}(I)} \hom_M(x(i), y_k(j))\right) \\
&\cong \lim_k \hom_{M^I}(x, \Delta(y_k)),
\end{aligned}$$

using the fact that small limits commute with small limits and hence in particular equalizers and small products. It follows that $\Delta(y) \cong \lim_k \Delta(y_k)$, as desired. The proof that Δ preserves small colimits is left as an exercise. □

EXERCISE 4.5. Show that Δ preserves small colimits.

DEFINITION 4.6. We will call the right adjoint to Δ the limit functor $\lim_I$, while the left adjoint is the colimit functor colim_I.

LEMMA 4.7. *The pairs of adjoint functors*

$$\Delta : M \rightleftarrows M^I_{\mathrm{inj}} : \lim_I$$

and

$$\mathrm{colim}_I : M^I_{\mathrm{proj}} \rightleftarrows M : \Delta$$

are Quillen pairs.

PROOF. Note that Δ preserves pointwise weak equivalences, pointwise fibrations, and pointwise cofibrations. □

DEFINITION 4.8. We will write holim_I for $\mathbf{R}\lim_I$ and $\mathrm{hocolim}_I$ for $\mathbf{L}\,\mathrm{colim}_I$, and call theese the homotopy limit and homotopy colimit functors.

EXERCISE 4.9. Let I be the small category $\bullet \leftarrow \bullet \rightarrow \bullet$, which classifies pushouts. To compute the homotopy pushout $x \leftarrow y \rightarrow z$ in M, we must take an cofibrant replacement $x' \leftarrow y' \rightarrow z'$ in M^I_{proj}, and then we can compute the categorical pushout of the new diagram. Describe the cofibrant objects of M^I. Show that the homotopy pushout can be computed as the pushout of $x' \leftarrow y' \rightarrow z'$ where x' and y' are cofibrant and $y' \rightarrow z'$ is a cofibration. Show however that such diagrams are not in general cofibrant in M^I_{proj}.

PROPOSITION 4.10. *Right derived functors of right Quillen functors commute with homotopy limits and left derived functors of left Quillen functors commute with homotopy colimits.*

PROOF. We prove the result for right Quillen functors and homotopy limits. Suppose that we have a Quillen adjunction:

$$F : M \leftrightarrows N : G.$$

It is easy to check that this induces a Quillen adjunction $F^I : M^I_{\text{inj}} \leftrightarrows N^I_{\text{inj}} : G^I$. Indeed, it is enough to check that F^I preserves cofibrations and acyclic cofibrations, but these are defined pointwise in the injective model category structure, so the fact that F is a left Quillen functor implies that F^I is as well. Consider the following diagram

$$\begin{array}{ccc} M^I_{\text{inj}} & \xrightarrow{F^I} & N^I_{\text{inj}} \\ \uparrow{\scriptstyle\Delta} & & \uparrow{\scriptstyle\Delta} \\ M & \xrightarrow{F} & N \end{array}$$

of left Quillen functors. This diagram commutes on the level of underlying categories; picking appropriate fibrant replacements to compute the right adjoints, the right derived versions of the functors commute which induces a commutative diagram

$$\begin{array}{ccc} \mathrm{Ho}(M^I_{\text{inj}}) & \xrightarrow{\mathbf{L}F^I} & \mathrm{Ho}(N^I_{\text{inj}}) \\ \uparrow{\scriptstyle\mathbf{L}\Delta} & & \uparrow{\scriptstyle\mathbf{L}\Delta} \\ \mathrm{Ho}(M) & \xrightarrow{\mathbf{L}F} & \mathrm{Ho}(N) \end{array}$$

of left adjoints on the level of homotopy categories. This means means that the diagram

$$\begin{array}{ccc} \mathrm{Ho}(M^I_{\text{inj}}) & \xleftarrow{\mathbf{R}G^I} & \mathrm{Ho}(N^I_{\text{inj}}) \\ {\scriptstyle\mathbf{R}\,\mathrm{holim}_I}\downarrow & & \downarrow{\scriptstyle\mathbf{R}\,\mathrm{holim}_I} \\ \mathrm{Ho}(M) & \xleftarrow{\mathbf{R}G} & \mathrm{Ho}(N) \end{array}$$

of right adjoints commutes. □

We are now in a position to give examples.

EXERCISE 4.11. One should be careful when trying to commute homotopy limits or colimits using the above proposition — the functors must be derived. Construct an example using a morphism of commutative rings $R \to S$, the functor $\otimes_R S : \mathrm{Ch}^{\geq 0}_R \to \mathrm{Ch}^{\geq 0}_S$, and the mapping cone of an R-module $M \to N$ thought of as chain complexes concentrated in a single degree to show that preservation of homotopy colimits fail if $\otimes_R S$ is not derived. Hint: see the example of mapping cones worked out in Example 4.12.

EXAMPLE 4.12. Let A be an associative ring, and consider $\mathrm{Ch}^{\geq 0}_A$, the category of non-negatively graded chain complexes equipped with the projective model category structure. Let M be an A-module viewed as a chain complex concentrated in degree 0, and let $N_\bullet$ be a chain complex. The actual pushout of a map $M \to N_\bullet$ along $M \to 0$ is just the cokernel of the map of complexes. If $N_\bullet = 0$, this cokernel is zero. However, by the recipe above, we should replace 0 with a quasi-isomorphic fibrant model $P_\bullet$ with a map $M \to P_\bullet$ that is a cofibration. A functorial choice turns out to be the cone on the identity of M. This is the complex $M \xrightarrow{\mathrm{id}_M} M$ with M placed in degrees 1 and 0. This time, when we take the cokernel, we get the complex $M[1]$. This confirms what everyone wants: that $M \to 0 \to M[1]$ should be a distinguished triangle in the derived category of A, which is what is needed to to have long exact sequences in homology.

Let us now turn to examples in $\mathbb{A}^1$-homotopy theory. The following proposition gives a way of constructing many examples of homotopy pushouts in the category Spc_S and is a consequence of the characterization of fibrant objects in Spc_S.

PROPOSITION 4.13. *If S is a noetherian scheme of finite Krull dimension, then an elementary distinguished (Nisnevich) square*

$$\begin{array}{ccc} U \times_X V & \longrightarrow & V \\ \downarrow & & \downarrow{\scriptstyle p} \\ U & \xrightarrow{\ i\ } & X, \end{array}$$

in Sm_S thought of as a diagram of simplicial presheaves is a homotopy pushout in Spc_S.

PROOF. Since the Nisnevich topology is subcanonical (it is coarser than the étale topology which is subcanonical) we may regard these squares as diagrams in Spc_S via the Yoneda embedding (or, rather, its simplicial analogue — we think of schemes as sheaves of discrete simplicial sets). Let X be a space, i.e., a fibrant object of Spc_S. Proposition 3.53 tells us that applying X to an elementary distinguished square gives rise to a homotopy pullback square. This verifies the universal property for a homotopy pushout. □

One problem with the category of schemes, as mentioned above, is that it lacks general colimits, even finite colimits. In particular, general quotient spaces do not exist in Sm_S.

DEFINITION 4.14. For the purposes of this paper, the **quotient** X/Y of a map $X \to Y$ of schemes in Sm_S is always defined to be the homotopy cofiber of the map in $\mathrm{Spc}_S^{\mathbb{A}^1}$. Recall that the homotopy cofiber is the homotopy pushout of $\star \leftarrow X \to Y$. Note that since localization is a left adjoint, this definition agrees up to homotopy with the $\mathbb{A}^1$-localization of the homotopy cofiber computed in Spc_S by Proposition 4.10.

EXAMPLE 4.15. Proposition 4.13 implies that in the situation of an elementary distinguished square, the natural map

$$\frac{V}{U \times_X V} \to \frac{X}{U}$$

is an $\mathbb{A}^1$-local weak equivalence. To see this, we see that Proposition 4.13 gives a Nisnevich local weak equivalence of the cofibers of the top and bottom horizontal arrows; since $\mathrm{L}_{\mathbb{A}^1}$ is a left adjoint, we see that it preserves cofibers and thus gives rise to the desired $\mathbb{A}^1$-local weak equivalence.

EXAMPLE 4.16. A particularly important example of a quotient or homotopy cofiber is the **suspension** of a pointed object X in $\mathrm{Spc}_{S,\star}^{\mathbb{A}^1}$. This is simply the homotopy cofiber of $X \to \star$, or in other words, the homotopy pushout of the diagram

$$\begin{array}{ccc} X & \longrightarrow & * \\ \big\downarrow & & \\ * & & \end{array}$$

which we denote by ΣX. See Section 4.6 for one use of the construction.

4.2. $\mathbb{A}^1$-homotopy fiber sequences and long exact sequences in homotopy sheaves.

DEFINITION 4.17. Let $X \to Y$ be a map of pointed objects in a model category. The **homotopy fiber** F is the homotopy pullback of $\star \to Y \leftarrow X$. In general, if $F \to X \to Y$ is a sequence of spaces and if F is weak equivalent to the homotopy fiber of $X \to Y$, then we call $F \to X \to Y$ a homotopy fiber sequence.

Recall that in ordinary algebraic topology, given a homotopy fiber sequence

$$F \to X \to Y$$

of pointed spaces, there is a long exact sequence

$$\cdots \pi_{n+1}Y \to \pi_n F \to \pi_n X \to \pi_n Y \to \pi_{n-1}F \to \cdots$$

of homotopy groups, where we omit the basepoint for simplicity. Exactness should be carefully interpreted for $n = 0, 1$, when these are only pointed sets or not-necessarily-abelian groups. For details, consult Bousfield and Kan [**BK72**, Section IX.4.1].

DEFINITION 4.18. The **Nisnevich homotopy sheaf** $\pi_n^{\mathrm{Nis}}(X)$ of a pointed object X of Spc_S is the Nisnevich sheafification of the presheaf

$$U \mapsto [S^n \wedge U_+, X]_{\mathrm{Nis},\star}.$$

DEFINITION 4.19. The **$\mathbb{A}^1$-homotopy sheaf** $\pi_n^{\mathbb{A}^1}(X)$ of a pointed object X of $\mathrm{Spc}_S^{\mathbb{A}^1}$ is the Nisnevich sheafification of the presheaf

$$U \mapsto [S^n \wedge U_+, X]_{\mathbb{A}^1,\star}.$$

EXERCISE 4.20. Show that if X is weakly equivalent to $\mathrm{L}_{\mathbb{A}^1}\mathrm{L}_{\mathrm{Nis}}X$, where X is a pointed simplicial presheaf, then the natural map $\pi_n^{\mathrm{Nis}}(X) \to \pi_n^{\mathbb{A}^1}(X)$ is an isomorphism of Nisnevich sheaves.

The following result is a good illustration of the theory we have developed so far.

PROPOSITION 4.21. *Let $F \to X \to Y$ be a homotopy fiber sequence in $\mathrm{Spc}_S^{\mathbb{A}^1}$. Then, there is a natural long exact sequence*

$$\cdots \to \pi_{n+1}^{\mathbb{A}^1}Y \to \pi_n^{\mathbb{A}^1}F \to \pi_n^{\mathbb{A}^1}X \to \pi_n^{\mathbb{A}^1}Y \to \cdots$$

of Nisnevich sheaves.

PROOF. The forgetful functor $\mathrm{Spc}_S^{\mathbb{A}^1} \to \mathrm{sPre}(\mathrm{Sm}_S)$ is a right adjoint, and hence it preserves homotopy fiber sequences. It follows from the fact that fibrations are defined as object-wise fibrations that there is a natural long exact sequence of homotopy presheaves. Since sheafification, and in particular Nisnevich sheafification, is exact [**TS14**, Tag 03CN], the claim follows. □

REMARK 4.22. We caution the reader that although the functor: $\mathrm{L}_{\mathbb{A}^1}\mathrm{L}_{\mathrm{Nis}}$: $\mathrm{sPre}(\mathrm{Sm}_S) \to \mathrm{Spc}_S^{\mathbb{A}^1}$ preserves homotopy colimits, it is not clear that resulting homotopy colimit diagram in $\mathrm{Spc}_S^{\mathbb{A}^1}$ possess any exactness properties. To be more explicit, let $i : \mathrm{Spc}_S^{\mathbb{A}^1} \to \mathrm{sPre}(\mathrm{Sm}_S)$ be the forgetful functor. Suppose that we have a homotopy cofiber sequence: $X \to Y \to Z$ in $\mathrm{sPre}(\mathrm{Sm}_S)$, then the it is not clear that $i\mathrm{L}_{\mathbb{A}^1}\mathrm{L}_{\mathrm{Nis}}(Z)$ is equivalent to the cofiber of $i\mathrm{L}_{\mathbb{A}^1}\mathrm{L}_{\mathrm{Nis}}(X) \to i\mathrm{L}_{\mathbb{A}^1}\mathrm{L}_{\mathrm{Nis}}(Y)$ since we are composing a Quillen left adjoint with a Quillen right adjoint. Consequently, long exact sequences which arise out of cofiber sequences (such as mapping into Eilenberg-MacLane spaces which produces the long exact sequences in ordinary cohomology) in $\mathrm{sPre}(\mathrm{Sm}_S)$ will not apply to this situation.

4.3. The $\mathrm{Sing}^{\mathbb{A}^1}$-construction. While the process of Nisnevich localization, which produces objects of Spc_S, is familiar from ordinary sheaf theory, the localization $\mathrm{L}_{\mathbb{A}^1} : \mathrm{Spc}_S \to \mathrm{Spc}_S^{\mathbb{A}^1}$ is more difficult to grasp concretely. This section describes one model for the localization functor $\mathrm{L}_{\mathbb{A}^1}$.

Consider the cosimplicial scheme $\Delta^\bullet$ where

$$\Delta^n = \mathrm{Spec}\ k[x_0, \dots, x_n]/(x_0 + \cdots + x_1 = 1)$$

with the face and degeneracy maps familiar from the standard topological simplex. The scheme Δ^n is a closed subscheme of $\mathbb{A}^{n+1}$ isomorphic to $\mathbb{A}^n$, the ith coface map $\partial_j : \Delta^n \to \Delta^{n+1}$ is defined by setting $x_j = 0$, and the ith codegeneracy $\sigma_i : \Delta^n \to \Delta^{n-1}$ is given by summing the ith and $i+1$st coordinates.

DEFINITION 4.23. Let X be a simplicial presheaf. We define the simplicial presheaf $\mathrm{Sing}^{\mathbb{A}^1} X := |X(- \times \Delta^\bullet)|$. This gives the **singular construction** functor

$$\mathrm{Sing}^{\mathbb{A}^1} : \mathrm{sPre}(\mathrm{Sm}_S) \to \mathrm{sPre}(\mathrm{Sm}_S).$$

We will also write $\mathrm{Sing}^{\mathbb{A}^1}$ for the restriction of the singular construction to $\mathrm{Spc}_S \subseteq \mathrm{sPre}(\mathrm{Sm}_S)$.

REMARK 4.24. Since geometric realizations do not commute in general with homotopy limits, there is no reason to expect $\mathrm{Sing}^{\mathbb{A}^1}$ to preserve the property of being Nisnevich-local. At heart this is the reason for both the subtlety and the depth of motivic homotopy theory.

From the above remark it is thus useful to introduce a new terminology: we say that a simplicial presheaf X is **$\mathbb{A}^1$-invariant** if $X(U) \to X(U \times \mathbb{A}^1)$ is a weak homotopy equivalence of simplicial sets for every U in Sm_S.

THEOREM 4.25. *Let S be a base separated noetherian scheme and X a simplicial presheaf. Then,*

(1) $\mathrm{Sing}^{\mathbb{A}^1} X$ *is $\mathbb{A}^1$-invariant, and*
(2) *the natural map $g : X \to \mathrm{Sing}^{\mathbb{A}^1} X$ induces a weak equivalence*

$$\mathrm{map}(\mathrm{Sing}^{\mathbb{A}^1} X, Y) \to \mathrm{map}(X, Y)$$

for any $\mathbb{A}^1$-invariant simplicial presheaf Y.

PROOF. For $i = 0, \dots, n$ we have maps $\theta_i : \mathbb{A}^{n+1} \simeq \Delta^{n+1} \to \mathbb{A}^n \simeq \Delta^n \times_S \mathbb{A}^1$ corresponding to a "simplicial decomposition" of $\Delta^n \times_S \mathbb{A}^1$ made up of Δ^{n+1}'s (see, for example, [**MVW06**, Figure 2.1]). For an arbitrary S-scheme U, the θ_i maps induce a morphism of cosimplicial schemes

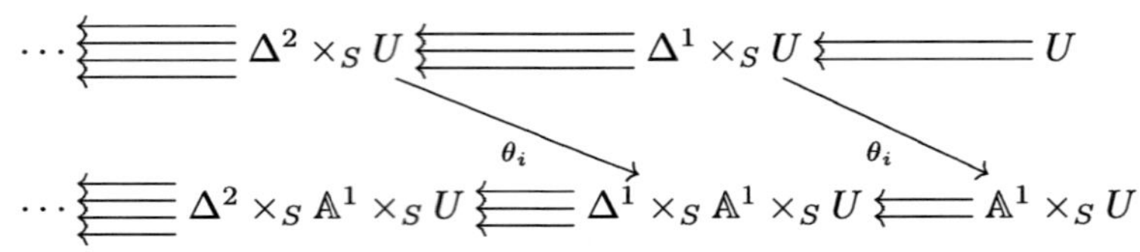

such that, upon applying a simplicial presheaf X, we get a simplicial homotopy [**Wei94**, Section 8.3.11] between the maps $\partial_0^*, \partial_1^* : \mathrm{Sing}^{\mathbb{A}^1} X(U \times \mathbb{A}^1) \to \mathrm{Sing}^{\mathbb{A}^1} X(U)$ induced by the 0 and 1-section respectively. Hence, as shown in the exercise below, $\mathrm{Sing}^{\mathbb{A}^1} X$ is $\mathbb{A}^1$-invariant.

Observe that the functor $U \mapsto X(U \times_S \Delta^n)$ is the same as the functor $U \mapsto \mathrm{map}(U \times_S \Delta^n, X)$. We have a natural map $X \simeq \mathrm{map}(\Delta^0, X) \to \mathrm{map}(\Delta^n, X)$ for each n, so we think of the map $X \to \mathrm{Sing}^{\mathbb{A}^1} X$ as the canonical map from the zero simplices.

To check the second claim, it is enough to prove that for all $n \geq 0$, we have a weak equivalence

$$\mathrm{map}(X, Y) \to \mathrm{map}(\mathrm{map}(\Delta^n, X), Y)$$

whenever Y is $\mathbb{A}^1$-invariant. Furthermore, $\mathrm{map}(\Delta^n, X) \simeq \mathrm{map}(\Delta^1, \mathrm{map}(\Delta^{n-1}, X))$ so by induction we just need to prove the claim for $n = 1$. To do so, we claim that the map $f : X \to \mathrm{map}(\Delta^1, X)$ induced by the projection $\mathbb{A}^1 \to S$ is an $\mathbb{A}^1$-homotopy equivalence, from which we conclude the desired claim from Exercise 3.63.

There is a map $g : \mathrm{map}(\Delta^1, X) \to X$ induced by the zero section, from which we automatically have $f \circ g = \mathrm{id}$. We then have to construct an $\mathbb{A}^1$-homotopy between $g \circ f$ and $\mathrm{id}_{\mathrm{map}(\Delta^1, X)}$ so we look for a map $H : \mathrm{map}(\Delta^1, X) \times \mathbb{A}^1 \to \mathrm{map}(\Delta^1, X)$. By adjunction this is the same data as a map $H : \mathrm{map}(\Delta^1, X) \to \mathrm{map}(\Delta^1 \times \Delta^1, X)$. To construct this map we use the multiplication map $\mathbb{A}^1 \times \mathbb{A}^1 \to \mathbb{A}^1$, $(x, y) \mapsto xy$ from which it is easy to see that $H^*(\mathrm{id} \times \partial_0)^* = \mathrm{id}$ and $H^*(\mathrm{id} \times \partial_1) = g \circ f$. □

EXERCISE 4.26. If X is a simplicial presheaf, then X is $\mathbb{A}^1$-invariant if and only if for any $U \in \mathrm{Sm}_S$ the morphisms $\partial_0^*, \partial_1^* : X(U \times_S \mathbb{A}^1) \to X(U)$ induced by the 0 and 1-sections are homotopic. Hint: use again the multiplication map $\mathbb{A}^1 \times \mathbb{A}^1 \to \mathbb{A}^1, (x, y) \mapsto xy$ as a homotopy. See [**MVW06**, Lemma 2.16].

We conclude from the above results that $\mathrm{Sing}^{\mathbb{A}^1} X$ is $\mathbb{A}^1$-invariant and, furthermore, X and $\mathrm{Sing}^{\mathbb{A}^1} X$ are $\mathbb{A}^1$-weak equivalent which means, more explicitly, that they become weakly equivalent in $\mathrm{Spc}_S^{\mathbb{A}^1}$ after applying $\mathrm{L}_{\mathbb{A}^1}\mathrm{L}_{\mathrm{Nis}}$.

THEOREM 4.27. *The functor* $\mathrm{L}_{\mathbb{A}^1}\mathrm{L}_{\mathrm{Nis}} : \mathrm{sPre}(\mathrm{Sm}_S) \to \mathrm{Spc}_S^{\mathbb{A}^1}$ *is equivalent to the countable iteration* $(\mathrm{L}_{\mathrm{Nis}} \mathrm{Sing}^{\mathbb{A}^1})^{\circ \mathrm{N}}$.

PROOF. Let $\Phi = \mathrm{L}_{\mathrm{Nis}} \mathrm{Sing}^{\mathbb{A}^1}$, so that the theorem claims that $\Phi^{\circ \mathrm{N}} \simeq \mathrm{L}_{\mathrm{Nis}}$. We first argue that $(\mathrm{L}_{\mathrm{Nis}} \mathrm{Sing}^{\mathbb{A}^1})^{\circ \mathrm{N}} X$ is fibrant in $\mathrm{Spc}_S^{\mathbb{A}^1}$ for any X in $\mathrm{sPre}(\mathrm{Sm}_S)$. We must simply check that it is Nisnevich and $\mathbb{A}^1$-local. To check that it is Nisnevich local, write

$$\Phi^{\circ \mathrm{N}}(X) \simeq \underset{n \to \infty}{\mathrm{hocolim}} (\mathrm{L}_{\mathrm{Nis}} \mathrm{Sing}^{\mathbb{A}^1})^{\circ n}(X),$$

a filtered homotopy colimit of Nisnevich local presheaves of spaces. It hence suffices to show that the forgetful functor $\mathrm{Spc}_S \to \mathrm{sPre}(\mathrm{Sm}_S)$ preserves filtered homotopy colimits. However, since the sheaf condition is checked on the finite homotopy limits induced from the elementary distinguished squares by Proposition 3.53, and since filtered homotopy limits commute with finite homotopy limits, the result is immediate. At this point we must be honest and point out that the main reference we know for the commutativity of finite homotopy limits and filtered homotopy colimits, namely [**Lur09**, Proposition 5.3.3.3], is for ∞-categories rather than model categories. However, since homotopy limits and colimits in combinatorial simplicial model categories (such as all model categories in this paper) agree with the corresponding ∞-categorical limits and colimits by [**Lur09**, Section 4.2.4], this should be no cause for concern.

To check that $(\mathrm{L}_{\mathrm{Nis}}\operatorname{Sing}^{\mathbb{A}^1})^{\circ\mathbb{N}}X$ is $\mathbb{A}^1$-local, note that we can write

$$\Phi^{\circ\mathbb{N}}X \simeq \operatorname*{hocolim}_{n\to\infty}(\operatorname{Sing}^{\mathbb{A}^1}\mathrm{L}_{\mathrm{Nis}})^{\circ n}\left(\operatorname{Sing}^{\mathbb{A}^1}X\right),$$

a filtered homotopy colimit of $\mathbb{A}^1$-invariant presheaves by Theorem 4.25. But, filtered homotopy colimits of $\mathbb{A}^1$-invariant presheaves are $\mathbb{A}^1$-invariant. Since $\Phi^{\circ\mathbb{N}}(X)$ is Nisnevich local and $\mathbb{A}^1$-invariant, it is $\mathbb{A}^1$-local.

Thus, we have seen that $\Phi^{\circ\mathbb{N}}$ does indeed take values in the fibrant objects of $\mathrm{Spc}_S^{\mathbb{A}^1}$. Finally, we claim that it suffices to show that $\Phi \simeq \mathrm{L}_{\mathrm{Nis}}\operatorname{Sing}^{\mathbb{A}^1}$ preserves $\mathbb{A}^1$-local weak equivalences. Indeed, if this is the case, then so does $\Phi^{\circ\mathbb{N}}$, which will show that

$$\Phi^{\circ\mathbb{N}}(X) \simeq \Phi^{\circ\mathbb{N}}(\mathrm{L}_{\mathbb{A}^1}\mathrm{L}_{\mathrm{Nis}}X) \simeq \mathrm{L}_{\mathbb{A}^1}\mathrm{L}_{\mathrm{Nis}}X,$$

since it is clear that $X \simeq \Phi(X)$ when X is $\mathbb{A}^1$-local. For the remainder of the proof, write $\mathrm{map}(-,-)$ for the mapping spaces in $\mathrm{sPre}(\mathrm{Sm}_S)$. We want to show that

$$\mathrm{map}(\Phi(X),Y) \simeq \mathrm{map}(X,Y)$$

for all $\mathbb{A}^1$-local objects Y of $\mathrm{sPre}(\mathrm{Sm}_S)$. But,

$$\begin{aligned}\mathrm{map}(\Phi(X),Y) &\simeq \mathrm{map}_{\mathrm{Spc}_S}(\mathrm{L}_{\mathrm{Nis}}\operatorname{Sing}^{\mathbb{A}^1}X,Y)\\ &\simeq \mathrm{map}(\operatorname{Sing}^{\mathbb{A}^1}X,Y)\end{aligned}$$

since Y is in particular Nisnevich local. As the singular construction functor $\operatorname{Sing}^{\mathbb{A}^1}$ is a homotopy colimit, it commutes with homotopy colimits. Since $X \simeq \operatorname{hocolim}_{U\to X}U$, where the colimit is over maps from smooth S-schemes U, it follows that it is enough to show that

$$\mathrm{map}(\operatorname{Sing}^{\mathbb{A}^1}U,X) \simeq \mathrm{map}(U,Y)$$

for U a smooth S-scheme and Y an $\mathbb{A}^1$-local presheaf. To prove this, it is enough in turn to show that

$$\mathrm{map}(U(-\times\mathbb{A}^n),Y) \simeq \mathrm{map}(U,Y),$$

where $U(-\times\mathbb{A}^n)$ is the presheaf of spaces $V \mapsto U(V\times\mathbb{A}^n)$. Note that because there is an S-point of $\mathbb{A}^n$, the representable presheaf U is a retract of $U(-\times\mathbb{A}^n)$, so it suffices to show that

$$\pi_0\mathrm{map}(U(-\times\mathbb{A}^n),Y) \cong \pi_0\mathrm{map}(U,Y),$$

or even just that the map

$$\pi_0\mathrm{map}(U,Y) \to \pi_0\mathrm{map}(U(-\times\mathbb{A}^n),Y)$$

induced by an S-point of $\mathbb{A}^n$ is a surjection. Now, $U(-\times\mathbb{A}^n) \simeq \operatorname{hocolim}_{V\times\mathbb{A}^n\to U} V\times\mathbb{A}^n$, so

$$\pi_0\mathrm{map}(U(-\times\mathbb{A}^n), Y) \cong \pi_0 \lim_{V\times\mathbb{A}^n\to U} \mathrm{map}(V\times\mathbb{A}^n, Y) \cong \pi_0 \lim_{V\times\mathbb{A}^n\to U} \mathrm{map}(V, Y),$$

the last weak equivalence owing to the fact that Y is $\mathbb{A}^1$-local. This limit can be computed as $\lim_{V\times\mathbb{A}^n\to U}\pi_0\mathrm{map}(V,Y)$ since π_0 commutes with all colimits (being left adjoint to the inclusion of discrete spaces in all spaces). Picking an S-point of $\mathbb{A}^n$ gives a compatible family

$$\lim_{V\to U}\pi_0\mathrm{map}(V,Y) \cong \pi_0\mathrm{map}(U,Y),$$

giving a section of the natural map $\pi_0\mathrm{map}(U,Y) \to \lim_{V\times\mathbb{A}^n\to U}\pi_0\mathrm{map}(V,Y)$. □

From this description, we get a number of non-formal consequences.

COROLLARY 4.28. *The $\mathbb{A}^1$-localization functor commutes with finite products.*

PROOF. Both $\mathrm{Sing}^{\mathbb{A}^1}$ (being a sifted colimit) and $\mathrm{L}_{\mathrm{Nis}}$ have this property. For $\mathrm{L}_{\mathrm{Nis}}$ the fact is clear because it is the left adjoint of a geometric morphism of ∞-topoi and hence left exact (see [**Lur09**]), while for $\mathrm{Sing}^{\mathbb{A}^1}$ we refer to [**ARV10**]. Alternatively, it is easy to check directly that the singular construction commutes with finite products and it is shown in [**MV99**, Theorem 1.66] that Nisnevich localization commutes with finite products. (Note that since finite products and finite homotopy products agree, it is easy to transfer the Morel-Voevodsky proof along the Quillen equivalences necessary to bring it over to our model for $\mathrm{Spc}_S^{\mathbb{A}^1}$.) □

The corollary is important in proving that certain functors which are symmetric monoidal on the level of presheaves, remain symmetric monoidal after $\mathbb{A}^1$-localization.

DEFINITION 4.29. If $X \in \mathrm{sPre}(\mathrm{Sm}_S)$, then X is **$\mathbb{A}^1$-connected** if the canonical map $X \to S$ induces an isomorphism of sheaves $\pi_0^{\mathbb{A}^1}X \to \pi_0^{\mathbb{A}^1}S = \star$. We say that X is **naively-$\mathbb{A}^1$-connected** if the canonical map $\mathrm{Sing}^{\mathbb{A}^1} X \to S$ induces an isomorphism $\pi_0^{\mathrm{Nis}}\,\mathrm{Sing}^{\mathbb{A}^1} X \to \pi_0^{\mathrm{Nis}}S = \star$.

COROLLARY 4.30 (Unstable $\mathbb{A}^1$-connectivity theorem). *Suppose that X is a simplicial presheaf on* Sm_S. *The canonical morphism $X \to \mathrm{L}_{\mathbb{A}^1}\mathrm{L}_{\mathrm{Nis}}X$ induces an epimorphism $\pi_0^{\mathrm{Nis}}X \to \pi_0^{\mathrm{Nis}}\mathrm{L}_{\mathbb{A}^1}\mathrm{L}_{\mathrm{Nis}}X = \pi_0^{\mathbb{A}^1}X$. Hence, if $\pi_0^{\mathrm{Nis}}X = \star$, then X is $\mathbb{A}^1$-connected.*

PROOF. By Theorem 3.36, it follows that $X \to \mathrm{L}_{\mathrm{Nis}}X$ induces isomorphisms on homotopy sheaves $\pi_0^{\mathrm{Nis}}X \to \pi_0^{\mathrm{Nis}}\mathrm{L}_{\mathrm{Nis}}X$. Hence, using the fact that sheafification preserves epimorphisms and Theorem 4.27, it suffices to show that $\pi_0X(U) \to \pi_0\,\mathrm{Sing}^{\mathbb{A}^1} X(U)$ is surjective for all $X \in \mathrm{sPre}(\mathrm{Sm}_S)$ and all $U \in \mathrm{Sm}_S$. To do so, we note that $\pi_0\,\mathrm{Sing}^{\mathbb{A}^1} X(U)$ is calculated as π_0 of the bisimplicial set $X_\bullet(U\times\Delta^\bullet)$. This is in turn calculated as the coequalizer of the diagram

$$\pi_0X(U\times_S\mathbb{A}^1) \rightrightarrows \pi_0X(U),$$

where the maps are induced by $\Delta^\bullet$ and thus we get the desired surjection. □

Consequently, to determine if a simplicial presheaf is $\mathbb{A}^1$-connected, it suffices to calculate its sheaf of "naive" $\mathbb{A}^1$-connected components. We will use this observation later to prove that SL_n is $\mathbb{A}^1$-connected.

COROLLARY 4.31. *If $X \in \mathrm{sPre}(\mathrm{Sm}_S)$, then the natural morphism $\pi_0^{\mathrm{Nis}} \mathrm{Sing}^{\mathbb{A}^1}(X) \to \pi_0^{\mathbb{A}^1} X$ is an epimorphism. Hence if X is naively $\mathbb{A}^1$-connected, then it is $\mathbb{A}^1$-connected.*

PROOF. Since the natural map $X \to \mathrm{Sing}^{\mathbb{A}^1} X$ is an $\mathbb{A}^1$-local weak equivalence by Theorem 4.25, we deduce that $\mathrm{L}_{\mathbb{A}^1}\mathrm{L}_{\mathrm{Nis}} X \simeq \mathrm{L}_{\mathbb{A}^1}\mathrm{L}_{\mathrm{Nis}}\,\mathrm{Sing}^{\mathbb{A}^1} X$, so we may apply Theorem 4.30 to $\mathrm{Sing}^{\mathbb{A}^1} X$ to get the desired conclusion. □

4.4. The sheaf of $\mathbb{A}^1$-connected components. The 0-th $\mathbb{A}^1$-homotopy sheaf, or the **sheaf of $\mathbb{A}^1$-connected components**, admits a simple interpretation: it is the Nisnevich sheafification of the presheaf $U \mapsto [U_+, X]_{\mathbb{A}^1} \simeq [U_+, \mathrm{L}_{\mathbb{A}^1}\mathrm{L}_{\mathrm{Nis}} X]_s$. With this description, we may perform some calculations whose results deviate from our intuition from topology.

DEFINITION 4.32. Let X be an S-scheme. We say that X is **$\mathbb{A}^1$-rigid** if $\mathrm{L}_{\mathbb{A}^1} X \simeq X$ in Spc_S. Concretely, this condition amounts to saying that $X(U \times_S \mathbb{A}^1_S) \simeq X(U)$ for any finitely presented smooth S-scheme U.

EXERCISE 4.33. Let k be a field. Prove that the following k-schemes are all $\mathbb{A}^1$-rigid:

(1) $\mathbb{G}_m$;
(2) smooth projective k-curves of positive genus;
(3) abelian varieties.

In fact, if S is a reduced scheme of finite Krull dimension, show that $\mathbb{G}_m$ is rigid in $\mathrm{Spc}_S^{\mathbb{A}^1}$.

PROPOSITION 4.34. *Let X be an $\mathbb{A}^1$-rigid S-scheme. Then $\pi_0^{\mathbb{A}^1}(X) \simeq X$ as Nisnevich sheaves, and $\pi_n^{\mathbb{A}^1}(X) = 0$ for $n > 0$.*

PROOF. The homotopy set $U \mapsto [U, X]_{\mathbb{A}^1} \simeq [U, X]_s = \pi_0(\mathrm{map}_{\mathrm{Spc}_S}(U, X))$ is equivalent to the set of S-scheme maps from U to X as U and X are discrete simplicial sets. Hence this presheaf is equivalent to the presheaf represented by X which is already a Nisnevich sheaf on Sm_S. Now, $[S^n \wedge U_+, X]_s = [S^n, \mathrm{map}_{\mathrm{Spc}_S}(U_+, X)]_{\mathrm{sSets}}$, which is trivial since the target is a discrete simplicial set. Since X is $\mathbb{A}^1$-rigid we see that $[S^n \wedge U_+, X]_s \cong [S^n \wedge U_+, X]_{\mathbb{A}^1}$, and thus the sheafification is also trivial. □

EXERCISE 4.35. Let $\mathrm{Sm}_S^{\mathbb{A}^1} \hookrightarrow \mathrm{Sm}_S$ be the full subcategory spanned by $\mathbb{A}^1$-rigid schemes. Then the natural functor $\mathrm{Sm}_S^{\mathbb{A}^1} \to \mathrm{Spc}_S^{\mathbb{A}^1}$ which is the composite of $\mathrm{L}_{\mathbb{A}^1} : \mathrm{Spc}_S \to \mathrm{Spc}_S^{\mathbb{A}^1}$ and the Yoneda embedding is fully faithful. In other words, two $\mathbb{A}^1$-rigid schemes are isomorphic as schemes if and only if they are $\mathbb{A}^1$-equivalent.

4.5. The smash product and the loops-suspension adjunction. Let us begin with some recollection about smash products in simplicial sets. Let $(X, x), (Y, y)$ be two pointed simplicial sets, then we can form the smash product

$(X,x) \wedge (Y,y)$ which is defined to be the pushout:

$$\begin{array}{ccc} (X,x) \vee (Y,y) & \longrightarrow & (X,x) \times (Y,y) \\ \downarrow & & \downarrow \\ \star & \longrightarrow & (X,x) \wedge (Y,y) \end{array}$$

The functor $-\wedge(X,x)$ is then a left Quillen endofunctor on the category of simplicial sets by the following argument: if $(Z,z) \to (Y,y)$ is a cofibration of simplicial sets, then we note that $(X,x) \wedge (Z,z) \to (X,x) \wedge (Y,y)$ is cofibration since cofibrations are stable under pushouts. The case of acyclic cofibrations is left to the reader. The right adjoint to $- \wedge (X,x)$ is given by the pointed mapping space $\mathrm{map}_\star(X,-)$ which is given by the formula:

$$\mathrm{map}_\star(X,Y)_n \cong \mathrm{Hom}_{\mathrm{sSets}_\star}(X \wedge \Delta^n_+, Y).$$

To promote the smash product to the level of simplicial presheaves, we first take the pointwise smash product, i.e., if $(X,x),(Y,y)$ are objects in $\mathrm{sPre}(\mathrm{Sm}_S)_\star$, then we form the smash product $(X,x) \wedge (Y,y)$ as the simplicial presheaf:

$$U \mapsto (X,x)(U) \wedge (Y,y)(U).$$

An analogous pointwise formula is used for the pointed mapping space functor.

PROPOSITION 4.36. *The Quillen adjunction*

$$(X,x) \wedge - : \mathrm{sPre}(\mathrm{Sm}_S)_\star \to \mathrm{sPre}(\mathrm{Sm}_S)_\star : \mathrm{map}_\star(X,-)$$

descends to a Quillen adjunction:

$$(X,x) \wedge - : \mathrm{Spc}^{\mathbb{A}^1}(S)_\star \to \mathrm{Spc}^{\mathbb{A}^1}(S)_\star : \mathrm{map}_\star(X,-)$$

and thus there are natural isomorphisms categories:

$$[(X,x) \wedge^{\mathbf{L}} (Z,z),(Y,y)]_{\mathbb{A}^1} \simeq [(X,x), \mathbf{R}\mathrm{map}_\star((X,x),(Y,y))]_{\mathbb{A}^1}$$

for $X,Y,Z \in \mathrm{Spc}^{\mathbb{A}^1}_S$.

PROOF. The question of whether a monoidal structure defined on the underlying category of a model category descends to a Quillen adjunction with mapping spaces as its right adjoint is answered in the paper of Schwede and Shipley [**SS00**]. The necessary conditions are checked in [**DRØ03**, Section 2.1]. □

Now, recall that the suspension of (X,x) is calculated either as the homotopy cofiber of the canonical morphism $(X,x) \to \star$ or, equivalently, as $S^1 \wedge (X,x)$ (check this!), while the loop space is calculated as the homotopy pullback of the diagram

$$\begin{array}{ccc} \Omega(X,x) & \longrightarrow & \star \\ \downarrow & & \downarrow \\ \star & \longrightarrow & (X,x) \end{array}$$

or, equivalently, as $\mathrm{map}_\star(S^1,(Y,y))$. Consequently:

COROLLARY 4.37. *For any objects* $(X,x),(Y,y) \in \mathrm{Spc}^{\mathbb{A}^1}(\mathrm{Sm}_S)_\star$,*there is an isomorphism*

$$[\mathbf{L}\Sigma(X,x),(Y,y)]_{\mathbb{A}^1} \simeq [(X,x), \mathbf{R}\Omega(Y,y)]_{\mathbb{A}^1}.$$

4.6. The bigraded spheres. We will now delve into some calculations in $\mathbb{A}^1$-homotopy theory. More precisely, these are calculations in the pointed category $\mathrm{Spc}_{S,\star}^{\mathbb{A}^1}$. We use the following conventions for base points of certain schemes which will play a major role in the theory.

(1) $\mathbb{A}^1$ is pointed by 1
(2) $\mathbb{G}_m$ is pointed by 1.
(3) $\mathbb{P}^1$ is pointed by ∞.
(4) X_+ denotes X with a disjoint base point for a space X.

In particular, we only write pointed objects as (X, x) when the base points are not the ones indicated above. We also note that the forgetful functors $\mathrm{Spc}_{S,\star} \to \mathrm{Spc}_S$ and $\mathrm{Spc}_{S,\star}^{\mathbb{A}^1} \to \mathrm{Spc}_S^{\mathbb{A}^1}$ preserve and detect weak equivalences. Hence when the context is clear, we will say Nisnevich or $\mathbb{A}^1$-weak equivalence as opposed to *pointed* Nisnevich or $\mathbb{A}^1$-weak equivalence.

REMARK 4.38. In many cases, base points of schemes are negotiable in the sense that there is an explicit pointed $\mathbb{A}^1$-local weak equivalence between (X, x) and (X, y) for two base points x, y. For example $(\mathbb{P}^1, \infty)$ is $\mathbb{A}^1$-equivalent in the pointed category to $(\mathbb{P}^1, x)$ for any other point $x \in \mathbb{P}^1$ via an explicit $\mathbb{A}^1$-homotopy.

Of course, if one takes a cofiber of pointed schemes (or even simplicial presheaves), the cofiber is automatically pointed: if $X \to Y \to X/Y$ is a cofiber sequence, then X/Y is pointed by the image of Y.

The first calculation one encounters in $\mathbb{A}^1$-homotopy theory is the following.

LEMMA 4.39. *In $\mathrm{Spc}_{S,\star}^{\mathbb{A}^1}$, there are $\mathbb{A}^1$-weak equivalences* $\Sigma(\mathbb{G}_m, 1) \simeq (\mathbb{P}^1, \infty) \simeq \mathbb{A}^1/(\mathbb{A}^1 - \{0\})$.

PROOF. Consider the distinguished Nisnevich square

$$\begin{array}{ccc} \mathbb{G}_m & \longrightarrow & \mathbb{A}^1 \\ \downarrow & & \downarrow \\ \mathbb{A}^1 & \longrightarrow & (\mathbb{P}^1, 1) \end{array}$$

in Sm_S. By Proposition 4.13, this can be viewed as a homotopy pushout in $\mathrm{Spc}_{S,\star}$ as well. Since the localization functor $\mathrm{Spc}_{S,\star} \to \mathrm{Spc}_{S,\star}^{\mathbb{A}^1}$ is a Quillen left adjoint, it commutes with homotopy colimits, and in particular with homotopy pushouts. Therefore, when viewed in $\mathrm{Spc}_{S,\star}^{\mathbb{A}^1}$ the square above is a homotopy pushout. However, since $\mathbb{A}^1 \simeq \star$ in the $\mathbb{A}^1$-homotopy theory, it follows that $\Sigma\mathbb{G}_m \simeq (\mathbb{P}^1, \infty)$ (by contracting both copies of $\mathbb{A}^1$ and noting that $(\mathbb{P}^1, 1) \simeq (\mathbb{P}^1, \infty)$) or $\Sigma\mathbb{G}_m \simeq \mathbb{A}^1/(\mathbb{A}^1 - \{0\})$ (by contracting one of the copies of $\mathbb{A}^1$). $\square$

The above calculation justifies the idea that in $\mathbb{A}^1$-homotopy theory there are two kinds of circles: the simplicial circle S^1 and the "Tate" circle $\mathbb{G}_m$. The usual convention (which matches up with the grading in motivic cohomology) is to define

$$S^{1,1} = \mathbb{G}_m,$$

and

$$S^{1,0} = S^1.$$

Consequently, by the lemma, we have an $\mathbb{A}^1$-weak equivalence $S^{2,1} \simeq \mathbb{P}^1$.

Now, given a pair a, b of non-negative integers satisfying $a \geq b$, we can define $S^{a,b} = \mathbb{G}_m^{\wedge b} \wedge (S^1)^{\wedge(a-b)}$. In general, there is no known nice description of these motivic spheres. However, the next two results give two important classes of exceptions.

PROPOSITION 4.40. *In* $\mathrm{Spc}_{S,\star}^{\mathbb{A}^1}$, *there are* $\mathbb{A}^1$*-weak equivalences* $S^{2n-1,n} \simeq \mathbb{A}^n - \{0\}$ *for* $n \geq 1$.

PROOF. The case $n = 1$ is Lemma 4.39; we need to do the $n = 2$ case and then perform induction. Specifically, the claim for $n = 2$ says that $\mathbb{A}^2 - \{0\} \simeq S^1 \wedge (\mathbb{G}_m^{\wedge 2})$. First, observe that we have a homotopy push-out diagram:

$$\begin{array}{ccc} \mathbb{G}_m \times \mathbb{G}_m & \longrightarrow & \mathbb{G}_m \times \mathbb{A}^1 \\ \downarrow & & \downarrow \\ \mathbb{G}_m \times \mathbb{A}^1 & \longrightarrow & \mathbb{A}^2 - \{0\}, \end{array}$$

from which we conclude that $\mathbb{A}^2 - \{0\}$ is calculated as the homotopy push-out of

$$\mathbb{G}_m \leftarrow \mathbb{G}_m \times \mathbb{G}_m \rightarrow \mathbb{G}_m.$$

On the other hand we may calculate this homotopy push-out using the diagram

$$\begin{array}{ccccc} \star & \longleftarrow & \star & \longrightarrow & \star \\ \uparrow & & \uparrow & & \uparrow \\ \mathbb{G}_m & \longleftarrow & \mathbb{G}_m \vee \mathbb{G}_m & \longrightarrow & \mathbb{G}_m \\ \downarrow & & \downarrow & & \downarrow \\ \mathbb{G}_m & \longleftarrow & \mathbb{G}_m \times \mathbb{G}_m & \longrightarrow & \mathbb{G}_m. \end{array}$$

Taking the homotopy push-out across the horizontal rows gives us $\star \leftarrow \star \rightarrow \mathbb{A}^2 - \{0\}$; taking this homotopy push-out gives back $\mathbb{A}^2 - \{0\}$. On the other hand, taking the homotopy push-put across the vertical rows give us $\star \leftarrow \mathbb{G}_m \wedge \mathbb{G}_m \rightarrow \star$ which calculates the homotopy push-out $S^1 \wedge (\mathbb{G}_m \wedge \mathbb{G}_m)$.

Let us now carry out the induction. We have a distinguished Nisnevich square

$$\begin{array}{ccc} \mathbb{A}^{n-1} - \{0\} \times \mathbb{G}_m & \longrightarrow & \mathbb{A}^n \times \mathbb{G}_m \\ \downarrow & & \downarrow \\ \mathbb{A}^{n-1} - \{0\} \times \mathbb{A}^1 & \longrightarrow & \mathbb{A}^n - \{0\}, \end{array}$$

from which we conclude that $\mathbb{A}^n - \{0\}$ is calculated as the homotopy push-out of

$$\mathbb{A}^{n-1} - \{0\} \leftarrow \mathbb{A}^{n-1} - \{0\} \times \mathbb{G}_m \rightarrow \mathbb{G}_m.$$

Hence we can set-up an analogous diagram:

$$\begin{array}{ccccc}
\star & \longleftarrow & \star & \longrightarrow & \star \\
\uparrow & & \uparrow & & \uparrow \\
\mathbb{A}^{n-1}-\{0\} & \longleftarrow & \mathbb{A}^{n-1}-\{0\}\vee\mathbb{G}_m & \longrightarrow & \mathbb{G}_m \\
\downarrow & & \downarrow & & \downarrow \\
\mathbb{A}^{n-1}-\{0\} & \longleftarrow & \mathbb{A}^{n-1}-\{0\}\times\mathbb{G}_m & \longrightarrow & \mathbb{G}_m
\end{array}$$

to conclude as in the base case that $S^1\wedge((\mathbb{A}^{n-1}-\{0\})\wedge\mathbb{G}_m)\simeq\mathbb{A}^n-\{0\}$. □

COROLLARY 4.41. *In* $\mathrm{Spc}_{S,\star}^{\mathbb{A}^1}$ *there are* $\mathbb{A}^1$*-weak equivalences* $\mathbb{A}^n/\mathbb{A}^n-\{0\}\simeq S^n\wedge\mathbb{G}_m\simeq S^{2n,n}$ *for* $n\geq 1$.

PROOF. The homotopy cofiber of the inclusion $\mathbb{A}^n-0\hookrightarrow\mathbb{A}^n$ is calculated as the homotopy pushout

$$\begin{array}{ccc}
\mathbb{A}^n-\{0\} & \longrightarrow & \mathbb{A}^n \\
\downarrow & & \downarrow \\
\star & \longrightarrow & \mathbb{A}^n/\mathbb{A}^n-\{0\}.
\end{array}$$

In $\mathrm{Spc}_S^{\mathbb{A}^1}$, this cofiber can be calculated as the homotopy pushout

$$\begin{array}{ccc}
\mathbb{A}^n-\{0\} & \longrightarrow & \star \\
\downarrow & & \downarrow \\
\star & \longrightarrow & \mathbb{A}^n/\mathbb{A}^n-\{0\}.
\end{array}$$

Therefore $\mathbb{A}^n/\mathbb{A}^n-\{0\}\simeq S^1\wedge(\mathbb{A}^n-\{0\})\simeq S^1\wedge(S^{2n-1,n})$ by Proposition 4.40. □

REMARK 4.42. In [**ADF**], the authors study the question of when the motivic sphere $S^{a,b}$ is $\mathbb{A}^1$-weak equivalent to a smooth scheme. Proposition 4.40 shows that this is the case for $S^{2n-1,n}\simeq\mathbb{A}^n-\{0\}$. Asok, Doran, and Fasel prove that it is also the case for $S^{2n,n}$, which they show is $\mathbb{A}^1$-weak equivalent to to the affine quadric with coordinate ring

$$k[x_1,\ldots,x_n,y_1,\ldots,y_n,z]/\left(\sum_i x_iy_i-z(1+z)\right)$$

when $S=\operatorname{Spec}k$ for a commutative ring k. They also show that $S^{a,b}$ is *not* $\mathbb{A}^1$-weak equivalent to a smooth affine scheme if $a>2b$. Conjecturally, the only motivic spheres $S^{a,b}$ admitting smooth models are those above, when $(a,b)=(2n-1,n)$ or $(a,b)=(2n,n)$.

REMARK 4.43. If we impose only Nisnevich descent rather than Nisnevich hyperdescent, the results in this section remain true. This might provide one compelling reason to do so. For details, see [**AHW15a**].

4.7. Affine and projective bundles.

PROPOSITION 4.44. *Let $p : E \to X$ be a Nisnevich-locally trivial affine space bundle. Then, $E \to X$ is an $\mathbb{A}^1$-weak equivalence.*

PROOF. Pick a Zariski cover $\mathfrak{U} := \{U_\alpha\}$ of X that trivializes E. Suppose that $\check{C}(U)_\bullet$ is the Čech nerve of the cover, then we have a weak equivalence

$$\operatorname*{hocolim}_{\Delta^{op}} \check{C}(U)_\bullet \simeq X$$

in Spc_S and an $\mathbb{A}^1$-weak equivalence

$$\operatorname*{hocolim}_{\Delta^{op}} \check{C}(U)_\bullet \times_X \mathbb{A}^n \simeq E$$

in $\mathrm{Spc}_S^{\mathbb{A}^1}$. But now, we have an levelwise-$\mathbb{A}^1$-weak equivalence of simplicial objects

$$\check{C}(U)_\bullet \times_X \mathbb{A}^n \to \check{C}(U)_\bullet.$$

Hence, the homotopy colimits are equivalent by construction. □

Note that the above proposition covers a larger class of morphisms than just vector bundles $p : E \to X$. For these, the homotopy inverse of the projection map is the zero section as per Exercise 3.62.

We obtain immediate applications of this proposition in the form of certain presentations of $\mathbb{A}^n - 0$ in terms of a homogeneous space and an affine scheme in the $\mathbb{A}^1$-homotopy category. We leave the proofs to the reader.

COROLLARY 4.45. *Let $n \geq 2$, and let $\mathrm{SL}_n \to \mathbb{A}^n - \{0\}$ be the map defined by taking the last column of a matrix in SL_n. In $\mathrm{Spc}_S^{\mathbb{A}^1}$, this map factors through the cofiber $\mathrm{SL}_n \to \mathrm{SL}_n / \mathrm{SL}_{n-1}$, and the map $\mathrm{SL}_n / \mathrm{SL}_{n-1} \to \mathbb{A}^n - \{0\}$ is an $\mathbb{A}^1$-weak equivalence.*

COROLLARY 4.46. *Let S be the spectrum of a field, and give $\mathbb{A}^{2n}$ the coordinates $x_1, \ldots, x_n, y_1, \ldots, y_n$. Consider the quadric $Q_{2n-1} = V(x_1y_1 + \cdots + x_ny_n = 1)$. The map $Q_{2n-1} \to \mathbb{A}^n - \{0\}$ induced by the projection to the x-coordinates is an $\mathbb{A}^1$-weak equivalence.*

Note that some authors might write Q_{2n} for what we have written Q_{2n-1}.

Now we will use the above proposition to deduce results about projective bundles. We will then recover yet another presentation of the spheres $S^{2n,n}$. Furthermore we will also introduce an important construction on vector bundles that we will encounter later.

DEFINITION 4.47. If $\nu : E \to X$ is a vector bundle, then the Thom space $\mathrm{Th}(\nu)$ of E (sometimes also written $\mathrm{Th}(E)$) is defined as the cofiber

$$E/(E - X),$$

where the embedding of X into E is given by the zero section.

The Thom space construction plays a central role in algebraic topology and homotopy theory, and is intimately wrapped up in computations of the bordism ring for manifolds and in the representation of homology classes by manifolds [**Tho54**].

EXAMPLE 4.48. Let S be a base scheme, then $\mathbb{A}^n_S \to S$ is a trivial vector bundle over S. The Thom space of the trivial rank n vector bundle is then by definition $\frac{\mathbb{A}^n}{\mathbb{A}^n - \{0\}}$. From Proposition 4.41 we conclude that Thom space in this case is given by $S^{2n,n}$.

In topology, one has a weak homotopy equivalence: $\mathbb{CP}^n/\mathbb{CP}^{n-1} \simeq S^{2n}$, thanks to the standard cell decomposition of projective space. One of the benefits of having this decomposition is that for a suitable class of generalized cohomology theories, the complex orientable theories, there exists a theory of Chern classes similar to the theory in ordinary cohomology. We would like a similar story in $\mathbb{A}^1$-homotopy theory, and this indeed exists.

EXERCISE 4.49. Let $E \to X$ be a trivial rank n vector bundle, then there is an $\mathbb{A}^1$-weak equivalence: $\mathrm{Th}(E) \simeq \mathbb{P}^{1^{\wedge n}} \wedge X_+$. Hint: use Corollary 4.41.

PROPOSITION 4.50. *Suppose that $E \to X$ is a vector bundle and $\mathbb{P}(E) \to \mathbb{P}(E \oplus \mathcal{O})$ is the closed embedding at infinity. Then, there is an $\mathbb{A}^1$-weak equivalence*

$$\frac{\mathbb{P}(E \oplus \mathcal{O})}{\mathbb{P}(E)} \to \mathrm{Th}(E).$$

PROOF. Throughout, X is identified with its zero section for ease of notation. Observe that we have a morphism $X \to E \to \mathbb{P}(E \oplus \mathcal{O})$ where the first map is the closed embedding of X via the zero section and the second map is the embedding complementary to the embedding $\mathbb{P}(E) \to \mathbb{P}(E \oplus \mathcal{O})$ at infinity. We also identify X in $\mathbb{P}(E \oplus \mathcal{O})$ via this embedding. Hence, there is an elementary distinguished square

$$\begin{array}{ccc} E - X & \longrightarrow & \mathbb{P}(E \oplus \mathcal{O}) - X \\ \downarrow & & \downarrow \\ E & \longrightarrow & \mathbb{P}(E \oplus \mathcal{O}), \end{array}$$

which means that we have an weak equivalence of simplicial presheaves:

$$\mathrm{Th}(E) \simeq \frac{\mathbb{P}(E \oplus \mathcal{O})}{\mathbb{P}(E \oplus \mathcal{O}) - X}.$$

We have a map $\frac{\mathbb{P}(E\oplus\mathcal{O})}{\mathbb{P}(E)} \to \frac{\mathbb{P}(E\oplus\mathcal{O})}{\mathbb{P}(E\oplus\mathcal{O})-X}$ because $\mathbb{P}(E)$ avoids the embedding of X described above. This is the map that we want to be an $\mathbb{A}^1$-weak equivalence, so it suffices to prove that we have an $\mathbb{A}^1$-weak equivalence $\mathbb{P}(E) \to \mathbb{P}(E \oplus \mathcal{O}) - X$. The lemma below shows that the map is indeed the zero section of an affine bundle, and so we are done by Proposition 4.44. □

LEMMA 4.51. *Let X be a scheme, $p : E \to X$ a vector bundle with s its zero section. Consider the open embedding $j : E \to \mathbb{P}(E \oplus \mathcal{O}_X)$ and its closed complement $i : \mathbb{P}(E) \to \mathbb{P}(E \oplus \mathcal{O}_S)$. In this case, there is a morphism*

$$q : \mathbb{P}(E \oplus \mathcal{O}_X) \setminus j(s(X)) \to \mathbb{P}(E)$$

such that $q \circ i = id$ and q is an $\mathbb{A}^1$-bundle over $\mathbb{P}(E)$.

PROOF. Recall that to give a morphism $T \to \mathbb{P}(E)$ over X, one must give a morphism $h : T \to X$ and a surjection $h^*(E) \to \mathcal{L}$ where $\mathcal{L}$ is a line bundle on T. Now $\bar{p} : \mathbb{P}(E \oplus \mathcal{O}_X) \to X$ has a universal line bundle $\mathcal{L}_{univ}$ and a universal quotient map $\bar{p}^*(E \oplus \mathcal{O}_X) \simeq \bar{p}^*(E) \oplus \mathcal{O}_{\mathbb{P}(E\oplus\mathcal{O}_X)} \to \mathcal{L}_{univ}$. Restricting to the first factor gives us a map $\bar{p}^*(E) \to \mathcal{L}_{univ}$ and hence a rational map $t : \mathbb{P}(E \oplus \mathcal{O}_X) \dashrightarrow \mathbb{P}(E)$. Over a point of X, t is given by projection onto E-coordinates. Hence, this map is well-defined away from $j(s(X))$, so we get a morphism $q : \mathbb{P}(E \oplus \mathcal{O}_X) \setminus j(s(X)) \to \mathbb{P}(E)$; by construction $q \circ i = \mathrm{id}_{\mathbb{P}(E)}$. To check the last claim, since it is local on the base, one may assume that $E \cong \mathcal{O}_X^{n+1}$, so we are looking at $\mathbb{P}_X^{n+1} \setminus X \to \mathbb{P}^n$. In coordinates X embeds as $[0 : \ldots 0 : 1]$, and the map is projection onto the first n coordinates, which is an $\mathbb{A}^1$-bundle. □

COROLLARY 4.52. *There are $\mathbb{A}^1$-weak equivalences $\mathbb{P}^n/\mathbb{P}^{n-1} \simeq S^{2n,n}$ for $n \geq 1$ when S is noetherian of finite Krull dimension.*

5. Classifying spaces in $\mathbb{A}^1$-homotopy theory

One of the main takeaways from Section 2 is that one can go very far using homotopical methods to study topological vector bundles on CW complexes. The key inputs in this technique are the existence of the Postnikov tower and knowledge of the homotopy groups of the classifying spaces BGL_n in low degrees. In this section, we will give a sampler of the techniques involved in accessing the $\mathbb{A}^1$-homotopy sheaves of the classifying spaces BGL_n. In the end will identify a "stable" range for these homotopy sheaves, which will naturally lead us to a discussion of algebraic K-theory in the next section.

As usual, S is a quasi-compact and quasi-separated unless stated otherwise.

5.1. Simplicial models for classifying spaces.

DEFINITION 5.1. Let τ be a topology (typically this will be Zariski, Nisnevich or étale) on Sm_S, and let G a τ-sheaf of groups. A τ-G-torsor over $X \in \mathrm{Sm}_S$ is the data of a τ-sheaf of sets P on Sm_S, a right action $a : P \times G \to P$ of G on P, and a G-equivariant morphism $\pi : P \to X$ (where X has the trivial G-action) such that

(1) the morphism $(\pi, a) : P \times G \to P \times_X P$ is an isomorphism, and
(2) there exists a τ-cover $\{U_i \to X\}_{i \in I}$ of X such that $U_i \times_X P \to U_i$ has a section for all $i \in I$.

Let G be a τ-sheaf of groups. Consider the simplicial presheaf $\mathrm{E}G$ described section-wise in the following way: $\mathrm{E}G_n(U) =: G(U)^{\times n+1}$ with the usual faces and degeneracies. We write $\mathrm{E}_\tau G$ as a fibrant replacement in the model category $\mathrm{L}_\tau(\mathrm{sPre}(\mathrm{Sm}_S))$.

PROPOSITION 5.2. *There is a weak equivalence $\mathrm{E}_\tau G \simeq \star$ in $\mathrm{L}_\tau\mathrm{sPre}(\mathrm{Sm}_S)$.*

PROOF. The fact that each $\mathrm{E}G(U)$ is contractible is standard: the diagonal morphism: $G(U) \to G(U) \times G(U)$ produces an extra degeneracy. See Goerss and Jardine [**GJ99**, Lemma III.5.1]. Thus, $\mathrm{E}G \to \star$ is a weak equivalence in $\mathrm{sPre}(\mathrm{Sm}_S)$. Since localization ($\tau$-sheafification) preserves weak equivalences, it follows that $\mathrm{E}_\tau G$ is contractible. □

There is a right G-action on $\mathrm{E}G$ by letting G act on the last coordinate in each simplicial degree. The level-wise quotient is the simplicial presheaf we christen $\mathrm{B}G$. We write $\mathrm{B}_\tau G$ for a fibrant replacement in the model category $\mathrm{L}_\tau\mathrm{sPre}(\mathrm{Sm}_S)$. We would like to make sense of $\mathrm{B}_\tau G$ as a simplicial presheaf classifying τ-G-torsors.

DEFINITION 5.3. Let $\mathrm{BTors}_\tau(G)$ be the simplicial presheaf which assigns to $U \in \mathrm{Sm}_S$ the nerve of the groupoid of G-torsors on U and to a morphism $f : U' \to U$ a map of simplicial presheaves $\mathrm{BTors}_\tau(G)(U) \to \mathrm{BTors}_\tau(G)(U')$ induced by pullback.

REMARK 5.4. The above definition is valid by the work of Hollander [**Hol08**, Section 3]; the functor that assigns to U the nerve of the groupoid of G-torsors over U does not have strictly functorial pullbacks and thus one needs to appeal to some rectification procedure.

The following proposition is well known.

PROPOSITION 5.5. *The simplicial presheaf* $\mathrm{BTors}_\tau(G)$ *is* τ*-local.*

PROOF. This follows from the local triviality condition and the fact that we can construct τ-G-torsors by gluing; see, for example, [**Vis05**]. □

Let $U \in \mathrm{Sm}_S$, we denote by $\mathrm{H}^1_\tau(U, G)$ be the (non-abelian) cohomology set of τ-G-bundles on U. More precisely, we set

$$\mathrm{H}^1_\tau(U, G) = \pi_0(\mathrm{BTors}_\tau(G)(U)).$$

PROPOSITION 5.6. *Let G be a τ-sheaf of groups, then there is a natural weak equivalence*

$$\mathrm{B}_\tau G \to \mathrm{BTors}_\tau(G).$$

Hence, for all $U \in \mathrm{Sm}_S$, there is a natural isomorphism $\pi_0(\mathrm{B}_\tau G(U)) \cong \mathrm{H}^1_\tau(U; G)$ and a natural weak equivalence $\mathbf{R}\Omega\mathrm{B}_\tau G(U) \simeq G(U)$.

PROOF. A proof is given in [**MV99**, section 4.1], we also recommend [**AHW15b**, Lemma 2.2.2] and the references therein. Let us sketch the main ideas. To define a map to $\mathrm{BTors}_\tau(G)$, we can first define a map $\mathrm{B}G \to B\mathrm{Tors}_\tau(G)$ of presheaves and then use the fact that the target is τ-local to get a map $\mathrm{B}_\tau G \to B\mathrm{Tors}_\tau(G)$. The former map is given by sending the unique vertex of $\mathrm{B}G(U)$ to the trivial G-torsor over U. Since G-torsors with respect to τ are τ-locally trivial, we conclude that the map must be a τ-local weak equivalence. The fact that $\mathrm{B}_\tau G$ is fibrant is by definition, and for $\mathrm{B}_\tau G$ it follows from [**AHW15b**, Lemma 2.2.2]. The second part of the assertion then follows by definition, and the standard fact that loop space of the nerve of a groupoid is homotopy equivalent to the automorphism group of a fixed object. □

Many interesting objects in algebraic geometry, such as Azumaya algebras and the associated PGL_n-torsors, are only étale locally trivial. The classifying spaces of these torsors are indeed objects of $\mathbb{A}^1$-homotopy theory as we shall explain. We can consider $\mathrm{Sm}_{S,\text{ét}}$, the full subcategory of the big étale site over S spanned by smooth S-schemes. Completely analogous to the Nisnevich case, one can develop étale-$\mathbb{A}^1$-homotopy theory by the formula $\mathrm{Spc}^{\mathbb{A}^1}_{S,\text{ét}} = \mathrm{L}_{\mathbb{A}^1}\mathrm{L}_{\text{ét}}\mathrm{sPre}(\mathrm{Sm}_S)$.

THEOREM 5.7. *The morphism of sites: $\pi : \mathrm{Sm}_{S,\text{ét}} \to \mathrm{Sm}_{S,\mathrm{Nis}}$ induced by the identity functor induces a Quillen pair*

$$\pi^* : \mathrm{Spc}^{\mathbb{A}^1}_S \leftrightarrows \mathrm{Spc}^{\mathbb{A}^1}_{S,\text{ét}} : \pi_*,$$

and hence an adjunction

$$\mathbf{L}\pi^* : \mathrm{Ho}(\mathrm{Spc}^{\mathbb{A}^1}_S) \leftrightarrows \mathrm{Ho}(\mathrm{Spc}^{\mathbb{A}^1}_{S,\text{ét}}) : \mathbf{R}\pi_*$$

on the level of homotopy categories.

PROOF. Since our categories are constructed via Bousfield-localization of $\mathrm{sPre}(\mathrm{Sm}_S)$, the universal property tells us that to define a Quillen pair

$$\pi^* : \mathrm{Spc}_S^{\mathbb{A}^1} \leftrightarrows \mathrm{Spc}_{S,\text{ét}}^{\mathbb{A}^1} : \pi_*$$

it suffices to define a Quillen pair:

$$\pi^* : \mathrm{sPre}(\mathrm{Sm}_S) \leftrightarrows \mathrm{Spc}_{S,\text{ét}}^{\mathbb{A}^1} : \pi_*$$

such that $\pi^*(i)$ is a weak equivalence for i belonging to the class of Nisnevich hypercovers and $\mathbb{A}^1$-weak equivalences. However, the model category $\mathrm{Spc}_{S,\text{ét}}^{\mathbb{A}^1}$ is also constructed via Bousfield localization, so we use the Quillen pair from this Bousfield localization. But, it is clear that the identity functor $\pi^* : \mathrm{sPre}(\mathrm{Sm}_S) \to \mathrm{sPre}(\mathrm{Sm}_S)$ takes Nisnevich hypercovers to étale hypercovers and the morphisms $X \times_S \mathbb{A}^1 \to X$ to $X \times_S \mathbb{A}^1$. Hence, the Quillen pair exists by the universal property of Bousfield localization. □

PROPOSITION 5.8. *There are natural isomorphisms of Nisnevich sheaves* $\pi_0^{\mathrm{Nis}}(\mathbf{R}\pi_*\mathrm{B}_{\text{ét}}G) \simeq \mathrm{H}^1_{\text{ét}}(-;G)$ *and* $\pi_1^{\mathrm{Nis}}(\mathbf{R}\pi_*\mathrm{B}_{\text{ét}}G) \simeq G$, *where the étale sheaf of groups G is considered as a Nisnevich sheaf.*

PROOF. By adjunction, $[U, \mathbf{R}\pi_*\mathrm{B}_{\text{ét}}G]_{\mathbb{A}^1} \simeq [\mathbf{L}\pi^*U, \mathrm{B}_{\text{ét}}G]_{\mathbb{A}^1} \cong \mathrm{H}^1_{\text{ét}}(U,G)$. To see the π_1-statement, we note that $\mathbf{R}\pi_*$ is a right Quillen functor and hence commutes with homotopy limits. Since the loop space is calculated via a homotopy limit, we have that $\Omega\mathbf{R}\pi_*\mathrm{B}_{\text{ét}}G \simeq \mathbf{R}\pi_*\Omega\mathrm{B}_{\text{ét}}G \simeq \mathbf{R}\pi_*G$, as desired. □

EXAMPLE 5.9. Let $\mathrm{G} = \mathrm{GL}_n, \mathrm{SL}_n$ or Sp_{2n}; these are the special groups in the sense of Serre. In this case, any étale-G-torsor is also a Zariski-locally trivial and hence a Zariski-G-torsor (or a Nisnevich-G-torsor). One way to say this in our language is to consider the Quillen adjunction

$$\pi^* : \mathrm{L}_{\mathrm{Nis}}(\mathrm{sPre}(\mathrm{Sm}_S)) \leftrightarrows \mathrm{L}_{\text{ét}}(\mathrm{sPre}(\mathrm{Sm}_S)) : \mathbf{R}\pi_*.$$

Then there is a unit map $\mathrm{B}_{\mathrm{Nis}}G \to \mathbf{R}\pi_*\pi^*\mathrm{B}_{\text{ét}}G$, which is an weak equivalence in the cases above.

5.2. Some calculations with classifying spaces. We are now interested in the $\mathbb{A}^1$-homotopy sheaves of classifying spaces. The first calculation is a direct consequence of the unstable-$\mathbb{A}^1$-0-connectivity theorem. We work over an arbitrary Noetherian base in this section, unless specified otherwise.

PROPOSITION 5.10. *If G is a Nisnevich sheaf of groups, then* $\pi_0^{\mathbb{A}^1}(\mathrm{B}G) = \star$.

PROOF. By Theorem 4.30, it suffices to prove that $\pi_0^{\mathrm{Nis}}(\mathrm{B}G)$ is trivial. Note that this is the sheafification of the functor $U \mapsto \mathrm{H}^1_{\mathrm{Nis}}(U,G)$. The claim follows from the fact that we are considering G-torsors which are Nisnevich-locally trivial. □

REMARK 5.11. Let G be an étale sheaf of groups. If we replace $\mathrm{B}G$ by $\mathbf{R}\pi_*\mathrm{B}_{\text{ét}}G$, then the above result will *not* hold unless étale G-torsors are also Nisnevich locally trivial. This is not the case for example for PGL_n. For more about $\mathrm{B}_{\text{ét}}\,\mathrm{PGL}_n$, see [**Aso13**, Corollary 3.16].

In order to proceed further, we need a theorem of Asok-Hoyois-Wendt [**AHW15b**].

THEOREM 5.12 ([**AHW15b**]). *If $X \to Y \to Z$ is a fiber sequence in* $\mathrm{sPre}(\mathrm{Sm}_S)$ *such that Z satisfies affine Nisnevich excision and $\pi_0(Z)$ satisfies affine $\mathbb{A}$-invariance, then $X \to Y \to Z$ is an $\mathbb{A}^1$-fiber sequence.*

COROLLARY 5.13. *If $\mathrm{H}^1_{\mathrm{Nis}}(-, G)$ is $\mathbb{A}^1$-invariant, then the sequence $G \to \mathrm{E}G \to \mathrm{B}G$ is an $\mathbb{A}^1$-fiber sequence.*

From now on to the end of this section, we will need the base scheme to be a field (although we can do better — see the discussions in [**AHW15b**]) in order to utilize $\mathbb{A}^1$-invariance of various cohomology sets and apply Theorem 5.12 above. As a first example, we let T be a split torus over a field k.

PROPOSITION 5.14. *Let T be a split torus over a field k. If $P \to X$ is a T-torsor with a k-point $x : \mathrm{Spec}\, k \to P$, then we have a short exact sequence*

$$1 \to \pi_1^{\mathbb{A}^1}(P, x) \to \pi_1^{\mathbb{A}^1}(X, x) \to T.$$

PROOF. We need to check that $\pi_0(\mathrm{B}T)$ is $\mathbb{A}^1$-invariant. Recall that a split torus over a field simply means that it is isomorphic over k to products of $\mathbb{G}_m$, and so $\pi_0(\mathrm{B}T) \cong \mathrm{Pic}(-)^{\oplus n}$ where n is the number of copies of $\mathbb{G}_m$. Therefore it is indeed $\mathbb{A}^1$-invariant on smooth k-schemes. This shows that T is an $\mathbb{A}^1$-rigid scheme over k, hence $\pi_0^{\mathbb{A}^1}(T) \simeq T$ and the higher homotopy groups are zero by Proposition 4.34, giving us the short exact sequence above. □

REMARK 5.15. The result is true in greater generality for not-necessarily-split tori with some assumptions on the base field, see [**Aso11**] for details.

5.3. BGL **and** BSL. In our classification of vector bundles, on affine schemes, we need to calculate the homotopy sheaves of BGL_n. We use the machinery above to highlight two features of this calculation. First, just like in topology, we may reduce the calculation of homotopy sheaves of BGL_n to that of BSL_n, save for π_1. Secondly, the $\mathbb{A}^1$-homotopy sheaves of BSL_n stabilize: for each i, $\pi_i^{\mathbb{A}^1}(\mathrm{BSL}_n)$ is independent of the value of n as n tends to ∞.

PROPOSITION 5.16. *Let S be a regular noetherian affine scheme of finite Krull dimension, and suppose that the Bass-Quillen conjecture holds for smooth schemes of finite presentation over S. The space SL_n in $\mathrm{Spc}_S^{\mathbb{A}^1}$ is $\mathbb{A}^1$-connected and BSL_n is $\mathbb{A}^1$-1-connected, i.e. $\pi_1^{\mathbb{A}^1}(\mathrm{BSL}_n) = \star$.*

PROOF. We show that the sheaf $\pi_0^{\mathbb{A}^1}(\mathrm{SL}_n)$ is trivial by showing that the stalks of $\pi_0^{\mathbb{A}^1}(\mathrm{SL}_n)$ are trivial. To show this it suffices by Theorem 4.31 to show that for any henselian local ring R,

$$[\mathrm{Spec}\ R, \mathrm{Sing}^{\mathbb{A}^1}(\mathrm{SL}_n)]_s = \star$$

(i.e. the set of naive $\mathbb{A}^1$-homotopy classes is trivial), where we view $\mathrm{Spec}\, R$ as an object of $\mathrm{Spc}_S^{\mathbb{A}^1}$ via the functor of points it represents.

In fact we will prove the above claim for R, any local ring. We want to connect any matrix $M \in \mathrm{SL}_n(R)$ to the identity via a chain of naive $\mathbb{A}^1$-homotopies. Let $\mathfrak{m}$ be the maximal ideal of R, and let $k = R/\mathfrak{m}$ be the residue field. The subgroup $\mathrm{E}_n(k) \subseteq \mathrm{SL}_n(k)$ generated by the elementary matrices is actually all of $\mathrm{SL}_n(k)$, so we can write $\overline{M}$, the image of M in $\mathrm{SL}_n(k)$, as a product of elementary matrices. Recall that an elementary matrix in $\mathrm{SL}_n(k)$ is the identity matrix except for a single off-diagonal entry. Since we can lift each of these to $\mathrm{SL}_n(R)$, we can write $M = EN$, where E is a product of elementary matrices in $\mathrm{SL}_n(R)$ and

$$N = \mathrm{I}_n + P,$$

where $P = (p_{ij}) \in \mathrm{M}_n(\mathfrak{m})$ is a matrix with entries in $\mathfrak{m}$. Note that the condition that $N \in \mathrm{SL}_n(R)$ means that we can solve for p_{11}. Indeed,

$$1 = \det(N) = (1 + p_{11})|C_{11}| - p_{12}|C_{12}| + \cdots + (-1)^n p_{1n}|C_{1n}|,$$

where C_{ij} is the ijth minor of N. Each p_{1r} is in $\mathfrak{m}$ for $2 \leq r \leq n$. Hence, $1 - n = (1 + p_{11})|C_{11}|$, where $n \in \mathfrak{m}$. Since $1 - n$ and $1 + p_{11}$ are units, $|C_{11}|$ must be a unit in R as well. Thus, we can solve

$$p_{11} = \frac{1-n}{|C_{11}|} - 1.$$

Now, define a new matrix $Q = (q_{ij})$ in $\mathrm{M}_n(\mathfrak{m}[t])$ by $q_{ij} = tp_{ij}$ unless $(i, j) = (1, 1)$, in which case set q_{11} so that $\det(1 + Q) = 1$, using the formula above. Then, we see that $Q(0) = \mathrm{I}_n$, while $Q(1) = P$. It follows that $1 + Q$ defines an explicit homotopy from I_n to $N = \mathrm{I}_n + P$. It follows that M is $\mathbb{A}^1$-homotopic to a product of elementary matrices. Since each elementary matrix is $\mathbb{A}^1$-homotopic to I_n, we have proved the claim.

Now, by Theorem 5.12, $\mathrm{SL}_n \to \mathrm{ESL}_n \to \mathrm{BSL}_n$ is an $\mathbb{A}^1$-fiber sequence due to the fact that SL_n-torsors are $\mathbb{A}^1$-invariant on smooth affine schemes (since GL_n-torsors are $\mathbb{A}^1$-invariant on smooth affine schemes). Therefore we have an exact sequence:

$$\pi_1^{\mathbb{A}^1}(\mathrm{ESL}_n) \to \pi_1^{\mathbb{A}^1}(\mathrm{BSL}_n) \to \pi_0^{\mathbb{A}^1}(\mathrm{SL}_n).$$

The left term is $\star$ since ESL_n is simplicially (and hence $\mathbb{A}^1$-)contractible and the right term is a singleton due to the first part of this proposition. □

EXERCISE 5.17. Prove the following statements when S satisfies the hypotheses of the previous theorem. For $i > 2$, $\pi_i^{\mathbb{A}^1}(\mathrm{B}\mathbb{G}_m) = 0$. For $i = 1$, the sheaf of groups $\pi_i^{\mathbb{A}^1}(\mathrm{B}\mathbb{G}_m) \cong \mathbb{G}_m$. Finally, $\pi_0^{\mathbb{A}^1}(\mathrm{B}\mathbb{G}_m) = \star$. Hint: use the $\mathbb{A}^1$-rigidity of $\mathbb{G}_m$ and Theorem 5.12.

PROPOSITION 5.18. *Let S be a regular noetherian affine scheme of finite Krull dimension, and suppose that the Bass-Quillen conjecture holds for smooth schemes of finite presentation over S. For $i > 1$, the map $\mathrm{SL}_n \to \mathrm{GL}_n$ induces an isomorphism $\pi_i^{\mathbb{A}^1}(\mathrm{BSL}_n) \to \pi_i^{\mathbb{A}^1}(\mathrm{BGL}_n)$.*

PROOF. By Theorem 5.12, the sequence $\mathrm{BSL}_n \to \mathrm{BGL}_n \to \mathrm{B}\mathbb{G}_m$ induces a long exact sequence of $\mathbb{A}^1$-homotopy sheaves and the result for $i > 1$ follows from the above proposition above. However we note that the case of $\pi_1^{\mathbb{A}^1}$ is different: we have an exact sequence $\pi_1^{\mathbb{A}^1}(\mathrm{BSL}_n) \to \pi_1^{\mathbb{A}^1}(\mathrm{BGL}_n) \to \pi_1^{\mathbb{A}^1}(\mathrm{B}\mathbb{G}_m) \to \pi_0^{\mathbb{A}^1}(\mathrm{BSL}_n)$; The groups on the right are zero by Proposition 5.10, and the group on the left is zero by Proposition 5.16. □

Recall from Corollary 4.45 that we have an $\mathbb{A}^1$-weak equivalence: $\mathrm{SL}_{n+1} / \mathrm{SL}_n \to \mathbb{A}^{n+1} - \{0\}$ for $n \geq 1$. Moreover, $\mathbb{A}^{n+1} - \{0\}$ is $\mathbb{A}^1$-weak equivalent to $(S^1)^{\wedge n} \wedge \mathbb{G}_m^{\wedge n+1}$. Our intuition from topology suggests therefore that $\mathrm{SL}_{n+1} / \mathrm{SL}_n$ should be $(n-1)$-$\mathbb{A}^1$-connected. This is indeed the case but it relies on a difficult theorem of Morel, the unstable $\mathbb{A}^1$-connectivity theorem [**Mor12**, Theorem 6.38]. That theorem uses an $\mathbb{A}^1$-homotopy theoretic version of Hurewicz theorem and of $\mathbb{A}^1$-homology sheaves, which are defined not by pointwise sheafification but instead using the so-called $\mathbb{A}^1$-derived category.

We may apply Theorem 5.12 to the fiber sequence of simplicial presheaves: $\mathrm{SL}_{n+1} / \mathrm{SL}_n \to \mathrm{BSL}_n \to \mathrm{BSL}_{n+1}$ to see that this is also an $\mathbb{A}^1$-fiber sequence. We have thus proved the following important stability result.

THEOREM 5.19 (Stability). *Let S be a regular noetherian affine scheme of finite Krull dimension, and suppose that the Bass-Quillen conjecture holds for smooth schemes of finite presentation over S. Let $i > 0$ and $n \geq 1$. The morphism*

$$\pi_i^{\mathbb{A}^1}(\mathrm{BSL}_n) \to \pi_i^{\mathbb{A}^1}(\mathrm{BSL}_{n+1})$$

is an epimorphism if $i \leq n$ and an isomorphism if $i \leq n-1$.

Setting $\mathrm{GL} = \operatorname{colim}_{n\to\infty} \mathrm{GL}_n$ and similarly for SL, we obtain the following corollary.

COROLLARY 5.20. *Let S be a regular noetherian affine scheme of finite Krull dimension, and suppose that the Bass-Quillen conjecture holds for smooth schemes of finite presentation over S. For $i \geq 2$, we have $\pi_i^{\mathbb{A}^1}(\mathrm{BSL}) \simeq \pi_i^{\mathbb{A}^1}(\mathrm{BGL})$.*

6. Representing algebraic K-theory

One reason to contemplate the $\mathbb{A}^1$-homotopy category is the fact that many invariants of schemes are $\mathbb{A}^1$-invariant; one important example is algebraic K-theory, at least for regular schemes. The goal of this section is to prove the representability of algebraic K-theory in $\mathbb{A}^1$-homotopy theory and identify its representing space when the base scheme S is regular and noetherian. One important consequence is an identification of the $\mathbb{A}^1$-homotopy sheaves of the classifying spaces BGL_n's in the stable range, which plays a crucial role in the classification of algebraic vector bundles via $\mathbb{A}^1$-homotopy theory. Indeed, it turns out that the relationship between algebraic K-theory and these classifying spaces is just like what happens in topology — the latter assembles into a representing space for the former. This is perhaps a little surprising as one way to define the algebraic K-theory of rings is via the complicated $+$-construction which alters the homotopy type of $\mathrm{BGL}_n(R)$ rather drastically. The key insight is that the $\mathrm{Sing}^{\mathbb{A}^1}$ construction is an alternative to the $+$-construction in nice cases, which leads to the identification of the representing space in $\mathrm{Spc}^{\mathbb{A}^1}(S)_\star$.

Throughout this section, we let S be a fixed *regular* noetherian scheme of finite Krull dimension. An argument using Weibel's homotopy invariant K-theory will yield a similar result over an arbitrary noetherian base, but for homotopy K-theory; for details see [**Cis13**].

6.1. Representability of algebraic K-theory. The first thing we note is that representability of algebraic K-theory in the $\mathbb{A}^1$-homotopy category itself is a formal consequence of basic properties of algebraic K-theory.

PROPOSITION 6.1. *Let S be a regular noetherian scheme of finite Krull dimension. Then, the K-theory space functor $\mathcal{K}$ is a fibrant object of $\mathrm{Spc}^{\mathbb{A}^1}_{S,\star}$. In particular, there are natural isomorphisms*

$$\mathrm{K}_i(X) \cong [\Sigma^i_+ X, \mathcal{K}]_{\mathbb{A}^1}$$

for all finitely presented smooth S-schemes X and all $i \geq 0$.

PROOF. It is enough to show that $\mathcal{K}$ is an $\mathbb{A}^1$-local object of $\mathrm{sPre}(\mathrm{Sm}_S)_\star$. For this, we must show that $\mathcal{K}$ is both a Nisnevich-local object and satisfies $\mathbb{A}^1$-homotopy invariance. The first property follows from [**TT90**, Proposition 6.8]. The second property is proved in [**TT90**, Theorem 10.8] for the K-theory spectra. Since $\mathbf{R}\Omega^\infty$ is a Quillen right adjoint, it preserves homotopy limits, and hence $\mathcal{K}$ also satisfies descent. □

Therefore algebraic K-theory is indeed representable in $\mathrm{Spc}^{\mathbb{A}^1}_{S,\star}$ by an object which we denote by $\mathcal{K}$. This argument is purely formal. Next, we need to get a better grasp of the representing object $\mathcal{K}$. To do so, we need some review on H-spaces.

DEFINITION 6.2. Let X be a simplicial set. We say that X is an H-space if it has a map $m : X \times X \to X$ and a point $e \in X$ which is a homotopy identity, that is, the maps $m(e,-), m(-,e) : X \to X$ are homotopic to the identity map.

EXERCISE 6.3. Prove that the fundamental group of any H-space is always abelian.

DEFINITION 6.4. Let X be a homotopy commutative and associative H-space. A group completion of X is an H-space Y together with an H-map $X \to Y$ such that

(1) $\pi_0(X) \to \pi_0(Y)$ is a group completion of the abelian monoid $\pi_0(X)$, and
(2) for any commutative ring R, the homomorphism $\mathrm{H}_*(X;R) \to \mathrm{H}_*(Y;R)$ is a localization of the graded commutative ring $\mathrm{H}_*(X;R)$ at the multiplicative subset $\pi_0(X) \subset \mathrm{H}_0(X,R)$.

We denote by X^{gp} the group completion of X.

There is a simple criterion for checking if a commutative and associative H-space is indeed its own group completion.

DEFINITION 6.5. Let X be an H-space. We say that X is group-like if the monoid $\pi_0(X)$ is a group.

The following proposition is standard. See [**MS75**] for example. A specific model of the group completion of X is ΩBX when X is homotopy commutative and associative [**Seg74**].

PROPOSITION 6.6. *Let X be a homotopy commutative and associative H-space, then the group completion of X is unique up to homotopy and further, if X is group-like, then X is weakly equivalent to its own group-completion.*

EXAMPLE 6.7. Let R be an associative ring. We have maps $m : \mathrm{GL}_n(R) \times \mathrm{GL}_m(R) \to \mathrm{GL}_{m+n}(R)$ defined by block sum. This map is a group homomorphism and thus induces a map

$$m : \left(\coprod_{n\geq 0} \mathrm{BGL}_n(R)\right)^{\times 2} \to \coprod_{n \geq 0} \mathrm{BGL}_n(R).$$

One easily checks that this is indeed a homotopy associative and homotopy commutative H-space.

REMARK 6.8. On the other hand we have the group $\mathrm{GL}(R) = \operatorname{colim} \mathrm{GL}_d(R)$ where the transition maps are induced by adding a single entry "1" at the bottom right corner. We can take $\mathrm{BGL}(R)$, the classifying space of R. This space is *not* an H-space: its fundamental group is $\mathrm{GL}(R)$ which is not an abelian group. For it to have any chance of being an H-space we need to perform the $+$-construction of Quillen which kills off a perfect normal subgroup of the fundamental group of a space and does not alter homology. For details see [**Wei13**, Section IV.1]. One key property of the $+$-construction that we will need is the following theorem of Quillen.

THEOREM 6.9 (Quillen). *Let R be an associative ring with unit, the map $i : \mathrm{BGL}(R) \to \mathrm{BGL}(R)^+$ is universal for maps into H-spaces. In other words for each map $f : \mathrm{BGL}(R) \to H$ where H is an H-space, there is a map $g : \mathrm{BGL}(R)^+ \to H$ such that $f \simeq g \circ i$ and the induced map on homotopy groups is independent of g.*

PROOF. See [**Wei13**, Section IV.1 Theorem 1.8] and the references therein. □

Having this construction, the two spaces we discussed are intimately related.

THEOREM 6.10 (Quillen). *Let R be an associative ring with unit, then the group completion of $\coprod \mathrm{BGL}_n(R)$ is weakly equivalent to $\mathbb{Z} \times \mathrm{BGL}(R)^+$.*

See [**Wei13**] for a proof. The plus construction alters the homotopy type of a space rather drastically. There are other models for the plus construction like Segal's $\Omega\mathrm{B}$ construction mentioned above. The $\mathrm{Sing}^{\mathbb{A}^1}$-construction turns out to provide another model, as we explain in the next section.

6.2. Applications to representability. The following theorem was established by Morel and Voevodsky [**MV99**], although a gap was pointed out by Schlichting and Tripathi [**ST15**], who also provided a fix.

THEOREM 6.11. *Let S be a regular noetherian scheme of finite Krull dimension. The natural map $\mathbb{Z} \times \mathrm{BGL} \to \mathcal{K}$ in $\mathrm{Spc}_S^{\mathbb{A}^1}$ is an $\mathbb{A}^1$-local weak equivalence.*

There is an $\mathbb{A}^1$-local weak equivalence

$$\left(\coprod_n \mathrm{BGL}_n\right)^{\mathrm{gp}} \simeq \mathcal{K}$$

from Theorem 6.10 since the $+$-construction is one way to obtain $\mathrm{K}(R)$ by Quillen. (Note that sheafification takes care of the fact that there might be non-trivial finitely generated projective R-modules.) Hence, to prove the theorem, we must construct an $\mathbb{A}^1$-weak equivalence

$$\mathbb{Z} \times \mathrm{BGL} \simeq \left(\coprod_n \mathrm{BGL}_n\right)^{\mathrm{gp}}.$$

We have already mentioned that BGL is not an H-space, so that group completion will not formally lead to a weak equivalence. It is rather the $\mathrm{Sing}^{\mathbb{A}^1}$-construction which leads to an H-space structure on the $\mathbb{A}^1$-localization of $\mathbb{Z} \times \mathrm{BGL}$.

LEMMA 6.12. *If R is a commutative ring, then $\mathrm{Sing}^{\mathbb{A}^1} \mathrm{BGL}(R)$ is an H-space.*

PROOF. See [**Wei13**, Exercise IV.11.9]. □

PROPOSITION 6.13. *If R is a commutative ring, the natural map*

$$\mathrm{Sing}^{\mathbb{A}^1}\mathrm{BGL}(R) \to \mathrm{Sing}^{\mathbb{A}^1}\mathrm{BGL}(R)^+$$

is a weak equivalence.

PROOF. The map is a homology equivalence since each $\mathrm{BGL}(\Delta^n_R) \to \mathrm{BGL}(\Delta^n_R)^+$ is a homology equivalence. Since both sides are group-like H-spaces they are nilpotent, so the fact that the map is a homology equivalence implies that it is a weak homotopy equivalence. □

PROOF OF THEOREM 6.11. More precisely, we claim that we have a weak equivalence of simplicial presheaves $\mathrm{L}_{\mathbb{A}^1}\mathrm{L}_{\mathrm{Nis}}(\mathbb{Z}\times\mathrm{BGL}) \to \mathcal{K}$. Since both simplicial presheaves are, in particular, Nisnevich local we need only check on stalks. Therefore we need only check that $\mathrm{Sing}^{\mathbb{A}^1}(\mathbb{Z}\times\mathrm{BGL}(R)) \simeq \mathcal{K}(R)$ for R a regular noetherian local ring because after this, further application of $\mathrm{Sing}^{\mathbb{A}^1}$ does not change the stalk by $\mathbb{A}^1$-homotopy invariance. but this follows from our work above since $\mathcal{K}(R) \simeq \mathrm{K}_0(R)\times\mathrm{BGL}(R)^+$. Since $\mathcal{K}$ is $\mathbb{A}^1$-invariant on smooth affine schemes, the natural map

$$|\mathrm{K}_0(R)\times\mathrm{BGL}(\Delta^\bullet_R)^+| \to \mathrm{K}_0(R)\times\mathrm{BGL}(R)^+$$

is a weak equivalence, and the result follows from Quillen. □

REMARK 6.14. One can further prove that BGL is represented in $\mathrm{Spc}^{\mathbb{A}^1}_{S,\star}$ by the Grassmanian schemes. In order to do this, Morel and Voevodsky used an elegant model for classifying spaces in [**MV99**], also considered by Totaro [**Tot99**].

As a corollary, we get a calculation of the stable range of the $\mathbb{A}^1$-homotopy sheaves of BGL_n and BSL_n.

COROLLARY 6.15. *Let $i > 1$ and $n \geq 1$. Then if $i \leq n-1$, we have isomorphisms*

$$\pi_i^{\mathbb{A}^1}\mathrm{BSL}_n \cong \pi_i^{\mathbb{A}^1}(\mathrm{BGL}_n) \cong \mathrm{K}_i.$$

PROOF. This follows from the stable range results in Theorem 5.19 and Corollary 5.20. □

7. Purity

In this section, we prove the purity theorem. The theorem has its roots in the following theorem from étale cohomology: suppose that k is an algebraically closed field with characteristic prime to an integer n and $Z \hookrightarrow X$ is a regular closed immersion of k-varieties. Suppose further that Z is of pure codimension c in X. Then, for any locally constant sheaf of $\mathbb{Z}/n$-modules $\mathcal{F}$ there is a canonical isomorphism

$$g : \mathrm{H}^{r-2c}_{\text{ét}}(Z, \mathcal{F}(-n)) \to \mathrm{H}^r_Z(X, \mathcal{F}),$$

the **purity isomorphism**. Here $\mathrm{H}^r_Z(X,-)$ is the étale cohomology of X with supports on Z, which is characterized as the group fitting into the long exact sequence

$$\cdots \to \mathrm{H}^r_Z(X,\mathcal{F}) \to \mathrm{H}^r_{\text{ét}}(X,\mathcal{F}) \to \mathrm{H}^r_{\text{ét}}(X - Z,\mathcal{F}) \to \mathrm{H}^{r+1}_Z(X,\mathcal{F}) \to \cdots .$$

Substituting the isomorphism above into the long exact sequence we obtain the **Gysin sequence**

$$\cdots \to \mathrm{H}^{r-2c}_{\text{ét}}(Z;\mathcal{F}(-c)) \to \mathrm{H}^{r}_{\text{ét}}(X;\mathcal{F}) \to \mathrm{H}^{r}_{\text{ét}}(X-Z;\mathcal{F}) \to \mathrm{H}^{r+1-2c}_{\text{ét}}(Z;\mathcal{F}(-c)) \to \cdots .$$

The Gysin sequence is extremely useful for calculation: the naturality of the long exact sequence and purity isomorphism leads to calculations of Frobenius weights of smooth varieties U by embedding them into a smooth projective variety $U \hookrightarrow X$ whose complement is often a normal crossing divisor [**Del75**].

In topology, the Gysin sequence is also available and is deduced in the following way. Suppose that $Z \hookrightarrow X$ is a closed immersion of smooth manifolds of (real) codimension c and ν_Z is the normal bundle of Z in X. The tubular neighborhood theorem identifies the Thom space of ν_Z with the cofiber of $X - Z \to X$, i.e. there is a weak homotopy equivalence

$$\mathrm{Th}(\nu_Z) \simeq \frac{X}{X-Z}.$$

One then proves that there is an isomorphism

$$\tilde{\mathrm{H}}^{i-c}(Z;k) \to \tilde{\mathrm{H}}^{i}(\mathrm{Th}(\nu_Z);k)$$

in reduced singular cohomology with coefficients in a field k. In fact, this last isomorphism is true if we replace ordinary singular cohomology with any complex-oriented cohomology theory [**May99**]. Therefore, the crucial step is identifying the cofiber $\frac{X}{X-Z}$ with the Thom space of the normal bundle. In this light, the purity theorem in $\mathbb{A}^1$-homotopy theory may be interpreted as a kind of tubular neighborhood theorem.

We will now prove the crucial purity theorem of Morel and Voevodsky [**MV99**, Theorem 2.23]. We benefited from unpublished notes of Asok and from the exposition of Hoyois in [**Hoy17**, Section 3.5] in the equivariant case. We follow the latter closely below. The discussion in this section is valid for S a quasi-compact quasi-separated base scheme S.

DEFINITION 7.1. A **smooth pair** over a scheme S is a closed embedding $i : Z \hookrightarrow X$ of finitely presented smooth S-schemes. We will often write such a pair as (X,Z), omitting reference to the map i. The smooth pairs over S form a category $\mathrm{Sm}^{\mathrm{pairs}}_S$ in which the morphisms $(X,Z) \to (X',Z')$ are pullback squares

$$\begin{array}{ccc} Z & \longrightarrow & X \\ \downarrow & & \downarrow \\ Z' & \longrightarrow & X'. \end{array}$$

A morphism $f : (X,Z) \to (X',Z')$ of smooth pairs is **Nisnevich** if $f : X \to X'$ is étale and if $f^{-1}(Z') \to Z'$ is an isomorphism.

We will need following local characterization of smooth pairs.

PROPOSITION 7.2. *Let $i : Z \to X$ be a smooth pair over a quasi-compact and quasi-separated scheme S. Assume that the codimension of i is c along Z. Then, there is a Zariski cover $\{U_i \to X\}_{i\in I}$ and a set of étale morphisms $\{U_i \to \mathbb{A}^{n_i}_S\}_{i\in I}$ such that the smooth pair $U_i \times_X Z \to U_i$ is isomorphic to the pullback of the inclusion of a linear subspace $\mathbb{A}^{n_i-c}_S \to \mathbb{A}^{n_i}_S$ for all $i \in I$.*

PROOF. See [**Gro63**, Théorème II.4.10]. □

Certain moves generate all smooth pairs, which lets one prove statements for all smooth pairs by checking them locally, checking that they transport along Nisnevich morphisms of smooth pairs, and checking that they hold for zero sections of vector bundles.

LEMMA 7.3. *Suppose that* **P** *is a property of smooth pairs over a quasi-compact and quasi-separated scheme* S *satisfying the following conditions:*

(1) *if* (X, Z) *is a smooth pair and if* $\{U_i \to X\}$ *is a Zariski cover such that* **P** *holds for* $(U_{i_1} \times_X \cdots \times_X U_{i_n}, Z \times_X U_{i_1} \times_X \cdots \times_X U_{i_n})$ *for all tuples* $i_1, \ldots, i_n \in I$, *then* **P** *holds for* (X, Z);
(2) *if* $(V, Z) \to (X, Z)$ *is a Nisnevich morphism of smooth pairs, then* **P** *holds for* (V, Z) *if and only if* **P** *holds for* (X, Z);
(3) **P** *holds for all smooth pairs of the form* $(\mathbb{A}^n_Z, Z)$.

Then, **P** *holds for all smooth pairs over* S.

PROOF. By (1), it suffices to check that **P** is true Zariski-locally on X. Pick a Zariski cover $\{U_i \to X\}$ satisfying the conclusion of Proposition 7.2. Thus, the problem is reduced to showing that if $(X, Z) \to (\mathbb{A}^n, \mathbb{A}^m)$ is a map of smooth pairs with $X \to \mathbb{A}^n$ étale, then **P** holds for (X, Z). Indeed, all pairs $(U_{i_1} \times_X \cdots \times_X U_{i_n}, Z \times_X U_{i_1} \times_X \cdots \times_X U_{i_n})$ have this form by our choice of cover. The rest of the argument follows [**MV99**, Lemma 2.28]. Form the fiber product $X \times_{\mathbb{A}^n} (Z \times_S \mathbb{A}^c)$, where $c = n - m$, and $Z \times_S \mathbb{A}^c \to \mathbb{A}^n$ is the product of the maps $Z \to \mathbb{A}^m$ and $\mathbb{A}^c \xrightarrow{\mathrm{id}} \mathbb{A}^c$. Since $Z \to \mathbb{A}^m$ is étale, we see that $Z \times_{\mathbb{A}^m} Z \subseteq X \times_{\mathbb{A}^n} (Z \times \mathbb{A}^c)$ is the disjoint union of Z and some closed subscheme W. Let $U = X \times_{\mathbb{A}^n} (Z \times_S \mathbb{A}^c) - W$. The projection maps induce Nisnevich maps of pairs $(U, Z) \to (X, Z)$ and $(U, Z) \to (Z \times_S \mathbb{A}^c, Z)$. By (3), **P** holds for $(Z \times_S \mathbb{A}^c, Z)$ and hence for (U, Z) by (2), and hence for (X, Z) by (2) again. □

DEFINITION 7.4. A morphism $(X, Z) \to (X', Z')$ of smooth pairs over S is **weakly excisive** if the induced square

$$\begin{array}{ccc} Z & \longrightarrow & X/(X - Z) \\ \downarrow & & \downarrow \\ Z' & \longrightarrow & X'/(X' - Z') \end{array}$$

is homotopy cocartesian in $\mathrm{Spc}_S^{\mathbb{A}^1}$.

The following exercise is used in the proof of the purity theorem.

EXERCISE 7.5 ([**Hoy17**, Lemma 3.19]). Let $(X, Z) \xrightarrow{f} (X', Z') \xrightarrow{g} (X'', Z'')$ be composable morphisms of smooth pairs over S. Prove the following statements.

(1) If f is weakly excisive, then g is weakly excisive if and only if $g \circ f$ is weakly excisive.
(2) If g and $g \circ f$ are weakly excisive, and if $g : Z' \to Z''$ is an $\mathbb{A}^1$-local weak equivalence, then f is weakly excisive.

Finally, we come to the purity theorem itself.

THEOREM 7.6 (Purity theorem [**MV99**, Theorem 2.23]). *Let $Z \hookrightarrow X$ be a closed embedding in Sm_S where S is quasi-compact and quasi-separated. If $\nu_Z : \mathrm{N}_X Z \to Z$ is the normal bundle to Z in X, then there is an $\mathbb{A}^1$-local weak equivalence*

$$\frac{X}{X-Z} \to \mathrm{Th}(\nu_Z)$$

which is natural in $\mathrm{Ho}(\mathrm{Spc}_S^{\mathbb{A}^1})$ for smooth pairs (X, Z) over S.

PROOF. First, we construct the map. Consider the construction

$$\mathrm{D}_Z X = \mathrm{Bl}_{Z\times_S\{0\}}(X \times_S \mathbb{A}^1) - \mathrm{Bl}_{Z\times_S\{0\}}(X \times_S \{0\}),$$

which is natural in smooth pairs (X, Z). The fiber of $\mathrm{D}_Z X \to \mathbb{A}^1$ at $\{0\}$ is the complement $\mathbb{P}(\mathrm{N}_Z X \oplus \mathcal{O}_Z) - \mathbb{P}(\mathrm{N}_Z X)$, which is naturally isomorphic to the vector bundle $\mathrm{N}_Z X$. Hence, by taking the zero section at $\{0\}$, we get a closed embedding $Z \times_S \mathbb{A}^1 \to \mathrm{D}_Z X$. The fiber at $\{0\}$ of $(\mathrm{D}_Z X, Z \times_S \mathbb{A}^1)$ is $(\mathrm{N}_Z X, Z)$, while the fiber at $\{1\}$ is (X, Z). Thus, there are morphisms of smooth pairs

$$(X, Z) \xrightarrow{i_1} (\mathrm{D}_Z X, Z \times_S \mathbb{A}^1) \xleftarrow{i_0} (\mathrm{N}_Z X, Z),$$

and it is enough to prove that i_1 and i_0 are weakly excisive for all smooth pairs (X, Z). Indeed, in that case there are natural $\mathbb{A}^1$-weak equivalences $X/(X-Z) \simeq \mathrm{D}_Z X/(\mathrm{D}_Z X - Z \times_S \mathbb{A}^1) \simeq \mathrm{N}_Z X/(\mathrm{N}_Z X - Z) = \mathrm{Th}(\nu_Z)$ because the cofiber of $Z \to Z \times_S \mathbb{A}^1$ is contractible.

Let **P** hold for the smooth pair (X, Z) if and only if i_0 and i_1 are excisive. We show that **P** satisfies conditions (1)-(3) of Lemma 7.3.

Let $\{U_i \to X\}_{i\in I}$ be a Zariski cover of X, and let $(U_{i_1,\ldots,i_n}, Z_{i_1,\ldots,i_n}) \to (X, Z)$ be the induced morphisms of smooth pairs. For Suppose that **P** holds for each $(U_{i_1,\ldots,i_n}, Z_{i_1,\ldots,i_n})$. Then, there is a diagram

$$\begin{array}{ccc} |Z_\bullet| & \xrightarrow{i_1} & |U_\bullet/(U_\bullet - Z_\bullet)| \\ \downarrow & & \downarrow \\ |Z_\bullet \times_S \mathbb{A}^1| & \xrightarrow{i_1} & |\mathrm{D}_{Z_i} U_i/(D_{Z_i} U_i - Z_i \times_S \mathbb{A}^1)| \end{array}$$

of geometric realizations. However, this is the geometric realization of a simplicial cocartesian square by hypothesis, so it is itself cocartesian. The same argument works for i_0, so we see that **P** satisfies (1).

Consider a Nisnevich morphism $(V, Z) \to (X, Z)$ of smooth pairs, and consider the diagram

$$\begin{array}{ccccc} (V, Z) & \xrightarrow{i_1} & (\mathrm{D}_Z V, Z \times_S \mathbb{A}^1) & \xleftarrow[i_0]{} & (\mathrm{N}_Z V, Z) \\ \downarrow & & \downarrow & & \downarrow \\ (X, Z) & \xrightarrow{i_1} & (\mathrm{D}_Z X, Z \times_S \mathbb{A}^1) & \xleftarrow[i_0]{} & (\mathrm{N}_Z X, Z). \end{array}$$

We leave it as an easy exercise to the reader to show using Exercise 7.5 that (2) will follow if the vertical arrows are all weakly excisive. But, the vertical maps are all Nisnevich morphisms. So, it is enough to check that Nisnevich morphisms $(V, Z) \to (X, Z)$ of smooth pairs are weakly excisive. Let U be the complement of

Z in X. By hypothesis, the diagram

$$\begin{array}{ccc} U \times_X V & \longrightarrow & V \\ \downarrow & & \downarrow \\ U & \longrightarrow & X \end{array}$$

is an elementary distinguished square, and hence a homotopy cocartesian square in $\mathrm{Spc}_S^{\mathbb{A}^1}$ by Proposition 4.13. In particular, the cofiber of $V/(U \times_X V) \to X/U$ is contractible. Since the cofiber of $Z \to Z$ is obviously contractible, this proves $(V, Z) \to (X, Z)$ is weakly excisive.

To complete the proof, we just have to show that (3) holds. Of course, in the situation $(\mathbb{A}^n_Z, Z)$ of (3) we can prove the main result of the theorem quite easily. However, the structure of the proof requires us to check weak excision for i_0 and i_1. For this we can immediately reduce to the case where $Z = S$, which we omit from the notation for the rest of the proof. The blowup $\mathrm{Bl}_{\{0\}}(\mathbb{A}^n \times \mathbb{A}^1)$ is the total space of an $\mathbb{A}^1$-bundle over $\mathbb{P}^n$, and the image of $\mathrm{Bl}_{\{0\}}(\mathbb{A}^n)$ in $\mathbb{P}^n$ is a hyperplane $\mathbb{P}^{n-1} \subseteq \mathbb{P}^n$. Hence, there is a morphism of pairs

$$(\mathrm{D}_{\{0\}}(\mathbb{A}^n), \{0\} \times \mathbb{A}^1) \xrightarrow{f} (\mathbb{A}^n, \{0\}).$$

Since $\mathrm{D}_{\{0\}}(\mathbb{A}^n) \to \mathbb{A}^n$ is the total space of an $\mathbb{A}^1$-bundle, this morphism is weakly excisive. The composition of f with i_1 is the identity on $(\mathbb{A}^n, \{0\})$ and hence is weakly excisive as well. By Exercise 7.5(2), it follows that i_1 is weakly excisive. Similarly, $f \circ i_0$ is the identity on $(\mathrm{N}_{\{0\}}\mathbb{A}^n, \{0\}) \cong (\mathbb{A}^n, \{0\})$, so i_0 is weakly excisive, again by Exercise 7.5(2). □

8. Vista: classification of vector bundles

In this section we give a brief summary of how to use the theory developed above to give a straightforward proof of the classification of vector bundles on smooth affine curves and surfaces. We must understand how to compute $\mathbb{A}^1$-homotopy classes of maps to BGL_n so we can apply the Postnikov obstruction approach to the classification problem.

Recall from Theorem 4.27 that the $\mathbb{A}^1$-localization functor above may be calculated as a transfinite composite of $\mathrm{L}_{\mathrm{Nis}}$ and $\mathrm{Sing}^{\mathbb{A}^1}$. This process is rather unwieldy. However, things get better if this process stops at a finite stage, in particular suppose that $\mathcal{F}$ was already a Nisnevich-local presheaf and suppose that one can somehow deduce that $\mathrm{Sing}^{\mathbb{A}^1} X$ was already Nisnevich local, then one can conclude that $\mathrm{L}_{\mathbb{A}^1} X \simeq \mathrm{Sing}^{\mathbb{A}^1} X$ and therefore, using our formulae for mapping spaces in the $\mathbb{A}^1$-homotopy category, one concludes that

$$[U, X]_{\mathbb{A}^1} \cong [U, \mathrm{L}_{\mathbb{A}^1} X]_s \simeq \pi_0 \,\mathrm{Sing}^{\mathbb{A}^1} X(U).$$

In general, this does not work.

REMARK 8.1. Work of Balwe-Hogadi-Sawant **[BHS15]** constructs explicit smooth projective varieties X over $\mathbb{C}$ for which $\mathrm{Sing}^{\mathbb{A}^1} X$ is *not* Nisnevich-local, so extra conditions must be imposed to calculate the $\mathbb{A}^1$-homotopy classes of maps naively. However, there is often an intimate relation between naive $\mathbb{A}^1$-homotopies and genuine ones: Cazanave constructs in **[Caz12]** a monoid structure on $\pi_0 \operatorname{Sing}^{\mathbb{A}^1}(\mathbb{P}^1_k)$ and proves that the map $\pi_0 \operatorname{Sing}^{\mathbb{A}^1}(\mathbb{P}^1_k) \to [\mathbb{P}^1_k, \mathbb{P}^1_k]_{\mathbb{A}^1}$ is group completion, with the group structure on the target induced by the $\mathbb{A}^1$-weak equivalence $\mathbb{P}^1 \simeq S^1 \wedge \mathbb{G}_m$.

EXERCISE 8.2. Even for fields, one can show that the sets of isomorphism classes of vector bundles over the simplest non-affine scheme $\mathbb{P}^1_k$ are not $\mathbb{A}^1$-invariant. Construct (e.g. write down explicit transition functions) a vector bundle over $\mathbb{P}^1 \times_k \mathbb{A}^1$ that restricts to $\mathcal{O}(0) \oplus \mathcal{O}(0)$ on $\mathbb{P}^1 \times \{1\}$ and $\mathcal{O}(1) \oplus \mathcal{O}(-1)$ on $\mathbb{P}^1 \times \{0\}$ for a counter-example.

REMARK 8.3. In [**AD08**], examples are given of smooth $\mathbb{A}^1$-contractible varieties with families of non-trivial vector bundles of any given dimension. Were vector bundles to be representable in $\mathrm{Spc}^{\mathbb{A}^1}$, such pathologies could not occur. These varieties are non-affine.

As the exercise and remark show, the only hope for computing vector bundles as $\mathbb{A}^1$-homotopy classes of maps to BGL_n is to restrict to affine schemes, but even there it is not at all obvious that this is possible, as the map $\mathrm{BGL}_n \to \mathrm{L}_{\mathbb{A}^1}\mathrm{BGL}_n$ is not a simplicial weak equivalence. Remarkably, despite this gulf, Morel and later Asok-Hoyois-Wendt showed that for smooth affine schemes one can compute vector bundles in this way. In fact, this follows from a much more general and formal result, which we now explain.

We say that a presheaf F of sets on Sm_S satisfies **affine $\mathbb{A}^1$-invariance** if the pullback maps $F(U) \to F(U \times_S \mathbb{A}^1)$ are isomorphisms for all finitely presented smooth affine S-schemes U. Note that we say that an S-scheme is affine if $U \to \operatorname{Spec} \mathbb{Z}$ is affine, so that $U = \operatorname{Spec} R$ for some commutative ring R.

THEOREM 8.4 ([**AHW15a**]). *Let S be a quasi-compact and quasi-separated scheme. Suppose that X is a simplicial presheaf on Sm_S. Assume that $\pi_0(X)$ is affine $\mathbb{A}^1$-invariant and that X satisfies affine Nisnevich excision. For all affine schemes U in Sm_S, the canonical map*

$$\pi_0(X)(U) \to [U, X]_{\mathbb{A}^1}$$

is an isomorphism.

SKETCH PROOF. The key homotopical input to this theorem is the π_*-Kan condition, which ensures that homotopy colimits of simplicial diagram commutes over pullbacks [**BF78**]. This condition was first used in this area by Schlichting [**Sch15**]. It provides a concrete criterion to check if the functor $\operatorname{Sing}^{\mathbb{A}^1}(F)$ restricted to smooth affine schemes is indeed Nisnevich local (to make this argument precise, the key algebro-geometric input is the equivalence between the Nisnevich cd-structure and the affine Nisnevich cd-structure defined above [**AHW15a**, Proposition 2.3.2] when restricted to affine schemes).

More precisely, for any elementary distinguished square

$$\begin{array}{ccc} U \times_X V & \longrightarrow & V \\ \downarrow & & \downarrow p \\ U & \xrightarrow{\;i\;} & Y \end{array}$$

we have a homotopy pullback square

$$\begin{array}{ccc} X(Y \times \mathbb{A}^n) & \longrightarrow & X(V \times \mathbb{A}^n) \\ \downarrow & & \downarrow \\ X(U \times \mathbb{A}^n) & \longrightarrow & X(U \times_X V \times \mathbb{A}^n) \end{array}$$

of simplicial sets for all $n \geq 0$.

The π_*-Kan condition applies with the hypothesis that $\pi_0(X)$ satisfies affine $\mathbb{A}^1$-invariance and we may conclude that taking $\mathrm{hocolim}_{\Delta^{\mathrm{op}}}$ of the above squares preserve pullbacks and therefore we conclude that $\mathrm{Sing}^{\mathbb{A}^1}(X)$ satisfies affine Nisnevich excision.

Applying the above proposition, we have that for any affine U,

$$\mathrm{Sing}^{\mathbb{A}^1}(X)(U) \to \mathrm{L}_{\mathrm{Nis}} \mathrm{Sing}^{\mathbb{A}^1}(X)(U)$$

is a weak equivalence. Since the left hand side is $\mathbb{A}^1$-invariant, we conclude that the right hand side is $\mathbb{A}^1$-invariant; since being $\mathbb{A}^1$-invariant and Nisnevich local may be tested on affine schemes (by [**AHW15a**, Proposition 2.3.2]), we conclude that $\mathrm{L}_{\mathrm{Nis}} \mathrm{Sing}^{\mathbb{A}^1}(X) \simeq \mathrm{L}_{\mathbb{A}^1}(X)$. Taking π_0 of the weak equivalence above gets us the desired claim. □

COROLLARY 8.5 (Affine representability of vector bundles). *Let S be a regular noetherian affine scheme of finite Krull dimension, and suppose that the Bass-Quillen conjecture holds for smooth schemes of finite presentation over S. In this case, the natural map* $\mathrm{Vect}_r(U) \to [U, \mathrm{BGL}_r]_{\mathbb{A}^1}$ *is an isomorphism for all $U \in \mathrm{Sm}_S^{\mathrm{Aff}}$ and all $r \geq 0$.*

The Jouanoulou-Thomason homotopy lemma states that, up to $\mathbb{A}^1$-homotopy, we may replace a smooth scheme with an affine one.

THEOREM 8.6 ([**Jou73**] and [**Wei89**]). *Given a smooth separated scheme U over a regular noetherian affine scheme S, there exists an affine vector bundle torsor $\widetilde{U} \to U$ such that $\widetilde{U}$ is affine.*

PROOF. The point is that U is quasi-compact and quasi-separated and hence admits an ample family of line bundles (so U is divisorial) by [**71**, Proposition II.2.2.7]. The theorem now follows from [**Wei89**, Proposition 4.4]. □

This theorem lets us compute in some sense $[U, \mathrm{BGL}_n]_{\mathbb{A}^1}$ for any $U \in \mathrm{Sm}_S$, but it is not known at the moment what kind of objects these are on U.

One of the main features of $\mathbb{A}^1$-localization is the ability to employ topological thinking in algebraic geometry, if one is willing to work $\mathbb{A}^1$-locally. The homotopy sheaves $\pi_i^{\mathbb{A}^1}(X)$ are sometimes computable using input from both homotopy theory and algebraic geometry. At the same time, many algebro-geometric problems are inherently not $\mathbb{A}^1$-local in nature so one only gets an actual algebro-geometric theorem under certain certain conditions, as in Theorem 8.13 below. Let us first start with a review of Postnikov towers in $\mathbb{A}^1$-homotopy theory. Our main reference is [**AF14**], which in turn uses [**Mor12**], [**MV99**] and [**GJ99**].

Let G be a Nisnevich sheaf of groups and M a Nisnevich sheaf of abelian groups on which G acts (a G-module). In this case, G acts on the Eilenberg-Maclane sheaf $K(A, n)$, from which we may construct $K^G(A, n) := EG \times^G K(A, n)$. The first projection gives us a map $K^G(A, n) \to G$.

Of primary interest is the Nisnevich sheaf of groups $\pi_1^{\mathbb{A}^1}(Y)$ for some pointed $\mathbb{A}^1$-connected space Y. In this case, $\pi_1^{\mathbb{A}^1}(Y)$ acts on the higher homotopy sheaves $\pi_n^{\mathbb{A}^1}(Y)$ where $n \geq 2$.

THEOREM 8.7. *Let Y be a pointed $\mathbb{A}^1$-connected space. There exists a commutative diagram of pointed $\mathbb{A}^1$-connected spaces*

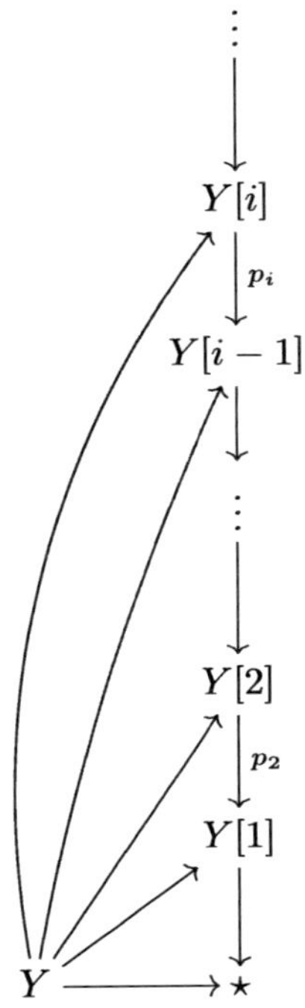

such that

(1) $Y[1] \simeq B\pi_1^{\mathbb{A}^1}(Y)$,
(2) $\pi_j^{\mathbb{A}^1} Y[i] = 0$ *for* $j > i$,
(3) *the map* $\pi_j^{\mathbb{A}^1} Y \to \pi_j^{\mathbb{A}^1} Y[i]$ *is an isomorphism of* $\pi_1^{\mathbb{A}^1} Y$*-modules for* $1 \leq j \leq i$,
(4) *the* $K(\pi_i^{\mathbb{A}^1} Y, i)$*-bundle* $Y[i] \to Y[i-1]$ *is a* **twisted principal fibration** *in the sense that there is a map*

$$k_i : Y[i-1] \to K^{\pi_1^{\mathbb{A}^1} Y}(\pi_i^{\mathbb{A}^1} Y, i+1)$$

such that $Y[i]$ *is obtained as the pullback*

$$\begin{array}{ccc} Y[i] & \longrightarrow & \mathrm{B}_{\mathrm{Nis}}\pi_1^{\mathbb{A}^1} Y \\ \downarrow & & \downarrow \\ Y[i-1] & \longrightarrow & K^{\pi_1^{\mathbb{A}^1} Y}(\pi_i^{\mathbb{A}^1} Y, i+1), \end{array}$$

(5) *and* $Y \to \lim_i Y[i]$ *is an* $\mathbb{A}^1$*-weak equivalence.*

The tower, which can be made functorial in Y, *is called the* **$\mathbb{A}^1$-Postnikov tower** *of* Y.

PROOF. This is left as an exercise, which basically amounts to Nisnevich sheafifying and $\mathbb{A}^1$-localizing the usual Postnikov tower. For an extensive discussion, see [**AF14**, Section 6] and the references therein. □

The point of the Postnikov tower is to make it possible to classify maps from X to Y by constructing maps inductively, i.e., starting with a map $X \to Y[1]$, lifting it to $X \to Y[2]$ while controlling the choices of lifts, and so on.

THEOREM 8.8. *Let S be a quasi-compact quasi-separated base scheme, and let X be a smooth noetherian S-scheme of Krull dimension at most d. Suppose that (Y, y) is a pointed $\mathbb{A}^1$-connected space. The natural map*

$$[X, Y]_{\mathbb{A}^1} \to [X, Y[i]]_{\mathbb{A}^1}$$

is an isomorphism for $i \geq d$ and a surjection for $i = d-1$.

PROOF. The first obstruction to lifting a map $X \to Y[i]$ to $X \to Y$ is the obstruction to lifting it to $X \to Y[i+1]$. This is classified by the k-invariant, and hence a class in $\mathrm{H}^{i+2}_{\mathrm{Nis},\pi_1^{\mathbb{A}^1}Y}(X, \pi_{i+1}^{\mathbb{A}^1}Y) = [X, K^{\pi_1^{\mathbb{A}^1}Y}(\pi_{i+1}^{\mathbb{A}^1}Y, i+2)]_{\mathbb{A}^1}$. One important feature of the theory intervenes at this point: the equivariant cohomology group $\mathrm{H}^{i+2}_{\mathrm{Nis},\pi_1^{\mathbb{A}^1}Y}(X, \pi_{i+1}^{\mathbb{A}^1}Y)$ can be identified with an ordinary Nisnevich cohomology group of a twisted form $(\pi_{i+1}^{\mathbb{A}^1}Y)_\lambda$ of $\pi_{i+1}^{\mathbb{A}^1}Y$ in Nisnevich sheaves on X:

$$\mathrm{H}^{i+2}_{\mathrm{Nis},\pi_1^{\mathbb{A}^1}Y}(X, \pi_{i+1}^{\mathbb{A}^1}Y) \cong \mathrm{H}^{i+2}_{\mathrm{Nis}}(X, (\pi_{i+1}^{\mathbb{A}^1}Y)_\lambda).$$

See [**Mor12**, Appendix B]. This group vanishes if $i+2 > d$, or $i+1 \geq d$, since the Nisnevich cohomological dimension of X is at most d by hypothesis. Thus, the map in the theorem is a surjection for $i \geq d-1$. The set of lifts, by the $\mathbb{A}^1$-fiber sequence induced from Theorem 8.7(4), is a quotient of $\mathrm{H}^{i+1}_{\mathrm{Nis},\pi_1^{\mathbb{A}^1}Y}(X, \pi_{i+1}^{\mathbb{A}^1}Y)$, which vanishes for the same reason as above if $i+1 > d$, or $i \geq d$. This completes the proof. □

We have an immediate consequence of the existence of the Postnikov towers as follows.

PROPOSITION 8.9. *If E is a rank $n > d$ vector bundle on a smooth affine d-dimensional variety X, then E splits off a trivial direct summand.*

PROOF. Using Proposition 5.18 and Theorem 5.19, we see that $\pi_i^{\mathbb{A}^1}\mathrm{BGL}_d \to \pi_i^{\mathbb{A}^1}\mathrm{BGL}_n$ is an isomorphism for $i \leq d-1$, and a surjection for $i = d$. By the representability theorem, E is represented by a map $X \to \mathrm{BGL}_n$ in the $\mathbb{A}^1$-homotopy category. Compose this map with $\mathrm{BGL}_n \to \mathrm{BGL}_n[d]$ to obtain $g : X \to \mathrm{BGL}_n[d]$. Note that E is uniquely determined by g by Theorem 8.8. It suffices to lift g to a map $h : X \to \mathrm{BGL}_d[d]$. The fiber of $\mathrm{BGL}_d[d] \to \mathrm{BGL}_n[d]$ is a $K(A, d)$-space for some Nisnevich sheaf A with an action of $\mathbb{G}_m = \pi_1^{\mathbb{A}^1}\mathrm{BGL}_d$. It follows that the obstruction to lifting g through $\mathrm{BGL}_d[d] \to \mathrm{BGL}_n[d]$ is a class of $\mathrm{H}^{d+1}_{\mathrm{Nis},\mathbb{G}_m}(X, A) = 0$. □

REMARK 8.10. As a Nisnevich sheaf of spaces, BGL_n is a $K(\pi, 1)$-sheaf in the sense that it has only one non-zero homotopy group. For the purposes of obstruction theory and classification theory this is not terribly useful as choosing a lift to the first stage of the Nisnevich-local Postnikov tower of BGL_n is equivalent to specifying a vector bundle. The process of $\mathbb{A}^1$-localization mysteriously acts as a prism that separates the single homotopy sheaf into an entire sequence (spectrum) of homotopy sheaves, allowing a finer step-by-step analysis.

PROPOSITION 8.11. *The first few* $\mathbb{A}^1$*-homotopy sheaves of* BGL_2 *are*

$$\pi_0^{\mathbb{A}^1}\mathrm{BGL}_2 = \star,$$
$$\pi_1^{\mathbb{A}^1}\mathrm{BGL}_2 \cong \mathbb{G}_m,$$
$$\pi_2^{\mathbb{A}^1}\mathrm{BGL}_2 \cong \mathrm{K}_2^{\mathrm{MW}},$$

where $\mathrm{K}_2^{\mathrm{MW}}$ *denotes the second unramified Milnor-Witt sheaf.*

PROOF. The $\mathbb{A}^1$-connectivity statement $\pi_0^{\mathbb{A}^1}\mathrm{BGL}_2 = \star$ follows from the fact that vector bundles are Zariski and hence Nisnevich locally trivial. The fact that $\pi_1\mathrm{BGL}_2 \cong \mathbb{G}_m$ follows from the stable range result that gives $\pi_1\mathrm{BGL}_2 \cong \pi_1\mathrm{BGL}_\infty \cong \mathrm{K}_1 \cong \mathbb{G}_m$, where the last two isomorphisms are explained in Section 5.3. The last follows from the $\mathbb{A}^1$-fiber sequence

$$\mathbb{A}^2 - \{0\} \to \mathrm{BGL}_1 \to \mathrm{BGL}_2,$$

the fact that BGL_1 is a $K(\mathbb{G}_m, 1)$-space, and Morel's result [**Mor12**, Theorem 6.40], which says that $\pi_1^{\mathbb{A}^1}\mathbb{A}^2 - \{0\} \cong \mathrm{K}_2^{\mathrm{MW}}$. □

Now, for any smooth scheme X and any line bundle $\mathcal{L}$ on X, there is an exact sequence of Nisnevich sheaves

$$0 \to \mathrm{I}^3(\mathcal{L}) \to \mathrm{K}_2^{\mathrm{MW}}(\mathcal{L}) \to \mathrm{K}_2 \to 0, \tag{2}$$

where the first and second terms are the $\mathcal{L}$-twisted forms (see [**Mor12**]). The sheaf $\mathrm{K}_2^{\mathrm{MW}}(\mathcal{L})$ controls the rank 2 vector bundles on X with determinant $\mathcal{L}$.

If X is a smooth affine surface, there is a bijection $[X, \mathrm{BGL}_2]_{\mathbb{A}^1} \to [X, \mathrm{BGL}_2[2]]_{\mathbb{A}^1}$, from which it follows that the rank 2 vector bundles on X with determinant $\mathcal{L}$ are classified by a quotient of

$$\mathrm{H}^2_{\mathrm{Nis},\mathbb{G}_m}(X, \mathrm{K}_2^{\mathrm{MW}}) \cong \mathrm{H}^2_{\mathrm{Nis}}(X, \mathrm{K}_2^{\mathrm{MW}}(\mathcal{L})).$$

In fact, we will see that the quotient is all of $\mathrm{H}^2_{\mathrm{Nis}}(X, \mathrm{K}_2^{\mathrm{MW}}(\mathcal{L}))$.

LEMMA 8.12. *If* X *is a smooth affine surface over a quadratically closed field* k*, then* $\mathrm{H}^n_{\mathrm{Nis}}(X, \mathrm{I}^3(\mathcal{L})) = 0$ *for* $n \geq 2$.

PROOF. This is [**AF14**, Proposition 5.2]. □

It follows from the lemma and the exact sequence (2) that the space of lifts is a quotient of $\mathrm{H}^2_{\mathrm{Nis}}(X, \mathrm{K}_2) \cong \mathrm{CH}^2(X)$, where the isomorphism is due to Quillen [**Qui73**] in the Zariski topology and Thomason-Trobaugh [**TT90**] in the Nisnevich topology. Now, looking at

$$\begin{aligned} &\mathbb{G}_m(X) \cong [X, K(\mathbb{G}_m, 0)]_{\mathbb{A}^1} \\ &\quad \to [X, K^{\mathbb{G}_m}(\mathrm{K}_2^{\mathrm{MW}}, 2)]_{\mathbb{A}^1} \to [X, \mathrm{BGL}_2[2]]_{\mathbb{A}^1} \to [X, \mathrm{BGL}_2[1]]_{\mathbb{A}^1} \cong \mathrm{Pic}(X), \end{aligned}$$

we see that the map $[X, K^{\mathbb{G}_m}(\mathrm{K}_2^{\mathrm{MW}}, 2)]_{\mathbb{A}^1} \to [X, \mathrm{BGL}_2[2]]_{\mathbb{A}^1}$ is injective because every element of $\mathbb{G}_m(X)$ extends to an automorphism of the vector bundle classified by $X \to \mathrm{BGL}_2$.

It follows that the map

$$\mathrm{Vect}_2(X) \to \mathrm{CH}^1(X) \times \mathrm{H}^2_{\mathrm{Nis}}(X, \mathrm{K}_2)$$

is a bijection. Thus, we have sketched a proof of the following theorem.

THEOREM 8.13. *Let X be a smooth affine surface over a quadratically closed field. Then, the map*

$$(c_1, c_2) : \mathrm{Vect}_2(X) \to \mathrm{CH}^1(X) \times \mathrm{CH}^2(X),$$

induced by taking the first and second Chern classes, is a bijection.

REMARK 8.14. To see that the natural maps involved are the Chern classes, as claimed, refer to [**AF14**, Section 6].

The fact that the theorem holds over quadratically closed fields is stronger than the previous results in this direction, which had been obtained without $\mathbb{A}^1$-homotopy theory.

Asok and Fasel have carried this program much farther in several papers, for instance showing in [**AF14**] that $\mathrm{Vect}_2(X) \cong \mathrm{CH}^1(X) \times \mathrm{CH}^2(X)$ when X is a smooth affine *three-fold* over a quadratically closed field. This theorem, which is outside the stable range, is much more difficult.

9. Further directions

In most of the exercises below, none of which are supposed to be easy, it will be useful to bear in mind the universal properties of $\mathrm{L}_{\mathrm{Nis}}$ and $\mathrm{L}_{\mathbb{A}^1}$.

EXERCISE 9.1. Use the formalism of model categories to construct topological and étale realization functors out of the $\mathbb{A}^1$-homotopy category $\mathrm{Spc}_S^{\mathbb{A}^1}$. Dugger's paper [**Dug01a**] on universal homotopy theories may come in handy. This problem is studied specifically in [**DI04**] and [**DI08**].

EXERCISE 9.2. Show that topological realization takes the motivic sphere $S^{a,b}$ where $a \leq b$ to the topological sphere S^a.

EXERCISE 9.3. Prove that complex topological K-theory is representable in $\mathrm{Spc}_S^{\mathbb{A}^1}$.

EXERCISE 9.4. Let $\mathbb{R}$ denote the field of real numbers. Construct a realization functor from $\mathrm{Spc}_{\mathbb{R}}^{\mathbb{A}^1}$ to the homotopy theory of $\mathbb{Z}/2$-equivariant topological spaces. Again, see [**DI04**].

EXERCISE 9.5. Construct a realization functor from $\mathrm{Spc}_S^{\mathbb{A}^1}$ to Voevodsky's category $\mathrm{DM}(S)$ of (big) motives over S. It will probably be necessary to search the literature for a model category structure for $\mathrm{DM}(S)$.

EXERCISE 9.6. Show that the realization functor from $\mathrm{Spc}_S^{\mathbb{A}^1}$ to Voevodsky's category factors through the *stable* motivic homotopy category obtained from $\mathrm{Spc}_S^{\mathbb{A}^1}$ by stabilizing with respect to $S^{2,1} \simeq \mathbb{P}^1$.

EXERCISE 9.7. Ayoub [**Ayo07**] has constructed a 6-functors formalism for stable motivic homotopy theory. Construct some functors between $\mathrm{Spc}_S^{\mathbb{A}^1}$ and $\mathrm{Spc}_U^{\mathbb{A}^1}$ when U is open in S and between $\mathrm{Spc}_S^{\mathbb{A}^1}$ and $\mathrm{Spc}_Z^{\mathbb{A}^1}$ when Z is closed in S.

References

[AR94] J. Adámek and J. Rosický, *Locally presentable and accessible categories*, London Mathematical Society Lecture Note Series, vol. 189, Cambridge University Press, Cambridge, 1994. MR1294136

[ARV10] J. Adámek, J. Rosický, and E. M. Vitale, *What are sifted colimits?*, Theory Appl. Categ. **23** (2010), No. 13, 251–260. MR2720191

[AM69] M. Artin and B. Mazur, *Etale homotopy*, Lecture Notes in Mathematics, No. 100, Springer-Verlag, Berlin-New York, 1969. MR0245577

[Aso11] A. Asok, *Remarks on $\mathbb{A}^1$-homotopy groups of smooth toric models*, Math. Res. Lett. **18** (2011), no. no. 2, 353–356. MR2784677

[Aso13] A. Asok, *Splitting vector bundles and $\mathbb{A}^1$-fundamental groups of higher-dimensional varieties*, J. Topol. **6** (2013), no. no. 2, 311–348. MR3065178

[AD08] A. Asok and B. Doran, *Vector bundles on contractible smooth schemes*, Duke Math. J. **143** (2008), no. no. 3, 513–530. MR2423761

[ADF] A. Asok, B. Doran, and J. Fasel, *Smooth models of motivic spheres and the clutching construction*, to appear in Int. Math. Res. Notices, available at `https://arxiv.org/abs/1408.0413`.

[AF14] A. Asok and J. Fasel, *A cohomological classification of vector bundles on smooth affine threefolds*, Duke Math. J. **163** (2014), no. no. 14, 2561–2601. MR3273577

[AHW15a] A. Asok, M. Hoyois, and M. Wendt, *Affine representability results in $\mathbb{A}^1$-homotopy theory I: vector bundles*, ArXiv e-prints (2015), available at `http://arxiv.org/abs/1506.07093`.

[AHW15b] A. Asok, M. Hoyois, and M. Wendt, *Affine representability results in $\mathbb{A}^1$-homotopy theory II: principal bundles and homogeneous spaces*, ArXiv e-prints (2015), available at `http://arxiv.org/abs/1507.08020`.

[Ayo07] J. Ayoub, *Les six opérations de Grothendieck et le formalisme des cycles évanescents dans le monde motivique. I* (French, with English and French summaries), Astérisque **No. 314** (2007), x+466 pp. (2008). MR2423375

[BHS15] C. Balwe, A. Hogadi, and A. Sawant, *$\mathbb{A}^1$-connected components of schemes*, Adv. Math. **282** (2015), 335–361. MR3374529

[71] *Théorie des intersections et théorème de Riemann-Roch* (French), Lecture Notes in Mathematics, Vol. 225, Springer-Verlag, Berlin-New York, 1971. Séminaire de Géométrie Algébrique du Bois-Marie 1966–1967 (SGA 6); Dirigé par P. Berthelot, A. Grothendieck et L. Illusie. Avec la collaboration de D. Ferrand, J. P. Jouanolou, O. Jussila, S. Kleiman, M. Raynaud et J. P. Serre. MR0354655

[Bot58] R. Bott, *The space of loops on a Lie group*, Michigan Math. J. **5** (1958), 35–61. MR0102803

[BF78] A. K. Bousfield and E. M. Friedlander, *Homotopy theory of Γ-spaces, spectra, and bisimplicial sets*, Geometric applications of homotopy theory (Proc. Conf., Evanston, Ill., 1977), II, Lecture Notes in Math., vol. 658, Springer, Berlin, 1978, pp. 80–130. MR513569

[BK72] A. K. Bousfield and D. M. Kan, *Homotopy limits, completions and localizations*, Lecture Notes in Mathematics, Vol. 304, Springer-Verlag, Berlin-New York, 1972. MR0365573

[BG73] K. S. Brown and S. M. Gersten, *Algebraic K-theory, I: Higher K-theories* (1973), 266–292. Lecture Notes in Math., Vol. 341. MR0347943

[Caz12] C. Cazanave, *Algebraic homotopy classes of rational functions* (English, with English and French summaries), Ann. Sci. Éc. Norm. Supér. (4) **45** (2012), no. no. 4, 511–534 (2013). MR3059240

[Cis13] D.-C. Cisinski, *Descente par éclatements en K-théorie invariante par homotopie* (French, with English and French summaries), Ann. of Math. (2) **177** (2013), no. no. 2, 425–448. MR3010804

[Del75] P. Deligne, *Proceedings of the International Congress of Mathematicians* (1975), 79–85. MR0432648

[Dug01a] D. Dugger, *Universal homotopy theories*, Adv. Math. **164** (2001), no. no. 1, 144–176. MR1870515

[Dug01b] D. Dugger, *The Zariski and Nisnevich descent theorems* (2001), available at `http://pages.uoregon.edu/ddugger/desc.html`.

[DHI04] D. Dugger, S. Hollander, and D. C. Isaksen, *Hypercovers and simplicial presheaves*, Math. Proc. Cambridge Philos. Soc. **136** (2004), no. no. 1, 9–51. MR2034012

[DI04] D. Dugger and D. C. Isaksen, *Topological hypercovers and $\mathbb{A}^1$-realizations*, Math. Z. **246** (2004), no. no. 4, 667–689. MR2045835

[DI08] D. Dugger and D. C. Isaksen, *Etale homotopy and sums-of-squares formulas*, Math. Proc. Cambridge Philos. Soc. **145** (2008), no. no. 1, 1–25. MR2431636

[DI10] D. Dugger and D. C. Isaksen, *The motivic Adams spectral sequence*, Geom. Topol. **14** (2010), no. no. 2, 967–1014. MR2629898

[DRØ03] B. I. Dundas, O. Röndigs, and P. A. Østvær, *Motivic functors*, Doc. Math. **8** (2003), 489–525. MR2029171

[DLØ+07] B. I. Dundas, M. Levine, P. A. Østvær, O. Röndigs, and V. Voevodsky, *Motivic homotopy theory*, Universitext, Springer-Verlag, Berlin, 2007. Lectures from the Summer School held in Nordfjordeid, August 2002. MR2334212

[DS95] W. G. Dwyer and J. Spaliński, *Homotopy theories and model categories*, Handbook of algebraic topology, North-Holland, Amsterdam, 1995, pp. 73–126. MR1361887

[GJ99] P. G. Goerss and J. F. Jardine, *Simplicial homotopy theory*, Progress in Mathematics, vol. 174, Birkhäuser Verlag, Basel, 1999. MR1711612

[GS07] P. Goerss and K. Schemmerhorn, *Model categories and simplicial methods*, Interactions between homotopy theory and algebra, Contemp. Math., vol. 436, Amer. Math. Soc., Providence, RI, 2007, pp. 3–49. MR2355769

[Gro63] A. Grothendieck, *Revêtements étales et groupe fondamental. Fasc. I: Exposés 1 à 5*, Séminaire de Géométrie Algébrique, vol. 1960/61, Institut des Hautes Études Scientifiques, Paris, 1963. MR0217087

[Hat02] A. Hatcher, *Algebraic topology*, Cambridge University Press, Cambridge, 2002. MR1867354

[Hir03] P. S. Hirschhorn, *Model categories and their localizations*, Mathematical Surveys and Monographs, vol. 99, American Mathematical Society, Providence, RI, 2003. MR1944041

[Hol08] S. Hollander, *A homotopy theory for stacks*, Israel J. Math. **163** (2008), 93–124.

[Hov99] M. Hovey, *Model categories*, Mathematical Surveys and Monographs, vol. 63, American Mathematical Society, Providence, RI, 1999. MR1650134

[Hoy17] M. Hoyois, *The six operations in equivariant motivic homotopy theory*, Adv. Math. **305** (2017), 197–279. MR3570135

[Hoy16] M. Hoyois, *A trivial remark on the Nisnevich topology* (2016), available at `http://math.mit.edu/~hoyois/papers/allagree.pdf`.

[Hus75] D. Husemoller, *Fibre bundles*, 2nd ed., Springer-Verlag, New York-Heidelberg, 1975. Graduate Texts in Mathematics, No. 20. MR0370578

[Jar87] J. F. Jardine, *Simplicial presheaves*, J. Pure Appl. Algebra **47** (1987), no. no. 1, 35–87. MR906403

[Jou73] J. P. Jouanolou, *Algebraic K-theory, I: Higher K-theories* (French) (1973), 293–316. Lecture Notes in Math., Vol. 341. MR0409476

[Lam06] T. Y. Lam, *Serre's problem on projective modules*, Springer Monographs in Mathematics, Springer-Verlag, Berlin, 2006. MR2235330

[Lev08] M. Levine, *Motivic homotopy theory*, Milan J. Math. **76** (2008), 165–199. MR2465990

[Lev16] M. Levine, *An overview of motivic homotopy theory*, Acta Math. Vietnam. **41** (2016), no. no. 3, 379–407. MR3534540

[LM07] M. Levine and F. Morel, *Algebraic cobordism*, Springer Monographs in Mathematics, Springer, Berlin, 2007. MR2286826

[Lin81] H. Lindel, *On the Bass-Quillen conjecture concerning projective modules over polynomial rings*, Invent. Math. **65** (1981/82), no. 2, 319–323. MR641133

[Lur09] J. Lurie, *Higher topos theory*, Annals of Mathematics Studies, vol. 170, Princeton University Press, Princeton, NJ, 2009. MR2522659

[Lur16] J. Lurie, *Spectral algebraic geometry*, 2016. `http://www.math.harvard.edu/~lurie/`.

[May99] J. P. May, *A concise course in algebraic topology*, Chicago Lectures in Mathematics, University of Chicago Press, Chicago, IL, 1999. MR1702278

[MVW06] C. Mazza, V. Voevodsky, and C. Weibel, *Lecture notes on motivic cohomology*, Clay Mathematics Monographs, vol. 2, American Mathematical Society, Providence, RI; Clay Mathematics Institute, Cambridge, MA, 2006. MR2242284

[MS75] D. McDuff and G. Segal, *Homology fibrations and the "group-completion" theorem*, Invent. Math. **31** (1975/76), no. 3, 279–284. MR0402733

[Mor06] F. Morel, *$\mathbb{A}^1$-algebraic topology*, International Congress of Mathematicians. Vol. II, Eur. Math. Soc., Zürich, 2006, pp. 1035–1059. MR2275634

[Mor12] F. Morel, $\mathbb{A}^1$*-algebraic topology over a field*, Lecture Notes in Mathematics, vol. 2052, Springer, Heidelberg, 2012. MR2934577

[MV99] F. Morel and V. Voevodsky, $\mathbf{A}^1$*-homotopy theory of schemes*, Inst. Hautes Études Sci. Publ. Math. **No. 90** (1999), 45–143 (2001). MR1813224

[Nis89] Ye. A. Nisnevich, *The completely decomposed topology on schemes and associated descent spectral sequences in algebraic K-theory*, Algebraic K-theory: connections with geometry and topology (Lake Louise, AB, 1987), NATO Adv. Sci. Inst. Ser. C Math. Phys. Sci., vol. 279, Kluwer Acad. Publ., Dordrecht, 1989, pp. 241–342. MR1045853

[Qui67] D. G. Quillen, *Homotopical algebra*, Lecture Notes in Mathematics, No. 43, Springer-Verlag, Berlin-New York, 1967. MR0223432

[Qui73] D. Quillen, *Algebraic K-theory, I: Higher K-theories* (1973), 85–147. Lecture Notes in Math., Vol. 341. MR0338129

[Qui76] D. Quillen, *Projective modules over polynomial rings*, Invent. Math. **36** (1976), 167–171. MR0427303

[Rob15] M. Robalo, *K-theory and the bridge from motives to noncommutative motives*, Adv. Math. **269** (2015), 399–550. MR3281141

[Sch15] M. Schlichting, *Euler class groups, and the homology of elementary and special linear groups*, ArXiv e-prints (2015), available at `https://arxiv.org/abs/1502.05424`.

[ST15] M. Schlichting and G. S. Tripathi, *Geometric models for higher Grothendieck-Witt groups in* $\mathbb{A}^1$*-homotopy theory*, Math. Ann. **362** (2015), no. no. 3-4, 1143–1167. MR3368095

[SS00] S. Schwede and B. E. Shipley, *Algebras and modules in monoidal model categories*, Proc. London Math. Soc. (3) **80** (2000), no. no. 2, 491–511. MR1734325

[Seg74] G. Segal, *Categories and cohomology theories*, Topology **13** (1974), 293–312. MR0353298

[Ser55] J.-P. Serre, *Faisceaux algébriques cohérents* (French), Ann. of Math. (2) **61** (1955), 197–278. MR0068874

[TS14] The Stacks Project Authors, *Stacks Project* (2014), available at `http://stacks.math.columbia.edu`.

[Sus76] A. A. Suslin, *Projective modules over polynomial rings are free* (Russian), Dokl. Akad. Nauk SSSR **229** (1976), no. 5, 1063–1066. MR0469905

[Swa98] R. G. Swan, *Néron-Popescu desingularization*, Algebra and geometry (Taipei, 1995), Lect. Algebra Geom., vol. 2, Int. Press, Cambridge, MA, 1998, pp. 135–192. MR1697953

[Tho54] R. Thom, *Quelques propriétés globales des variétés différentiables* (French), Comment. Math. Helv. **28** (1954), 17–86. MR0061823

[Tho66] E. Thomas, *Seminar on fiber spaces*, Lectures delivered in 1964 in Berkeley and 1965 in Zürich. Berkeley notes by J. F. McClendon. Lecture Notes in Mathematics, vol. 13, Springer-Verlag, Berlin-New York, 1966. MR0203733

[TT90] R. W. Thomason and T. Trobaugh, *Higher algebraic K-theory of schemes and of derived categories*, The Grothendieck Festschrift, Vol. III, Progr. Math., vol. 88, Birkhäuser Boston, Boston, MA, 1990, pp. 247–435. MR1106918

[Tot99] B. Totaro, *The Chow ring of a classifying space*, Algebraic K-theory (Seattle, WA, 1997), Proc. Sympos. Pure Math., vol. 67, Amer. Math. Soc., Providence, RI, 1999, pp. 249–281. MR1743244

[Vis05] A. Vistoli, *Grothendieck topologies, fibered categories and descent theory*, Fundamental algebraic geometry, Math. Surveys Monogr., vol. 123, Amer. Math. Soc., Providence, RI, 2005, pp. 1–104. MR2223406

[Voe98] V. Voevodsky, $\mathbf{A}^1$*-homotopy theory*, Proceedings of the International Congress of Mathematicians, Vol. I (Berlin, 1998), Doc. Math. **Extra Vol. I** (1998), 579–604. MR1648048

[Voe10a] V. Voevodsky, *Homotopy theory of simplicial sheaves in completely decomposable topologies*, J. Pure Appl. Algebra **214** (2010), no. no. 8, 1384–1398. MR2593670

[Voe10b] V. Voevodsky, *Unstable motivic homotopy categories in Nisnevich and cdh-topologies*, J. Pure Appl. Algebra **214** (2010), no. no. 8, 1399–1406. MR2593671

[Wei89] C. A. Weibel, *Homotopy algebraic K-theory*, Algebraic K-theory and algebraic number theory (Honolulu, HI, 1987), Contemp. Math., vol. 83, Amer. Math. Soc., Providence, RI, 1989, pp. 461–488. MR991991

[Wei94] C. A. Weibel, *An introduction to homological algebra*, Cambridge Studies in Advanced Mathematics, vol. 38, Cambridge University Press, Cambridge, 1994. MR1269324

[Wei13] C. A. Weibel, *The K-book*, Graduate Studies in Mathematics, vol. 145, American Mathematical Society, Providence, RI, 2013. An introduction to algebraic K-theory. MR3076731

Department of Mathematics, University of Illinois at Chicago, Chicago, Illinois 60607

Department of Mathematics, Northwestern University, Evanston, Illinois 60208